12/18/91

PERTURBATIONS

PERTURBATIONS
Theory and Methods

JAMES A. MURDOCK
Department of Mathematics
Iowa State University
Ames, Iowa

A Wiley-Interscience Publication
JOHN WILEY & SONS, INC.
New York / Chichester / Brisbane / Toronto / Singapore

Copyright © 1991 by John Wiley & Sons, Inc.

All rights reserved. Published simultaneously in Canada.

Reproduction or translation of any part of this work
beyond that permitted by Section 107 or 108 of the
1976 United States Copyright Act without the permission
of the copyright owner is unlawful. Requests for
permission or further information should be addressed to
the Permissions Department, John Wiley & Sons, Inc.

Library of Congress Cataloging in Publication Data:
Murdock, James A.
    Perturbations : theory and methods / James A. Murdock.

        p.   cm.
    "A Wiley-Interscience publication."
    Includes bibliographical references.
    ISBN 0-471-61294-4
    1. Perturbation (Mathematics)   I. Title.

QA871.M87   1991
515'.35--dc20   90-12944
                                                           CIP

Printed in the United States of America

10 9 8 7 6 5 4 3 2 1

To William B. Inhelder,
My High School Mathematics Teacher

CONTENTS

PREFACE

The beginning student of differential equations quickly exhausts the few
types which can be solved "in closed form," that is, using elementary functions: the first order exact, linear, and homogeneous equations; the higher
order linear equations with constant coefficients; the partial differential
equations that are reducible to these by separation of variables. After this,
there are several directions: advanced theory, approximation of solutions
on the digital computer by numerical analysis, approximation of solutions
by means of formulas. The last is the primary subject of this book. The
advantage of having an approximate formula for the solution of an equation, as opposed to having a computer program that generates numbers,
is that it is more easily possible to see the role of the different variables
and parameters in the solution: to recognize, for instance, the effect of a
friction parameter on the frequency of an oscillation. There are, of course,
advantages to numerical methods as well, one of the greatest being that
they apply over a wider range of the parameters. Perturbation methods, as
the name implies, are useful only when the equation to be solved is close
to ("is a perturbation of") a solvable equation.

Perturbation theory has the reputation of being a bag of tricks giving
formulas that often work but are seldom justifiable. It is viewed this way
by many who use it successfully, as well as by those who refuse to use
it because of its alleged lack of mathematical rigor. But one only needs
to remember that the definition of asymptotic series was due to Henri
Poincaré to realize that the separation of perturbation methods from other
approaches (such as the geometrical analysis of solution orbits), although
common in the classroom today, is a betrayal of the true nature of the
subject. Just as one should never feed a differential equation into a computer without having a feeling for the type of solutions to be expected

and what numerical difficulties they threaten, one should never (or almost never) apply a perturbation method without at the same time considering the existence and uniqueness of solutions, the solution geometry, possible bifurcations, and other factors that might affect the accuracy and meaning of the solutions.

On the other hand, to present perturbation theory in such a holistic context poses a difficulty. Because of the demand for courses in perturbation theory from the extremely applied end of the mathematical spectrum, and because of the ease of teaching a course that is almost entirely formal in content (a "cookbook" course), it has become popular to teach perturbation theory at a beginning graduate level to students who have not yet mastered, or even been exposed to, the methods of proof needed to establish the existence of the solutions to which approximations are being found. Some of the students in such a course, particularly those coming from outside the mathematics department, may never study these proof techniques. Does this mean that they should be denied an awareness of the extent to which mathematical theory interacts with practical computational methods, especially when it is this very theory that makes it possible to appreciate the conditions under which the methods may fail? On the contrary, it would seem that one of the most important mathematical skills to be acquired by a user of mathematics is precisely the ability to read a mathematical theorem and glean its significance for an application, without necessarily studying its proof. And for the prospective mathematician, who already has some appreciation of the need for proof, an exposure to problem solving based on theorems whose proof will be studied later can only help the student to feel "at home" when the proofs are finally encountered.

Therefore it is the premise of this book that even at a beginning level, perturbation theory should be presented not as an isolated collection of cookbook techniques but as a part of the mainstream of mathematics. This necessarily means that theorems covered in other mathematics courses will be encountered. Our approach is to make these theorems available, whether or not the student has seen them elsewhere, by a clear statement and intuitive discussion in an appendix. The mathematically sophisticated reader will of course not need these appendices, although even such a reader may find bits of intuitive insight fall into place in reading them; I certainly did, as I wrote them. For others, they will serve as examples of how to extract the meaning from a mathematical statement, or as a foretaste of things to be studied later.

Throughout this book, my effort has been to say the important things that no one ever seems to say, the simple insights that finally dawned on me after months (sometimes years) of working with a method and reading its standard literature; the things that made me think: "Why didn't they tell me that in the first place?" If my experience is at all typical, there must be many students who will welcome the discussions in this book, although others may find the book too "chatty" and wish for more formulas and fewer words. Some may feel that there are not enough solved computational examples. My response to this is one of personal temperament. I do not enjoy computation for its own sake, and whenever I attempt a problem, I end up spending two weeks searching into its meaning, its theoretical significance, possible variations, places where it might stretch the limits of existing theory, and so forth. Therefore I do not have a large stock of problems from which to draw for this book; almost every problem I have ever solved, if it had anything worthwhile to contribute, is included here. I can make no claim to the kind of computational virtuosity exhibited in, for instance, the books by Ali Nayfeh. Some of my examples are taken from his books, frequently with additional development. (For instance, the solvable triple-deck boundary value problem taken from Nayfeh and given here in Example 8.1.1 becomes unsolvable when changed to an initial value problem, Example 7.6.2.) This book is in many ways complementary to Nayfeh's books, and certainly does not replace them.

I have made no effort to distinguish between new and old results in the main text of the book. There are many elementary ideas here that I have never seen in print, which I had to develop in the same way as I would a research result, and yet I cannot claim them as my own, since many people must have known them without writing them out explicitly. An example is the discussion of what I call the "trade-off" property of Lindstedt series: A solution that is accurate to order $\mathcal{O}(\varepsilon^k)$ on a time interval of length $\mathcal{O}(1/\varepsilon)$ is also accurate to order $\mathcal{O}(\varepsilon^{k-j})$ on an interval of length $\mathcal{O}(1/\varepsilon^{1+j})$ for $0 < j < k$. This can be proved easily for the Lindstedt method, and yet fails for multiple scale methods in general. The literature is full of vague claims to this effect for both methods. I have not seen the proof for the Lindstedt case before, and yet it is too elementary to claim as new. On the other hand, I have recently given an example which I think makes clear for the first time that there are fundamental limitations on the attempt to extend the multiple scale method to longer intervals of time. This example (in a simplified form) is outlined in the text, without indicating it as mine. Each chapter does end with an annotated "Notes and References" section, which is intended to guide the student to related reading and occasionally to give historical information; the original attribution of results that are not in general circulation is given here when it is known to me. These notes are in no way intended to be exhaustive or up to the minute.

The book is divided into three parts, covering regular perturbation theory, oscillatory phenomena (with the Lindstedt, multiple scale, and averaging methods), and transition layer phenomena (initial layers, boundary layers, turning points, and such). The first two of these parts have been in preparation for a number of years and contain most of the original contributions; the second part, especially, is closely related to my own research and goes far enough to provide initial access to my papers. (But no farther; I have resisted the temptation to unbalance this book too greatly by presenting actual research here.) The third part contains new expositions of classical ideas but does not reach as close to current research as does the second. (The one relatively recent topic mentioned in this part, without much development, is canards.) Largely this is because of my own lack of extensive knowledge in transition layer phenomena. I have tried to learn, and since as a thinker I tend strongly toward the "foundational," I have worked hard at understanding the basics. The result is that students struggling with "matching" for the first time seem to find the approach taken here to be much less mysterious than the usual exposition; that is, if they are interested in more than computational facility.

Most of the notations used are self-explanatory. Boldface is used for vectors, although in advanced mathematics this is usually dispensed with, and students need not feel obligated to underline or otherwise designate vectors in their own work. All vectors are to be treated as column vectors, unless otherwise indicated, whenever matrix multiplication is involved, but we still write them as $\mathbf{x} = (x_1, \ldots, x_N)$ most of the time without indicating "transpose" or "col." A boldfaced function with a boldfaced subscript indicates the matrix of partial derivatives of the components of the function with respect to the components of the subscript; in each row of such a matrix, the component of the function is fixed and the component of the subscript varies. As a general rule, N and M are used for dimensions of vectors, n for the general term of a finite or infinite series, and k for the last term of a finite series. Other indices are used as needed; it should be clear when i is an index and when it is $\sqrt{-1}$.

There are a few textbooks that serve as general references for additional reading on most topics in this book. These will not be repeated in each "Notes and References" section, except to point out something particularly good for an individual topic. Therefore they will be listed here, with brief comments. The works of Ali Nayfeh are excellent references for the computational aspects of all parts of perturbation theory, but are not reliable in regard to the theory. The most useful are

Ali Nayfeh, *Introduction to Perturbation Techniques*, Wiley, New York, 1981

and

Ali Nayfeh, *Perturbation Methods*, Wiley, New York, 1973.

The former is an introduction to the basic techniques with many worked examples and is in most cases the best book for the student to consult when looking for additional exercises or for an alternative explanation of a topic that is found to be difficult. The latter is an encyclopedic reference covering almost all methods and giving many references. The principal "error" to watch out for in these books is in the interpretation of the big-oh symbol. An equation will be given containing a perturbation parameter ε and a constant control parameter λ, and it will be stated that when λ is $\mathcal{O}(1)$, one method should be used, but that this method "breaks down" when λ is $\mathcal{O}(\varepsilon)$, and then another method (perhaps a rescaling) should be used.

Of course, it is impossible for a constant to be $\mathcal{O}(\varepsilon)$, no matter how small it may be, so the statement is entirely meaningless. The correct statement is that the first method is uniformly valid for λ in any compact set (closed bounded interval) not containing zero, whereas the second is uniformly valid for λ in a shrinking ε-dependent interval whose length is $\mathcal{O}(\varepsilon)$. Once these matters are clearly understood—and they are explained at length in Section 1.7 below—it is easy to profit from the books by Nayfeh without becoming confused.

The book

J. Kevorkian and J. D. Cole, *Perturbation Methods in Applied Mathematics*, Springer-Verlag, New York, 1981

covers many of the topics in this book, emphasizing the multiple scale and matching methods. The range of applications given is much greater than in this book and includes many difficult examples with partial differential equations. Most of the work is formal, without error estimates, but it is quite carefully done. In a different spirit is

Donald R. Smith, *Singular Perturbation Theory*, Cambridge University Press, Cambridge, 1985.

This book emphasizes what we call direct error estimation and presents many up-to-date theorems about the degree of accuracy of multiple scale and boundary layer correction methods. (This book prefers the correction method to the essentially equivalent matching method for transition layer problems.) The last three books are good starting places for a student wishing to go beyond the present book, but they leave unsaid many of the first principles that are explained here. A reference that includes a variety of perturbation methods explained from a practical point of view using a variety of interesting examples is

Carl M. Bender and Steven A. Orszag, *Advanced Mathematical Methods for Scientists and Engineers*, McGraw-Hill, New York, 1978.

An extensive discussion of perturbation methods in partial differential equations (barely touched on in the present text) is given in Chapter 9 (all 214 pages of it!) of

Erich Zauderer, *Partial Differential Equations of Applied Mathematics*, Wiley, New York, 1983.

ACKNOWLEDGMENTS

This book would not have been possible without the teachers that encouraged and supported me over the years, and the students who used preliminary drafts of portions of the book in their classes. Among the teachers were William Inhelder, who knew his epsilons and deltas and believed that real mathematics belonged in high school, where it has now been replaced by mass-produced advanced placement calculus; Wendell Fleming, who thought undergraduates deserved differential forms and Lebesgue integrals; and Jürgen Moser, who introduced me to perturbation theory and criticized some of my early attempts at mathematical writing. Among the students, Chao-Pao Ho did the largest amount of proofreading.

The majority of the graphs (35) were done by Kurt Whitmore, an undergraduate at Iowa State University, using Mathematica and other software on a Macintosh SE/30 with a laser printer; a few were done in the same way by Tony Walker (15), Tom Bullers (9), and Jonathan Schultz (1), also undergraduates. Only Fig. 4.7.4 was drawn by hand.

JAMES A. MURDOCK

PERTURBATIONS

PART I

INTRODUCTION TO PERTURBATION THEORY

CHAPTER 1

ROOT FINDING

1.1. THE NATURE OF PERTURBATION THEORY

Perturbation theory first appeared in one of the oldest branches of applied mathematics: celestial mechanics, the study of the motions of the planets. From antiquity, various mathematical methods were used to describe these motions (as seen from earth), usually with no attempt to state their causes. After Newton's formulation of the law of gravity, it became possible to deduce the planetary motions from physical laws which were considered to be more fundamental. If only the sun and one planet are considered, the result is elliptical motion with the sun at a focus. However, this does not quite correspond to the actually observed motion. The explanation is that the planets exert gravitational forces on each other, and therefore *perturb*, that is, modify, their motions. Perturbation theory in its original sense refers to various ways of taking these modifications into account. In essence, one begins with the "unperturbed solution," that is, with purely elliptical motion, as a first approximation, then computes the forces which the planets would exert on each other if this unperturbed motion were correct, and then corrects the unperturbed solution accordingly. The first corrections are still not accurate, since their construction depended upon the unperturbed solution, and so a second set of corrections can be computed, and so on. The sum of the unperturbed solution and the sequence of corrections forms a series, and one hopes that a partial sum of a reasonable number of terms gives an adequate approximation to the motion for perhaps a few hundred years.

The scope of perturbation theory at the present time is much broader than its applications to celestial mechanics, but the main idea is the same.

One begins with a solvable problem, called the unperturbed or reduced problem, and uses the solution of this problem as an approximation to the solution of a more complicated problem that differs from the reduced problem only by some small terms in the equations. Then one looks for a series of successive corrections to this initial approximation, most often in the form of a power series (or some kind of modified power series) in a small quantity called the *perturbation parameter*. Finally one attempts to show that the use of only a few of these correction terms (usually only one or two) provides a useful approximate solution to the actual problem at hand.

The simplest problem which can be addressed by perturbation theory is that of finding roots of polynomials. This problem illustrates many of the important ideas: proper formulation of perturbation families; degenerate and nondegenerate cases; uniform and nonuniform solutions; rescaling coordinates; rescaling parameters. The problem is purely mathematical, so it is not necessary to address physical and mathematical issues simultaneously. There are no differential equations involved, so the only mathematical difficulties are those coming from perturbation theory itself. All in all, it is an ideal place to begin.

What does it mean to solve a polynomial equation by a perturbation method? Suppose the problem is

$$x^2 - 3.99x + 3.02 = 0. \tag{1.1.1}$$

Of course, this can be solved easily by the quadratic formula. But to approach it by perturbation theory, there are four steps:

1. The first is to notice that since $-3.99 = -4 + 0.01$ and $3.02 = 3 + 0.02$, equation (1.1.1) is almost the same as $x^2 - 4x + 3 = 0$, which can be solved easily by factoring: $(x - 1)(x - 3) = 0$, giving two roots $x^{(1)} = 1$, $x^{(2)} = 3$.

2. The second step is to create a family of problems intermediate between the easy, factorable problem and the original problem (1.1.1). This can be done by letting ε denote the small quantity 0.01, so that $-3.99 = -4 + \varepsilon$ and $3.02 = 3 + 2\varepsilon$; then (1.1.1) can be written

$$x^2 + (\varepsilon - 4)x + (3 + 2\varepsilon) = 0. \tag{1.1.2}$$

Now allow ε to vary. Then (1.1.2) is no longer a single equation, but a family of equations, one equation for each value of ε. When $\varepsilon = 0$, (1.1.2) reduces to the factorable problem, and when $\varepsilon = 0.01$, it is the "target problem" (1.1.1). For $0 < \varepsilon < 0.01$, it is midway between the two. Equation (1.1.2) is an example of a *perturbation family*, a family of problems depending on a small parameter ε which is easily solvable when $\varepsilon = 0$.

3. The third step is to find approximate solutions of (1.1.2), in the form of polynomials (truncated power series) in the small parameter ε. The method for finding these will be explained in the next section. In this example suitable solutions turn out to be

$$x^{(1)} \cong 1 + \frac{3}{2}\varepsilon + \frac{15}{8}\varepsilon^2,$$

$$x^{(2)} \cong 3 - \frac{5}{2}\varepsilon - \frac{15}{8}\varepsilon^2. \tag{1.1.3}$$

Evaluating these solutions at $\varepsilon = 0.01$ gives an approximate solution of the original problem (1.1.1), namely $x^{(1)} \cong 1.0151875$, $x^{(2)} \cong 2.9748125$.

4. The fourth step is, whenever possible, to say something about the amount of error in these approximations. We will leave this topic until Section 1.4.

This brief example already reveals a good deal about perturbation theory. First of all, the method can only be applied when the "target" problem is close to a solvable problem (that is, close to a problem solvable exactly or approximately by some method other than perturbation theory). A polynomial equation chosen at random can probably not be solved by perturbation theory, since it is unlikely to be close to a factorable polynomial.

Next, the example shows that in solving a problem by perturbation theory, one solves not only a single target problem such as (1.1.1), but every problem belonging to the perturbation family (1.1.2), as long as ε is "sufficiently small." A perturbation solution such as (1.1.3) is "valid"—in the loose sense of "gives a good approximation"—only for sufficiently small values of ε. The meaning of "sufficiently small" is not clear until step 4 (the error analysis) has been completed. It could turn out that $\varepsilon = 0.01$ is not sufficiently small to give a good solution in our problem. In that case we would not have solved the target problem (1.1.1) satisfactorily, although we might still have a good solution of (1.1.2) for smaller ε, perhaps $\varepsilon = 0.001$. On the other hand, (1.1.3) might be a reasonably accurate approximate solution as far out as $\varepsilon = 0.1$, or even 1 or 10. For these reasons it will become second nature to think of perturbation families, rather than individual problems, as the fundamental object of study in perturbation theory.

It is sometimes assumed in perturbation theory that $0 < \varepsilon < 1$, so that $\varepsilon^2 < \varepsilon$. Under this assumption, higher powers of ε decrease in size, which gives an intuitive justification for ignoring sufficiently high powers, for instance, terms of degree higher than ε^2 in (1.1.3). It is important to think of higher powers of a small parameter as less important. But how small ε must be cannot be stated in advance by an assumption such as $\varepsilon < 1$, since the influence of a given term such as ε^3 depends on the coefficient of that term. One approach to this question is to ask about the radius of convergence of a power series;

within its radius of convergence, partial sums of enough terms represent the function well, and of course, the radius of convergence can be greater or less than 1. This alone shows that there is no magic in assuming $\varepsilon < 1$, although the radius of convergence is not very important for the type of error estimates needed in this book.

It is useful to think of a perturbation family such as (1.1.2) as a *path* leading from the easy *reduced problem* to the given *target problem*. In order to clarify this idea of a "path," it is helpful to think of (1.1.1) as a special case of the general quadratic equation $ax^2 + bx + c = 0$. If a set of three perpendicular coordinate axes, labeled a, b, and c, is constructed in space, then each point (a, b, c) in this coordinate system corresponds to a specific quadratic equation; thus (1.1.1) corresponds to the point $(1, -3.99, 3.02)$, and the reduced problem corresponds to $(1, -4, 3)$. Then the three equations $a = 1$, $b = \varepsilon - 4$, and $c = 3 + 2\varepsilon$, are the parametric equations (with parameter ε) of a straight line joining the reduced and target problems. Typically, a problem to be solved by perturbation theory is initially stated in terms of certain *natural parameters* such as a, b, and c, which may have physical or mathematical significance. Then a path is chosen through the possible values of the natural parameters by introducing a perturbation parameter ε.

It is important to realize that perturbation families must be created; they do not arise automatically. There is no ε in the target problem (1.1.1); the ε has to be put in artificially. Instead of drawing a straight line from $(1, -4, 3)$ to $(1, -3.99, 3.02)$ we could have drawn a curve with different parametric equations. In problems arising from physics there are often good reasons to choose a certain variable as the perturbation parameter, but there are usually several possible choices, and the creation of good perturbation problems is an art rather than a science. As with any other form of mathematical modelling, it requires a good understanding both of the subject area from which the problem arises, and of the mathematical theory behind the solutions.

As an example of the choices involved in formulating a perturbation problem, suppose the "target problem" is

$$x^2 + 3.9x - 5.002 = 0. \tag{1.1.4}$$

This is close to $x^2 + 4x - 5 = (x - 1)(x + 5)$. Two possible perturbation families for (1.1.4) are

$$x^2 + (4 - \varepsilon)x - (5 + 2\varepsilon^3) = 0 \tag{1.1.5}$$

and

$$x^2 + (4 - \varepsilon)x - (5 + 0.02\varepsilon) = 0. \tag{1.1.6}$$

Both reduce to (1.1.4) when $\varepsilon = 0.1$, and they represent two different paths leading from the easy problem $x^2 + 4x - 5 = 0$ to the target prob lem (1.1.4). Which of them will yield a better solution? In the present case the issue is to be decided by error analysis alone, because the only basis for decision is numerical accuracy. If the problem comes from physics (or some other application), there could be reasons other than mere accuracy which suggest one choice rather than another. For instance, a certain nondimensional combination of physical variables (such as the Reynolds number or 1 over the Reynolds number) might present itself as a physically meaningful small quantity which almost begs to be chosen as ε. (The importance of nondimensional quantities will be explained in Section 2.3.) Even so, physical problems lend themselves to a variety of formulations, and the final choice should be affected by both mathematical and physical considerations.

In this chapter, our approach will be to take a perturbation family as given, and concentrate on the solution and its error analysis (steps 3 and 4). These results will be used to compare the error in alternative formulations such as (1.1.5) and (1.1.6). In Sections 1.6 and 1.7 we will see that the mathematical theory sometimes suggests a reformulation (called a rescaling) of an already given perturbation problem in order to improve the accuracy of the solutions. Therefore a poorly formulated problem need not lead to a dead end. Often a physical problem is stated at first in terms suited to the application, and rescaled several times in the course of analysis; then the solutions must be interpreted carefully to see how they apply to the original problem. The easy example of finding roots is a good introduction to these ideas.

A good metaphor for perturbation theory is the exploration of a jungle by helicopter. The jungle stands for all problems of a given type. One searches the jungle from the air, looking for clearings in which to land. These clearings are the easily solvable problems. After landing, the explorers cut a path into the jungle; this path, parameterized by ε, is the perturbation family. The explorers proceed until they reach an impenetrable swamp or are ambushed by a lion. At this point ε is too large and the perturbation method has failed. (Actually, lions live in plains and not jungles. But this book is about perturbation theory and not zoology.)

Exercises 1.1

1. Use the quadratic formula and a calculator to find the "exact" solutions of (1.1.1). (These are probably accurate to the number of digits given by the calculator, because the calculator carries extra "invisible" digits. On the other hand, the methods used by a calculator to extract square roots are approximation methods, so the calculator solution is not to be thought of as necessarily exact.) Compare these solutions with the solutions obtained by putting $\varepsilon = 0.01$ in (1.1.3). Also try using (1.1.3) with the ε^2 terms omitted.

2. Write two different perturbation families that might be suitable if the target problem is $x^2 + 10.1x + 21.03$.

3. The differential equation for the motion of a particle of mass m suspended from a spring with Hooke's law constant k and damping factor h is $md^2y/dt^2 + hdy/dt + ky = 0$. What are the natural parameters in this equation? Using ε to represent a small quantity, rewrite this equation to express each of the following situations:

   (a) The mass is small.

   (b) The damping is small.

   (c) $k = 2h$, and both k and h are small. Observe that in each case, you have made the natural parameters into functions of ε, although some of these functions are constant functions that do not actually depend upon ε. (In this problem we are ignoring the fact that the natural parameters in a physical problem should usually be nondimensionalized before introducing ε. This topic will be addressed in Chapter 2.)

1.2. FORMAL APPROXIMATIONS IN THE NONDEGENERATE CASE

Approximate solutions in applied mathematics are frequently constructed on the basis of guesswork, and justified later (if at all). The words "heuristic" and "formal" are used to describe this process. *Heuristic* reasoning is thinking that is imprecise, based on intuition, and aimed at discovery. For instance, one might say: "This problem concerns the interval $-a \leq x \leq a$ and the given information is symmetric around $x = 0$, so let's try a solution that is symmetric. That is, let's guess that the solution is an even function and so try a Fourier cosine series." *Formal* calculation is calculation based strictly on the *form* of the equations, that is, what the equations look like, without regard to whether the steps in the calculation are actually justified by mathematical theorems. As an example, in a formal argument one might differentiate an infinite series term by term without stopping to consider whether the resulting series converges. These types of exploratory thinking have led to important breakthroughs: It is only necessary to think of the "delta function," which at first had no rigorous justification (it is not a function in the usual sense) but is now well established in both mathematics and physics, having led to the creation of an entirely new subject called "generalized functions." On the other hand the literature of perturbation theory is full of erroneous results obtained by heuristic reasoning and formal calculation. It is important for the student to develop facility at exploration and discovery, on the one hand, and at justification of the discovered results, on the other.

Confusingly, the word "formal" is sometimes used in almost the opposite sense. A mathematician will sometimes say: "I have an informal (that is, somewhat sloppy, nonrigorous) argument for a certain result, but I haven't written down a formal proof." In this usage, a "formal proof" is one with all the details spelled out, with careful attention to logic. In both usages the word "formal" refers to "form," but in our case it means "looking only at the form *of the equations* and not worrying about the proof," while in the other case it means "looking very carefully at the form *of the proof*."

We will develop approximate solutions to the perturbation problem (1.1.2) by using two heuristic ideas:

(i) Try polynomials.
(ii) Ignore high powers of ε.

As a motivation for (ii), we have already remarked that if $0 < \varepsilon < 1$ then $\ldots \varepsilon^4 < \varepsilon^3 < \varepsilon^2 < \varepsilon < 1$. For instance, if $\varepsilon = 0.01$, then $\varepsilon^2 = 0.0001$ and $\varepsilon^3 = 0.000001$. If one is interested in a solution with three or four decimal places of accuracy, it seems reasonable to "ignore" ε^3, that is, to compute as though $\varepsilon^3 = 0$. (Remember, this is only heuristic reasoning: The actual error introduced by ignoring ε^3 remains to be determined and will vary from problem to problem. Recall the discussion about this in Section 1.1.) Idea (i) can be motivated by the fact that lots of functions can be represented by power series, and ignoring high powers in a power series leaves a polynomial. Combining ideas (i) and (ii), taking the "cutoff" at ε^3, suggests looking for approximate solutions to (1.1.2) in the form of quadratic polynomials:

$$x^{(1)} \cong 1 + a\varepsilon + b\varepsilon^2,$$
$$x^{(2)} \cong 3 + c\varepsilon + d\varepsilon^2. \tag{1.2.1}$$

The leading terms 1 and 3 in (1.2.1) are the known solutions of (1.1.2) when $\varepsilon = 0$. Substituting the first equation of (1.2.1) into (1.1.2) gives

$$(1 + a\varepsilon + b\varepsilon^2)^2 + (\varepsilon - 4)(1 + a\varepsilon + b\varepsilon^2) + (3 + 2\varepsilon) = 0.$$

The next step is to multiply this out and arrange it in powers of ε. Since this operation occurs frequently in perturbation theory, it is useful to be able to find the terms of any specific degree in such a product without having to carry out the complete multiplication; for instance, the coefficient of ε^2 in $(1 + a\varepsilon + b\varepsilon^2)^2 = (1 + a\varepsilon + b\varepsilon^2)(1 + a\varepsilon + b\varepsilon^2)$ is easily seen to be $a^2 + 2b$ since ε^2 can only arise by multiplying $a\varepsilon$ by itself or by multiplying 1 times $b\varepsilon^2$ (which can happen in two ways). Either by this kind of reasoning or by longhand algebra, the result is found to be

$$\varepsilon(-2a + 3) + \varepsilon^2(a + a^2 - 2b) + \varepsilon^3(2ab + b) + \varepsilon^4(b^2) = 0. \tag{1.2.2}$$

Since a polynomial in ε which equals zero for all ε must have every coefficient equal to zero, (1.2.2) can hold exactly only if

$$-2a + 3 = 0,$$
$$a + a^2 - 2b = 0,$$
$$2ab + b = 0,$$
$$b^2 = 0.$$

$$(1.2.3)$$

This is a system of four equations in two unknowns, which clearly has no solution. But this is just what should have been expected: (1.2.2) cannot hold exactly, since that would amount to demanding that (1.2.1) be an exact solution of (1.1.2). We only expect an approximate solution. To find one, use idea (ii) again. If the ε^3 and ε^4 terms in (1.2.2) are ignored, a and b need only satisfy the first two equations of (1.2.3). This leads at last to a solution, $a = 3/2$ and $b = 15/8$. A similar calculation for c and d completes the result stated previously in (1.1.3). (See Exercise 1.2.1.)

In presenting this approximate solution we have used the sign $\cong$, meaning "approximately equal to." In Section 1.8, several approximation symbols will be introduced. Among these, $\cong$ is the most general. When we write $x^{(1)} \cong 1 + 3\varepsilon/2 + 15\varepsilon^2/8$, it means merely that we are *proposing* $1 + 3\varepsilon/2 + 15\varepsilon^2/8$ as an approximation to the actual solution $x^{(1)}$. No error analysis has yet been performed, and the proposed approximation still could be a very bad one.

We have computed a tentative solution to (1.1.2), but will the same procedure work on every problem? Several examples are given in Exercise 1.2.2 which show that the answer is no: Sometimes the computations cannot be carried out because they lead to a division by zero. Therefore in order to understand when the method works, it is not enough to look at examples; it is necessary to study the general case. The purpose of the following analysis is to find out the conditions under which the computational procedures described above can be carried out. This is called a *formal analysis* of the method, because it deals only with the form of the solutions and not with error estimates. Such a formal analysis is an essential part of the justification of a perturbation method, but it should be kept in mind that the possibility of performing a calculation does not guarantee that the computation will be accurate.

The general problem, of which (1.1.2) is a special case, is the following: to find an approximate solution of

$$\varphi(x, \varepsilon) = 0 \qquad (1.2.4)$$

in the form

$$x \cong x_0 + \varepsilon x_1 + \varepsilon^2 x_2. \qquad (1.2.5)$$

Here φ is an arbitrary function of two variables, which is considered as given (and therefore known). The form of (1.2.5) indicates that we are still "ignoring" ε^3; the cutoff could be taken at any power of ε, but the more terms that are retained, the longer the calculations become. The first step is to substitute (1.2.5) into (1.2.4), obtaining

$$\varphi(x_0 + \varepsilon x_1 + \varepsilon^2 x_2, \varepsilon) \cong 0. \tag{1.2.6}$$

(The symbol $\cong$ is used because we cannot expect to achieve exactly zero.) The next step is to expand (1.2.6) in powers of ε. In any specific example where φ is a polynomial, this is just a matter of multiplying and collecting like terms, as was done in finding (1.2.2). In the general case the function φ has not been specified, and the only way to proceed is by a Taylor series. Since we are still reasoning formally and not rigorously, there is no need to pause over the justification of this step. It is convenient to have a name for the left hand side of (1.2.6) regarded as a function of ε, so we set $h(\varepsilon) := \varphi(x_0 + \varepsilon x_1 + \varepsilon^2 x_2, \varepsilon)$. (The symbol := means "is defined to equal.") Expanding in powers of ε and ignoring ε^3 leads to

$$\begin{aligned}
h(0) + h'(0)\varepsilon + \tfrac{1}{2}h''(0)\varepsilon^2 &= \varphi(x_0, 0) + \varepsilon\{\varphi_x(x_0, 0)x_1 + \varphi_\varepsilon(x_0, 0)\} \\
&\quad + \tfrac{1}{2}\varepsilon^2\{\varphi_{xx}(x_0, 0)x_1^2 + 2\varphi_{x\varepsilon}(x_0, 0)x_1 \\
&\quad + 2\varphi_x(x_0, 0)x_2 + \varphi_{\varepsilon\varepsilon}(x_0, 0)\} = 0.
\end{aligned} \tag{1.2.7}$$

The coefficients of (1.2.7) must equal zero, which means that x_0, x_1, and x_2 must satisfy

$$\varphi(x_0, 0) = 0,$$

$$x_1 = -\frac{\varphi_\varepsilon(x_0, 0)}{\varphi_x(x_0, 0)} = -\left.\frac{\varphi_\varepsilon}{\varphi_x}\right|_{(x_0,0)}, \tag{1.2.8}$$

$$x_2 = \left.\frac{-\varphi_{xx}\varphi_\varepsilon^2 + 2\varphi_{x\varepsilon}\varphi_\varepsilon\varphi_x - \varphi_{\varepsilon\varepsilon}\varphi_x^2}{2\varphi_x^3}\right|_{(x_0,0)}.$$

The first equation of (1.2.8) says that x_0 must be a solution of the "reduced problem" obtained by setting $\varepsilon = 0$ in (1.2.4). In order to use perturbation theory, it is necessary to know the solution to the reduced problem. (Of course in the root-finding problem the reduced problem may have several solutions, and it is necessary to choose one before going on, or take each of them in turn.) Once a suitable value of x_0 is selected, the remaining equations in (1.2.8) determine x_1 and x_2. These equations clearly show the conditions that must be satisfied in order for the calculation to be possible: The indicated first and second partial derivatives of φ must exist, and the quantity $\varphi_x(x_0, 0)$ must not be zero. Notice that both denominators in (1.2.8) are nonzero if this condition is met. If (1.2.5) is carried to higher order, and the coefficients $(x_3, x_4, \ldots)$ are computed, it is found that the denominators always contain $\varphi_x(x_0, 0)$ raised to some power. No other factors arise in the denominators which could equal zero, so it is not necessary to impose any further restrictions beyond the fundamental assumption that $\varphi_x(x_0, 0) \neq 0$. This assumption defines what is called the *nondegenerate case*. This completes the formal analysis of the simplest perturbation method for root finding. The results can be stated as a theorem:

Theorem 1.2.1. *Let $\varphi(x, \varepsilon)$ be a function having partial derivatives up to second order. Let x_0 satisfy $\varphi(x_0, 0) = 0$ and $\varphi_x(x_0, 0) \neq 0$. Then there is a unique quadratic polynomial $x_0 + \varepsilon x_1 + \varepsilon^2 x_2$ which is an exact solution of (1.2.7) and is therefore regarded as a formal approximate solution of $\varphi(x, \varepsilon) = 0$ for small ε. The coefficients x_1 and x_2 are given by (1.2.8). Similar formal approximations of higher order can be computed if φ has partial derivatives of higher order.*

For the mathematician, this theorem raises more questions than it answers. For instance, the *existence problem*: Does there actually exist a true root close to the formal approximate root given by (1.2.8)? And the *error problem*: If the true root exists, how close is it to the approximate root? And perhaps the most important question of all, because it leads into a whole field called bifurcation and singularity theory: What is the deeper significance of the nondegeneracy condition $\varphi_x(x_0, 0) \neq 0$, and what happens when it fails? All of these issues will be addressed in the remainder of this chapter.

One further comment is in order concerning the equations (1.2.8): *these equations are not intended to be used in actually solving specific problems!* Reliance on formulas is a way to avoid understanding, and the feeling that a problem should be solved by "plugging into" the appropriate formula should be resisted. Of course, if someone had fifty problems of the same sort to solve, a formula might be useful. (The quadratic formula is useful because quadratic equations come up frequently, but no one should learn the quadratic formula until they know how to solve a quadratic equation by completing the square. Otherwise they have been cheated out of the opportunity to understand what they are doing.) The purpose of this section is to introduce ideas that will occur in many contexts, and it is these ideas that are useful, not the final formulas. Besides, no one can remember formulas as complicated as (1.2.8). *The method to be used in solving specific problems is the method illustrated in equations (1.2.1) through (1.2.3).* The purpose of (1.2.8) is to show under what circumstances (namely, nondegeneracy) this method works (at least in a formal sense).

Calculations of the sort given in this section are usually expressed in terms of power series rather than polynomials. A *formal power series* in ε is an expression $\sum_{n=0}^{\infty} a_n \varepsilon^n$, considered merely as an expression, without regard to whether it converges or in any way represents a function. It is merely an infinite sum arranged by powers of ε. Two formal power series are called equal if they have equal coefficients. When only the first few terms of a formal power series are written, the following notation will be used:

$$\sum_{n=0}^{\infty} a_n \varepsilon^n = a_0 + a_1 \varepsilon + a_2 \varepsilon^2 + \mathcal{O}_F(\varepsilon^3).$$

The symbol $\mathcal{O}_F(\varepsilon^3)$ means "terms of formal order ε^3 or greater," that is, terms that contain powers of ε greater than or equal to 3. (The subscript F for "formal" is intended to contrast with the asymptotic order symbol $\mathcal{O}$, which will be introduced in Section 1.4.) A formal solution of $\varphi(x, \varepsilon) = 0$ can be found in the following way. Substitute a formal power series

$$x = \sum_{n=0}^{\infty} x_n \varepsilon^n = x_0 + x_1 \varepsilon + \mathcal{O}_F(\varepsilon^2) \tag{1.2.9}$$

into $\varphi(x, \varepsilon)$ and expand into a (formal) Taylor series

$$\varphi(x_0, 0) + \varepsilon \{ \varphi_x(x_0, 0) x_1 + \varphi_\varepsilon(x_0, 0) \} + \mathcal{O}_F(\varepsilon^2) = 0. \tag{1.2.10}$$

The series (1.2.9) is called a *formal power series solution* of $\varphi(x, \varepsilon) = 0$ if each coefficient of (1.2.10) is zero. This gives an infinite sequence of equations for x_n, the first three being (1.2.8). The full infinite series (1.2.9) can never be computed and used, and one is really only interested in partial sums of this series, which are just the polynomial approximations constructed earlier. It is often convenient to speak of infinite series as a way of remembering that the solutions can be carried out to any order, but perturbation theory is never actually concerned with the infinite series as such. (This is in contrast to the use of infinite series in, for instance, the theory of Bessel functions, where one might use the fact that the full infinite series converges to the Bessel function to prove an identity involving such functions.)

Exercises 1.2

1. Compute c and d in (1.2.1) by the method of this section.

2. Using the method of this section, attempt to compute approximate solutions to the following problems. (Some of them will not work.)

   (a) $x^2 + (10 + \varepsilon)x + (21 + 2\varepsilon) = 0$,
   (b) $x^2 + (6 + \varepsilon)x + (9 + 2\varepsilon) = 0$,
   (c) $(x - 2)^2(x - 3) + \varepsilon = 0$.

3. Compute third order solutions $x \cong x_0 + x_1 \varepsilon + x_2 \varepsilon^2 + x_3 \varepsilon^3$ for both roots of (1.1.5) and (1.1.6). Observe that the term $2\varepsilon^3$ in (1.1.5) does

not affect the calculation until the third order term is reached, whereas the term 0.02ε in (1.1.6) affects the first order term in the solution. (Contrast this with the following problem.)

4. Using the quadratic formula, show that the solutions of $x^2 - 2\varepsilon x + (1 - a^2)\varepsilon^2 = 0$ are (exactly) $x = (1 \pm a)\varepsilon$. Therefore the quantity a, which enters the perturbation problem at the second order, already affects the solution at the first order, in contrast to what is expected after solving the previous problem. Observe that this problem is degenerate and the previous one is nondegenerate. (For advanced students: Can you formulate and prove a theorem about circumstances under which a term of a given order affects the solution only at that order and beyond?)

1.3. EXAMPLES, COMPARISONS, AND WARNING SIGNS

The previous section leaves us in this situation: We have a method for computing proposed approximate solutions of $\varphi(x, \varepsilon) = 0$ for ε near zero. But we have no idea how good these solutions are, or even if they correspond to actually existing solutions at all. Since this situation arises frequently in perturbation theory, it is worth pausing here to experience the feeling of being in this position. A variety of attitudes are possible. If the problem arises from an immediate engineering necessity, it is quite appropriate to take our tentative formulas into the laboratory, compare them with experimental data, and if they hold up, put them to use. In this case, the mathematics is being used to generate a rule-of-thumb formula which is then evaluated by practical experience. If the formula works, the principal question will be: For what range of ε does it give good results? This too will be settled by use. There is no danger in this, unless someone puts exaggerated faith in the unproved result merely because it has an aura of mathematical reliability.

The extreme opposite attitude is that what has been accomplished up to this point is precisely nothing. We have a formula of unknown accuracy which is supposed to approximate a solution that may not even exist. We would have done just as well to consult a witch doctor! There is no value—according to this view—to any approximate formula unless it is accompanied by an existence theorem proving that there really is a solution, and an error estimate comparing the exact solution and its approximation.

It would be nice if it were always possible to satisfy these rigorous demands. It is possible in the present case and will be done in the next section. It is also possible in most problems involving ordinary differential equations, as will be seen in later chapters. For problems involving partial differential equations, the theory is much less complete, and what theory exists is more difficult; no justification for partial differential equations examples will be attempted in this book. But there are many cases where approximate solutions of practical value exist although justifications have not as yet been found. It is therefore worth asking: What can be done to test the validity of an approximation, besides laboratory work or a complete mathematical theory?

There are two answers. Sometimes there exist simple problems, belonging to the class covered by a particular perturbation method, which can also be solved exactly. A comparison between the exact and approximate solutions to such a problem can give clues as to the validity of the perturbation method. A second possibility is to compare a perturbation solution (of a problem that cannot be solved exactly) with a numerical solution (obtained by digital computer or, in simple cases, pocket calculator). The numerical solution is also approximate, and perhaps of unknown error, but a concurrence between two approximate solutions obtained by greatly different methods lends weight to both. In our present case of $\varphi(x, \varepsilon) = 0$, quadratic equations provide examples which can be solved exactly, by the quadratic formula. (Cubics and quartics can also be solved exactly, but this is much harder.) Other examples can be solved numerically by the bisection method or Newton's method.

In this section we will investigate several such examples in order to gain insight into our problem and also into the nature of perturbation methods in general. These examples will also illustrate the geometrical significance of the nondegeneracy hypothesis and introduce the idea of *bifurcation*, which is important in the degenerate case.

Example 1.3.1

The first example,

$$\varphi(x, \varepsilon) = x^2 - 1 + \varepsilon = 0, \qquad (1.3.1)$$

is easily understood geometrically as well as algebraically. The graph of $y = \varphi(x, 0) = x^2 - 1$ has x intercepts at ± 1 (Fig. 1.3.1). As ε increases, this curve is simply raised until at $\varepsilon = 1$ these intercepts coalesce and, for larger ε, disappear. (Of course, varying ε has this effect only because ε is just added to the equation. When ε enters in more complicated ways, it is impossible to visualize the result of changing ε, except to say that the graph of $y = \varphi(x, 0)$ gradually becomes distorted as ε increases.) The exact algebraic solution

$$x^{(1)} = f(\varepsilon) = +\sqrt{1 - \varepsilon} \qquad (1.3.2)$$

becomes complex for $\varepsilon > 1$. (Unless otherwise stated we will take the position, appropriate for most applied problems, that complex solutions "do not exist.") The exact solution (1.3.2) can be expanded in a Taylor series $x^{(1)} = 1 - \varepsilon/2 - \varepsilon^2/8 + \cdots$, and the reader should verify that this expansion is the same as the formal series solution generated by the method of the last section. Since the exact solution is known for this problem,

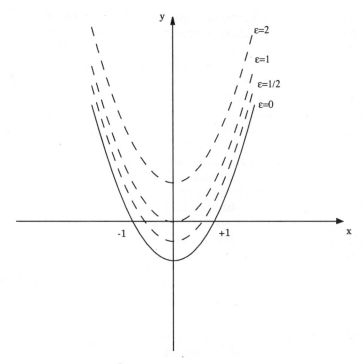

FIG. 1.3.1. Graphs of $y = x^2 - 1 + \varepsilon$. As ε increases, the roots coalesce and then disappear. The pair bifurcation takes place at $\varepsilon = 1$.

Taylor's Theorem gives an easy way to find an upper bound for the error in the approximate solutions. For instance, consider the first approximation $x^{(1)} \cong 1 - \varepsilon/2$ and define the *remainder* to be $R(\varepsilon) := \sqrt{1 - \varepsilon} - (1 - \varepsilon/2)$. Then the exact solution equals the approximation plus the remainder (or error), that is, $x^{(1)} = 1 - \varepsilon/2 + R$. Then Taylor's Theorem (see Appendix A) implies that $|R(\varepsilon)| \leq M\varepsilon^2/2$ for $|\varepsilon| \leq \varepsilon_0$, where $M = \max\{|f''(\varepsilon)| : |\varepsilon| \leq \varepsilon_0\}$. For any given value of ε_0 it is possible to compute M, since $f(\varepsilon)$ is known. For instance, for $\varepsilon_0 = 0.5$ and $\varepsilon_0 = 0.37$ the following results are obtained:

$$|R(\varepsilon)| \leq (0.353553391)\varepsilon^2 \quad \text{for} \quad |\varepsilon| \leq 0.5,$$
$$|R(\varepsilon)| \leq (0.249976503)\varepsilon^2 \quad \text{for} \quad |\varepsilon| \leq 0.37. \tag{1.3.3}$$

(The value $\varepsilon_0 = 0.5$ was chosen arbitrarily: $\varepsilon_0 = 0.37$ was chosen for later use in Section 1.4.) Of course, the results in (1.3.3) are possible only because we have used the quadratic formula to obtain the exact solution; this method is of no use in "real" perturbation problems (those for which the perturbation method is the only method available). In the next section we will derive similar (but less exact) estimates for this same problem without using the exact solution; the method used there is applicable to "real" perturbation problems, at least in principle, although it is rather difficult to apply. As a final remark about the numerical accuracy of the perturbation method for this problem, Table 1.3.1 gives data (obtained by pocket calculator) showing that the actual errors are smaller than the upper bounds given by (1.3.3). For values of ε from 0.01 to 3.00, the table gives the exact solution, the first approximation, the error, and the two upper bounds for the error from (1.3.3). (For ε_0 outside the range of applicability of these upper bounds, the entry "not relevant" appears in the table.) Notice that the approximate solution continues to exist for $\varepsilon > 1$, although there is no actual solution (in the real number system) in this case. *The existence of a formal "approximate solution" does not imply that there really is a solution to which it is an approximation.* This is the reason for the attention that must be paid to the existence of solutions throughout this book.

We have shown both geometrically (from Fig. 1.3.1) and algebraically (from equation (1.3.2)) that the number of solutions of (1.3.1) changes from two to one and then to zero as ε increases through the critical value $\varepsilon = 1$. In Fig. 1.3.2 we have graphed the two solutions for x as functions of ε; the top half of the parabola is the graph of (1.3.2), and the bottom half uses the negative square root. The points $(\varepsilon, x) = (0, \pm 1)$ are the starting points for the perturbation solutions, which (to first order) are shown as dotted lines. It is clear from this figure that the approximate solutions are a good approximation only for small ε, and that for $\varepsilon > 1$ the approximate solutions continue to exist although the actual solutions do not. A warning: Don't confuse Fig. 1.3.1 with Fig. 1.3.2. The former is a graph of y against x for $\varepsilon = 0$, with $y = \varphi(x, 0)$, and only the x-intercepts represent roots. The latter is a graph of x against ε and all points represent roots. It is coincidental that in this simple problem both graphs are parabolas; in more complicated examples they can have quite different shapes. One pesky detail is that although the root-finding problem

TABLE 1.3.1

ε	$\sqrt{1-\varepsilon}$	$1 - \frac{1}{2}\varepsilon$	Error	$(0.249976503)\varepsilon^2$	$(0.353553391)\varepsilon^2$
0.01	0.994987437	0.995	0.000012563	0.000024998	—
0.05	0.974679435	0.975	0.000320566	0.000624941	—
0.10	0.948683298	0.950	0.001316702	0.002499765	—
0.30	0.836660027	0.870	0.013339974	0.022497885	—
0.37	0.793725393	0.815	0.021274607	0.034221783	—
0.40	0.774596669	0.800	0.025403331	not relevant	0.056568543
0.50	0.707106781	0.750	0.042893219	not relevant	0.088388348
1.00	0	0.500	0.500000000	not relevant	not relevant
2.00	complex	0	complex	not relevant	not relevant
3.00	complex	-0.50	complex	not relevant	not relevant

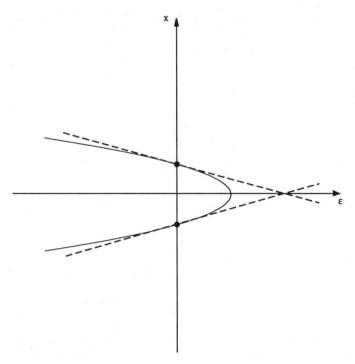

FIG. 1.3.2. The bifurcation diagram for $x^2 - 1 + \varepsilon$. The dotted straight lines are the first order perturbation approximations to the roots through the two starting points. These do not correctly detect the pair bifurcation point.

is generally written as $\varphi(x, \varepsilon) = 0$, the bifurcation diagram is customarily plotted in the (ε, x) plane (that is, the coordinates are reversed). To avoid confusion we will always speak of "the point $(x, \varepsilon) = (0, 1)$" or "the point $(\varepsilon, x) = (1, 0)$" rather than merely "the point $(0,1)$."

Any change in the number or character of solutions when a parameter is varied is called a *bifurcation*. The parameter that is varied is called the *bifurcation parameter*; here, the bifurcation parameter is also the perturbation parameter, but this is not always the case. The graph of solutions versus the bifurcation parameter (Fig. 1.3.2) is called the *bifurcation diagram*. The *bifurcation point* is $(\varepsilon, x) = (1, 0)$. Certain common bifurcation diagrams have been given names descriptive of their shape (such as the "pitchfork bifurcation" which will appear in Section 1.5). The type of bifurcation illustrated in Fig. 1.3.2, in which two solutions coalesce into one and then disappear as the bifurcation parameter is varied, is so common that it has several names in the literature. When it appears in differential equations it is often called a *saddle-node bifurcation*. In other settings it can be called a *limit point bifurcation* or a *tangent bifurcation*. These names refer to special features of the different applications and none of them is suitable in all circumstances. In this book the term *pair bifurcation* will be used to describe this situation.

Example 1.3.2

The next example is closely related to the first:

$$\varphi(x,\varepsilon) = x^2 + \varepsilon = 0. \tag{1.3.4}$$

The graph for $\varepsilon = 0$ and the bifurcation diagram are given in Figs. 1.3.3 and 1.3.4.

The exact solutions $x = \pm\sqrt{-\varepsilon}$ exist for $\varepsilon < 0$ but not for $\varepsilon > 0$. It is not possible to construct perturbation solutions for this example by the method of Section 1.2, because it is a degenerate problem: The starting point is $(x,\varepsilon) = (0,0)$, and $\varphi_x(0,0) = 0$. This fact can be looked at in several ways. First, $\varphi_x(0,0) = 0$ means that the graph of $y = \varphi(x,0)$ in Fig. 1.3.3 is tangent to the x axis at its intercept and is therefore easily pulled away from the x axis by a perturbation so that there is no intercept, or, alternatively, it is pushed across the x axis creating two intercepts.

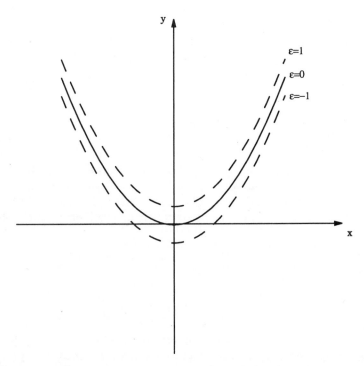

FIG. 1.3.3. $y = x^2 + \varepsilon$ for various ε. The bifurcation is at $\varepsilon = 0$. The reduced equation has a double root at $x = 0$: it is a degenerate problem.

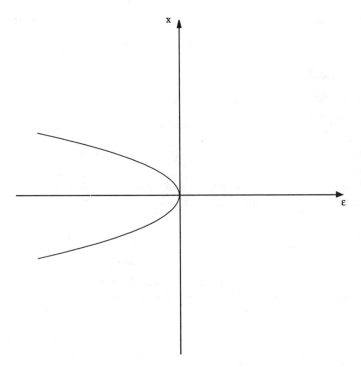

FIG. 1.3.4. The bifurcation diagram for $x^2 + \varepsilon = 0$. The bifurcation point is also the starting point. This is typical for degenerate problems.

This means that *the starting point $(\varepsilon, x) = (0,0)$ is at the same time a bifurcation point*, in this case a pair bifurcation point, in Fig. 1.3.4. In algebraic terms, $x = 0$ *is a double root of $\varphi(x,0) = 0$*, and in general, the nondegeneracy condition $\varphi_x(x_0, 0) \neq 0$ is equivalent to x_0 being a simple (not multiple) root if $\varphi(x,0)$ is a polynomial. In analytic terms, the exact solution $\sqrt{-\varepsilon}$ *cannot be expanded in a Taylor series* in ε because the derivatives do not exist at $\varepsilon = 0$. All of these phenomena are typical of the degenerate case, which will be studied in Sections 1.5 and 1.6.

Example 1.3.3

The equation

$$\varphi(x, \varepsilon) = x^2 + (\varepsilon - 2.01)x + 1.01 = 0 \qquad (1.3.5)$$

is one for which the method we have developed seems to give entirely unsatisfactory results. When $\varepsilon = 0$ the equation factors into $(x-1)(x-1.01) = 0$, with roots $x^{(1)} = 1$ and $x^{(2)} = 1.01$. To find the continuation of the first root, try $x^{(1)}(\varepsilon) = 1 + x_1\varepsilon + \mathcal{O}_F(\varepsilon^2)$. Calculation yields $x_1 = 100$, so

$$x^{(1)} \cong 1 + 100\varepsilon. \qquad (1.3.6)$$

Putting $\varepsilon = 0.1$ in this formula gives $x^{(1)} \cong 11$, whereas the actual value (from the quadratic formula) is complex; with $\varepsilon = -0.1$, the proposed approximation gives $x^{(1)} \cong -9$, whereas in fact $x^{(1)} = 0.734$ to three decimal places. It will follow from the theory in the next section that (1.3.6) does in fact give a good approximation for sufficiently small ε but that $\varepsilon = 0.1$, which is "sufficiently small" for Example 1.3.1, is not nearly small enough for the present problem. There are several "warning signs" present in this problem (or in its approximate solution) that provide a partial explanation of the very large errors which appear for ε as small as 0.1. Before giving these indicators, it should be emphasized that these considerations are not in any way conclusive. For each of the following danger signals, examples can be found in which the warning is present but the approximate solution is quite accurate, and other examples in which inaccuracy is present without warning. Nothing but a complete error analysis along the lines of Section 1.4 is able to resolve fully the question of accuracy, because inaccuracy results from an interaction of factors: The reduced problem must be somehow "sensitive" or unstable, and the perturbation must be one which reacts with this instability to produce a misleading solution. The following "warning signs" each focus attention on only one aspect of this complex story. With this understood, the danger signals are:

1. The reduced problem is nearly degenerate, in two senses:
   (a) The roots (1 and 1.01) are close (nearly a double root);
   (b) $\varphi_x(1, 0) = -0.01$ is nearly zero.
2. The coefficient of ε in the approximate solution (1.3.6) is large.
3. The residual $\rho := \varphi(-9, -0.1) = 101$ is large.

We will discuss each of these in turn.

1a. A graph of $y = \varphi(x, 0)$ is shown in Fig. 1.3.5a. The graph is very nearly tangent to the x axis at $x = 1$, so it does not take a great disturbance to pull it away from the x axis altogether, as happens for $\varepsilon = 0.1$, giving a complex solution. However, the fact that the two roots are close is not proof that the problem is sensitive in this respect. More important is the fact that the graph is close to the x axis between the roots. Figure 1.3.5b shows a graph that is not sensitive in spite of close roots, and Fig. 1.3.5c shows one that is, although the roots are far apart.

1b. An argument can be made that when $\varphi_x(x_0, 0)$ is close to zero, approximate solutions built according to (1.2.8) are likely to be inaccurate. Since $\varphi_x(x_0, 0)$ occurs in the denominator of each coefficient x_n, these coefficients will tend to be large when $\varphi_x(x_0, 0)$ is small, unless the numerators also happen to be small. In our case, $x_1 = 100$, which is rather large, and it is likely that x_2 is also large. Now, the argument goes, Taylor's theorem tells us that the error in any truncation of a Taylor series has a form which strongly resembles (but is not identical to) the first omitted

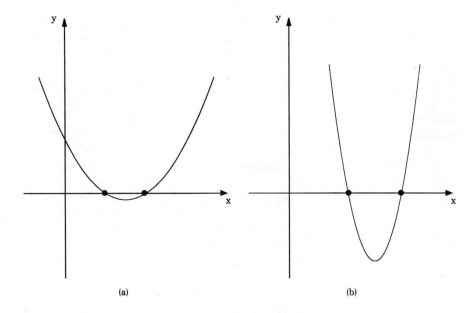

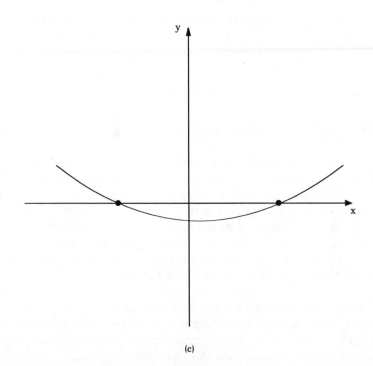

FIG. 1.3.5. (a) Two nearby simple roots that are almost a double root and are easily destroyed by perturbation. (b) Two nearby roots that are not easily destroyed. (c) Two distant roots that are easily destroyed.

term. Thus a clue to the size of the error of $x_0 + x_1\varepsilon$ is the size of $x_2\varepsilon^2$. If $\varphi_x(x_0, 0)$ is small, then not only x_2, but $x_2\varepsilon^2$, is probably large, since the small quantity $\varphi_x(x_0, 0)$ occurs cubed in the denominator of x_2 and ε^2 is only squared. This reasoning is not to be taken too seriously, because the first omitted term is not always a good indication of the numerical size of the error (even when that term is asymptotically the dominant part of the error as $\varepsilon \to 0$, as we will prove to be the case in the next section). The real lesson to be learned here is that an approximation method usually does not fail *suddenly* in a case such as degeneracy; it usually becomes less accurate gradually as the condition for failure (in this case $\varphi_x = 0$) is approached.

2. The fact that $x_1 = 100$ is large is closely connected, as we have just seen, with the fact that $\varphi_x(1, 0)$ is small. However, quite apart from this, the appearance of large coefficients in an approximation such as this is frequently taken as a danger sign in itself. The argument goes like this: In setting up $x \cong x_0 + x_1\varepsilon + x_2\varepsilon^2 + \cdots$ for small ε, the intention is that each additional term should be a small correction to what has gone before. If one term turns out to be comparable in magnitude to the preceding term, then the entire heuristic foundation of the method has failed and there is no longer any reason to believe that the solution is accurate. This is exactly what happens in the case of $1 + 100\varepsilon$ for $\varepsilon = -0.1$; we might expect that $1 + 100\varepsilon$ is a good approximation for $\varepsilon = 0.001$, where $100\varepsilon = 0.1$ is small compared to the previous term 1, but not for $\varepsilon = -0.1$, where $100\varepsilon = -10$. However persuasive this argument might sound, the fact remains that the accuracy of an approximation depends ultimately on the error term, not on the size relationships between terms that appear in the approximation itself. The best justification for the "large coefficient warning sign" is probably this: A large coefficient may signal the presence of *a mechanism that causes* large coefficients. (An example of such a mechanism is the small denominator in 1b.) If such a mechanism is present, it may cause other large coefficients in omitted terms, and these in turn may signal a large error. The following example shows that this does not always happen: $\varphi(x, \varepsilon) = 0.01(x - 1) - \varepsilon = 0$ has the exact solution $x = 1 + 100\varepsilon$, valid for all ε, with a large coefficient and, nevertheless, zero error. (The same example has warning sign 1b and nevertheless has zero error.)

3. Suppose that we have calculated the approximation $x^{(1)} \cong -9$ when $\varepsilon = -0.1$, and (without knowing the exact solution) wish to put this to a test. Since the exact solution should make $\varphi = 0$, a reasonable test would be whether $\varphi(-9, -0.1)$ is close to zero. Calculating this, we find the answer to be 101, which can hardly be considered small. What we have just calculated is called a *residual*; it is the amount by which the approximate solution fails to satisfy the equation that it should. The residual $\rho := \varphi(x_{\text{approx}}, \varepsilon)$ should be sharply distinguished from the error or *remainder* $R := x_{\text{exact}} - x_{\text{approx}}$. Figure 1.3.4 shows that ρ can be large and R small or vice versa, but also suggests that these extremes only occur when the slope of φ is either very

large or very small: In most cases the residual bears some relationship to the remainder, and if the residual seems large it is reasonable to worry about the accuracy of the solution. Some methods of error estimation make explicit use of the residual, together with other factors, in finding error bounds. (See Section 1.4.) Our method clearly fails for (1.3.5), unless we are content with extremely small ε indeed. What are we to do? If our "target problem" is (1.3.5) with $\varepsilon = -0.1$, that is, if we wish to solve

$$x^2 - 2.11x + 1.01 = 0 \qquad (1.3.7)$$

and problems close to it, then (1.3.5) is a poorly formulated perturbation problem to accomplish this task. (Recall here the discussion in Section 1.1 about the arbitrariness in the introduction of ε.) A clue about how to proceed comes from our observation that (1.3.7) is close to degenerate. Formulation (1.3.5) ignores this. It is better to "build in" the fact that the problem is nearly degenerate by creating a perturbation family in which the reduced problem is degenerate, for instance

$$\tilde{\varphi}(x, \varepsilon) = (x - 1)^2 - 1.1\varepsilon x + 0.1\varepsilon = 0, \qquad (1.3.8)$$

which reduces to (1.3.7) when $\varepsilon = 0.1$. This problem, of course, cannot be solved by the methods of Section 1.2, but when it is solved by methods suitable for degenerate problems the result is satisfactory. (See Exercise 1.5.4 and Section 1.7.)

The moral, perhaps, is to face difficulties head on (admit the problem is nearly degenerate and set it up as such) rather than flee from them (by setting up a nondegenerate but highly sensitive problem like (1.3.5).) Another example of the same sort to be encountered later is that a nearly resonant nonlinear oscillator should be treated as a perturbation of a resonant one, even though the problem seems more difficult when posed this way.

Example 1.3.4

Consider the fifth degree polynomial

$$x(x - 1)(x - 2)(x - 3)(x - 4) + \varepsilon(x^2 + x + 1) = 0. \qquad (1.3.9)$$

We have chosen a fifth degree polynomial because it is not possible (according to algebraic field theory) to solve the general fifth degree polynomial by a "closed form" expression using radicals; therefore there is no choice but to use an approximation method. Three such methods suggest themselves: Newton's method, the bisection method, and perturbation theory. Focusing on the root that reduces to $x = 1$ when $\varepsilon = 0$, we have solved the problem by all three methods, with results presented in Table 1.3.2. The perturbation solution used there is the second order approximation

$$x \cong 1 + \frac{2}{3}\varepsilon + \frac{19}{27}\varepsilon^2. \qquad (1.3.10)$$

TABLE 1.3.2

Parameter	Perturbation Method	Newton's Method	Error
0.11	1.0810	1.0618	+0.0200
0.12	1.0901	1.0683	+0.0218
0.17	1.1337	1.1040	+0.0297
0.30	1.2633	1.2591	+0.0042
0.32	1.2854	1.3308	−0.0454
0.3222	1.2879	1.3574	−0.0695
0.3223	1.2880	1.3610	−0.0730
0.3224	1.2881	−0.0129	n/a

This expression was evaluated by computer and rounded to four decimal places for many more values of ε than shown here; the values selected for the table were chosen to illustrate features that will be discussed below.

The Newton's method solutions were obtained in the following way. For $\varepsilon = 0.1$, the first guess taken for the root was $x = 1$. Then Newton's method was iterated until there was no change in the first four decimal places of the approximate root. Next, ε was increased by 0.01, and Newton's method was repeated as before, except that for this and all subsequent stages the starting guess for the root was taken to be the (approximate) root calculated for the previous value of ε.

On the first run it was found that the root jumped from 1.3308 at $\varepsilon = 0.32$ to −0.0132 at $\varepsilon = 0.33$. (This behavior is not unexpected, for reasons to be explained below.) To explore this further, on the second run after reaching $\varepsilon = 0.32$ the step size in ε was reduced to 0.0001; the jump in the root then occurred (as shown in the table) between $\varepsilon = 0.3222$ and 0.3223.

After obtaining these results, they were checked by using the bisection method to find roots between 1 and 1.5 for the specific values of ε shown in the table. Again the aim was four-place accuracy, and in all cases the bisection method agreed with Newton's method to four places, except for $\varepsilon = 0.3223$. At $\varepsilon = 0.3222$, two roots were found, 1.3574 (as shown in the table) and 1.3772. For $\varepsilon = 0.3223$ these were closer, 1.3610 and 1.3735. For $\varepsilon = 0.3224$, no roots were found in this vicinity.

Now it is clear what is happening at this point. A pair bifurcation is taking place; two roots are merging and then disappear as ε is increased. The bisection method, since it is searching for roots in a given interval, finds none after the bifurcation, whereas Newton's method simply converges to a different root, which is not a continuation of the root $x = 1$ of the reduced problem, but of one of the other roots (apparently $x = 0$ in this case). In fact it is easy to predict that this bifurcation will take place, since the reduced polynomial is negative between the two roots at 1 and

2 and the perturbation is positive there. Therefore increasing ε will raise the graph until these two roots coalesce and vanish.

As we have seen in other examples, the perturbation solution does not reveal the bifurcation behavior; the formula (1.3.10) gives a value of the root for any ε. The fourth column of Table 1.3.2, labelled "error," gives the difference between the perturbation solution and the numerical solution; in the last line, where there is no actual solution, the notation "n/a" means "not applicable." Notice that the error begins positive, then increases, but finally decreases again and becomes negative. Of course, the asymptotic error bound increases with ε in a monotone fashion, but the actual error may always be less than the error estimate. (In fact, the table together with the intermediate value theorem implies that there is a value of ε between 0.3 and 0.32 where the error of the perturbation approximation is zero!)

1.4. JUSTIFICATION AND ERROR ESTIMATES

It is time to turn to the question of justifying the approximate solutions found by the method of Section 1.2. The examples in Section 1.3 ought to have convinced you that an "approximate solution" may exist when there is no actual solution to which it corresponds (Example 1.2.1 when $\varepsilon > 1$) and that an "approximate solution" may actually give a very bad approximation to the actual solution (Example 1.3.3). Therefore it is not just an academic matter to worry about the justification of these methods, and the only way to truly settle them is by rigorous mathematical proof. Every student of perturbation theory should at least be aware of the statement and meaning of the major definitions and theorems of the subject, although some students may not need to become involved in the details of the proofs. In this book certain fundamental mathematical theorems are stated carefully in the appendices and then used freely in the text. The extent to which you (the individual student) become involved in these theorems will depend upon your mathematical background. If you have already studied real analysis (with some point set topology) and have had a graduate course in ordinary differential equations, the theorems in the appendices will be familiar to you, and you will have studied their proofs. If not, the appendices attempt to make the theorems intuitively reasonable, and you can accept them without proof, provided you make the effort to understand precisely what they say. In this case, you will begin to see why these theorems are useful and will be ready to appreciate them if you do encounter their proofs in other courses.

For the root-finding problem $\varphi(x, \varepsilon) = 0$ in the nondegenerate case, the existence question is settled by the *implicit function theorem*, one of the major theorems of calculus. Information about this theorem, together with proof sketches, is given in Appendix B. In the special case with which we are concerned here, the theorem may be stated as follows:

Theorem 1.4.1. *Suppose $\varphi(x, \varepsilon)$ has continuous partial derivatives φ_x and φ_ε, and suppose*

$$\varphi(x_0, 0) = 0 \quad and \quad \varphi_x(x_0, 0) \neq 0. \tag{1.4.1}$$

Then the starting solution x_0 has a unique continuation to some interval about $\varepsilon = 0$; that is, there exist a constant $\varepsilon_0 > 0$ and a unique function $x = f(\varepsilon)$ defined for $|\varepsilon| < \varepsilon_0$ such that

$$f(0) = x_0 \quad and \quad \varphi(f(\varepsilon), \varepsilon) = 0. \tag{1.4.2}$$

Furthermore f has a continuous derivative, and if all partial derivatives of φ (including mixed partials) of total order $\leq r$ exist and are continuous then $f', f'', \ldots, f^{(r)}$ exist and are continuous.

In stating theorems of this sort, it is necessary to state technical hypotheses about the number of continuous derivatives that a function has. Most of the time it is reasonable to ignore these hypotheses and merely assume that all functions in question are infinitely differentiable; the last line of Theorem 1.4.1 implies that if φ has partial derivatives of all orders, then so does f. (You may want to think of this as "$r = \infty$," although it is more correct to say that "the statement holds for all integers r" since there is, of course, no such thing as $f^{(\infty)}$.)

Theorem 1.4.1 guarantees the existence of a solution for some interval $|\varepsilon| < \varepsilon_0$. In Example 1.3.1, ε_0 can be taken to be 1 (or any positive number less than 1). Theorem 1.4.1 does not give a method for finding the "radius" ε_0 of the interval of existence, but merely states that there is such an interval; this is typical of theoretical results in perturbation theory. The phrase "so-and-so is true for sufficiently small ε" is often used in place of the more exact expression "there exists an ε_0 such that so-and-so is true for $|\varepsilon| \leq \varepsilon_0$." (Sometimes, instead of this, sufficiently small ε means $0 \leq \varepsilon \leq \varepsilon_0$ or even $0 < \varepsilon < \varepsilon_0$. The meaning is usually clear from the context.) In this book, proclaimed theorems (with theorem numbers, appearing in italics) will be stated in their precise form. Outside of such theorem statements, almost every statement or equation involving ε is only intended to hold for sufficiently small ε, and it is tedious to repeat this constantly, so it will only be mentioned occasionally when it is important to have it in mind. (The frequency of such statements will decrease as the book goes on and you become used to thinking in this way.) As an example of the general principle that any statement involving ε is only required to hold for sufficiently small ε, observe that it is not even necessary to assume in Theorem 1.4.1 that the function $\varphi(x, \varepsilon)$ is defined for all ε. The theorem is still true if φ is only defined in some interval $|\varepsilon| < \varepsilon^*$. In this case the ε_0 occurring in the theorem cannot be larger than ε^*, and may be smaller.

Having settled the existence question, we go on to consider the error of the approximations constructed in Section 1.2. For simplicity, let us suppose that $\varphi(x, \varepsilon)$ is a polynomial in x and ε, or at least a function that is infinitely differentiable. Then Theorem 1.4.1 guarantees that f is infinitely differentiable (this means that it has derivatives of all orders), and Taylor's theorem (Appendix A, Theorem A.1) implies that $f(\varepsilon)$ can be approximated by a Taylor polynomial of any desired order. That is, choose a positive integer k and let

$$p_k(\varepsilon) := f(0) + f'(0)\varepsilon + \frac{1}{2}f''(0)\varepsilon^2 + \cdots + \frac{1}{k!}f^{(k)}(0)\varepsilon^k. \qquad (1.4.3)$$

Then for any ε_1 in the interval $0 < \varepsilon_1 < \varepsilon_0$ there is a constant $c_1 > 0$ such that

$$|f(\varepsilon) - p_k(\varepsilon)| \leq c_1 \varepsilon^{k+1} \qquad \text{for} \quad |\varepsilon| \leq \varepsilon_1. \qquad (1.4.4)$$

There are two reasons for giving this estimate on a smaller interval $|\varepsilon| \leq \varepsilon_1$ than the full interval $|\varepsilon| < \varepsilon_0$ on which $f(\varepsilon)$ is known to exist. One is the practical reason that taking a smaller interval may give a better (smaller) value for c_1; this is seen already in (1.3.3). The other reason is that in the proof of Taylor's theorem, c_1 involves the maximum of $f^{(k+1)}(\varepsilon)$ over $|\varepsilon| \leq \varepsilon_1$. Such a maximum is only guaranteed to exist for a closed bounded interval of ε, and the interval $|\varepsilon| < \varepsilon_0$ given by the implicit function theorem is open. (For instance, in Example 1.3.1 with $\varepsilon_0 = 1$, the function $f(\varepsilon)$ exists at $\varepsilon = 1$, but the first derivative is already unbounded there.)

The inequality (1.4.4) is an error estimate for the Taylor polynomial approximation of $f(\varepsilon)$. It is not immediately clear that the same error estimate applies to the polynomials constructed by the method of Section 1.2. In order to show this, it is necessary to show that in fact the polynomials obtained by that method are exactly the same as the Taylor polynomials. One way to do this is to compute the Taylor polynomials $p_k(\varepsilon)$ in terms of φ and to compare the results with Section 1.2. We will now carry this out for the case $k = 2$. Beginning with $\varphi(f(\varepsilon), \varepsilon) = 0$, differentiate twice with respect to ε:

$$\varphi_x f' + \varphi_\varepsilon = 0,$$
$$\varphi_{xx}(f')^2 + \varphi_x f'' + 2\varphi_{x\varepsilon}f' + \varphi_{\varepsilon\varepsilon} = 0. \qquad (1.4.5)$$

Set $\varepsilon = 0$ and $f(0) = x_0$, and solve for $f'(0)$ and $f''(0)$:

$$f'(0) = -\left.\frac{\varphi_\varepsilon}{\varphi_x}\right|_{(x_0, 0)},$$
$$f''(0) = \left.\frac{-\varphi_{xx}\varphi_\varepsilon^2 + 2\varphi_{x\varepsilon}\varphi_x\varphi_\varepsilon - \varphi_{\varepsilon\varepsilon}\varphi_x^2}{\varphi_x^3}\right|_{(x_0, 0)}. \qquad (1.4.6)$$

Compare (1.4.6) with (1.2.8); it is easy to see that $p_2(\varepsilon) = f(0) + f'(0)\varepsilon + f''(0)\varepsilon^2/2 = x_0 + x_1\varepsilon + x_2\varepsilon^2$. Thus $p_2(\varepsilon)$ coincides with the second order approximation found heuristically in Section 1.2. This computation can be done to any order. (See Exercise 1.4.1.) We have assumed that φ is infinitely differentiable. If φ has only r continuous derivatives, one can only get as far as $k = r - 1$ in constructing $p_k(\varepsilon)$, saving the last derivative to get the estimate (1.4.4). These calculations prove the following theorem:

Theorem 1.4.2. *Under the same hypotheses as Theorem 1.4.1, the function $f(\varepsilon)$ has a polynomial approximation*

$$f(\varepsilon) \cong x_0 + x_1\varepsilon + \cdots + x_k\varepsilon^k$$

for each $k \leq r - 1$ which can be found as in Section 1.2. The remainder $R(\varepsilon)$ defined by

$$f(\varepsilon) = x_0 + x_1\varepsilon + \cdots + x_k\varepsilon^k + R(\varepsilon)$$

satisfies an estimate of the form

$$|R(\varepsilon)| \leq c_1\varepsilon^{k+1} \quad \text{for} \quad |\varepsilon| \leq \varepsilon_1 \quad\quad (1.4.7)$$

for any $\varepsilon_1 < \varepsilon_0$. The constant c_1 may depend upon the choice of ε_1, but once an ε_1 is fixed, c_1 does not depend in any further way upon ε.

Our proof of Theorem 1.4.2 was based upon (1.4.6) and essentially involves rederiving the results of Section 1.2 by a more rigorous method. There is another way to prove Theorem 1.4.2. One begins with (1.4.3) and (1.4.4), which show that there exists a polynomial approximation satisfying the desired error estimates. Knowing that such a polynomial exists, it may be written as $x_0 + x_1\varepsilon + x_2\varepsilon^2 + \cdots + x_k\varepsilon^k$, where the x_n are to be determined. This is the same starting point as in Section 1.2 (see, for instance, (1.2.5)), except that in Section 1.2 this form of solution was merely conjectured and now it is known to exist. From here, one proceeds through the steps leading from (1.2.5) to (1.2.8), *showing that these steps are in fact legitimate operations when performed on a polynomial satisfying (1.4.4)*. In Section 1.2 these steps were carried out carelessly; we are now proposing to carry them out rigorously. It should be clear that if this can be done, we will have another proof of Theorem 1.4.2. The details of this proof require various theorems that will be presented in Section 1.8, and the proof itself will be assigned to the reader as Exercise 1.8.6.

Inequality (1.4.7) is usually expressed in words as "the error is of the order of the first omitted term," and is abbreviated $R(\varepsilon) = \mathcal{O}(\varepsilon^{k+1})$. The approximate solution is then written $f(\varepsilon) = x_0 + x_1\varepsilon + \cdots + x_k\varepsilon^k + \mathcal{O}(\varepsilon^{k+1})$. The symbol $\mathcal{O}$, read "big-oh of" or "order of", always refers to an estimate for the *error* in an approximate solution, as opposed to $\mathcal{O}_F$, which refers not to the actual error (the difference between exact and approximate solutions) but to the omitted terms in a formal solution. Thus if we say $f(\varepsilon) \cong x_0 + \varepsilon x_1$, it merely means $x_0 + \varepsilon x_1$ is proposed as an approximation to $f(\varepsilon)$. If we write $f(\varepsilon) = x_0 + \varepsilon x_1 + \mathcal{O}_F(\varepsilon^2)$, it means $x_0 + \varepsilon x_1$ are the first two terms of a formal solution continuing with terms in ε^2 and beyond. But if we say $f(\varepsilon) = x_0 + \varepsilon x_1 + \mathcal{O}(\varepsilon^2)$, it means that an actual estimate of the difference between $f(\varepsilon)$ and $x_0 + \varepsilon x_1$ has been performed; there are constants c_1 and ε_1 such that $|f(\varepsilon) - x_0 - \varepsilon x_1| \leq c_1 \varepsilon^2$ for $|\varepsilon| \leq \varepsilon_1$. The exact rules governing the $\mathcal{O}$ symbol will be given in Section 1.8; until then we will state all estimates in their precise form such as (1.4.7) as well as give their $\mathcal{O}$ equivalents.

> Perhaps a warning is in order to the beginning student. In many calculus texts, part of the theorem called the "alternating series test" says that for convergent alternating series satisfying certain conditions the error in taking a partial sum is *less than* the first omitted term. In Theorem 1.4.2, we only claim that the error is *of the order of* the first omitted term, in the sense made precise by (1.4.7). The first omitted term is $x_{k+1}\varepsilon^{k+1}$, and x_{k+1} might be quite different numerically from c_1. The point is that there is an upper bound for the error which involves the same power of ε as the first omitted term, and therefore in some sense "goes to zero at the same rate" as the first omitted term, when ε approaches zero. A great deal more will be explained about the concept of asymptotic order in Section 1.8.)

Theorem 1.4.2 involves two constants ε_1 and c_1, but provides no way of computing these constants. We will now show, by way of an example, how suitable values of ε_1 and c_1 can be obtained. For most problems of actual interest, it is very difficult to carry out these calculations, and it is almost never done. In practice, a result such as Theorem 1.4.2 provides a reason to trust the approximations to some extent, and then one can turn to numerical or experimental data for more detailed information about accuracy and range of validity. But it is worthwhile to see how concrete error estimates can be found.

The example we will work with is (1.3.1), $x^2 - 1 + \varepsilon = 0$. The approximate solution $x^{(1)} \cong 1 - \varepsilon/2$ has error estimates (1.3.3) found using the quadratic formula. In order to proceed without the quadratic formula, and so illustrate a method usable in general, the first step is to obtain an equation satisfied by the error R. Putting $x^{(1)} = 1 - \varepsilon/2 + R$ into (1.3.1) leads to

$$\Phi(R, \varepsilon) = R^2 + 2R - \varepsilon R + \tfrac{1}{4}\varepsilon^2 = 0. \qquad (1.4.8)$$

When $\varepsilon = 0$ this has two roots, $R^{(1)} = 0$ and $R^{(2)} = -2$; clearly the zero root is the difference between $x^{(1)}$ and $1 - \varepsilon/2$ when $\varepsilon = 0$, while $R^{(2)}$ is the difference between $x^{(2)}$ and $1 - \varepsilon/2$. So it is $R^{(1)}(\varepsilon)$, the continuation of the zero root, that we wish to study. Since $\Phi(0, 0) = 0$ and $\Phi_R(0, 0) = 2 \neq 0$, Theorem 1.4.1 guarantees the existence of $R^{(1)}(\varepsilon)$, which from here on will be called simply $R(\varepsilon)$. Implicit differentiation of (1.4.8) gives

$$R' = \frac{2R - \varepsilon}{2(2 + 2R - \varepsilon)}, \qquad R'' = \frac{-2}{(2 + 2R - \varepsilon)^3}. \qquad (1.4.9)$$

Since $R(0) = 0$, it follows from the first equation that $R'(0) = 0$. Taylor's theorem then implies $|R(\varepsilon)| \leq M\varepsilon^2/2$, where M is an upper bound for $|R''|$ on the interval from 0 to ε. Such an upper bound can be obtained from the second equation of (1.4.9).

First, observe that there is a line in the (ε, R) plane on which the denominator of R'' vanishes. (See Fig. 1.4.1). Choose a box $|\varepsilon| \leq \varepsilon_1$, $|R| \leq R_1$ around $(0,0)$ that misses this line. Let M be the maximum of $|2/(2 + 2R - \varepsilon)^3|$ over this box; it can be seen by inspection that for this example, the maximum occurs at the lower right hand corner, so $M = 2/(2 - 2R_1 - \varepsilon_1)^3$. (In general it can be difficult to find this maximum. One must examine critical points of the function in the interior of the box, as well as its restriction to the edges.) Then it follows that $|R(\varepsilon)| \leq M\varepsilon^2/2$ as long as $(\varepsilon, R(\varepsilon))$ remains in the box. In other words

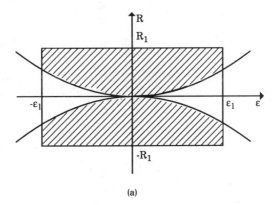

(a)

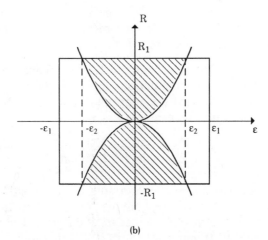

(b)

FIG. 1.4.1. If $|\varepsilon|$ is less than ε_1 in figure (a) or less than ε_2 in figure (b), then the remainder R is confined to the unshaded area between the parabolas.

$(\varepsilon, R(\varepsilon))$ is excluded from the shaded regions in Fig. 1.4.1. There are two possibilities. Either the parabolas $\pm M\varepsilon^2/2$ intersect the sides of the box (Fig. 1.4.1a) or the top and bottom (Fig. 1.4.1b). In the former case the inequality $|R(\varepsilon)| \leq M\varepsilon^2/2$ is valid for $|\varepsilon| \leq \varepsilon_1$, and we have found suitable constants (ε_1 and $c_1 := M/2$) for Theorem 1.4.2 (with $k = 1$). In the second case, $|R(\varepsilon)| \leq M\varepsilon^2/2$ is only valid for $|\varepsilon| \leq \varepsilon_2$, the point where the parabola leaves the box. Again we have found suitable constants (ε_2 and $M/2$). (We have assumed that $R(\varepsilon)$ exists for $|\varepsilon| \leq \varepsilon_1$ in the first case and for $|\varepsilon| \leq \varepsilon_2$ in the second. But it follows from the proof of the implicit function theorem that $R(\varepsilon)$ can cease to exist only by becoming infinite or by having a vertical tangent, and neither of these can occur in the box. So this assumption is correct.)

If we take $\varepsilon_1 = 0.5$ and $R_1 = 0.05$, then $M/2 = (1.4)^{-3} = 0.364431487$. The parabola $M\varepsilon^2/2$ crosses the top of the box at $\varepsilon_2 = 0.370405184$. This gives the estimate

$$|R(\varepsilon)| \leq (0.364431487)\varepsilon^2 \quad \text{for} \quad |\varepsilon| \leq 0.370405184.$$

Since the result is not good out to 0.5, it is better to try again with $\varepsilon_1 = 0.37$ and $R_1 = 0.05$. This gives $M/2 = (1.53)^{-3} = 0.279206618$ and

$$|R(\varepsilon)| \leq (0.279206618)\varepsilon^2 \quad \text{for} \quad |\varepsilon| \leq 0.37. \tag{1.4.10}$$

This estimate is slightly weaker than the second part of (1.3.3).

Exercises 1.4

1. Compute the next equation (for x_3) in (1.2.8) and the next equation (for $f'''(0)$) in (1.4.6). Show that the results are consistent with each other and with Taylor's theorem.

2. Does Theorem 1.4.2 apply to Example 1.3.3? If Theorem 1.4.2 is supposed to "justify" the approximations in this problem, is the theorem contradicted by the fact that the approximations are so bad? (The next exercise continues with this example.)

3. Try to carry out an error analysis of Example 1.3.3 along the lines of our error analysis for Example 1.3.1. What can you say from the denominator of R'' about the size of ε_1, and what does this tell you about why the approximations in Example 1.3.3 are bad? (It is difficult to continue the error analysis beyond this point, because it is not clear at what point the maximum value M occurs and a good deal of calculation is required to find it. If you want to try it, the maximum can be found by looking for critical points in the interior of the box and boundary critical points on the four sides.)

4. Review the concepts of *remainder* and *residual* at the end of Section 1.3. Then examine (1.4.8) and determine how the residual appears in it. This shows that the residual affects the remainder, but is not the only thing that influences it.

1.5. THE METHOD OF UNDETERMINED GAUGES

One of the most widely used heuristic methods in perturbation theory, but one which does not seem to have a generally accepted name, is what we will call the "method of undetermined gauges." Briefly stated, the method is to seek a solution in the form

$$x \cong x_0 \delta_0(\varepsilon) + x_1 \delta_1(\varepsilon) + \cdots + x_k \delta_k(\varepsilon) \qquad (1.5.1)$$

in place of the less general polynomial form $x \cong x_0 + x_1 \varepsilon + \cdots + x_k \varepsilon^k$; the function $\delta_n(\varepsilon)$, called gauges, must be determined sequentially along with the coefficients x_n. This method, in combination with others, is essential for complicated problems such as arise in fluid flow, where it is quite impossible to guess in advance what gauge functions will be required. We will illustrate the use of this method by finding approximate roots of polynomials in the degenerate case, and also by finding roots which are unbounded as $\varepsilon \to 0$. Although the method is powerful as an exploratory device, it has both practical and theoretical drawbacks: The computations can become longer than necessary, and they are difficult to justify. A more satisfactory method for these problems will be presented in the next section, using insights gained from the examples studied here.

Why is it reasonable to expect that a degenerate root-finding problem might have a solution in the form (1.5.1)? Consider the degenerate problem $x^2 - 10x + 25 - 4\varepsilon - 4\varepsilon^2 = 0$, with exact solutions $x = 5 \pm 2\sqrt{\varepsilon + \varepsilon^2}$. These solutions cannot be expanded in a power series in ε, because they are not differentiable at $\varepsilon = 0$, but they can easily be expanded in power series in $\sqrt{\varepsilon}$. For instance, $x^{(1)} = 5 + 2\sqrt{\varepsilon + \varepsilon^2} = 5 + 2\sqrt{\varepsilon}\sqrt{1 + \varepsilon} = 5 + 2\varepsilon^{1/2} + \varepsilon^{3/2} + \mathcal{O}(\varepsilon^{5/2})$, using the "binomial" Taylor expansion of $\sqrt{1 + \varepsilon}$. This has the form (1.5.1), with $x_0 = 5$, $\delta_0 = 1$, $x_1 = 2$, $\delta_1 = \varepsilon^{1/2}$, and so on. The need for $\varepsilon^{1/2}$ can be thought of as follows: The root $x^{(1)}(\varepsilon)$ departs from 5 as ε increases from 0 at a faster than linear rate, whereas any power series $5 + a\varepsilon + b\varepsilon^2 + \cdots$ would be able to express only a linear rate of divergence (in the term $a\varepsilon$) modified by even smaller corrections ($b\varepsilon^2$ and so forth). The gauge function $\varepsilon^{1/2}$ expresses exactly the rate of divergence from the unperturbed solution needed for this particular problem, but other problems will have their own rates of divergence, and so the gauges cannot be imposed in advance. (The term "rate" is being used loosely here and does not mean derivative. In fact $dx^{(1)}/d\varepsilon = \infty$ at $\varepsilon = 0$, and the gauge function $\varepsilon^{1/2}$ gives more precise information than this about the rate of divergence.)

The only assumption to be placed on the gauge functions $\delta_n(\varepsilon)$ in advance is that they are arranged in "order of importance." In other words, the lowest order approximation should be $x \cong x_0\delta_0(\varepsilon)$; the term $x_1\delta_1(\varepsilon)$ should be a smaller correction (small, that is, in absolute value), $x_2\delta_2(\varepsilon)$ should be smaller still, and so forth, at least when ε is sufficiently small. This will be the case, provided $\delta_{n+1}(\varepsilon)/\delta_n(\varepsilon) \to 0$ as $\varepsilon \to 0$ for each n; this condition will be expressed as $\delta_{n+1} \ll \delta_n$. Notice that $\ll$ is quite different from $<$; it is neither stronger nor weaker, just different. For instance, in the case of polynomial gauges $\delta_n(\varepsilon) = \varepsilon^n$ one has $\varepsilon^0 \gg \varepsilon^1 \gg \varepsilon^2 \gg \cdots$, but the similar statement $|\varepsilon^0| > |\varepsilon^1| > |\varepsilon^2| > \cdots$ is not always true, but holds only for $|\varepsilon| < 1$. On the other hand, $|2\varepsilon| > |\varepsilon|$ for all $\varepsilon \neq 0$, but $2\varepsilon \gg \varepsilon$ is false. Loosely speaking, $\delta(\varepsilon) \ll \delta'(\varepsilon)$ means that δ can be ignored when δ' is present (δ is much smaller than δ' for small ε). The symbol $\delta \approx \delta'$ means δ and δ' are of comparable magnitude, in the sense that their ratio tends to a nonzero (and noninfinite) limit as $\varepsilon \to 0$. These symbols (and others related to them) will be defined more precisely in Section 1.8. Since nothing in this section is intended to be rigorous, the ideas will be developed only through examples.

It is generally assumed that none of the coefficients x_n in (1.5.1) are zero, since if one of them is zero the corresponding gauge function can simply be omitted from the sequence. The following terminology is then used. The approximation $x_0\delta_0(\varepsilon)$ is called the *zeroth order, leading order,* or *one-term* approximation. Equation (1.5.1) with $k = 1$ is called the *first order* or *two-term* approximation, and so forth. The naming of these approximations differs in different applications; for instance, what we call the zeroth approximation is sometimes called the first. Our convention (explained in more detail in Section 5.1, where the distinctions become more important) is that the zeroth approximation is the solution of the reduced problem, the first approximation is the first improvement of this, and so forth. Until Chapter 5 the first term of the approximate solution always coincides with the solution of the reduced problem, and so it is the zeroth approximation.

Example 1.5.1

Consider the problem

$$\varphi(x, \varepsilon) = (x - 2)^2(x - 3) + \varepsilon = 0, \tag{1.5.2}$$

which has already been encountered in Exercise 1.2.2(c). The reduced problem has roots $x^{(1)} = x^{(2)} = 2$, $x^{(3)} = 3$. Since 3 is a simple root, that is, $\varphi(3, 0) = 0$ and $\varphi_x(3, 0) \neq 0$, the root $x^{(3)}(\varepsilon)$ near 3 can be approximated by a polynomial in ε, as in Sections 1.2 and 1.3, with the result that $x^{(3)} = 3 - \varepsilon + \mathcal{O}(\varepsilon^2)$. But $\varphi_x(2, 0) = 0$, so the continuation problem for $x^{(1)}$ and $x^{(2)}$ is degenerate and cannot be solved by these methods. (If you try $x^{(1)} \cong 2 + x_1\varepsilon$, solving for x_1 is impossible without dividing by zero; see (1.2.8).) So it is reasonable to try a solution in the form

$$x \cong 2 + x_1\delta_1(\varepsilon) + x_2\delta_2(\varepsilon). \qquad (1.5.3)$$

It is not necessary to use an unknown gauge in the first term since the starting solution 2 is known and we only need to find small corrections. (In other words $\delta_0(\varepsilon) = 1$ and $x_0 = 2$ in (1.5.1).)

If (1.5.3) is substituted into (1.5.2) and the result is multiplied out, one obtains a confusing assortment of terms involving δ_1^3, $\delta_1^2\delta_2$, and other combinations of the unknown gauges, and it is difficult to decide which terms are most significant. It is much simpler to find the first approximation first, then the second, and so forth, to as many terms as needed. Substituting

$$x \cong 2 + x_1\delta_1 \qquad (1.5.4)$$

into (1.5.2) leads to the fairly simple equation

$$-x_1^2\delta_1^2 + x_1^3\delta_1^3 + \varepsilon \cong 0. \qquad (1.5.5)$$

Three functions of ε appear: δ_1^2, δ_1^3, and ε. The gauge function δ_1 is unknown, but it is small ($\delta_1 \ll 1$) and so $\delta_1^3 \ll \delta_1^2$. It is reasonable to ignore the term $x_1^3\delta_1^3$ in comparison with $-x_1^2\delta_1^2$ and simplify (1.5.5) to

$$-x_1^2\delta_1^2 + \varepsilon \cong 0. \qquad (1.5.6)$$

The two remaining order functions δ_1^2 and ε must now be compared. If the δ_1^2 term were dominant we could drop the ε and conclude $x_1 = 0$, but this would lead to no useful correction term in (1.5.4). If the ε term were dominant, we could ignore the δ_1^2 term and conclude that $\varepsilon \cong 0$; this is not startling news, but leaves us with no way to find x_1. The only way to make progress is to assume that δ_1^2 and ε are of equal significance for small ε, that is, $\delta_1^2 \approx \varepsilon$. The simplest way to achieve this is to set $\delta_1^2 = \varepsilon$ and so define

$$\delta_1(\varepsilon) := \varepsilon^{1/2}. \qquad (1.5.7)$$

This is not the only possible choice for δ_1; any gauge function of comparable magnitude as $\varepsilon \to 0$ could be used, such as $\delta_1' := 2\varepsilon^{1/2} + \varepsilon\ln\varepsilon$. (Here $\varepsilon\ln\varepsilon$ is small compared to $2\varepsilon^{1/2}$ for small ε, that is, $\varepsilon\ln\varepsilon \ll 2\varepsilon^{1/2}$, and therefore $\delta_1' \approx \delta_1$.) But (1.5.7) is the simplest choice and allows the approximate solution to be constructed, for when (1.5.7) is put into (1.5.6) we find $x_1 = \pm 1$, giving two solutions for (1.5.4):

$$x^{(1)} \cong 2 + \varepsilon^{1/2},$$
$$x^{(2)} \cong 2 - \varepsilon^{1/2}. \qquad (1.5.8)$$

These are the formal first approximations to the two roots of (1.5.2) near 2. The form of (1.5.8) immediately suggests that these roots are real for small positive ε and complex (hence "do not exist") for small negative ε. In other words, it appears that (as in Example 1.3.2) the starting point for this degenerate problem is a pair bifurcation point. The method of rescaled coordinates (in the next section) will make it obvious that this conjecture is true (see Exercise 1.6.2), but the present method does not rule out the possibility that higher order corrections to (1.5.8) are complex even for positive ε.

To compute the second approximation for $x^{(1)}$, substitute

$$x \cong 2 + \varepsilon^{1/2} + x_2 \delta_2(\varepsilon) \tag{1.5.9}$$

into (1.5.2), obtaining

$$-2x_2(\varepsilon^{1/2}\delta_2) - x_2^2\delta_2^2 + \varepsilon^{3/2}$$
$$+ 3x_2(\varepsilon\delta_2) + 3x_2^2(\varepsilon^{1/2}\delta_2^2) + x_2^3\delta_2^3 \cong 0. \tag{1.5.10}$$

Examining the gauges appearing in (1.5.10), one can see that $\delta_2^3 \ll \delta_2^2$, $\varepsilon^{1/2}\delta_2^2 \ll \delta_2^2$, and $\varepsilon\delta_2 \ll \varepsilon^{1/2}\delta_2$. Omitting the *subdominant terms* (those with the clearly smaller gauges) leaves

$$-2x_2(\varepsilon^{1/2}\delta_2) - x_2^2\delta_2^2 + \varepsilon^{3/2} \cong 0, \tag{1.5.11}$$

in which nothing can be said about the relative dominance of the three terms until δ_2 is chosen. For this sample problem we will consider in detail the possibilities for δ_2 in order to arrive at a rule for choosing gauges easily. First, consider the choices of δ_2 which make two terms of (1.5.11) equal in significance. These are the so-called *distinguished gauges* for the problem. If $\varepsilon^{1/2}\delta_2 = \delta_2^2$, then $\delta_2 = \varepsilon^{1/2}$; if $\varepsilon^{1/2}\delta_2 = \varepsilon^{3/2}$, then $\delta_2 = \varepsilon$; if $\delta_2^2 = \varepsilon^{3/2}$, then $\delta_2 = \varepsilon^{3/4}$. Arranging these distinguished gauges in decreasing order, we have $\varepsilon^{1/2} \gg \varepsilon^{3/4} \gg \varepsilon$. In relation to these there are seven possibilities for δ_2:

$$
\begin{array}{lll}
\text{(i)} & \delta_2 \gg \varepsilon^{1/2} & \\
\text{(ii)} & \delta_2 \approx \varepsilon^{1/2} & \\
\text{(iii)} & \varepsilon^{1/2} \gg \delta_2 \gg \varepsilon^{3/4} & \\
\text{(iv)} & \delta_2 \approx \varepsilon^{3/4} & \text{(1.5.12)} \\
\text{(v)} & \varepsilon^{3/4} \gg \delta_2 \gg \varepsilon & \\
\text{(vi)} & \delta_2 \approx \varepsilon & \\
\text{(vii)} & \varepsilon \gg \delta_2 & \\
\end{array}
$$

Cases (i) and (ii) can be rejected at once because δ_2 is too large: The philosophy behind the method requires $\delta_2 \ll \delta_1 = \varepsilon^{1/2}$. Case (iii) makes

$\varepsilon^{1/2}\delta_2 \gg \delta_2^2 \gg \varepsilon^{3/2}$, so that the dominant term in (1.5.11) is the first term, and (ignoring the others) $x_2 = 0$. This means that such a δ_2 leads to no improvement in (1.5.9), and we reject this case. In case (iv) we take $\delta_2 := \varepsilon^{3/4}$ for simplicity, and find that while this makes the second and third terms of (1.5.11) equal in significance (of order $\varepsilon^{3/2}$), they are not the dominant terms. Again $x_2 = 0$ and this case also is to be rejected. In case (v) the first term is still dominant and the result is the same. As δ_2 is decreased to ε (case (vi)), the first term, which has been dominant until now, comes into balance with the third term, and the two together dominate the middle term. Taking

$$\delta_2 := \varepsilon \qquad (1.5.13)$$

and discarding the middle term gives $-2x_2 + 1 = 0$ or $x_2 = 1/2$. Therefore

$$x^{(1)} \cong 2 + \varepsilon^{1/2} + \tfrac{1}{2}\varepsilon. \qquad (1.5.14)$$

Continuing through the list of possibilities, in case (vii) the dominant term in (1.5.11) is the third, and x_2 cannot be found. Only one choice of δ_2 is successful, namely (1.5.13). (More accurately: Any $\delta_2' \approx \varepsilon$ could be used, but ours is the simplest.) A similar calculation gives the second order correction to $x^{(2)}$, and the final results for this problem are

$$x^{(1)} \cong 2 + \varepsilon^{1/2} + \tfrac{1}{2}\varepsilon,$$
$$x^{(2)} \cong 2 - \varepsilon^{1/2} + \tfrac{1}{2}\varepsilon, \qquad (1.5.15)$$
$$x^{(3)} \cong 3 - \varepsilon.$$

We have carried each solution to order ε, which requires the second approximation for $x^{(1)}$ and $x^{(2)}$ but only the first for $x^{(3)}$, since that uses only polynomial gauges.

It would be tedious to work through such a long list of possible gauges every time a problem is solved. Fortunately there is a rule of thumb which makes this unnecessary. Any *significant gauge* (that is, a gauge which will yield a solution) must meet three conditions:

(I) δ_n must be of higher order than the gauges $\delta_1, \dots, \delta_{n-1}$ previously used in the same solution.

(II) δ_n must be a distinguished gauge, defined above as one which makes at least two terms balance in the equation for x_n. (This equation should be simplified by omitting subdominant terms before calculating distinguished gauges.)

(III) The terms which are brought into balance by condition (II) must become the dominant terms.

In (1.5.12) condition (I) rules out the cases (i) and (ii); (II) eliminates (i), (iii), (v), and (vii); and (III) eliminates (iv). So the procedure is simply to set the gauges in any two terms equal and see if the resulting δ_n makes those terms dominant and is of higher order than previously used gauges. All pairs of terms must be checked even if one significant gauge has already been found, because it is possible when a problem has multiple solutions that there will be more than one significant gauge. Various possibilities will be illustrated in the following examples and in the exercises. (The rule that all combinations must be checked applies only after all terms that are clearly subdominant have been removed. Thus there is no need to try $\varepsilon^{1/2}\delta_2 = \delta_2^3$ in (1.5.10) because δ_2^3 is less significant than δ_2^2 regardless of how δ_2 is chosen. It is only necessary to try combinations occurring in (1.5.11).)

Example 1.5.2

$$x^3 + \varepsilon x + 2\varepsilon^2 = 0. \tag{1.5.16}$$

The reduced problem $x^3 = 0$ has a triple root $x = 0$. It is therefore to be expected that the full equation has three roots near $x = 0$, which we seek in the form

$$x \cong 0 + x_1\delta_1. \tag{1.5.17}$$

Substituting this into (1.5.16) gives

$$x_1^3\delta_1^3 + x_1(\varepsilon\delta_1) + 2\varepsilon^2 \cong 0; \tag{1.5.18}$$

there are no subdominant terms to discard at this stage. The distinguished gauges are found by setting $\delta_1^3 = \varepsilon\delta_1$, $\delta_1^3 = \varepsilon^2$, and $\varepsilon\delta_1 = \varepsilon^2$; solving for δ_1 gives $\varepsilon^{1/2}$, $\varepsilon^{2/3}$, and ε. If $\delta_1 := \varepsilon^{2/3}$, the first and third terms in (1.5.18) are of equal order ε^2, but they are not dominant because the middle term is of lower (more significant) order $\varepsilon^{5/3}$. So the gauge $\delta_1 := \varepsilon^{2/3}$, although distinguished, is not significant and should be discarded. The remaining gauges $\varepsilon^{1/2}$ and ε are significant. Taking $\delta_1 := \varepsilon^{1/2}$ and discarding ε^2 turns (1.5.18) into $x_1^3 + x_1 = 0$, with three roots $x_1 = 0$, i, $-i$. If $x_1 = 0$, then no correction term is generated in (1.5.17); this means that $\varepsilon^{1/2}$ is not the leading term in the solution for that root. Setting $x_1 = 0$ aside, the other values for x_1 give useful solutions

$$x^{(1)} \cong i\sqrt{\varepsilon},$$
$$x^{(2)} \cong -i\sqrt{\varepsilon}. \tag{1.5.19}$$

Since $\varepsilon^{1/2} = \sqrt{\varepsilon}$ is pure imaginary when $\varepsilon < 0$, these approximate solutions are real in that case. For real solutions, it is better to rewrite these as

$$x^{(1)} \cong \sqrt{-\varepsilon},$$
$$x^{(2)} \cong -\sqrt{-\varepsilon}, \tag{1.5.20}$$

which avoids imaginary coefficients. The symbol $\sqrt{-\varepsilon}$ of course denotes the positive square root of $-\varepsilon$ when ε is negative. This can be thought of as choosing $\delta_1 := \sqrt{-\varepsilon}$ instead of $\sqrt{\varepsilon}$; if this choice of δ_1 is used in (1.5.17) and (1.5.18), it turns out that $x_1 = 0, 1, -1$ instead of $0, i, -i$, leading directly to (1.5.20). The method of undetermined gauges only required that $\delta_1 \approx \varepsilon^{1/2}$, and this can be achieved as well by using $\sqrt{-\varepsilon}$ as $\sqrt{\varepsilon}$. The same decision must be made for any gauge which is an even root, and the most convenient choice, whenever possible, is the one that gives real coefficients.

We have not yet used the other significant gauge, $\delta_1 := \varepsilon$. Using this in (1.5.18) and discarding δ_1^3 gives $x_1 = -2$, so the third solution is

$$x^{(3)} \cong -2\varepsilon. \tag{1.5.21}$$

This example shows how more than one significant gauge may be necessary to find all the solutions to a given problem. The gauge $\delta_1 = \varepsilon^{1/2}$ gives a hint of the third root $x^{(3)}$ in the form of $x_1 = 0$, which we discarded; if we had written $x^{(3)} \cong 0\varepsilon^{1/2}$ along with (1.5.19) or (1.5.20) it would not have been wrong, to this order. But the true leading order of $x^{(3)}$ is ε, not $\varepsilon^{1/2}$, and the leading term (1.5.21) is found using that gauge. The gauge $\delta_1 := \varepsilon$, taken by itself, does not even hint at the presence of the roots (1.5.20) since there is only one solution for x when that gauge is used. It is, of course, possible to compute higher order terms for these solutions, but this will be easier by the method of the next section.

The full bifurcation diagram for this example is shown in Fig. 1.5.1a. This diagram can be deduced by a trick in this particular example: Equation (1.5.16) is a cubic in x, but it also happens to be a quadratic in ε. If it is rewritten as $2\varepsilon^2 + x\varepsilon + x^3 = 0$ and solved for ε by the quadratic formula, the result is $\varepsilon = \left(-x \pm \sqrt{x^2 - 8x^3}\right)/4$. The discriminant $x^2 - 8x^3$ is nonnegative for $x \leq 1/8$ and vanishes at $x = 1/8$ and $x = 0$, so the graph has two points for each $x \leq 1/8$ except at $x = 1/8$ and $x = 0$, where there is only one. The solid arcs on the graph crossing at $(\varepsilon, x) = (0, 0)$ are meant to indicate the portion of the full bifurcation diagram which we were able to detect by perturbation theory. The arc with a vertical tangent consists of $x^{(1)}$ and $x^{(2)}$, the solutions which are real only for $\varepsilon < 0$, and the arc which crosses it is $x^{(3)}$. Once again, the starting point for a degenerate problem has turned out to be a bifurcation point, but this time not a pair bifurcation; the configuration near $(0,0)$, with three solutions on one side and one on the other, is called a *pitchfork bifurcation* because of its shape. Perturbation theory only detects behavior for small ε; there is no hint in our results (1.5.20) and (1.5.21) to tell us that $x^{(1)}$ and $x^{(3)}$ join at $\varepsilon = -1/32$ in a pair bifurcation and cease to exist for smaller ε. Figure 1.5.1b shows the incorrect bifurcation diagram obtained by using the approximations beyond their range of validity.

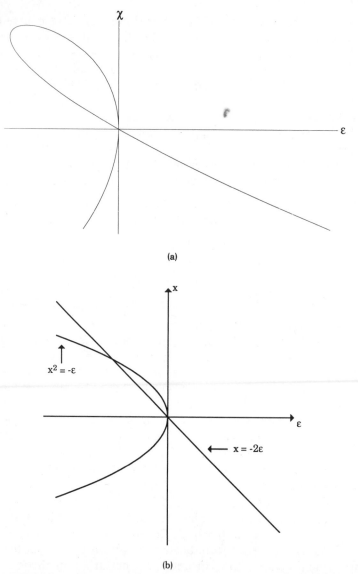

(a)

(b)

FIG. 1.5.1. Bifurcation diagram for Example 1.5.2. (a) The exact bifurcation diagram. (b) The approximate bifurcation diagram obtained by perturbation theory. It is accurate for small ε but does not properly represent the pair bifurcation that occurs when ε is sufficiently large and negative.

Example 1.5.3

$$(x - 1)^3 + \varepsilon x = 0. \qquad (1.5.22)$$

The reduced problem again has a triple root $x = 1$. Setting $x \cong 1 + x_1 \delta_1$ in the full problem gives $x_1^3 \delta_1^3 + x_1 (\varepsilon \delta_1) + \varepsilon \cong 0$. There are three distinguished gauges ($\varepsilon^{1/3}$, $\varepsilon^{1/2}$, and 1) but only one of these is significant: $\delta_1 := \varepsilon^{1/3}$. This leads easily to $x_1^3 + 1 = 0$, which has one real root $x_1 = -1$ and two complex roots. Since $\varepsilon^{1/3}$, being an odd root, is always real, there is no hope of obtaining a real root from these two. Therefore the problem has only one real root given to two terms by

$$x^{(1)} \cong 1 - \varepsilon^{1/3}. \tag{1.5.23}$$

(Of course, this existence statement is merely heuristic at this point.) Notice that this time the starting point, although degenerate, is not a bifurcation point for real solutions, although it becomes one if complex solutions are allowed.

We will begin to find the next correction to this solution in order to show that computations using the method of undetermined gauges can become unnecessarily complicated. Substituting $x \cong 1 - \varepsilon^{1/3} + x_2\delta_2$ into (1.5.22) leads to

$$3x_2(\varepsilon^{2/3}\delta_2) - 3x_2^2(\varepsilon^{1/3}\delta_2^2) + x_2^3\delta_2^3 + x_2(\varepsilon\delta_2) - \varepsilon^{4/3} \cong 0.$$

Here $\varepsilon\delta_2 \ll \varepsilon^{2/3}\delta_2$ and therefore $\varepsilon\delta_2$ is subdominant and may be omitted.

Equating the other gauges in pairs leads to six distinguished gauges for δ_2. Testing all of these to find the significant one (only one can be significant, since there is only a single root whose expansion starts off like (1.5.23)) is laborious, and quite unnecessary, since proceeding in a different way makes it immediately clear what gauge should appear next. This will be done in the next section.

Example 1.5.4

$$(x - 1)^4 - 2\varepsilon(x - 1)^2 - \varepsilon^2(x - 1)^2 + \varepsilon^2 - 2\varepsilon^3 + \varepsilon^4 = 0. \tag{1.5.24}$$

In this problem most details will be left to the reader. The quadruple root $x = 1$ when $\varepsilon = 0$ suggests $x \cong 1 + x_1\delta_1$. The only distinguished gauge is $\delta_1 := \varepsilon^{1/2}$; it makes three terms have equal dominance, and $x_1^4 - 2x_1^2 + 1 = 0$. This has two double roots, $x_1 = \pm 1$. Since (1.5.24) is known to have four roots (including complex roots and counting multiplicity) by the fundamental theorem of algebra, it seems that each double root for x_1 must be used twice, giving

$$x^{(1)} \cong 1 + \varepsilon^{1/2},$$
$$x^{(2)} \cong 1 + \varepsilon^{1/2},$$
$$x^{(3)} \cong 1 - \varepsilon^{1/2},$$
$$x^{(4)} \cong 1 - \varepsilon^{1/2}. \tag{1.5.25}$$

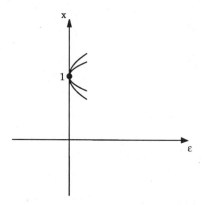

FIG. 1.5.2 The bifurcation diagram for Example 1.5.4.

Since these roots are real for $\varepsilon > 0$, it is correct to use $\varepsilon^{1/2}$ rather than $\sqrt{-\varepsilon}$. The roots $x^{(1)}$ and $x^{(2)}$ become distinct at the next stage of the calculation. Putting $x \cong 1 + \varepsilon^{1/2} + x_2 \delta_2$ into (1.5.24) leads to four distinguished gauges $(\varepsilon^{1/2}, \varepsilon^{3/4}, \varepsilon^{5/6}$, and ε) from which $\varepsilon^{1/2}$ is eliminated by rule (I), and $\varepsilon^{3/4}$ and $\varepsilon^{5/6}$ by rule (III), leaving $\delta_2 := \varepsilon$ as the only significant gauge. Then $x_2 = \pm\sqrt{3}/2$. A similar calculation for $x^{(3)}$ and $x^{(4)}$ completes the picture:

$$x^{(1)} \cong 1 + \varepsilon^{1/2} + \frac{\sqrt{3}}{2}\varepsilon,$$

$$x^{(2)} \cong 1 + \varepsilon^{1/2} - \frac{\sqrt{3}}{2}\varepsilon,$$

$$x^{(3)} \cong 1 - \varepsilon^{1/2} + \frac{\sqrt{3}}{2}\varepsilon,$$

$$x^{(4)} \cong 1 - \varepsilon^{1/2} - \frac{\sqrt{3}}{2}\varepsilon.$$

(1.5.26)

The bifurcation diagram is shown in Fig. 1.5.2; there is apparently no standard name for this bifurcation.

Example 1.5.5

The final example is of a somewhat different character. According to the fundamental theorem of algebra (or even, in this case, the quadratic formula), the problem

$$\varphi(x, \varepsilon) = \varepsilon x^2 + x + 1 = 0 \qquad (1.5.27)$$

has two roots for $\varepsilon \neq 0$. But the reduced problem $x + 1 = 0$ has only one. Since $\varphi(-1, 0) = 0$ and $\varphi_x(-1, 0) \neq 0$, the implicit function theorem (Section 1.4) applies, and there is a unique continuation of this root, namely $x^{(1)} = -1 - \varepsilon - 2\varepsilon^2 + \mathcal{O}(\varepsilon^3)$. The second solution $x^{(2)}(\varepsilon)$ exists only for $\varepsilon \neq 0$ and cannot approach a finite limit as $\varepsilon \to 0$. (If there were such a limit, it would be a second zero of $\varphi(x, 0)$, since φ is continuous.) We will look for this solution in the form (1.5.1), where this time we cannot take $\delta_0 = 1$. Substituting $x \cong x_0 \delta_0$ into (1.5.27) leads to $x_0^2 (\varepsilon \delta_0^2) + x_0 \delta_0 + 1 = 0$. The distinguished gauges are $\delta_0 = \varepsilon^{-1}$ (from $\varepsilon \delta_0^2 = \delta_0$), $\delta_0 = \varepsilon^{-1/2}$ (from $\varepsilon \delta_0^2 = 1$), and $\delta_0 = 1$. The significant gauges are ε^{-1} and 1, since $\varepsilon^{-1/2}$ renders the first and third terms of equal order but the middle term is dominant. The gauge $\delta_0 = 1$ gives the bounded root $x^{(1)}$ again, so we take $\delta_0 := \varepsilon^{-1}$. Then $x_0 = -1$, and, continuing, $\delta_1 = 1$, $x_1 = 1$, $\delta_2 = \varepsilon$, and $x_2 = 1$, so

$$x^{(2)} \cong -\frac{1}{\varepsilon} + 1 + \varepsilon. \tag{1.5.28}$$

We will see in the next section that this has error $\mathcal{O}(\varepsilon^2)$. The root approaches $-\infty$ as $\varepsilon \to 0^+$ and $+\infty$ as $\varepsilon \to 0^-$. See Fig. 1.5.3 for the full bifurcation diagram; the solid part of the graph shows what is detected by perturbation theory. This example shows that perturbation theory is not

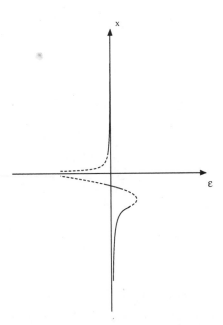

FIG. 1.5.3. The bifurcation diagram for Example 1.5.5. The solid portions (where ε is small) indicate the behavior detectable by perturbation theory.

always limited to the continuation of solutions that exist for $\varepsilon = 0$. In this instance, if one were to say that (1.5.28) holds "for sufficiently small ε," it would have to mean $0 < |\varepsilon| < \varepsilon_0$.

Exercises 1.5

1. (a) Sketch the graph of $y = (x - 2)^2(x - 3)$. Use this graph to study the roots of (1.5.2) for all real ε. Sketch the bifurcation diagram in the (ε, x) plane. (Use Example 1.3.1 for guidance.) Compare the result with (1.5.8).

   (b) Again using the graph of $y = (x - 2)^2(x - 3)$, sketch the bifurcation diagram of $(x - 2)^2(x - 3) + \varepsilon^2 = 0$ for all real ε. Notice that this diagram is symmetrical about the x axis. Find the values of ε at the two pair bifurcation points.

2. In (1.5.12) we discarded case (ii) because $\delta_2 \approx \delta_1$. Show that if we substitute $\delta_2 := \varepsilon^{1/2}$ into (1.5.11) the resulting values for x_2 change (1.5.9) back into (1.5.8). This shows that no progress can be made by reusing a previous gauge.

3. Compute $x^{(2)}$ and $x^{(3)}$ in Example 1.5.1 to obtain (1.5.15).

4. Apply undetermined gauges to (1.3.8). Using the second approximation (that is, up through δ_2), obtain the value 0.738 for the smaller root (in comparison with an exact solution of 0.734...) when $\varepsilon = 0.1$. This confirms that (1.3.8) is a better formulation of (1.3.7) than is (1.3.5), which gave the faulty value of -9. (Another method is given in Section 1.7.)

5. Find three approximate roots of $x^3 - 2x^2 + x - \varepsilon^2 = 0$. Compute them to sufficient accuracy to determine whether they increase or decrease as ε increases from zero. Sketch the bifurcation diagram in the (ε, x) plane for small ε. The bifurcation point at $(\varepsilon, x) = (0, 1)$ is called a *transcritical bifurcation*. (See also Exercise 1.6.3.)

6. Find the five roots of $x^5 + \varepsilon x^2 - \varepsilon^2 = 0$ approximately. (Two choices of δ_1 are necessary.) How many real roots are there for $\varepsilon > 0$? For $\varepsilon < 0$? Sketch the bifurcation diagram for small ε. (See also Exercise 1.6.4.)

1.6. RESCALED COORDINATES

The method of undetermined gauges was successful, in the last section, in constructing formal solutions to a variety of problems to which the implicit function theorem does not apply. But two drawbacks to the method are apparent: The calculations can become very long, especially when there are many distinguished gauges that must be checked for significance; and there is no obvious way to justify the results and equip them with error estimates. Both of these drawbacks can be corrected with a simple modification of the method. The basic idea will be used repeatedly in many contexts: When an approximate solution is proposed, try using the formula that gives the approximate solution as a change of variables. If the approximate solution is a good one, using it as a change of variables should change the original problem to a simpler one. A change of variables needs no justification; if I were to fall asleep at my desk and dream of a good change of variables, and upon awakening, find that it simplified my problem, it would not matter that it had come from a dream. So using a conjectured approximate solution as a change of variables is a good way to "cash in" on what has been discovered heuristically, without having to worry about justifying it.

To illustrate the idea, consider Example 1.5.3,

$$\varphi(x, \varepsilon) = (x - 1)^3 + \varepsilon x = 0. \tag{1.6.1}$$

The method of undetermined gauges called for a first approximation in the form $x \cong 1 + x_1 \varepsilon^{1/3}$, where x_1 turned out to be -1. (See (1.5.23).) Instead of treating x_1 as a constant to be determined, let us consider it as a new variable y and regard the equation

$$x = 1 + y\varepsilon^{1/3} \tag{1.6.2}$$

not as an approximate solution, but as a change of variables from the old variable x to the new variable y. Notice that the equality sign in (1.6.2) is an exact equality, not $\cong$ as in the method of undetermined gauges. Upon substituting (1.6.2) into (1.6.1) we obtain $\varepsilon y^3 + \varepsilon + \varepsilon^{4/3}y = 0$. Dividing by ε, and setting

$$\mu := \varepsilon^{1/3}, \tag{1.6.3}$$

gives

$$\psi(y, \mu) = y^3 + 1 + \mu y = 0. \tag{1.6.4}$$

This is a new equation in y and μ which is completely equivalent to (1.6.1) by way of (1.6.2) and (1.6.3). It is important in this kind of calculation not to discard any subdominant terms: We are looking for an exact copy of the original problem in new variables, not an approximation. (The approximation stage will come later.)

Examining (1.6.4), the reduced problem $y^3 + 1$ is seen to have three simple roots, two of them complex. The real solution -1 satisfies $\psi(-1, 0) = 0$ and $\psi_y(-1, 0) \neq 0$. The change of variables from x and ε to y and μ has changed a degenerate problem into a nondegenerate problem. Thus according to Theorems 1.4.1 and 1.4.2 there exists a unique solution to (1.6.4) which reduces to -1 when $\mu = 0$, and this solution can be expanded in the form

$$y^{(1)} = -1 + y_1\mu + y_2\mu^2 + \cdots + y_k\mu^k + \mathcal{O}(\mu^{k+1}) \tag{1.6.5}$$

for any k. Recall that $\mathcal{O}(\mu^{k+1})$ stands for a rigorous error estimate of the form $|y^{(1)} - y^{(1)}_{\text{approx}}| \leq c\mu^{k+1}$ for $|\mu| \leq \mu_0$. The coefficients $y_1, \ldots, y_k$ can be found by routine computation; in particular, there is no need to find any undetermined gauges, since we know in advance that integer powers of μ suffice. (We did, of course, need to go through the first step of undetermined gauges to obtain $\delta_1 = \varepsilon^{1/3}$ in order to set up (1.6.2). It is only the computation of later gauges $\delta_2, \delta_3, \ldots$ that is no longer necessary. Recall from Example 1.5.3 that there were six distinguished gauges for δ_2, so we have eliminated a good deal of work.) When (1.6.5) is substituted into (1.6.2) using (1.6.3), the solution in the original variables is seen to have the form

$$x^{(1)} = 1 - \varepsilon^{1/3} + y_1\varepsilon^{2/3} + y_2\varepsilon + \cdots + y_k\varepsilon^{(k+1)/3} + \mathcal{O}(\varepsilon^{(k+2)/3}). \quad (1.6.6)$$

Such a solution is called a *fractional power series* or *Puiseux series*. It is a special case of (1.5.1) in which all the gauges are fractional powers of ε with a fixed denominator (in this case 3), or equivalently, integral powers of a fixed fractional power of ε (in this case $\varepsilon^{1/3}$).

The change of variables (1.6.2) is called a *rescaling* of x around $x = 1$. When written in the form

$$y := \frac{x - 1}{\varepsilon^{1/3}} \quad (1.6.7)$$

it is clear that the distance of each point x from 1 is being stretched or dilated by the factor $\varepsilon^{-1/3}$, which is large when ε is small. This can be regarded as a change in the unit of length used to measure distance. Whereas x represents a fixed scale independent of ε, y represents distance measured from $x = 1$ using a short interval, whose length is $\varepsilon^{1/3}$ in the original units, as the new unit of length. Since this unit of length is short, measurements using it come out large; thus a small interval around $x = 1$ is magnified into a larger interval in y. For instance, the interval $|x - 1| \leq \varepsilon^{1/3}$ is magnified to $|y| \leq 1$. Recalling the discussion in Section 1.1 about the difficulty of properly formulating a perturbation problem, it is possible to take the following point of view: (1.6.1) is a degenerate problem because the "wrong" coordinate (or the "wrong" unit of length) was chosen initially. According to this attitude, rescaling is a way of correcting this initial mistake and finding the "proper" variables in which to formulate the problem. This can be a helpful way of thinking, provided it is understood that "proper" only means "mathematically convenient." No set of variables is actually "wrong," and in some important methods (notably the methods of multiple scales and matching) it is convenient to use several scalings of the same physical quantity at the same time.

Why does rescaling change a degenerate problem to a nondegenerate problem? The original problem (1.6.1) has a triple root $x = 1$ when $\varepsilon = 0$. As ε increases from zero, three roots (two of them complex) diverge from the single value $x = 1$ at a "rate" governed by the gauge $\varepsilon^{1/3}$. After dilating the coordinate around $x = 1$ by $\varepsilon^{-1/3}$ (see (1.6.7)), equation (1.6.4) is found, which has three simple roots for $\varepsilon = 0$. Thus dilating the variables at exactly the right rate (the rate at which the roots approach each other as $\varepsilon \to 0$) keeps them apart as $\varepsilon \to 0$, so that in the limit the roots are simple and not multiple.

As a second example of the method of rescaled coordinates, consider Example 1.5.2,

$$x^3 + \varepsilon x + 2\varepsilon^2 = 0. \tag{1.6.8}$$

To determine the three roots by undetermined gauges required two choices for δ_1, namely, $\delta_1 = \sqrt{-\varepsilon}$ and $\delta_1 = \varepsilon$. It turns out that in the method of rescaled coordinates, the scale $\sqrt{-\varepsilon}$ suffices for all three roots. Setting

$$\mu = \sqrt{-\varepsilon}, \qquad x = \mu y \tag{1.6.9}$$

in (1.6.8)—here $x = \mu y$ is perhaps better thought of as $x = 0 + \mu y$, to remember that we are rescaling around $x = 0$, the root of the reduced equation—results in

$$y^3 - y + 2\mu = 0 \tag{1.6.10}$$

after removing a factor of μ^3. The reduced equation of (1.6.10), $y^3 - y = 0$, has three simple roots, $y = \pm 1$ and $y = 0$, which therefore have continuations in the form of power series in μ:

$$y^{(1)} = 1 - \mu + \mathcal{O}(\mu^2),$$
$$y^{(2)} = -1 - \mu + \mathcal{O}(\mu^2), \tag{1.6.11}$$
$$y^{(3)} = 0 + 2\mu + \mathcal{O}(\mu^2).$$

In view of (1.6.9) these solutions take the following form in the original variables:

$$x^{(1)} = \sqrt{-\varepsilon} + \varepsilon + \mathcal{O}(|\varepsilon|^{3/2}),$$
$$x^{(2)} = -\sqrt{-\varepsilon} + \varepsilon + \mathcal{O}(|\varepsilon|^{3/2}), \tag{1.6.12}$$
$$x^{(3)} = 0 - 2\varepsilon + \mathcal{O}(|\varepsilon|^{3/2}).$$

Notice that (1.6.10) has solutions that are real for all real μ near zero, but that when these solutions are applied to the original problem, two of them (at least) are real only when ε is negative. (There is no contradiction here: According to (1.6.9), μ itself is real only when $\varepsilon < 0$.) The reality of the third root will be discussed below.

It is unusual that $x^{(3)}$ can be captured using the scale $\sqrt{-\varepsilon}$; typically, a given scale will expose only those roots that have that scale as their δ_1 gauge under the method of undetermined gauges. But the reason why one scale suffices in this problem is not far to seek. Equation (1.6.8) has three real roots diverging from $x = 0$ as ε deceases from zero, two at the "rate" $\sqrt{-\varepsilon}$ and one at the slower rate ε. Dilating the coordinate at the rate $\sqrt{-\varepsilon}$ separates the two roots that diverge at that rate and leaves the third root to approach zero by itself; thus all three roots are separated. If there were a fourth root having rate ε, the third and fourth roots would not be separated by this scaling, and it would be necessary to use the scale ε to resolve these roots. In our present problem, it is still possible to use this scale. Set

$$\mu := \varepsilon, \qquad y := x/\mu, \tag{1.6.13}$$

and obtain

$$y + 2 + \mu y^3 = 0. \tag{1.6.14}$$

The reduced problem has only one root, which is simple and has the continuation $y = -2 + \mathcal{O}(\mu)$, or

$$x^{(3)} = -2\varepsilon + \mathcal{O}(\varepsilon^2). \tag{1.6.15}$$

This is clearly the same root as $x^{(3)}$ in (1.6.12). The roots $x^{(1)}$ and $x^{(2)}$ are not captured under this rescaling because in the rescaled coordinate they become unbounded as $\varepsilon \to 0$.

Although the second scaling (1.6.13) is in some sense unnecessary, it does provide additional information about $x^{(3)}$ that is not available from (1.6.12). Not only does (1.6.15) give a sharper error estimate ($\mathcal{O}(\varepsilon^2)$ in place of $\mathcal{O}(|\varepsilon|^{3/2})$), but also using the scale $\mu = \varepsilon$ reveals that the higher order expansion of $x^{(3)}$ will never involve fractional powers of ε. In addition, we gain a rigorous proof that $x^{(3)}$ is real for all ε near zero, not merely for $\varepsilon < 0$ as in the case of $x^{(1)}$ and $x^{(2)}$. Namely, Theorem 1.4.1 guarantees that the solution of (1.6.14) is real for μ near zero, and μ (this time) is just ε. Thus, by using both the $\sqrt{-\varepsilon}$ and ε scales, we have achieved a proof of the pitchfork bifurcation shown in Fig. 1.5.1, which was previously obtained by a trick suitable only to this particular problem.

It is not our purpose to classify all the possibilities that can arise in using rescaled coordinates for root finding. The exercises will explore some of them, such as what to do when the first rescaling fails to produce a nondegenerate problem (answer: Rescale again), and how to rescale a problem with unbounded roots. Our goal has been to introduce the fundamental ideas of gauges and scales which appear throughout perturbation theory in a variety of forms. We will continue this program in the next section by introducing the ideas of control parameters and uniformity.

Exercises 1.6

1. Compute y_1 and y_2 in (1.6.5) and put the result into (1.6.6).

2. Use rescaled coordinates to find the roots $x^{(1)}$ and $x^{(2)}$ in Example 1.5.1 and to prove rigorously that the starting point for these roots is a pair bifurcation. (See the remarks following equations (1.5.8).)

3. Repeat Exercise 1.5.5 using rescaled coordinates. Establish the transcritical bifurcation rigorously. That is, show that there exist two solutions on either side of the bifurcation point at which the solutions coincide.

4. Repeat Exercise 1.5.6 using rescaled coordinates.

5. (a) Repeat Example 1.5.4 using rescaled coordinates. You will find that rescaling separates the quadruple root into two double roots. Each of these can in turn be separated by another rescaling. Compare the scales needed with the gauges used in the example.

   (b) Solve this problem again by using the second order undetermined gauge solutions to define new coordinates instead of using two successive rescalings. Hint: One of the new coordinates y is given by $x = 1 + \sqrt{\varepsilon} + y\varepsilon$.

6. Use rescaling to solve (a) Example 1.5.5 and (b) the similar problem $\varepsilon x^3 + x^2 - 4 = 0$.

7. At the end of Example 1.3.3, it is stated that a nearly degenerate problem should be reformulated as a degenerate problem. In the present section, degenerate problems are reformulated as nondegenerate problems. If this seems contradictory, re-read both discussions and explain how they are related. (Both ideas will be used in Section 1.7.)

1.7. PARAMETERS, UNIFORMITY, AND RESCALING

The problem

$$\varphi(x, \tau, \varepsilon) = (x - 1)(x - \tau) + \varepsilon x$$
$$= x^2 + (\varepsilon - \tau - 1)x + \tau = 0 \tag{1.7.1}$$

is of the same form as those in previous sections except that it contains an unspecified constant τ. This constant can be called a *control parameter* to distinguish it from a *coordinate* such as x and from a *perturbation parameter* such as ε. If $\tau \neq 1$, the problem is nondegenerate and may be solved by the method of Section 1.2; the coefficients x_n in the solution will, of course, depend upon τ. (This solution will be given below in (1.7.2) and (1.7.10).) If $\tau = 1$, the problem is degenerate and the method of Section 1.6 applies; this solution will not contain τ because it is valid only for $\tau = 1$. It may seem that this is all there is to say about the matter, but in fact there is a great deal lurking behind this simple problem. We have seen in Example 1.3.3 that nearly degenerate problems tend to suffer from poor error estimates; in fact, the present problem reduces to that example when $\tau = 1.01$. Therefore it is to be expected that the solution constructed under the nondegenerate assumption $\tau \neq 1$ will actually become less useful as $\tau \to 1$ and should be replaced, for τ close enough to 1, by a solution resembling that of the degenerate problem $\tau = 1$. In discussing Example 1.3.3, we suggested reformulating the nearly degenerate perturbation problem in such a way that the new problem is actually degenerate. At that time we proposed (1.3.8) as such a reformulation (for $\tau = 1.01$), and in Exercise 1.5.4, you have solved (1.3.8) and found that, indeed, it gives a useful solution. But (1.3.8) was "made up" arbitrarily. In this section we will present a systematic method for reformulating problems that contain a control parameter when that parameter is close to a value for which the problem is degenerate. But first we will study the nondegenerate case $(\tau \neq 1)$ in more detail to see exactly how its accuracy breaks down as $\tau \to 1$. This will lead us to the concept of uniformity (and nonuniformity) of approximations, which is a central concept for the remainder of this book. Almost every advanced method in perturbation theory is introduced because solutions constructed by an easier method are nonuniform. Our first example of nonuniform solutions, and how to improve them, is (1.7.1) with τ near 1.

Treating τ as a constant unequal to 1, problem (1.7.1) for $\varepsilon = 0$ has two simple roots: $x^{(1)} = 1$ and $x^{(2)} = \tau$. That is, $\varphi(1, \tau, 0) = 0$, $\varphi_x(1, \tau, 0) \neq 0$, $\varphi(\tau, \tau, 0) = 0$, and $\varphi_x(\tau, \tau, 0) \neq 0$. It follows from Theorems 1.4.1 and 1.4.2 that both solutions have unique continuations for ε near zero in the form $x = x_0 + x_1\varepsilon + x_2\varepsilon^2 + \mathcal{O}(\varepsilon^3)$. Calculation (Exercise 1.7.1a) reveals the continuation of $x^{(1)}$ to be

$$x^{(1)}(\tau, \varepsilon) = 1 - \frac{1}{1-\tau}\varepsilon - \frac{\tau}{(1-\tau)^3}\varepsilon^2 + \mathcal{O}(\varepsilon^3). \tag{1.7.2}$$

Here the precise meaning of $\mathcal{O}(\varepsilon^3)$ is as follows: There exists a constant ε_0 such that $x^{(1)}(\tau, \varepsilon)$ exists for $|\varepsilon| \leq \varepsilon_0$, and for any ε_1 in the interval $0 < \varepsilon_1 < \varepsilon_0$ there is a c such that

$$\left| x^{(1)}(\tau, \varepsilon) - \left(1 - \frac{1}{1-\tau}\varepsilon - \frac{\tau}{(1-\tau)^3}\varepsilon^2 \right) \right| \leq c\varepsilon^3 \quad \text{for} \quad |\varepsilon| \leq \varepsilon_1. \tag{1.7.3}$$

In the following discussion we will ignore ε_0 and emphasize the role of ε_1 and c, since our focus is not the interval of existence but the interval in which the error estimate (1.7.3) holds. (But ε_0 will reappear when we formulate the result carefully in Theorem 1.7.1.)

The results in the last paragraph depend strongly on the first eight words: "Treating τ as a constant unequal to 1." If the value of τ is changed, Theorems 1.4.1 and 1.4.2 must be applied again with the new value of τ, as though an entirely new problem were being solved. Equations (1.7.2) do not change, except for the value of τ, but the constants ε_1 and c in (1.7.3) may take different values. Therefore if τ is to be taken as a *variable* unequal to 1, rather than as a *constant* unequal to 1, then (1.7.3) should be restated as follows: There exist *functions* $\varepsilon_1(\tau)$ and $c(\tau)$ defined for all $\tau \neq 1$ such that

$$\left| x^{(1)}(\tau, \varepsilon) - \left(1 - \frac{1}{1-\tau}\varepsilon - \frac{\tau}{(1-\tau)^3}\varepsilon^2 \right) \right| \leq c(\tau)\varepsilon^3 \qquad \text{for} \quad |\varepsilon| \leq \varepsilon_1(\tau).$$
$$(1.7.4)$$

Now the important question is: What happens to $c(\tau)$ and $\varepsilon_1(\tau)$ as $\tau \to 1$? In fact the worst possible things happen: $c(\tau) \to \infty$ and $\varepsilon_1(\tau) \to 0$. (See Exercise 1.7.2.) As τ approaches 1, the interval of ε for which the solution is valid shrinks, and the error bound becomes larger. This is expressed by saying that the "big-oh" equations (1.7.2) *do not hold uniformly* for τ near 1.

Away from $\tau = 1$ it can be shown that these difficulties do not arise: $c(\tau)$ remains bounded, and $\varepsilon_1(\tau)$ stays away from zero. More precisely: If an interval $\tau_1 \leq \tau \leq \tau_2$ not containing $\tau = 1$ is chosen, there will be a finite upper bound c for $c(\tau)$ on this interval and a positive lower bound ε_1 for $\varepsilon_1(\tau)$, which means that the original error bound (1.7.3) holds for all $|\varepsilon| \leq \varepsilon_1$ and all τ in $\tau_1 \leq \tau \leq \tau_2$. This fact—that the constants c and ε_1 can actually be taken as constants independent of τ, provided that the range of τ is suitably restricted—is expressed by saying that the "big-oh" equation (1.7.2) holds *uniformly* with respect to τ for $\tau_1 \leq \tau \leq \tau_2$.

Precise definitions of uniformity will be given in the next section. It is more important now to develop a feeling for the idea. Roughly speaking, one should always think of $\mathcal{O}(\varepsilon^n)$ as standing for "something less than $c\varepsilon^n$ in absolute value." Here of course the number c must never depend upon ε; the upper bound must depend on ε only by the indicated power. If the quantity (usually an error) that is denoted by the big-oh symbol depends only on ε, there is nothing more to say. But if that quantity depends on other variables (in our case, τ), then the "constant" c may also depend on these other variables. In this case, c is "constant" only with respect to changes in ε, but changes if the other variables are changed. So, when faced with the symbol $\mathcal{O}(\varepsilon^n)$ standing for something that depends on other variables besides ε, it is always necessary to ask: *Is it possible* to choose c to be independent of the other variables, or *is it necessary* to allow c to depend on these variables? In the first case (when the constant is "really constant") the big-oh symbol is said to *hold uniformly* or be *uniformly valid*. In the second case (when there is *no way* to find a c that is "really constant"), the big-oh symbol is called *nonuniform*. (The real danger in nonuniformity is that c might approach infinity as the variables are allowed to change, making the error estimate meaningless. If c is bounded, it can be replaced by a constant upper bound and then the big-oh symbol holds uniformly after all.) The discussion in this paragraph has focused on the constant

c, but the same things must be said for ε_1 (except that the danger here is in approaching zero rather than infinity). Uniformity requires that *both* constants c and ε_1 be truly constant, independent of all variables in the problem, and not only independent of ε.

It often happens that a big-oh symbol holds uniformly over a limited part of the range of the variables. In the example we have been considering, the technical jargon is to say that (1.7.2) holds "uniformly for τ in compact subsets not including $\tau = 1$." (A subset of the real line is compact if and only if it is closed and bounded.) This means: *First* choose any compact subset such as a closed bounded interval $\tau_1 \leq \tau \leq \tau_2$ not containing $\tau = 1$; *then* there exist constants c and ε_1 such that (1.7.3) holds for $\tau_1 \leq \tau \leq \tau_2$ and $|\varepsilon| \leq \varepsilon_1$. Any such compact set can be chosen, but different sets will lead to different constants c and ε_1. (In the language of the previous paragraph, the constants are "really constant," but only after restricting to a compact set.) There are no constants which work on an interval such as $0 \leq \tau < 1$, since this interval is not compact, and there are no constants which work on $0 \leq \tau \leq 2$, since this subset, although compact, includes 1. Figure 1.7.1 shows a typical *region of uniformity* for this problem in the (ε, τ) plane. Two intervals of τ have been chosen, one above $\tau = 1$ and one below; together these make a compact set not containing $\tau = 1$. For this range of τ, there exists a constant ε_1, which determines the vertical edges of the shaded region ($|\varepsilon| \leq \varepsilon_1$), and a constant c such that $c\varepsilon^3$ bounds the error

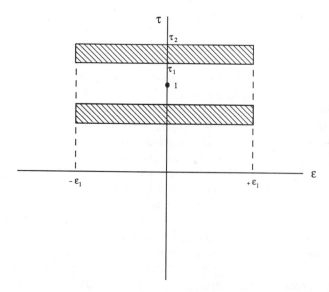

FIG. 1.7.1. A domain of uniformity for the straightforward perturbation solution of (1.7.1).

of our approximation as long as (ε, τ) belongs to the shaded region. If the initial choice of the two τ-intervals is changed so that the horizontal edges of the rectangles are brought closer to the line $\tau = 1$, then ε_1 (which determines the width of the rectangles) is expected to decrease and the constant c to increase. This diagram gives a good feeling for the type of validity possessed by the solutions (1.7.2).

The alert reader will have noticed that none of the rather complicated claims made in the previous paragraphs have been proved, even for the single example (1.7.1). The following theorem gives a general version of the result; the proof, which uses the implicit function theorem and a certain amount of point set topology, is given in Appendix B, Theorem B.3.

Theorem 1.7.1. *Suppose $\varphi(x, \tau, \varepsilon)$ has continuous partial derivatives up to order r and suppose that for τ in a certain interval $\tau_1 \leq \tau \leq \tau_2$ (or more generally, in a certain compact set) the reduced problem $\varphi(x, \tau, 0) = 0$ has a solution $x = x_0(\tau)$, that is,*

$$\varphi(x_0(\tau), \tau, 0) = 0 \quad for \quad \tau_1 \leq \tau \leq \tau_2. \tag{1.7.5}$$

Suppose that $x_0(\tau)$ is a continuous function of τ. Suppose also that for each τ, the solution of the reduced problem is nondegenerate:

$$\varphi_x(x_0(\tau), \tau, 0) \neq 0 \quad for \quad \tau_1 \leq \tau \leq \tau_2. \tag{1.7.6}$$

Then there exists a constant ε_0 (independent of τ) and a unique function $x = f(\tau, \varepsilon)$ defined for $|\varepsilon| < \varepsilon_0$ and $\tau_1 \leq \tau \leq \tau_2$ such that

$$\varphi(f(\tau, \varepsilon), \tau, \varepsilon) = 0 \quad and \quad f(\tau, 0) = x_0(\tau). \tag{1.7.7}$$

Furthermore the function $f(\tau, \varepsilon)$ has continuous derivatives up to order r, and there exist functions $x_1(\tau), \ldots, x_{r-1}(\tau)$ such that for each $k \leq r - 1$,

$$f(\tau, \varepsilon) = x_0(\tau) + x_1(\tau)\varepsilon + \cdots + x_k(\tau)\varepsilon^k + R_{k+1}(\tau, \varepsilon), \tag{1.7.8}$$

where the remainder $R_{k+1}(\tau)$ is $\mathcal{O}(\varepsilon^{k+1})$ uniformly for $\tau_1 \leq \tau \leq \tau_2$; that is, given $0 < \varepsilon_1 < \varepsilon_0$ there exists a constant c_{k+1} such that

$$|R_{k+1}(\tau, \varepsilon)| \leq c_{k+1}\varepsilon^{k+1} \quad for \quad \tau_1 \leq \tau \leq \tau_2 \quad and \quad |\varepsilon| \leq \varepsilon_1. \tag{1.7.9}$$

In order to apply this theorem to (1.7.1), we can select for $x_0(\tau)$ either of the two roots of the reduced problem. Taking $x_0(\tau) \equiv 1$ (that is, the constant function equal to 1 for all τ), the hypothesis (1.7.6) holds, provided the interval $\tau_1 \leq \tau \leq \tau_2$ does not contain 1. Then the conclusion (1.7.9), with $k = 2$, is exactly the kind of uniformity that we have claimed for (1.7.2). As another application of the theorem, the second root of the reduced problem for (1.7.1) is $x_0(\tau) = \tau$. Then (1.7.5) and (1.7.6) hold, so the approximations for $x^{(2)}$ are uniformly valid, provided (again) that the interval $\tau_1 \le \tau \le \tau_2$ does not include 1. Calculation (Exercise 1.7.1(b)) shows that

$$x^{(2)}(\tau, \varepsilon) = \tau + \frac{\tau}{1 - \tau}\varepsilon + \frac{\tau}{(1 - \tau)^3}\varepsilon^2 + \mathcal{O}(\varepsilon^3). \qquad (1.7.10)$$

This completes the study of (1.7.1) in the region where these solutions are uniform. As we have said, much of the effort in perturbation theory is directed to finding remedies for nonuniformity, and we now turn to this question.

Since these solutions for $x^{(1)}$ and $x^{(2)}$, constructed assuming $\tau \ne 1$, turn out to be highly inaccurate for τ near 1, it is necessary to formulate problem (1.7.1) in a different way to handle this case. Following the suggestion at the end of Example 1.3.3, the idea is to "build in" the fact that τ is near 1. Since ε denotes a small quantity, it is natural to express the idea that $\tau - 1$ is small by writing it as a constant times ε, that is, to write $\tau = 1 + \sigma\varepsilon$. Our experience with rescaling (in the last section) suggests that we generalize this a bit by writing

$$\tau = 1 + \sigma\varepsilon^\lambda, \qquad (1.7.11)$$

where λ is a positive constant to be determined. (We could generalize even further and put $\tau = 1 + \sigma\delta(\varepsilon)$ with a completely undetermined gauge δ, but it turns out that a fractional power of ε is sufficient.) There appears to be something paradoxical about equation (1.7.11), since until now τ has been regarded as a parameter which can be varied independently of ε, and in (1.7.11) it seems to have become a function of ε. The appearance of paradox disappears if we think of (1.7.11) as defining a new control parameter

$$\sigma := \frac{\tau - 1}{\varepsilon^\lambda}, \qquad (1.7.12)$$

which is a function of the two independent variables τ and ε. Then (1.7.11) is the inverse relationship, expressing τ as a function of σ and ε. This is another example of exploiting a good heuristic idea by using it as a change of variables. The original idea was to express the closeness of τ to 1 by writing (1.7.11). In order to use this idea without having to justify it, simply define a new variable by (1.7.12).

When (1.7.11) is substituted into (1.7.1), the result is

$$\psi(x, \sigma, \varepsilon) = (x - 1)(x - 1 - \sigma\varepsilon^\lambda) + \varepsilon x = 0. \qquad (1.7.13)$$

We will solve (1.7.13), after finding the best choice of λ, and then use (1.7.11) to translate the results into the original variable τ. The roles of τ and σ are almost entirely symmetrical: Substituting (1.7.12) into (1.7.13) changes it back into (1.7.1). The only asymmetry occurs when $\varepsilon = 0$; in this case (1.7.12) is not defined. This asymmetry has the following effect: Any specific equation in the family (1.7.13) with specified values of ε and σ can also be found in the family (1.7.1) with the same ε and with τ given by (1.7.11), but when ε is set equal to 0 in (1.7.1), the resulting equations (for various values of τ) cannot be found in the family of problems given by (1.7.13).

The procedure leading from (1.7.1) to (1.7.13) is called *rescaling the control parameter*; it is to be distinguished from the similar procedure of the last section, called *rescaling the coordinate*. Problem (1.7.13) is not fully specified until λ is determined. But it is clearly degenerate, with a double root at $x = 1$ when $\varepsilon = 0$. Therefore it is to be solved by a rescaling of the coordinate x around $x = 1$, which we express as

$$x = 1 + y\varepsilon^{\nu} \qquad (1.7.14)$$

for a positive constant ν. (Again we could use an undetermined gauge $\Delta(\varepsilon)$, but a fractional power of ε turns out to work, and it is simpler to assume that form from the beginning.) We will determine the rescaling factors ε^{λ} and ε^{ν} simultaneously. Substituting (1.7.14) into (1.7.13) gives

$$y^2\varepsilon^{2\nu} - \sigma y\varepsilon^{\nu+\lambda} + \varepsilon + y\varepsilon^{\nu+1} = 0. \qquad (1.7.15)$$

The principle governing the choice of ν and λ is that the reduced problem should have two simple roots. The reduced problem is obtained by factoring out and removing the highest possible power of ε from (1.7.15) and then setting $\varepsilon = 0$. The terms which survive are precisely the dominant terms of (1.7.15), that is, the term or terms with the lowest power of ε after λ and ν are fixed. The term $y\varepsilon^{\nu+1}$ is always subdominant (to the term ε), so it will not appear in the new reduced problem. The term y^2 must be one of the dominant terms, or else the reduced problem will not have two roots. There must be at least one other dominant term surviving along with y^2, or else the two roots of the reduced problem will not be distinct. A little experimentation shows that it is possible to balance out these terms in several ways:

(i)   Since both ν and λ are at our disposal, we can make all three terms (other than $y\varepsilon^{\nu+1}$) equally dominant by setting $2\nu = \nu + \lambda = 1$, which gives $\nu = \lambda = 1/2$. After dividing by ε this leaves

$$y^2 - \sigma y + 1 + y\varepsilon^{1/2} = 0. \qquad (1.7.16)$$

This is the choice which we will ultimately pursue below.

(ii) We could make the terms $y^2\varepsilon^{2\nu}$ and ε dominant by setting $2\nu = 1$ and $\nu + \lambda > 1$, that is, $\nu = 1/2$ and $\lambda > 1/2$. This possibility (with $\lambda = 1$) will be followed up in Exercise 1.7.4.

(iii) We could make the terms $y^2\varepsilon^{2\nu}$ and $\sigma y\varepsilon^{\nu+\lambda}$ dominant by making $2\nu = \nu + \lambda < 1$. This implies $\nu = \lambda$, and the condition $2\nu < 1$ can be achieved by taking, for example, $\nu = \lambda = 1/3$.

Of these three choices, only (i) sets a unique value for both ν and λ, and thus $\nu = \lambda = 1/2$ gives a set of *distinguished scales*. (Compare the notion of distinguished gauges in Section 1.5.) This in itself is not decisive, but there is another factor in favor of (i): Its reduced problem, $y^2 - \sigma y + 1 = 0$, is more complicated, or "richer," than the reduced problems of (ii) or (iii), which are $y^2 + 1 = 0$ and $y^2 - \sigma y = 0$. This is an advantage because the solution of the reduced problem is the starting point for the approximations, and a "richer" reduced problem means that more of the terms in the equation are contributing to the determination of the starting solutions. This should (one hopes, heuristically) make the starting solutions better and leave less to be corrected in the higher order terms of the solution. Of course, all three ways to rescale this problem give a nondegenerate reduced problem and hence lead to valid conclusions. If the "richer" formulation (1.7.16) were too difficult to solve, one of the others might be preferable. Exercise 1.7.4 illustrates the way in which conclusions drawn from a "less rich" version are consistent with the "richer" conclusions but less precise. (It should be noticed that each choice of λ and ν leads to a different definition of σ and y; so if two rescalings are to be compared, different letters should be used for each. In Exercise 1.7.4 we use $\tilde{\sigma}$ and $\tilde{y}$. Otherwise, results obtained by the two rescalings will appear to conflict.)

Having decided to use $\nu = \lambda = 1/2$, it is best (since $1/2$ is an even root) to break the problem into the cases $\varepsilon > 0$ and $\varepsilon < 0$, using $\mu = \sqrt{-\varepsilon}$ in the latter case. Thus the entire discussion up to now of (1.7.1) for τ near 1 can be summarized as follows:

If $\varepsilon > 0$:

original eq.	$(x - 1)(x - \tau) + \varepsilon x = 0$	
rescaling	$x = 1 + y\mu, \quad \tau = 1 + \sigma\mu, \quad \mu^2 = \varepsilon$	(1.7.17)
rescaled eq.	$y^2 + (\mu - \sigma)y + 1 = 0;$	

If $\varepsilon < 0$:

original eq.	$(x - 1)(x - \tau) + \varepsilon x = 0$	
rescaling	$x = 1 + y\mu, \quad \tau = 1 + \sigma\mu, \quad \mu^2 = -\varepsilon$	(1.7.18)
rescaled eq.	$y^2 - (\mu + \sigma)y - 1 = 0.$	

Problem (1.7.18) for $\varepsilon < 0$ is easier, so it will be treated first. For any fixed σ, the reduced equation $y^2 - \sigma y - 1 = 0$ has two simple roots $y = \left(\sigma \pm \sqrt{\sigma^2 + 4}\right)/2$ with continuations $y = \left(\sigma \pm \sqrt{\sigma^2 + 4}\right)/2 + \mathcal{O}(\mu)$. According to Theorem 1.7.1 (with σ playing the role of τ), the error estimate $\mathcal{O}(\mu)$ holds uniformly for σ in any closed bounded interval, since no values of σ lead to double roots. Therefore given $\sigma_0 > 0$ there exist $\varepsilon_1 > 0$ and $c > 0$ such that the error is bounded by $c\mu$ for $|\sigma| \leq \sigma_0$ and $-\varepsilon_1 \leq \varepsilon \leq 0$.

We must now interpret this result in the original variables. We have

$$x^{(1)} = 1 + \left(\frac{\sigma - \sqrt{\sigma^2 + 4}}{2}\right)\sqrt{-\varepsilon} + \mathcal{O}(\varepsilon)$$

$$= 1 + \frac{(\tau - 1) - \sqrt{(\tau - 1)^2 - 4\varepsilon}}{2} + \mathcal{O}(\varepsilon),$$

$$x^{(2)} = 1 + \left(\frac{\sigma + \sqrt{\sigma^2 + 4}}{2}\right)\sqrt{-\varepsilon} + \mathcal{O}(\varepsilon) \qquad (1.7.19)$$

$$= 1 + \frac{(\tau - 1) + \sqrt{(\tau - 1)^2 - 4\varepsilon}}{2} + \mathcal{O}(\varepsilon),$$

where the $\mathcal{O}(\varepsilon)$ terms are bounded by $c|\varepsilon|$ for $|\sigma| \leq \sigma_0$ and $-\varepsilon_1 \leq \varepsilon \leq 0$. The restriction on σ must be expressed in terms of τ. Figure 1.7.2 shows the graph of $\tau = 1 + \sigma\sqrt{-\varepsilon}$, for $\varepsilon < 0$ and for a fixed σ, in the (ε, τ) plane. It is half of a parabola; as σ is varied, the arc moves but remains a half parabola tangent to the τ axis at 1, except for $\sigma = 0$, when it is the half-line $\tau = 1$, $\varepsilon < 0$. Figure 1.7.3 shows the region $|\sigma| \leq \sigma_0$, $-\varepsilon_1 \leq \varepsilon \leq 0$. It is in this region that (1.7.19) is uniformly valid. The "opening" of the region can be increased by increasing σ_0, at the (probable) expense of decreasing ε_1 and increasing c.

The second expression for each root in (1.7.19) requires a warning. It is

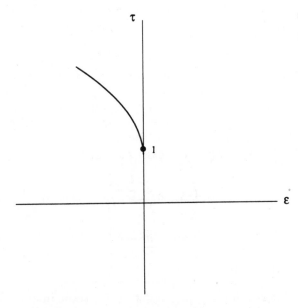

FIG. 1.7.2 A curve of fixed σ in the (ε, τ) plane.

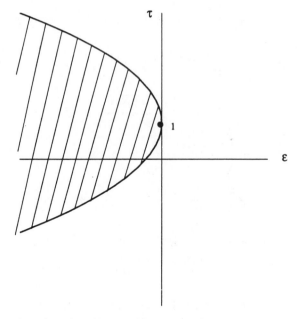

FIG. 1.7.3. A domain of uniformity for the perturbation solution of (1.7.1) obtained by rescaling, for negative ε. It is a parabolic region bounded by curves of constant σ.

tempting to expand the square root in powers of ε, and absorb all but the constant term into the $\mathcal{O}(\varepsilon)$. But to do this destroys the uniform validity in the region in question. In fact to do this is precisely to undo the rescaling and return to the form (1.7.2) which is uniformly valid in the region of Fig. 1.7.1 rather than 1.7.3. (See Exercise 1.7.1(d).) The distinction between (1.7.2) and (1.7.19) is often expressed in the following language: (1.7.2) is "expanded in ε with τ held constant" and (1.7.19) is "expanded in ε with σ held constant."

Turning now to the case (1.7.17) of $\varepsilon > 0$, the reduced rescaled equation has roots $y = (\sigma \pm \sqrt{\sigma^2 - 4})/2$. These roots are real and distinct only when $|\sigma| > 2$. Thus by the implicit function theorem (Theorems 1.4.1 and 1.4.2), for any fixed σ in this range there exist real roots

$$x^{(1)} = 1 + \left(\frac{\sigma - \sqrt{\sigma^2 - 4}}{2} \right) \sqrt{\varepsilon} + \mathcal{O}(\varepsilon),$$

$$x^{(2)} = 1 + \left(\frac{\sigma + \sqrt{\sigma^2 - 4}}{2} \right) \sqrt{\varepsilon} + \mathcal{O}(\varepsilon) \tag{1.7.20}$$

for ε in some interval $0 \le \varepsilon \le \varepsilon_1(\sigma)$, that is, for some interval along an arc of fixed σ outside the shaded region in Fig. 1.7.4. (The shaded region is

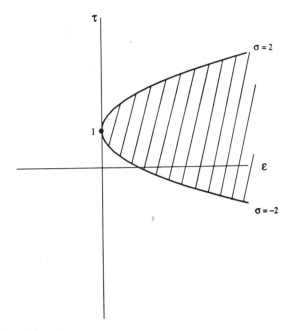

FIG. 1.7.4. The region in which the approximate solutions of (1.7.1) obtained by rescaling are complex. The boundary curves are the half-parabolas $\sigma = \pm 2$. This is not exactly the region where the true solutions are complex.

$|\sigma| \leq 2$.) The endpoint ε_1 of the interval of validity depends upon σ, and it is expected that $\varepsilon_1(\sigma) \to 0$ as $\sigma \to \pm 2$; similarly, the constant $c(\sigma)$ in the error estimate can approach infinity as $\sigma \to \pm 2$. Because the arc of validity along the curve $\sigma = $ constant shrinks as $\sigma \to \pm 2$, we cannot claim that the unshaded portion of Fig. 1.7.4 is the exact region in which solutions are real, although we can say that solutions are real *for a while* along each arc $\sigma = $ constant with $|\sigma| > 2$. In order to find specific regions in which the roots are real, we may turn to Theorem 1.7.1. Choosing σ_1 and σ_2 so that $2 < \sigma_1 < \sigma_2$ or $\sigma_1 < \sigma_2 < -2$, there exists ε_1 and c such that the roots are real and (1.7.20) holds for $0 \leq \varepsilon \leq \varepsilon_1$ and $\sigma_1 \leq \sigma \leq \sigma_2$ with uniform error bound $c\varepsilon$. See Fig. 1.7.5 for a graph of such a region.

The approximations constructed in this section can be carried to any order, with uniform error estimates in the same types of regions shown in Figs. 1.7.1, 1.7.3, and 1.7.5. But these improvements can never lead to any improvement in finding the boundary curve between real and complex solutions. The curve $|\sigma| = 2$ is only an approximation to this boundary. Two ways to determine the exact boundary are explored in Exercise 1.7.3. For our purposes the chief importance of this exercise is to make clear the need for careful interpretation of asymptotic results. It would be easy to arrive at the erroneous conclusion that for $\varepsilon > 0$, the roots are real if and only if $|\sigma| \geq 2$. Only by paying close attention to the exact meaning of the

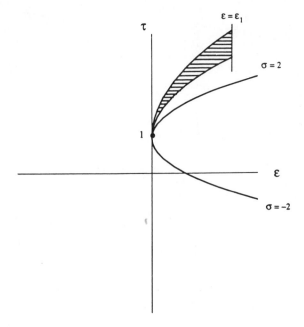

FIG. 1.7.5. A region of uniformity in which the approximate solutions are real and the exact solutions are also guaranteed to be real.

implicit function theorem can correct conclusions be drawn.

In summary, there are two approximate solutions to (1.7.1). The first consists of (1.7.2) and (1.7.10), and is valid when τ is not too close to 1. The second consists of (1.7.19) and (1.7.20). Typical domains of uniform validity for these solutions are sketched in Figs. 1.7.1–1.7.5, but the information contained in these sketches is incomplete because it does not include the value of c, which generally remains unknown. Therefore there is no exact rule for deciding which of the two approximate solutions is better at a specific point (ε, τ). The commonly accepted rule of thumb is that if τ is close enough to 1 that $(\tau - 1)$ is "comparable in magnitude" to $\varepsilon^{1/2}$, so that σ (defined by (1.7.12) with $\lambda = 1/2$) comes out to be "of the order of magnitude 1," then the second solution should be used. This amounts to choosing one of the parabolas of constant σ not too far from $\sigma = 1$ (whatever one thinks of as "order of magnitude 1") in the (ε, τ) plane as the boundary separating the regions in which the two solutions should be used. In other words, according to this rule of thumb the first (not rescaled) solution should be used outside of the shaded region in Fig. 1.7.3 (for a suitable choice of the boundary curve) and the second (rescaled) solution inside it. But this rule of thumb is not a rigorous mathematical statement. For instance, consider a point outside this shaded region but close to the point $(0, 1)$ and just to the left of the τ axis. Although this point does not belong to the shaded region of Fig. 1.7.3, it also does not belong to the shaded region of Fig. 1.7.1, so it appears not to be covered by either solution. But such a point can be included in either shaded region, either by increasing the "opening" of the parabola or by bringing the rectangles closer to $\tau = 1$. Either approach means increasing the constant in the appropriate error estimate, and which approach is better depends upon how large the constant becomes in each case, which is not in general known. To confuse the issue further, it is common in the literature to express the usual rule of thumb in big-oh notation. Thus, one will see statements such as "use the rescaled solution when $\tau - 1 = \mathcal{O}(\varepsilon^{1/2})$, that is, when $\sigma = \mathcal{O}(1)$." Strictly speaking, such a statement is nonsense. We have seen that if τ is regarded as a constant, then σ is a function of ε, and vice versa. If τ is constant, then $\tau - 1$ cannot be $\mathcal{O}(\varepsilon^{1/2})$; on the other hand, if σ is constant, then σ is automatically $\mathcal{O}(1)$ regardless of how large or small it is. Any author who writes a statement of this kind is confusing the big-oh symbol (asymptotic order) with the idea of order of magnitude and is using big-oh as though it meant approximate equality.

We will conclude this section by relating the rescaling (1.7.11) to the discussion of natural parameters and perturbation families in Section 1.1. Recall that for quadratic equations $ax^2 + bx + c = 0$, the natural parameters might be taken as a, b, and c. Then the perturbation family (1.7.1) can be specified as $a = 1$, $b = \varepsilon - \tau - 1$, $c = \tau$. Recall that this is to be regarded as a curve parameterized by ε, with the reduced problem corresponding to the point $(a, b, c) = (1, -\tau - 1, \tau)$. Since a has the fixed value 1, these curves (for various τ) can be drawn in the (b, c) plane, and they are straight horizontal lines (Fig. 1.7.6). Each solid dot in this figure represents a problem solvable by factoring, and (1.7.1) appears to be a suitable perturbation family for studying solutions near one of these dots on the same horizontal line. But the dot $(-2, 1)$ on the line $c = 1$ corresponds to a reduced problem with a double root. In a neighborhood of this point, the perturbation family (1.7.13) gives better solutions. This family, for $\varepsilon \geq 0$,

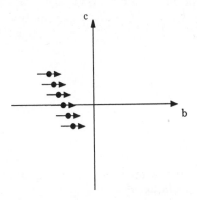

FIG. 1.7.6. The unrescaled perturbation problems for (1.7.1) for various τ, represented as curves in the space of natural parameters.

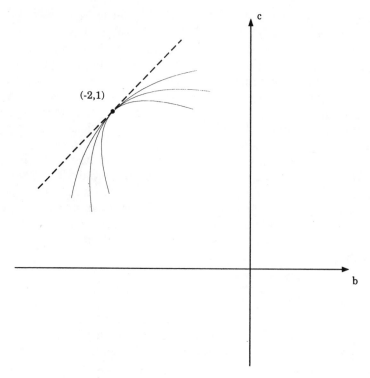

FIG. 1.7.7. The rescaled perturbation problems for (1.7.1) for various σ, represented as curves in the space of natural parameters.

is described by $a = 1$, $b = \varepsilon - 2 - \sigma\varepsilon^{1/2}$, $c = 1 + \sigma\varepsilon^{1/2}$. These curves, for various σ, are parabolas through $(-2, 1)$, sketched in Fig. 1.7.7. The fact that all these curves pass through the degenerate point reflects the fact that we wanted to make the nearly degenerate problems (those that pass near the bad point on horizontal lines in Fig. 1.7.6) into ones that are actually degenerate. Thus the rescaling process is really a major change in the type of path that we travel through the space of possible problems.

Exercises 1.7

1. Calculate:

   (a) the coefficients in (1.7.2);
   (b) the coefficients in (1.7.10);
   (c) the next term (the ε term) for each equation in (1.7.19), using the form that contains σ;
   (d) express the result of part c in terms of τ, expand in powers of ε, and compare with (1.7.2).

2. Show that $c(\tau) \to \infty$ as $\tau \to 1$ in (1.7.4). Hint: Use the quadratic formula to find $x^{(1)}(\tau, \varepsilon)$ exactly and substitute this into (1.7.4).

3. (a) Use the quadratic formula to determine an equation of the form $\chi(\varepsilon, \tau) = 0$ for the exact boundary curve in the ε, τ plane between the region where solutions of (1.7.1) are real and where they are complex. Show that the curve $\sigma = \pm 2$ drawn in Fig. 1.7.4 is an approximation to this curve.

   (b) To find the boundary between real and complex solutions without using the quadratic formula, carry out the following steps. Write the equations $\varphi(x, \tau, \varepsilon) = 0$ and $\varphi_x(x, \tau, \varepsilon) = 0$, with φ given by (1.7.1). These equations hold simultaneously at the points where $\varphi = 0$ has a double root, and hence at the points where two complex roots coalesce and become real. Eliminate x between these equations and recover the equation $\chi(\varepsilon, \tau) = 0$ found in part (a). (This method can be used for higher order equations where the quadratic formula is not available, but eliminating x may be difficult. The multiple variable implicit function theorem can be used to study the solution set of $\varphi = 0$, $\varphi_x = 0$.)

4. Study (1.7.1) in the nearly degenerate case by using $\lambda = 1$ in (1.7.11). That is, set $\tau = 1 + \tilde{\sigma}\varepsilon$. Determine suitable $\tilde{y}$ and $\tilde{\mu}$, discussing cases $\varepsilon < 0$ and $\varepsilon > 0$. Sketch the regions of uniformity. What can you say about the regions where solutions are real or complex? Observe that this gives much less precise information than taking $\lambda = 1/2$. Make sure you understand why the results of this problem do not contradict the results obtained in the text. (If you are puzzled about the region of complexity, consider how the curves of constant $\tilde{\sigma}$ in the ε, τ plane lie in relation to the curve $\sigma = \pm 2$ in Fig. 1.7.4 and remember that along each curve $\tilde{\sigma} = $ constant the implicit function theorem gives results only for a short distance which can depend upon $\tilde{\sigma}$.)

5. $(x)(x - 2)(x - \tau) + \varepsilon x + \varepsilon^2 = 0$ has three roots, $x \cong 0, 2$, and τ.

   (a) If τ is a constant $\neq 0$, find an approximation to the root $x \cong 0$ with error $\mathcal{O}(\varepsilon^3)$. For what intervals of τ is the error estimate $\mathcal{O}(\varepsilon^3)$ uniformly valid?

   (b) If τ is a constant $\neq 2$, find an approximation to the root $x \cong 2$ with error $\mathcal{O}(\varepsilon^2)$. For what intervals of τ is this uniform?

   (c) If τ is a constant $\neq 0$ and $\neq 2$, find an approximation to the root $x \cong \tau$ with error $\mathcal{O}(\varepsilon^2)$. For what intervals of τ is this uniform?

   (d) Find expansions for the roots $x \cong 0$ and $x \cong \tau$ valid when $\tau \cong 0$ with error $\mathcal{O}(\varepsilon^2)$. Determine the shape of the regions in the ε, τ plane for which the estimate holds uniformly.

   (e) Find approximations to the roots $x \cong 2$ and $x \cong \tau$ valid when $\tau \cong 2$, with error $\mathcal{O}(\varepsilon)$. Treat the cases $\varepsilon > 0$ and $\varepsilon < 0$. Sketch the region in the (ε, τ) plane near $(\varepsilon, \tau) = (0, 2)$ for which your approximate solutions are real. Do you think the exact solutions are real in this region? For $\varepsilon < 0$, find the shape of regions where your approximations are uniformly valid.

1.8. APPROXIMATIONS AND SERIES

Two forms of approximations have occurred so far in this chapter: polynomials and a generalization of polynomials in which the powers are replaced by arbitrary gauge functions. It is time now to explain in greater detail these and other forms of approximations used in later chapters, and the rules by which they operate. But first it is helpful to classify the various types of variables which can occur in a problem and its solutions. These are divided into dependent and independent variables, and also into coordinates and parameters.

Consider a differential equation with initial conditions:

$$a\frac{d^2y}{dx^2} + b\frac{dy}{dx} + cy = \varepsilon f(x,y),$$

$$y(0) = \alpha, \qquad \frac{dy}{dx}(0) = \beta. \tag{1.8.1}$$

The variables $a, b, c, \alpha, \beta,$ and ε are *parameters*; these are "variables" which can be thought of as "constants," because fixing them does not change the character of the problem: For each set of values for these parameters, the problem becomes a specific second order differential equation. In contrast, the variables x and y cannot be fixed without making the problem meaningless, since there is no longer a differential equation if x and y cannot vary. These variables, the fundamental variables for the problem, are called *coordinates*. The solution of (1.8.1) has the form

$$y = y(x; a, b, c, \alpha, \beta; \varepsilon), \tag{1.8.2}$$

that is, the coordinate y is the *dependent variable*, while the coordinate x and the parameters $a, b, c, \alpha, \beta,$ and ε are *independent variables*. It is clear that parameters are always independent variables, whereas coordinates can be dependent or independent. The parameter ε is singled out in (1.8.2) as the *perturbation parameter*, because when $\varepsilon = 0$ a *reduced problem* is obtained which belongs to a solvable class (in this case, it is a homogeneous second order linear ordinary differential equation with constant coefficients). The other parameters, $a, b, c, \alpha,$ and β are *control parameters*. Sometimes some parameters are regarded as truly fixed constants and omitted from the solution function. For instance, if (1.8.1) is to be studied for various initial conditions but fixed $a, b,$ and c, the solution will be denoted $y = \varphi(x; \alpha, \beta; \varepsilon)$. In writing (1.8.2) we have used the common, but occasionally dangerous, practice of using the letter y to represent both a variable (on the left hand side) and a function (on the right). This causes no confusion unless several such functions are in use at the same time; in such a case we will use function letters such as $y = \varphi(x; a, b, c, \alpha, \beta; \varepsilon)$. An example occurs in Section 2.5, where $y = \varphi(x; \alpha, \beta; \varepsilon)$ is the solution of an initial value problem and $y = \psi(x; \alpha, \gamma; \varepsilon)$ is the solution of a boundary problem for the same differential equation.

Problem (1.8.1) is typical of the problems encountered in this book, except that the number of variables of each type may differ. A partial differential equation has more independent coordinates $x_1, \ldots, x_n$; a system of ordinary differential equations has more dependent coordinates $y_1, \ldots, y_m$. Usually such additional coordinates will be amalgamated into vectors $\mathbf{x} = (x_1, \ldots, x_n) \in \mathbf{R}^n$, $\mathbf{y} = (y_1, \ldots, y_m) \in \mathbf{R}^m$ so that the form

$$\mathbf{y} = \mathbf{y}(\mathbf{x}; \mathbf{p}; \varepsilon) \qquad (1.8.3)$$

can still be used to represent the solution; here $\mathbf{p} = (p_1, \ldots, p_\ell) \in \mathbf{R}^\ell$ is a vector of control parameters, but ε remains a scalar, as we will not deal with multiple perturbation parameters in this book (although this is an important subject of current research, under the name of *theory of unfoldings*). The problems of root finding studied in previous sections of this chapter have solutions of the form $x^{(i)}(\varepsilon)$ or $x^{(i)}(\tau, \varepsilon)$, where i denotes the specific root being considered; this has the form (1.8.3) with $x^{(i)}$ playing the role of $\mathbf{y}$, τ being $\mathbf{p}$ (these are both scalars, or vectors with one component), and $\mathbf{x}$ being absent (there is no independent coordinate).

In this chapter, most functions of ε have been defined in an interval of the form $|\varepsilon| < \varepsilon_0$. In later chapters this is not always the case. Frequently, functions of ε will only be defined (or will only be useful) in an interval of the form $0 \le \varepsilon < \varepsilon_0$, or even $0 < \varepsilon < \varepsilon_0$. (One example of this latter type was already given in Example 1.5.5, where a root approaches infinity as ε approaches zero. Other examples will occur in Chapters 7, 8, and 9.) In this section we will assume that all functions are defined for $0 < \varepsilon < \varepsilon_0$; it is easy to modify the definitions to handle other cases.

Several symbols are used in discussing approximations to functions such as (1.8.3). These include $\cong$, $\ll$, o, $\mathcal{O}$, $\mathcal{O}_F$, $\mathcal{O}_S$, $\sim$, and $\approx$. The symbol $\cong$ has been defined before; writing

$$\mathbf{y}(\mathbf{x}; \mathbf{p}; \varepsilon) \cong \hat{\mathbf{y}}(\mathbf{x}; \mathbf{p}; \varepsilon) \qquad (1.8.4)$$

merely means that the function $\hat{\mathbf{y}}$ is proposed (on some heuristic grounds) as an approximation to $\mathbf{y}$. Nothing is implied about the validity of the approximation.

A *gauge function* is a positive monotone function $\delta(\varepsilon)$ defined in some interval $0 < \varepsilon < \varepsilon_0$ of interest for a particular problem; the most common gauge functions (such as ε^ν for rational ν) are defined for all

$\varepsilon > 0$ (that is, "$\varepsilon_0 = \infty$"). "Monotone" means that a gauge function must be either increasing or decreasing throughout its domain (not sometimes increasing and sometimes decreasing). Gauge functions are used to measure the "size" of other functions "as $\varepsilon \to 0$." The following are the most important definitions:

$$f(\varepsilon) = o\big(\delta(\varepsilon)\big) \qquad \text{if there is a positive function } h(\varepsilon)$$

$$\text{and a constant } \varepsilon_1 \text{ with } 0 < \varepsilon_1 < \varepsilon_0$$

$$\text{such that } \lim_{\varepsilon \to 0^+} h(\varepsilon) = 0 \text{ and}$$

$$|f(\varepsilon)| \le h(\varepsilon)\delta(\varepsilon) \text{ for } 0 < \varepsilon \le \varepsilon_1; \qquad (1.8.5)$$

$$f(\varepsilon) = \mathcal{O}\big(\delta(\varepsilon)\big) \qquad \text{if there are constants } c \text{ and } \varepsilon_1$$

$$\text{with } 0 < \varepsilon_1 < \varepsilon_0 \text{ such that}$$

$$|f(\varepsilon)| \le c\delta(\varepsilon) \text{ for } 0 < \varepsilon < \varepsilon_1;$$

$$f(\varepsilon) = \mathcal{O}_S\big(\delta(\varepsilon)\big) \qquad \text{if } f = \mathcal{O}(\delta) \text{ and } f \text{ is not } o(\delta).$$

The symbols o and $\mathcal{O}$ are read "little-oh" and "big-oh" or "small order" and "large order." The symbol $\mathcal{O}_S$ is read "strict order," and means "big-oh but not little-oh." The symbol $\mathcal{O}_F(\delta)$ (encountered already in Section 1.2) is read "formal order" and is used in this book to denote an expression which appears to be $\mathcal{O}(\delta)$ because of its form, but for which the $\mathcal{O}$ statement has not been proved. Often, especially in applied literature, the symbol $\mathcal{O}$ is used with the meaning of $\mathcal{O}_F$. However, we will be careful to distinguish these.

These definitions can be stated in a simpler form under the assumption that the limit

$$L := \lim_{\varepsilon \to 0^+} |f(\varepsilon)|/\delta(\varepsilon) \qquad (1.8.6)$$

exists. In this case (1.8.5) is equivalent (see Exercise 1.8.1) to

$$f(\varepsilon) = o\big(\delta(\varepsilon)\big) \qquad \text{if} \quad L = 0;$$
$$f(\varepsilon) = \mathcal{O}\big(\delta(\varepsilon)\big) \qquad \text{if} \quad L < \infty; \qquad (1.8.7)$$
$$f(\varepsilon) = \mathcal{O}_S\big(\delta(\varepsilon)\big) \qquad \text{if} \quad 0 < L < \infty.$$

If the function $f(\varepsilon)$ is vector valued, $|\cdot|$ is replaced by $\|\cdot\|$ in definitions (1.8.5) and (1.8.7).

There are two additional symbols, $\ll$ and $\approx$, which are most commonly used in comparing two gauge functions. To say $\delta_2 \ll \delta_1$ is the same as to say $\delta_2 = o(\delta_1)$, and means that δ_2 is negligible compared to δ_1 when ε is sufficiently small. This symbol has already been used informally with this meaning in Section 1.5. The symbol $\delta_1 \approx \delta_2$ is defined to mean "$\delta_1 = \mathcal{O}(\delta_2)$ and $\delta_2 = \mathcal{O}(\delta_1)$" and is read "$\delta_1$ and δ_2 are asymptotically equivalent." If the limit of $\delta_1(\varepsilon)/\delta_2(\varepsilon)$ exists as $\varepsilon \to 0^+$, then $\delta_1 \approx \delta_2$ if and only if $\delta_1 = \mathcal{O}_S(\delta_2)$, but otherwise $\delta_1 \approx \delta_2$ is stronger. (See Exercise 1.8.1).

Suppose now that $f(x, \varepsilon)$ is a function depending upon another variable x in addition to the small parameter ε. There are two ways to adapt the definitions in (1.8.5) to this case. The first, and simplest, is to suppose that these conditions hold *for each fixed value* of x. This is called the *pointwise* interpretation of the big-oh and little-oh symbols. For instance, to say that $f(x, \varepsilon) = \mathcal{O}(\delta(\varepsilon))$ *pointwise for x in the interval $a \leq x \leq b$* means that for each x in the interval, there exist constants c and ε_1 such that $|f(x, \varepsilon)| \leq c\delta(\varepsilon)$ for $0 < \varepsilon \leq \varepsilon_1$. Since one fixes x before asserting the existence of c and ε_1, it may be necessary to choose these constants differently for different x. In other words, the "constants" are not constant unless x has been fixed first; actually, they must be thought of as varying with x. For the pointwise interpretation of the little-oh symbol, the function $h(\varepsilon)$ in (1.8.5) becomes a function $h(x, \varepsilon)$.

The other, and much more important, meaning that can be assigned to the big- and little-oh symbols when an additional variable x is present is called the *uniform* interpretation. Here, the constants c and ε_1 (and the function $h(\varepsilon)$) occurring in (1.8.5) are required to be independent of x, as long as x lies in a specified set. Since the big-oh case is the most important, we will present the following formal definition:

Definition 1.8.1. *The function $f(x, \varepsilon)$ satisfies the condition $f(x, \varepsilon) = \mathcal{O}(\delta(\varepsilon))$ uniformly for x in the interval $a \leq x \leq b$ if and only if there exist constants c and ε_1 such that $|f(x, \varepsilon)| \leq c\delta(\varepsilon)$ for all x in $a \leq x \leq b$ and for all ε in $0 < \varepsilon < \varepsilon_1$.*

It is often the case that a function $f(x, \varepsilon)$ is defined for x in an open interval or on the entire real line, and is uniformly $\mathcal{O}(\delta)$ for a certain gauge function δ on every finite closed subinterval of its domain, but not on its entire domain. In this case the following expression is used: $f(x, \varepsilon) = \mathcal{O}(\delta)$ *uniformly for x in compact subsets*. The meaning is: Given any compact subset K (such as a finite closed interval) of the domain of x, there exist the usual constants c and ε_1 such that $|f(x, \varepsilon)| \leq c\delta(\varepsilon)$ for all x in K and for all ε in $0 < \varepsilon \leq \varepsilon_1$. Here the constants c and ε_1 depend upon the compact set K, but once K is chosen, they do not depend any further on the value of x in K.

There are many possible combinations and variations of these ideas. As a typical illustration, suppose that a differential equation has a solution $y(x; \alpha, \beta; \varepsilon)$ defined for x in $0 \leq x \leq 1$ and for ε in $|\varepsilon| < \varepsilon_0$. Suppose that this solution has an approximation with error (or remainder) $R(x; \alpha, \beta; \varepsilon)$. It could then be the case that "$R = \mathcal{O}(|\varepsilon|^3)$ uniformly in x and uniformly for α and β in compact subsets." The complete "unwinding" of this phrase would be: Given any compact set of α and any compact set of β (for instance, given any closed bounded interval of these values), there exist constants c and ε_1 such that $|R(x; \alpha, \beta; \varepsilon)| \leq c|\varepsilon|^3$ for all $0 \leq x \leq 1$, for all α and β in the given compact sets, and for all ε in $|\varepsilon| \leq \varepsilon_1$. Notice that because ε is allowed to be negative here, we have written $\mathcal{O}(|\varepsilon|^3)$ instead of $\mathcal{O}(\varepsilon^3)$. In general, when negative ε are allowed one may write $\delta(|\varepsilon|)$ and keep the same definitions used above; alternatively, the definitions can be modified slightly. The reader is expected to be able to formulate appropriate variations on this theme as the need arises and to avoid mathematically "ungrammatical" versions which actually say nothing. (For instance, it means nothing to say "$|f(x, \varepsilon)| \leq c\delta(\varepsilon)$ for all x in compact subsets." This is meaningless because *every* x belongs to *some* compact subset; no restriction has been placed upon x until a particular compact subset has been specified. The expression "$\mathcal{O}(\delta)$ uniformly in compact subsets" is a technical phrase with a precise definition when taken as a whole. The expression "in compact subsets" cannot be lifted from this phrase and used by itself elsewhere.) Appendix G gives a more thorough discussion of the rules governing the meaning of mathematical statements containing the words "for every" and "there exists," along with the application of these rules to specific statements of uniformity properties. As a final remark, the word "uniform" sometimes seems to be used loosely, almost as a synonym for "valid"; it is a good habit never to say "uniformly" without saying, or at least thinking, "uniformly for certain variables in certain sets." And always remember that "uniformly" refers only to variables *other than* the perturbation parameter ε.

Because the symbol $\mathcal{O}$ is frequently misused, it is worthwhile to conclude this discussion by pointing out what $\mathcal{O}$ does *not* mean. Although the symbol is often read "of the order of," it must be remembered that this is "of the *asymptotic* order of" and not "of the *order of magnitude* of." For instance, according to (1.8.5), $f(\varepsilon) = \mathcal{O}(1)$ means that $|f(\varepsilon)|$ is bounded. Since any constant (regarded as a constant function of ε) is bounded, any constant is automatically $\mathcal{O}(1)$. Therefore $1,000,000 = \mathcal{O}(1)$, and also $0.00000001 = \mathcal{O}(1)$, in spite of the fact that these numbers are not at all of the *order of magnitude* of 1. (The order of magnitude of a number is the power of 10 that is needed to write it in "scientific notation," or the mantissa of its common logarithm.) One frequently encounters statements like this: Given the differential equation $y'' + 0.002y' + y = 0.1 \cos x$, take $\varepsilon = 0.1$, and then $0.002 = \mathcal{O}(\varepsilon^3)$, and so the problem should be written $y'' + 2\varepsilon^3 y' + y = \varepsilon \cos x$. Of course, this is nonsense. The perturbation problem we have just written down is one way of connecting the reduced problem $y'' + y = 0$ to the target problem, but there are many others; compare the discussion of (1.1.4) in Section 1.1. The fact that 0.002 is of the *order of magnitude* of $(0.1)^3$ does not mean that it must be written using the *asymptotic order* of ε^3 when creating a perturbation problem.

Enough language has now been developed to define the various types of approximations which will be used. Let $\delta(\varepsilon)$ be a gauge function. Any approximation $\mathbf{y}(\mathbf{x}; \mathbf{p}; \varepsilon) \cong \hat{\mathbf{y}}(\mathbf{x}; \mathbf{p}; \varepsilon)$ is called a *pointwise asymptotic approximation* of order $\delta(\varepsilon)$ if

$$\mathbf{y}(\mathbf{x}; \mathbf{p}; \varepsilon) = \hat{\mathbf{y}}(\mathbf{x}; \mathbf{p}, \varepsilon) + o\big(\delta(\varepsilon)\big) \tag{1.8.8}$$

as $\varepsilon \to 0$, pointwise in $\mathbf{x}$ and $\mathbf{p}$. Of course, equation (1.8.8) is a short way of writing $\mathbf{y}(\mathbf{x}; \mathbf{p}; \varepsilon) = \hat{\mathbf{y}}(\mathbf{x}; \mathbf{p}; \varepsilon) + R(\mathbf{x}; \mathbf{p}; \varepsilon)$ together with $R = o(\delta)$. The approximation is called a *uniform asymptotic approximation* (for $\mathbf{x}$ and $\mathbf{p}$ in specified sets) if the o symbol holds uniformly (in those sets).

Most asymptotic approximations are built up sequentially using a set of gauge functions $\delta_1(\varepsilon) \gg \delta_2(\varepsilon) \gg \cdots \gg \delta_k(\varepsilon)$. An approximation

$$\mathbf{y}(\mathbf{x}; \mathbf{p}; \varepsilon) \cong \mathbf{y}_0(\mathbf{x}; \mathbf{p})\delta_0(\varepsilon) + \cdots + \mathbf{y}_k(\mathbf{x}; \mathbf{p})\delta_k(\varepsilon) \tag{1.8.9}$$

is called a (finite) *asymptotic series* (pointwise or uniformly) provided that each of the following statements holds (pointwise or uniformly):

$$\mathbf{y}(\mathbf{x}; \mathbf{p}; \varepsilon) = \mathbf{y}_0(\mathbf{x}; \mathbf{p})\delta_0(\varepsilon) + o(\delta_0(\varepsilon))$$
$$\mathbf{y}(\mathbf{x}; \mathbf{p}; \varepsilon) = \mathbf{y}_0(\mathbf{x}; \mathbf{p})\delta_0(\varepsilon) + \mathbf{y}_1(\mathbf{x}; \mathbf{p})\delta_1(\varepsilon) + o(\delta_1(\varepsilon))$$
$$\vdots \tag{1.8.10}$$
$$\mathbf{y}(\mathbf{x}; \mathbf{p}; \varepsilon) = \sum_{n=0}^{k} \mathbf{y}_n(\mathbf{x}; \mathbf{p})\delta_n(\varepsilon) + o(\delta_k(\varepsilon)).$$

This situation is denoted by the symbol $\sim$:

$$\mathbf{y}(\mathbf{x}; \mathbf{p}; \varepsilon) \sim \mathbf{y}_0(\mathbf{x}; \mathbf{p})\delta_0(\varepsilon) + \cdots + \mathbf{y}_k(\mathbf{x}; \mathbf{p})\delta_k(\varepsilon). \tag{1.8.11}$$

The same symbol $\sim$ is used if the series on the right hand side is continued indefinitely as an infinite asymptotic series. Such a series is not assumed to converge, only to satisfy an infinite sequence of error estimates of the form (1.8.10). Each error term (or remainder) with fixed n has the behavior as $\varepsilon \to 0$ specified by $o(\delta_n(\varepsilon))$, but for fixed ε the remainders need not go to zero as $n \to \infty$. The word *expansion* is synonymous with *series*, and these words will be used interchangeably, both for finite and infinite series.

Uniformity of an asymptotic series refers to uniform validity of the error estimates in (1.8.10). There is a closely related, but weaker, notion which is sometimes treated as if it were equivalent to uniformity. Each term of (1.8.11) is clearly of the order indicated by its gauge, for fixed $\mathbf{x}$ and $\mathbf{p}$; that is,

$$\mathbf{y}_n(\mathbf{x}; \mathbf{p})\delta_n(\varepsilon) = \mathcal{O}(\delta_n(\varepsilon)) \tag{1.8.12}$$

always holds pointwise. If (for a given set of $\mathbf{x}$ and $\mathbf{p}$) equation (1.8.12) holds uniformly for each $n > 0$, the approximation (1.8.11) will be called *uniformly ordered* (for $\mathbf{x}$ and $\mathbf{p}$ in their respective sets). Notice that no requirement is placed on the leading term $n = 0$. An example of a series that is not uniformly ordered is (1.7.2),

$$x^{(1)}(\tau, \varepsilon) \sim 1 - \frac{1}{1-\tau}\varepsilon - \frac{\tau}{(1-\tau)^3}\varepsilon^2, \tag{1.8.13}$$

on an interval such as $1 < \tau < 2$. The ε term, for instance, is not uniformly $\mathcal{O}(\varepsilon)$ on this interval since there is no constant c such that $|\varepsilon/(1-\tau)| \leq c|\varepsilon|$; if there were, then $|1/(1-\tau)|$ would be bounded $(\leq c)$ on $1 < \tau < 2$. Other examples of series that are not uniformly ordered are series with "secular terms", discussed in detail in Chapter 4. A series which is not uniformly ordered is "disordered" in the sense that a term can be comparable in magnitude to an earlier term in the series, regardless of how small ε is. (Choose ε arbitrarily small and choose $\tau = 1 + \varepsilon$ in $1 < \tau < 2$; then the second term of (1.8.13) is equal to the first.) The leading term can become arbitrarily large without overwhelming any earlier term, which explains why no restriction is needed for the leading term.

Uniform ordering is often confused with uniformity. Since uniform ordering depends only on the terms appearing in the approximation, it is much easier to check that a series is uniformly ordered than to test it for uniformity, which requires an estimate for the error. The following theorem develops the relationship between the two concepts in greater detail. The most important part of the theorem is that uniformity implies uniform ordering but not conversely; in other words, uniformity is a stronger condition than uniform ordering. In practice this means that if the straightforward solution of a problem is not uniformly ordered then it cannot be uniformly valid; you should first look for an alternate method that gives a uniformly ordered solution, and only then should you bother with error estimates and trying to prove uniformity. Most of the important methods to be developed in this book, beginning with Chapter 3, are motivated by the search for uniformly ordered solutions; a well-conceived solution that is uniformly ordered most often turns out to be uniform as well, although the proof may be difficult, and in many cases (especially in partial differential equations) the proofs are not yet available. (On the other hand there exist perturbation series, actually arising in practical problems, which are uniformly ordered but not uniformly valid. This is the point that is not recognized in many textbooks. A specific example will be given in the last paragraph of Section 3.2 and in Exercise 3.2.4.)

Theorem 1.8.1. *(a) The series (1.8.11) is uniformly ordered (for* $\mathbf{x}$ *and* $\mathbf{p}$ *in specified sets) if and only if each coefficient* $\mathbf{y}_n(\mathbf{x}; \mathbf{p})$ *for* $n > 0$ *is bounded (for* $\mathbf{x}$ *and* $\mathbf{p}$ *in those sets). (b) If the series (1.8.11) is uniform, then it is uniformly ordered. (b') If a series is not uniformly ordered, it is not uniform. (c) If (1.8.11) is uniformly ordered and the last equation of (1.8.10) holds uniformly, then all of (1.8.10) holds uniformly and the series is uniform.*

Proof. Statement (a) is easy and is left to the reader. If (1.8.11) is uniform, that is, if (1.8.10) holds uniformly, a comparison of the first two equations shows that $\mathbf{y}_1(\mathbf{x}; \mathbf{p})\delta_1(\varepsilon) = o(\delta_0) - o(\delta_1) = o(\delta_0)$ uniformly; that is, $\mathbf{y}_1(\mathbf{x}; \mathbf{p})\delta_1(\varepsilon)/\delta_0(\varepsilon) \to 0$ uniformly as $\varepsilon \to 0$. This is possible only if

$y_1(x; p)$ is bounded. Proceeding similarly shows that each y_n is bounded except possibly for y_0, proving statement (b). (The example ε/p, which is its own one-term asymptotic expansion uniformly for $0 < p < 1$, shows that we cannot require boundedness of the leading term.) (b') is logically equivalent to (b), being the contrapositive, but is stated separately because of its usefulness. Suppose now that (1.8.11) is uniformly ordered. From this alone it does not follow that it is a uniform asymptotic series, because uniform ordering says nothing about the error. But if the last line of (1.8.10) holds, the rest follows. For instance,

$$\begin{aligned} y(x; p; \varepsilon) &= \sum_{n=0}^{k-1} y_n(x; p)\delta_n(\varepsilon) + \left\{ y_k(x; p)\delta_k(\varepsilon) + o\big(\delta_k(\varepsilon)\big) \right\} \\ &= \sum_{n=0}^{k-1} y_n(x; p)\delta_n(\varepsilon) + \mathcal{O}\big(\delta_k(\varepsilon)\big) \end{aligned} \tag{1.8.14}$$

uniformly (because of uniform ordering), and $\mathcal{O}(\delta_k)$ implies $o(\delta_{k-1})$. This is the next to the last line of (1.8.10), and the remaining lines are proved in the same way, working upward. This proves (c). ∎

Corollary 1.8.2. *If (1.8.10) holds pointwise or uniformly, then the error estimates can be improved from "little-oh of the last gauge used" to "big-oh of the next gauge," except possibly in the last line. That is, the following hold pointwise or uniformly:*

$$y(x; p; \varepsilon) = y_0(x; p)\delta_0(\varepsilon) + \mathcal{O}\big(\delta_1(\varepsilon)\big)$$

$$y(x; p; \varepsilon) = y_0(x; p)\delta_0(\varepsilon) + y_1(x; p)\delta_1(\varepsilon) + \mathcal{O}\big(\delta_2(\varepsilon)\big)$$

$$\vdots$$

$$y(x; p; \varepsilon) = \sum_{n=0}^{k-1} y_n(x; p)\delta_n(\varepsilon) + \mathcal{O}\big(\delta_k(\varepsilon)\big) \tag{1.8.15}$$

$$y(x; p; \varepsilon) = \sum_{n=0}^{k} y_n(x; p)\delta_n(\varepsilon) + o\big(\delta_k(\varepsilon)\big).$$

If the series (1.8.11) can be carried one step farther using some gauge $\delta_{k+1}(\varepsilon)$, the last line of (1.8.15) can also be improved to $\mathcal{O}(\delta_{k+1})$.

Proof. In the uniform case, since uniformity implies uniform ordering, the statements follow as in (1.8.14). The pointwise case is easy and is left to the reader; the point is that every series of the form (1.8.9) is "pointwise ordered" as long as the gauges satisfy $\delta_i \gg \delta_{i+1}$. ∎

The idea that the $n + 1$st term should be a uniformly small correction to the nth is occasionally seen stated in the form $\|y_{n+1}\|\delta_{n+1} / \|y_n\|\delta_n = o(1)$

uniformly as $\varepsilon \to 0$. (If y is a scalar, $\| \cdot \|$ may be omitted.) We will have no use for this condition, and it is in fact neither equivalent to uniformity nor to uniform ordering. The function $1/\tau^2 + \varepsilon/\tau$ is its own two-term asymptotic expansion pointwise on $0 < \tau < 1$ and satisfies the condition that $y_2\delta_2/y_1\delta_1 = \varepsilon\tau$ is uniformly $o(1)$. It is of course a uniform asymptotic *approximation* because its error is zero. But it is not a uniform asymptotic *series*, because its first term $1/\tau^2$ has error ε/τ which is not uniformly $o(1)$. By the same token the series is not uniformly ordered, because ε/τ is not uniformly $\mathcal{O}(\varepsilon)$. (This is a good example to make sure you understand all the definitions of this section; see Exercise 1.8.3.)

Although this book will never deal with *convergent series*, it is important to make certain that the difference between convergent and asymptotic series is understood. A convergent series must first of all have an infinite number of terms (although it is possible that they are all zero from some point on); to say that the series is convergent means that the nth partial sum (the sum of the first n terms) approaches a limit as n approaches infinity. An asymptotic series may have a finite number of terms, and even if it is infinite, one is never concerned with letting n become large. Instead, (1.8.10) indicates that the focus is on the error in approximating a function by a partial sum when the number of terms in the partial sum is held constant; this error should decrease at a specified rate as ε approaches zero, not as n approaches infinity. It is easiest to study this distinction in the case of Taylor series. Taylor's theorem with remainder (Theorem A.1) expresses the error as a function of both ε and n. It is always true that this error is $\mathcal{O}(\varepsilon^{n+1})$ for fixed n; this is an asymptotic statement. It is not always true that the error approaches zero for fixed ε as $n \to \infty$, although this is true for such familiar Taylor series as those for e^{ε}, $\sin\varepsilon$, and $\ln(1 + \varepsilon)$ on a suitable domain.

Since an asymptotic series need not converge (although it may), increasing the number of terms used in a partial sum of such a series may not increase the accuracy of the approximation, for a specific value of ε. For instance, in an asymptotic power series, the error after the first two terms (the constant term and the linear term) may be bounded by $10\varepsilon^2$; including the quadratic term might give an error bounded by $1000\varepsilon^3$. The asymptotic nature of the series specifies the powers in these error bounds, but not the coefficients. In this example, the linear approximation has error less than 0.1 when $\varepsilon = 0.1$; the error of the quadratic approximation is only bounded by 1, so the linear approximation is probably best. ("Probably," because it is still possible that the quadratic approximation actually has a smaller error than the linear one. We only have upper bounds for these errors, not lower bounds.) What one does gain by increasing the number of terms in an asymptotic series is an improvement in the *rate* at which the error goes to zero when ε is decreased. In our illustration, the linear approximation is better for $\varepsilon = 0.1$, but for $\varepsilon = 0.01$ the quadratic is better. One always gains accuracy by increasing n *and decreasing ε sufficiently*. On the other hand, for fixed ε there may be a certain number of terms which gives the best approximation possible, after which increasing the number of terms leads to a loss of accuracy. This happens whenever the asymptotic series is actually divergent; of course if it is convergent in addition to being asymptotic, one can only gain by taking more terms. All of these points are usually moot in the applications, since it becomes computationally very difficult to compute terms beyond the second or third. In the heroic days of celestial mechanics people sometimes computed eight or ten terms of such a series, and by doing algebra on a computer (symbolic manipulation) twenty-five or more terms have been computed in some cases, but this is still quite unusual and is only worthwhile under very special circumstances.

All of the asymptotic expansions considered until now have been of the form (1.8.11). Series of this type are not sufficient for all perturbation problems. A perturbation problem is called *regular* if its exact solution $y(x; p; \varepsilon)$ has an asymptotic expansion of the form (1.8.11) which holds uniformly for x and p in the desired domains. Otherwise the problem is called *singular*. The exact requirements for a regular problem vary from author to author, but it is not necessary to be precise, since the distinction between regular and singular perturbations is used only to sort problems according to their difficulty. To call a problem regular means at least this: that its solution has an expansion of the form (1.8.11) to some order k with respect to some gauges $\delta_0, \ldots, \delta_k$ which holds uniformly for x in the entire natural domain of the independent variables and for p in a specified subset of the parameter domain. Most authors require that this uniform expansion exist to all orders as an infinite asymptotic series.

Usually, a singular perturbation problem has an asymptotic series solution of the form (1.8.11) which holds pointwise in the full domain of x but not uniformly, although it may hold uniformly for x in certain subsets of the domain. In order to find approximate solutions of singular problems which are uniformly valid in the full domain of x, it is necessary to use expressions of the form

$$y(x; p; \varepsilon) \sim y_0(x; p; \varepsilon)\delta_0(\varepsilon) + \cdots + y_k(x; p; \varepsilon)\delta_k(\varepsilon) \qquad (1.8.16)$$

in which ε is allowed to enter through the coefficients y_n as well as the gauges. Such an approximation is called a *generalized asymptotic series* (or *expansion*); the symbol $\sim$ indicates that the same error estimates as in (1.8.10) are required to hold. In order that the coefficients do not subvert the meaning of the gauges, it is necessary to require at least that each $y_n(x; p, \varepsilon)$ remains bounded as $\varepsilon \to 0$ for fixed x and p; this guarantees that each term of (1.8.16) is pointwise $\mathcal{O}$ of its gauge. If each term after the leading term is uniformly $\mathcal{O}$ of its gauge, the series is termwise uniformly ordered and satisfies estimates like (1.8.15).

It is possible to interpret equation (1.7.19) as a generalized asymptotic expansion. Namely, if (1.7.12), with $\lambda = 1/2$, is substituted into the "sigma form" of (1.7.19), the coefficient of the gauge $\sqrt{-\varepsilon}$ becomes a function of ε. (In the "tau form" of (1.7.19) we have obscured this by multiplying the expression out.) The observation that this is a generalized expansion may help to understand why the "tau form" must not be re-expanded in powers of ε: This would not only undo the rescaling (as we pointed out), but also return to an ordinary expansion. (Of course, (1.7.19) *is* an ordinary expansion in terms of the rescaled parameter σ.) This example should make it clear that a generalized asymptotic expansion can have different uniformity properties from an ordinary expansion for the same problem.

Apart from this example, we will not encounter generalized asymptotic expansions until Chapter 3. In this book *asymptotic series* (or for emphasis, *ordinary asymptotic series*) always means a series of the form (1.8.11), and (1.8.16) will always be called a *generalized asymptotic series*; the phrase *perturbation series* will be used loosely to refer to both of these. This usage is common, although many books refer to both types as asymptotic series and distinguish the first (ordinary) type as a "Poincaré series."

The remainder of this section is devoted to theorems which hold for ordinary, but not for generalized, asymptotic expansions. These theorems play an essential role in regular perturbation theory (Chapter 2). For an example of their use, see Exercise 1.8.5. Although some of these theorems resemble theorems about power series in calculus texts, they are not the same (even when $\delta_n = \varepsilon^n$), because the proofs in most calculus texts are only valid for convergent series.

Theorem 1.8.3. *If a function* $y(x; p; \varepsilon)$ *has an ordinary asymptotic expansion using gauges* $\delta_0, \ldots, \delta_k$, *then that expansion is unique; that is, if* $y \sim \sum y_n \delta_n$ *and* $y \sim \sum \tilde{y}_n \delta_n$, *then* $y_n = \tilde{y}_n$ *for* $n = 0, \ldots, k$. *This is true even if the expansions are only valid pointwise in* x *and* p.

Proof. From the assumption that $y \sim \sum y_n \delta_n$ we will derive formulas for y_n. Since the same formulas apply to $\tilde{y}_n$, it follows that $y_n = \tilde{y}_n$. First divide both sides of $y = y_0 \delta_0 + o(\delta_0)$ by δ_0 and let $\varepsilon \to 0$. The result is

$$y_0(x; p) = \lim_{\varepsilon \to 0} \frac{y(x; p; \varepsilon)}{\delta_0(\varepsilon)}. \tag{1.8.17a}$$

This determines y_0 uniquely. Then divide both sides of $y = y_0 \delta_0 + y_1 \delta_1 + o(\delta_1)$ by δ_1 and let $\varepsilon \to 0$ to obtain

$$y_1(x; p) = \lim_{\varepsilon \to 0} \frac{y(x; p; \varepsilon) - y_0(x; p)\delta_0(\varepsilon)}{\delta_1(\varepsilon)}. \tag{1.8.17b}$$

Since y_0 is unique, this determines y_1 uniquely. The general formula is

$$y_n(x; p) = \lim_{\varepsilon \to 0} \frac{1}{\delta_n(\varepsilon)} \left\{ y(x; p; \varepsilon) - \sum_{i=0}^{n-1} y_i(x; p)\delta_i(\varepsilon) \right\}. \quad \blacksquare \tag{1.8.17c}$$

An expansion obtained by formulas (1.8.17) is sometimes called a *limit process expansion*. If $y(x; p; \varepsilon)$ is any function for which the limits (1.8.17) exist, the series $\sum y_n \delta_n$ defined from (1.8.17) will be asymptotic to y, at least pointwise (Exercise 1.8.3). To see that this is more general than a Taylor expansion, even when $\delta_n = \varepsilon^n$, consider the function

$$y(x; \varepsilon) = e^{\varepsilon x} + e^{-x/\varepsilon} \tag{1.8.18}$$

for $\varepsilon > 0$ and $x > 0$. This has no Taylor series expansion in ε, since it is not even defined for $\varepsilon = 0$. (It can be extended to $\varepsilon = 0$ by letting $\varepsilon \to 0^+$, resulting in $y(x; 0) = 1$, and a Taylor series of sorts can be defined using one-sided derivatives, but this is somewhat awkward.) But it is easy to calculate the limits in (1.8.17) using $\delta_n = \varepsilon^n$ and find an asymptotic expansion, which turns out to coincide with the Taylor series of $e^{\varepsilon x}$. (See Exercise 1.8.4.) A function such as $e^{-x/\varepsilon}$, which approaches zero so rapidly as $\varepsilon \to 0$ that its influence never shows up in an asymptotic power series, is called *transcendentally small*. This example shows that an asymptotic series does not determine a function uniquely, although the function does determine the series uniquely (provided the series is regular and the gauges are fixed.)

Addition of asymptotic series (using the same gauges), and multiplication by a constant, are always valid. Multiplication of two scalar asymptotic series is trickier. (For the vector case, see Exercise 1.8.9.) The simplest case is when the gauges satisfy $\delta_n \delta_m = \delta_{n+m}$; notice that this requires $\delta_0 = 1$. This condition holds for asymptotic power series ($\delta_n = \varepsilon^n$) and, more generally, for fractional power series with fixed denominator ($\delta_n = \varepsilon^{n/q}$). The following theorem covers the most important cases.

Theorem 1.8.4. *Suppose that*

$$\begin{aligned} u(\mathbf{x}; \mathbf{p}; \varepsilon) &\sim u_0(\mathbf{x}; \mathbf{p})\delta_0(\varepsilon) + \cdots + u_k(\mathbf{x}; \mathbf{p})\delta_k(\varepsilon), \\ v(\mathbf{x}; \mathbf{p}; \varepsilon) &\sim v_0(\mathbf{x}; \mathbf{p})\delta_0(\varepsilon) + \cdots + v_k(\mathbf{x}; \mathbf{p})\delta_k(\varepsilon) \end{aligned} \tag{1.8.19}$$

either pointwise or uniformly. Then

(a) $u + v \sim (u_0 + v_0)\delta_0 + \cdots + (u_k + v_k)\delta_k$ *pointwise or uniformly;*

(b) $cu \sim cu_0\delta_0 + \cdots + cu_k\delta_k$ *pointwise or uniformly;*

(c) $uv = (u_0\delta_0 + \cdots + u_k\delta_k)(v_0\delta_0 + \cdots + v_k\delta_k) + o(\delta_0\delta_k)$ *pointwise, and this holds uniformly, provided (1.8.19) holds uniformly and u_0 and v_0 are bounded;*

(d) *If $\delta_n\delta_m = \delta_{n+m}$ (and hence $\delta_0 = 1$), then*

$$uv \sim u_0v_0 + (u_0v_1 + u_1v_0)\delta_1 + \cdots + (u_0v_k + u_1v_{k-1} + \cdots + u_kv_0)\delta_k \tag{1.8.20}$$

pointwise, and this holds uniformly, provided (1.8.19) holds uniformly and u_0 and v_0 are bounded.

Proof. (a) and (b) are left to the reader. For (c), write $u = u_0\delta_0 + \cdots + u_k\delta_k + R_k$ and $v = v_0\delta_0 + \cdots + v_k\delta_k + S_k$. Multiplying these equations gives

$$uv = (u_0\delta_0 + \cdots + u_k\delta_k)(v_0\delta_0 + \cdots + v_k\delta_k)$$
$$+ \{(u_0\delta_0 + \cdots + u_k\delta_k)S_k + (v_0\delta_0 + \cdots + v_k\delta_k)R_k\}.$$

Since R_k and S_k are $o(\delta_k)$ and each δ_n is $\mathcal{O}(\delta_0)$ it follows that any product $\delta_n R_k$ or $\delta_n S_k$ is $o(\delta_0\delta_k)$. Therefore any product $u_n\delta_n R_k$ or $v_n\delta_n S_k$ is pointwise $o(\delta_0\delta_k)$ and is uniformly $o(\delta_0\delta_k)$, provided that the coefficient u_n or v_n is bounded (for **x** and **p** in the region of uniformity). If (1.8.19) holds uniformly, then all u_n and v_n for $n > 0$ are bounded by Theorem 1.8.1, so it is only necessary to assume this specifically for u_0 and v_0. Notice that statement (c) does not use the symbol $\sim$, because it is only an asymptotic approximation and not an asymptotic series. If $\delta_0 = 1$ and $\delta_n\delta_m = \delta_{n+m}$, (c) can be multiplied out to obtain

$$uv = u_0v_0 + (u_0v_1 + u_1v_0)\delta_1 + \cdots + (u_0v_k + \cdots + u_kv_0)\delta_k + o(\delta_k),$$

which is one of the estimates making up statement (d); the others follow by truncating this and estimating the error, again using boundedness of coefficients in the uniform case. ∎

Differentiation of an asymptotic series with respect to an independent variable (coordinate or parameter) is essential in finding heuristic solutions to most perturbation problems, but cannot be justified by a general theorem. The reason is that a statement such as $\mathbf{y} = \mathbf{y}_0\delta_0 + \mathbf{y}_1\delta_1 + \mathcal{O}(\delta_2)$ is actually an inequality, $\|\mathbf{y} - \mathbf{y}_0\delta_0 + \mathbf{y}_1\delta_1\| \leq c\delta_2$, and inequalities cannot be differentiated. (If $f(x) < g(x)$, i.e., if the graph of f lies beneath the graph of g, it does not follow that $f'(x) < g'(x)$, i.e., that the slope of f is less than that of g). The only case in which differentiation is easily justified is the case of Taylor series. This is sufficient for most regular problems (see Chapter 2); otherwise, differentiation of an asymptotic series must be regarded as a formal and heuristic process, justifiable only by an error estimate performed after the solution is obtained (Chapter 4).

Theorem 1.8.5. *Suppose that* $\mathbf{y}(\mathbf{x}; \mathbf{p}; \varepsilon)$ *has continuous partial derivatives of all orders* $\leq r$. *Then for each* $k \leq r - 2$, *the Taylor approximation*

$$\mathbf{y} = \mathbf{y}_0(\mathbf{x}; \mathbf{p}) + \mathbf{y}_1(\mathbf{x}; \mathbf{p})\varepsilon + \cdots + \mathbf{y}_k(\mathbf{x}; \mathbf{p})\varepsilon^k + \mathcal{O}(\varepsilon^{k+1}) \tag{1.8.21}$$

can be differentiated termwise to give

$$\frac{\partial \mathbf{y}}{\partial x_i} = \frac{\partial \mathbf{y}_0}{\partial x_i} + \frac{\partial \mathbf{y}_1}{\partial x_i}\varepsilon + \cdots + \frac{\partial \mathbf{y}_k}{\partial x_i}\varepsilon^k + \mathcal{O}(\varepsilon^{k+1}),$$

$$\frac{\partial \mathbf{y}}{\partial p_j} = \frac{\partial \mathbf{y}_0}{\partial p_j} + \frac{\partial \mathbf{y}_1}{\partial p_j}\varepsilon + \cdots + \frac{\partial \mathbf{y}_k}{\partial p_j}\varepsilon^k + \mathcal{O}(\varepsilon^{k+1}). \tag{1.8.22}$$

Proof. Taylor's theorem (Appendix A) implies that the error term in (1.8.21) can be written

$$\int_0^\varepsilon \frac{\partial^{k+1}}{\partial \varepsilon^{k+1}} \mathbf{y}(\mathbf{x}; \mathbf{p}; \varepsilon)\bigg|_{\varepsilon=s} \frac{(\varepsilon - s)^k}{k!} ds.$$

This can be differentiated under the integral sign to give the error in (1.8.22), and since the resulting integrand is continuous, the integral can be estimated as in the proof of Taylor's theorem. (Why does this theorem say $k \leq r - 2$ while Taylor's theorem only requires $k \leq r - 1$? Can you formulate a slightly weaker requirement on the derivatives of $\mathbf{y}$ under which the same proof is valid?) ∎

Exercises 1.8

1. (a) Show that each definition in (1.8.7) is equivalent to the corresponding definition in (1.8.5) in the case that the limit (1.8.6) exists. In addition, show that $\mathcal{O}_S$ and $\approx$ are the same when this limit exists.

   (b) Let $f(\varepsilon) = \varepsilon \sin(1/\varepsilon)$ and $g(\varepsilon) = \varepsilon$. Show that $f = \mathcal{O}_S(g)$ but that f is not $\approx g$. (Here $f(\varepsilon)$ is not a gauge function, but use the same definition of $\approx$ as we have given for gauge functions. After seeing this example, can you find a gauge function for which $\mathcal{O}_S$ and $\approx$ are not the same?)

2. Define from memory the following expressions, in the sense used in this book:

   > pointwise asymptotic approximation
   > uniform asymptotic approximation
   > pointwise asymptotic expansion (or series)
   > uniform asymptotic expansion
   > uniformly ordered asymptotic expansion
   > generalized asymptotic expansion
   > regular perturbation problem
   > singular perturbation problem

3. Let $f(\tau; \varepsilon) := 1/\tau^2 + \varepsilon/\tau$ for $0 < \tau < 1$, and consider this series as an approximation to itself: $f(\tau; \varepsilon) \sim 1/\tau^2 + \varepsilon/\tau$. Prove in complete detail, using only the definitions, that this is a uniform asymptotic approximation and a pointwise asymptotic series, but not a uniform asymptotic series and not uniformly ordered. Show that *on compact subsets of* τ it is uniformly ordered and is a uniformly asymptotic series.

4. Prove that for any function $\mathbf{y}(\mathbf{x}; \mathbf{p}; \varepsilon)$ the series defined by $\sum \mathbf{y}_n \delta_n$, with $\mathbf{y}_n$ given by (1.8.17), is pointwise asymptotic to $\mathbf{y}$, provided only that

the limits in (1.8.17) exist. Can you give a condition under which the series is uniformly asymptotic?

5. Show that the limit process power series expansion of (1.8.18) coincides with the Taylor series in ε of $e^{\varepsilon x}$. Hint: Use L'Hospital's rule to show that $\lim_{t \to \infty} t^n e^{-t} = 0$ for all n.

6. Use Theorems 1.8.3 and 1.8.4 to re-prove Theorem 1.4.2 without using (1.4.5). Hint: Theorem 1.4.1 and Taylor's theorem imply that an asymptotic power series solution exists, and Theorems 1.8.3 and 1.8.4 can be used to show that the procedures of Section 1.2 are a legal way of finding the coefficients.

7. Let $f(\varepsilon)$ have r continuous derivatives and let $g(\varepsilon)$ have an asymptotic expansion $g(\varepsilon) \sim b_0 + \cdots + b_k \varepsilon^k$ with $k \le r - 1$. Show how to compute a kth order asymptotic expansion of $f(g(\varepsilon))$ and verify the asymptotic estimates for this series. (In particular, this result will hold if g also has r continuous derivatives, but your proof should not use this assumption.)

8. Suppose that $y(x, \varepsilon)$ has an asymptotic expansion $y(x, \varepsilon) \sim y_0(x) + \varepsilon y_1(x) + \cdots + \varepsilon^k y_k(x)$ uniformly for x in compact intervals. Show that this series may be integrated with respect to x over any compact interval, that is,

$$\int_a^b y(x; \varepsilon)dx \sim \int_a^b y_0(x)dx + \cdots + \varepsilon^k \int_a^b y_k(x)dx.$$

(Hint: This is much easier than differentiation because inequalities can be integrated: If $f(x) \le g(x)$, then any definite integral of f is less than or equal to the corresponding integral of g.) Notice that there is no question of "uniformity with respect to x" in the conclusion since x is no longer present as a free variable; it appears only as a bound or "dummy" variable. Now show that

$$\int_0^x y(\xi; \varepsilon)d\xi \sim \int_0^x y_0(\xi)d\xi + \cdots + \varepsilon^k \int_0^x y_k(\xi)d\xi$$

uniformly for x in compact intervals. If the original asymptotic expansion for y holds uniformly for *all* x, explain why the last expansion does not (in general) hold uniformly for all x, but still only for x in compact intervals. (Remark: All of these statements continue to hold if **y** is a vector. If **x** is a vector, it is possible to integrate with respect to a component x_i or, using a multiple integral, over a compact region in **x**-space. If **y** also depends on additional variables **p** and if the original expansion holds uniformly for **p** in compact subsets, the same is true of the integrated expansions.)

9. Formulate and prove versions of Theorem 1.8.4 for a scalar-valued asymptotic series times a vector-valued one and for the inner product (dot product or scalar product) of two vector-valued series. (There is less

occasion to use the cross (or vector) product of two three-dimensional vector-valued asymptotic series, but a similar theorem could be proved for that case.)

1.9. NOTES AND REFERENCES

The material in this chapter stands at the center of a great deal of important mathematics. The subjects of *algebraic geometry, singularity theory, catastrophe theory*, and *bifurcation theory* all deal with the zero-sets of mappings from one real or complex vector space to another.

Algebraic geometry focuses on polynomial maps, usually in complex variables. The zero-set of such a mapping is called a *variety*, and the description of the portion of a variety in a neighborhood of a given point is called *local algebraic geometry*. When the mapping is from two-dimensional complex space to one-dimensional complex space, this means describing the set on which $\varphi(z, w) = 0$ near a point (z_0, w_0). Except for notation, this is the same as our problem $\varphi(x, \varepsilon) = 0$ if x and ε are allowed to be complex. The local description of such a variety can be achieved using fractional power series (Puiseux series). Our treatment emphasized asymptotic aspects (error estimates for truncations of the series), whereas algebraic geometry, as the name implies, is more concerned with geometric questions. An excellent introduction is

Robert J. Walker, *Algebraic Curves*, Princeton University Press, Princeton, N.J., 1950; also Dover, New York, 1962.

The point of view taken in the early part of this book is close to that of this chapter; more modern treatments rely heavily on abstract algebra (the theory of local rings and valuations) in place of fractional power series, since algebraic tools have been found to be powerful in higher dimensions. For a taste of this approach, one might (after reading part of Walker's book) look at

William Fulton, *Algebraic Curves: An Introduction to Algebraic Geometry*, Benjamin, New York, 1969.

Algebraic geometry is a good place to experience the real unity of all mathematics. The present chapter indicates some of the connections between algebraic geometry and applied mathematics; there are many well-known connections between algebraic geometry and topics such as number theory, complex analysis, topology, and on and on. For instance, in calculus one learns to integrate a function like $1/\sqrt{az^2 + bz + c}$ using trig functions, but if the quadratic is replaced by a cubic, the integration is not possible using elementary functions (it requires elliptic functions). The reason can be found by considering the zero-set (or variety) of the function $\varphi(z, w) = w^2 - (az^2 + bz + c)$. Topologically, this zero-set is a sphere (after adding "points at infinity"), whereas in the cubic case it becomes a torus, and this turns out to be what makes the difference. The question here involves the overall or *global* properties of the variety, not merely the local description using fractional power series, but there is nevertheless a far-reaching connection with the topic of this chapter. One should learn to search for such connections in order not to feel totally lost in the sea of modern mathematics. The particular topic just described may be found in the insightful book

Harvey Cohn, *Conformal Mappings on Riemann Surfaces*, McGraw-Hill, New York, 1967.

Bifurcation theory is concerned with changes in the number of solutions to an equation as a parameter is varied. Usually the equations are ordinary or partial differential equations, and the subject has a very applied flavor. A typical problem concerns the "buckling of a beam." A yardstick, for instance, in a vertical position with one end on the floor, will remain straight if the top end is pressed slightly downward, but if the pressure reaches a critical value the yardstick will bend to one side or the other. The position of the stick is an equilibrium solution of a partial differential equation, and the downward force is a parameter. For small values of the parameter there is one equilibrium solution (the vertical), for larger values there are three (two bent positions, one on either side, and the straight vertical position, which is now unstable and will not be physically observed). This is a *pitchfork bifurcation*. A technique called *Liapunov-Schmidt reduction* turns the partial differential equation into just the type of problem treated in this chapter, and the bifurcation can be established by the fractional power series or rescaling techniques developed here. We could, once again, employ the language of the previous paragraph and say that the pitchfork is a variety whose local geometry changes at the bifurcation point. But there are two important differences in outlook between bifurcation theory and algebraic geometry. The first is that in bifurcation theory only *real* solutions count. (For example, a pitchfork bifurcation has three solutions on each side of the critical value if complex solutions are counted, but this is not what is wanted in bifurcation theory.) The second is that in bifurcation theory it matters *on which side* of the critical value the solution branches exist. One wants to distinguish a transcritical bifurcation (two solutions on each side, crossing at the bifurcation point) from a pitchfork (one and three), but from the point of view of algebraic geometry they are the same: In both cases if the crossing point is deleted, four arcs remain. An introduction to bifurcation theory, at about the same mathematical level as this text, is

Gérard Iooss and Daniel D. Joseph, *Elementary Stability and Bifurcation Theory*, Springer-Verlag, New York, 1980.

In fact there are many examples near the beginning of this book that are nearly the same as examples in Walker (above), except for the two differences just noted, in spite of the fact that Iooss and Joseph is considered "applied" and Walker is "pure." At a somewhat higher level, the best existing book on bifurcation theory is probably

Martin Golubitsky and David G. Schaeffer, *Singularities and Groups in Bifurcation Theory*, Vol. I, Springer-Verlag, New York, 1985

and its companion volume

Martin Golubitsky, Ian Stewart, and David G. Schaeffer, *Singularities and Groups in Bifurcation Theory*, Vol. II, Springer-Verlag, New York, 1988.

These books make use of some of the algebraic and analytic tools of algebraic geometry, singularity theory, and catastrophe theory, in order to clarify the determination of complicated bifurcations for which the methods presented in this chapter are not sufficient.

Singularity theory and *catastrophe theory* can be described as specific aspects of *real* algebraic geometry that have flourished in recent years as a result of some new mathematical ideas first introduced by René Thom and then proved in detail by a number of others. The questions are too technical to discuss here, except to say that every mathematics graduate student should be aware of the debate over catastrophe theory in the late 1970s, one of the few occasions on which articles about mathematics appeared in the newspapers. The controversy concerned not the mathematics itself, which was sound, but some of the applications: It was claimed that catastrophe theory was a breakthrough as significant as calculus, and that it would do for biology, psychology, and social science what calculus did for physics. Eventually these claims were seen to be exaggerated, partly the result of enthusiastic mathematicians rushing into new applications without a full understanding of the difficulties in fields with which they were not familiar; it turned out to be impossible to verify many of the hypotheses needed for the application of the theorems. But catastrophe theory continues on, although stripped of some of its glamour, and serious applications have been and will continue to be found. For an introduction, one might consult

T. Poston and I. Stewart, *Catastrophe Theory and Its Applications*, Pitman, London, 1978.

For a partial account of the controversy, see

Gina Bari Kolata, Catastrophe theory: The emperor has no clothes, *Science* **196** (April 15, 1977), 287–351.

CHAPTER 2

REGULAR PERTURBATIONS

2.1. PERTURBED SECOND ORDER DIFFERENTIAL EQUATIONS

This chapter concerns the approximate solution of differential equations
of the form

$$a\frac{d^2y}{dx^2} + b\frac{dy}{dx} + cy = \varepsilon f\left(x, y, \frac{dy}{dx}, \varepsilon\right) \qquad (2.1.1a)$$

when the solutions $y(x)$ are to be studied over a finite interval $x_1 \leq x \leq x_2$.
Here a, b, and c are constants, independent of both x and ε, with $a \neq 0$.
Two important cases resembling this are excluded: problems which are
to be solved over an infinite interval such as $0 \leq x < \infty$, and problems
which "drop order" when $\varepsilon = 0$ (such as $\varepsilon d^2y/dx^2 + b\,dy/dx + cy = \varepsilon f$).
These problems are not regular perturbation problems, and are considered
in later chapters. Equation (2.1.1a) is usually written

$$ay'' + by' + cy = \varepsilon f(x, y, y', \varepsilon). \qquad (2.1.1b)$$

The independent variable will often be time, in which case the equation is
written

$$a\frac{d^2y}{dt^2} + b\frac{dy}{dt} + cy = \varepsilon f\left(t, y, \frac{dy}{dt}, \varepsilon\right) \qquad (2.1.2a)$$

or

$$a\ddot{y} + b\dot{y} + cy = \varepsilon f(t, y, \dot{y}, \varepsilon). \qquad (2.1.2b)$$

Sometimes part of the perturbation will be written on the left hand side. Thus $\ddot{y} + \varepsilon\dot{y} + y = \varepsilon y^2 \sin t$ is to be regarded as an instance of (2.1.2) with $b = 0$ and $f = y^2 \sin t - \dot{y}$, not with $b = \varepsilon$, which would violate the constancy of b. Perturbation methods for these problems depend upon the ability to solve both the reduced problem $ay'' + by' + cy = 0$ and linear inhomogeneous problems of the form $ay'' + by' + cy = g(x)$; these topics are reviewed briefly in Appendix C. In general, (2.1.1) and (2.1.2) are nonlinear, with the nonlinearities entering through f. But since the reduced problem is linear, (2.1.1) and (2.1.2) are called *nearly linear*.

The solution of a second order equation is not uniquely determined unless two "side conditions" are imposed, usually in the form of *initial conditions* or *boundary conditions*. An *initial value problem* has the form

$$a\ddot{y} + b\dot{y} + cy = \varepsilon f(t, y, \dot{y}, \varepsilon),$$
$$y(0) = \alpha, \tag{2.1.3}$$
$$\dot{y}(0) = \beta.$$

Problems of this type cannot usually be solved exactly using the known elementary functions; if they could, there would be no need to use perturbation methods. But before attempting a perturbation solution, it is worthwhile to try to find out if the solution actually exists. (Sometimes, especially in partial differential equations, a perturbation solution is attempted even though it is not known whether the solution exists, but then the results must be taken with a grain of salt.) The existence theory for solutions of initial value problems is summarized in Appendix D. Those who have not studied this theory must take the results "on faith," but fortunately this is not difficult, since anyone who has had a course in "elementary" differential equations ("elementary" meaning ones that *are* solvable with known functions) is already "brainwashed" to believe that initial value problems have solutions. The following theorem states the main facts needed for this chapter; it is proved by changing (2.1.3) into a system of two first order equations and then applying Theorem D.1.

Theorem 2.1.1. *Let f be defined for all t in a compact interval $0 \le t \le T$, for all y and $\dot{y}$, and for all ε near zero. Let f have continuous partial derivatives of all orders $\le r$. Let compact intervals A and B be specified for α and β. Then there exists $\varepsilon_0 > 0$ such that a solution*

$$y = \varphi(t; \alpha, \beta; \varepsilon) \tag{2.1.4}$$

of (2.1.3) exists for all $0 \le t \le T$, $|\varepsilon| < \varepsilon_0$, $\alpha \in A$, and $\beta \in B$. Furthermore this solution is unique, and φ is as smooth as f; that is, it has continuous partial derivatives of all orders $\le r$ with respect to all of its arguments t, α, β, and ε (including mixed partial derivatives).

It follows immediately from Theorem 2.1.1 and Taylor's theorem that for any $k \leq r - 1$, φ has an asymptotic approximation of the form

$$\varphi(t;\alpha,\beta,\varepsilon) = \sum_{n=0}^{k} \varphi_n(t;\alpha,\beta)\varepsilon^n + \mathcal{O}(\varepsilon^{k+1}) \qquad (2.1.5)$$

uniformly for $0 \leq t \leq T$ and for α, β in compact subsets. (That is, given $0 < \varepsilon_1 < \varepsilon_0$, there exists $c > 0$ such that the error in (2.1.5) is $\leq c\varepsilon^{k+1}$ for $|\varepsilon| \leq \varepsilon_1$, $0 \leq t \leq T$, $\alpha \in A$, and $\beta \in B$.) If f is infinitely differentiable, the approximation can be carried to any order:

$$\varphi(t;\alpha,\beta,\varepsilon) \sim \sum_{n=0}^{\infty} \varphi_n(t;\alpha,\beta)\varepsilon^n \qquad (2.1.6)$$

with the same type of uniformity. (Recall that $\sim$ here simply means that (2.1.5) holds for all k.) Since the existence of the asymptotic approximations is known in advance, and the correct gauges are known, this is a regular perturbation problem, and it only remains to explain how to compute the coefficients φ_n using the principles explained in Section 1.8. The details will be carried out in Section 2.4.

It should be added at once that for many initial value problems the solution obtained in this way is not adequate. Some physical problems naturally concern a finite interval of time; falling bodies or objects sliding down an incline, for instance, satisfy a differential equation until they collide with an obstacle. Other physical systems, particularly those undergoing oscillatory motion, continue to satisfy a differential equation for all time, and a solution that only retains its accuracy for a finite time interval is not entirely satisfactory. Therefore special methods have been developed to handle oscillatory problems, and some of these methods will be presented in Chapters 3, 5, and 6. Oscillatory initial value problems are not excluded from this chapter, but the results obtained here will not be the best possible for these problems.

Since the oscillatory case is rather important, it is worthwhile to have some technical definitions available from the beginning. There are three words to define, "oscillation," "oscillatory," and "oscillator," and these are not the same (as used in this book). An *oscillation* is a function (of t or x) which crosses the (t or x) axis infinitely many times. A differential equation is *oscillatory* if it has (at least some) solutions that are oscillations. We will, of course, be concerned primarily with perturbed second order differential equations. In order to decide whether a perturbation problem is oscillatory, it is sufficient to look at the reduced problem. The reduced problem of (2.1.2) can have oscillating solutions only if $b^2 - 4ac < 0$ (Exercise 2.1.4), and we will call (2.1.2) an *oscillatory problem* if it satisfies this condition, and a *nonoscillatory problem* if $b^2 - 4ac > 0$. (The case where it equals zero is a borderline case.) An *oscillator*, on the other hand, is a physical system that is designed to have oscillatory solutions, at least when the natural parameters are in a certain range; by extension, the word *oscillator* also applies to the differential equation for such a physical system. For our purposes (2.1.2) will be called an *oscillator* if $c/a > 0$, because in this case the c term represents a restoring force tending to pull the system towards equilibrium, while the a term is an inertia which tends to make it pass through equilibrium, leading to solutions that cross through equilibrium infinitely often (and are therefore oscillations). After dividing by a and redefining f, the general *nearly linear oscillator* can be written

$$\ddot{y} + 2h\dot{y} + k^2 y = \varepsilon f(t, y, \dot{y}, \varepsilon),$$
$$y(0) = \alpha, \qquad\qquad\qquad (2.1.7)$$
$$\dot{y}(0) = \beta.$$

The constants are denoted $2h$ and k^2 for convenience in solving the characteristic equation $r^2 + 2hr + k^2 = 0$. Of course writing k^2 emphasizes that the coefficient of y is positive in the case of an oscillator (after dividing by a). We will also assume (without loss of generality) that k itself is positive (that is, k is defined to be the positive square root of c/a).

It is important to realize that although (2.1.7) is called an *oscillator*, it is not always *oscillatory*, that is, its solutions are not always *oscillations*; the condition $c/a > 0$ for an oscillator is weaker than the condition $b^2 - 4ac < 0$ for an oscillatory problem. Remember that an oscillator is defined in physical terms. A spring–mass system (or an electric circuit with a capacitor) is naturally regarded as an oscillator, and it is not reasonable to insist that it no longer be called an oscillator when the damping is large enough to prevent oscillations. For instance, consider the case of $h > k$ in (2.1.7); when $\varepsilon = 0$, this is called an *overcritically damped linear oscillator* and has no solution which crosses the equilibrium position $y = 0$ more than once (see Appendix C). It is shown in Exercise 2.1.1 that there exists a change of variables which eliminates the $\dot{y}$ term in (2.1.7) and replaces the k^2 term with $k^2 - h^2$, so that whenever $|h| > k$, (2.1.7) is equivalent to a clearly nonoscillatory problem. For mathematical purposes it may be simplest to use this transformation at all times and so assume $b = 0$ in (2.1.1), but for physical problems it is best to preserve the damping term because of its obvious significance; that is, to leave (2.1.7) as it is and treat h as an adjustable parameter. Then it should be remembered that the character of the problem changes when $|h|$ crosses k. Any oscillatory problem with $h > 0$ is called *strongly damped*; if $h = 0$ but damping exists through the perturbation (that is, in a term such as $\varepsilon \dot{y}$), the problem is called *weakly damped*. Strong damping is *undercritical* if $0 < h < k$, *critical* if $h = k$, and *overcritical* if $h > k$. Weak damping is always undercritical because ε is small compared to k.

Boundary value problems are automatically confined to a finite interval of the independent variable, and so do not suffer from the difficulties we have been discussing for initial value problems. However, they have their own difficulties, which, again, are especially pronounced in the oscillatory case $c/a > 0$. Consider the nonoscillatory problem for $0 \le x \le L$,

$$y'' + 2hy' - k^2y = \varepsilon f(x, y, y', \varepsilon),$$
$$y(0) = \alpha, \tag{2.1.8}$$
$$y(L) = \gamma,$$

and the similar oscillatory problem (assuming $|h| < k$), differing only in the sign before the k^2,

$$y'' + 2hy' + k^2y = \varepsilon f(x, y, y', \varepsilon),$$
$$y(0) = \alpha, \tag{2.1.9}$$
$$y(L) = \gamma.$$

There is a striking difference between these problems even for $\varepsilon = 0$: The nonoscillatory problem has a unique solution (see Exercise 2.1.2), whereas the oscillatory problem may have one solution, infinitely many solutions, or else no solution at all, depending upon the relationship between h, k, and L. To see this, observe that the general solution of $y'' + 2hy' - k^2y = 0$ is $y = Ae^{r_1x} + Be^{r_2x}$, where $r_1, r_2 = -h \pm \sqrt{h^2 + k^2}$. Thus r_1 and r_2 are always distinct real numbers, so that the determinant of the equations

$$y(0) = A + B = \alpha,$$
$$y(L) = Ae^{r_1L} + Be^{r_2L} = \gamma \tag{2.1.10}$$

is nonzero, and these equations can always be solved for A and B. On the other hand the general solution of $y'' + 2hy' + k^2y = 0$ in the case $0 \le h < k$ (undercritical damping) is $y = e^{-hx}(A\cos\omega x + B\sin\omega x)$, where $\omega = \sqrt{k^2 - h^2}$, and the equations for A and B are

$$y(0) = A = \alpha,$$
$$y(L) = e^{-hL}(A\cos\omega L + B\sin\omega L) = \gamma. \tag{2.1.11}$$

Clearly $A = \alpha$, and B must be determined from

$$B\sin\omega L = \gamma e^{hL} - \alpha\cos\omega L.$$

This is uniquely solvable for B if and only if $\sin\omega L \ne 0$, that is, provided $L\sqrt{k^2 - h^2}$ is not an integral multiple of π; if $\sin\omega L = 0$, either every value of B is a solution, or none, depending upon whether the right hand side is zero. It is evident, then, that (2.1.8) is always a reasonable perturbation problem as it stands, whereas (2.1.9) is sometimes not. Boundary value problems for which the reduced problem has infinitely many solutions are called *degenerate*; a nonoscillatory problem is nondegenerate, whereas an oscillatory problem can be degenerate or nondegenerate (depending on the vanishing of a determinant, as seen above). For simplicity, we will focus only on nonoscillatory problems in Section 2.5, although nondegenerate oscillatory problems can be treated the same way. (Once again, as for initial value problems, an oscillator for which $h > k$ is not "truly oscillatory" because it is overcritically damped. Such problems cannot be degenerate. See Exercise 2.1.3).

In Section 2.6 we will deal with a special degenerate case of the oscillatory boundary value problem (2.1.9), namely

$$
\begin{aligned}
&y'' + \lambda y = \varepsilon f(y), \\
&y(0) = 0, \\
&y(1) = 0,
\end{aligned}
\tag{2.1.12}
$$

in which $h = \alpha = \gamma = 0$. The reduced problem of (2.1.12) is a familiar eigenvalue problem, the eigenvalues being values of λ for which nonzero solutions exist. The perturbed problem can be considered as a "nonlinear eigenvalue problem." Apart from this special case, we will confine our study of boundary value problems to the nonoscillatory cases.

For the nonoscillatory boundary value problem (2.1.8), there does exist a unique solution

$$
y = \psi(x; \alpha, \gamma; \varepsilon)
\tag{2.1.13}
$$

for small ε, and this solution is as smooth as f. Hence there exist asymptotic solutions, having the same form as (2.1.5) and (2.1.6) except that φ is replaced by ψ and β by γ. It is not hard to derive this existence result using tools already at hand: the implicit function theorem and the existence theorem for initial value problems (Theorem 2.1.1). This will be done in Section 5.2 by what is called the *shooting method*. The idea is to drop the second boundary condition $y(L) = \gamma$ from (2.1.8) and replace it temporarily by the initial condition $y'(0) = \beta$. Let $y = \varphi(x; \alpha, \beta; \varepsilon)$ denote the solution of the resulting initial value problem. Then φ is known to exist and to be smooth, and we can define the *shooting problem*: to find β such that

$$
\varphi(L; \alpha, \beta; \varepsilon) = \gamma.
\tag{2.1.14}
$$

See Fig. 2.1.1; β can be thought of as the "angle of fire" from α which causes the solution to arrive at γ when $x = L$. The implicit function theorem can be used to show that there is a smooth solution for β, thus proving existence and smoothness of (2.1.13). Therefore (2.1.8) is a regular perturbation problem. The details, and the computation of perturbation solutions, will be given in Section 2.5.

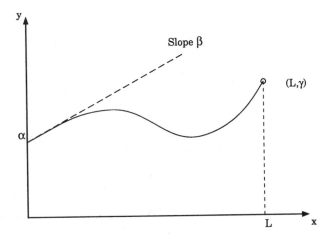

FIG. 2.1.1. The shooting method. With the left endpoint fixed at α, a correct choice of the initial slope β will cause the solution to reach the desired right endpoint at (L, γ).

In summary, there are three types of problems discussed in this chapter. Initial value problems (whether oscillatory or not, although better results in the oscillatory case will be obtained in later chapters) are treated in Section 2.4. Nonoscillatory boundary value problems are handled in Section 2.5. One special kind of oscillatory boundary value problem, called a nonlinear eigenvalue problem, is studied in Section 2.6. But before describing the perturbation methods for solving these three types of problems, we will digress in the next two sections to show how these problems arise in physical examples. In Section 2.2, one example of each type will be presented as it arises from fundamental physical laws. Then in Section 2.3 we return to the theme introduced at the beginning of the first chapter: How should a perturbation family be formulated? One of the fundamental techniques for finding perturbation families that are physically meaningful is *dimensional analysis*, and in Section 2.3 we introduce this idea and use it on each of the three examples.

Exercises 2.1

1. Show that the change of variables $u := e^{ht}y$ applied to (2.1.7) eliminates the damping term. (This is a special case of the *Sturm transformation*.)

2. Show by explicit computation that (2.1.8) has a unique solution when $\varepsilon = 0$. (See Appendix C.)

3. Show that (2.1.9) for $h > k$ has a unique solution when $\varepsilon = 0$. Do this in two ways: (a) directly, using Appendix C; (b) indirectly, using Exercise 2.1.1 and 2.1.2.

4. Check that $ay'' + by' + cy = 0$ has solutions with infinitely many zeroes if and only if $b^2 - 4ac < 0$.

2.2. PHYSICAL EXAMPLES

Example 2.2.1. The Nonlinear Spring or Duffing Equation

Consider a mass of M kilograms suspended from the ceiling by a spring which has length L meters when the mass is at rest. (In this section, upper-case letters will be used to denote dimensional quantities, that is, quantities measured using a system of units such as the metric system. The significance of units will be discussed in Section 2.3.) Let the spring move so that its displacement from equilibrium is $Y = Y(T)$ meters at time T seconds after the motion is started, with initial displacement $Y(0) = A$ meters and initial velocity $dY/dT(0) = B$ meters per second.

In elementary discussions the restoring force R is usually taken to be proportional to the displacement, according to Hooke's law, $R = -KY$. This is of course not correct, as there is some limit beyond which the spring cannot be compressed, but it is a fair approximation when the displacements are small compared to the unextended length L. To find a better approximation, R can be represented by a Taylor series in Y : $R(Y) = -(K_1 Y + K_2 Y^2 + K_3 Y^3 + \dots)$. The constant term K_0 is absent because $R(0) = 0$ since $Y = 0$ is the equilibrium point, and K_1 is the Hooke's law constant. Since most springs are nearly symmetrical in their response, K_2 is nearly zero; we will assume that there is a range of Y for which the K_1 and K_3 terms give a satisfactory approximation, and write $R(Y) = -(K_1 Y + K_3 Y^3)$. But before dispensing with the Taylor series, we pause to observe that each coefficient K_n is a dimensional quantity having the correct units to make $K_n Y^n$ have the units of force. Since force is measured in newtons, one newton being one kilogram meter per second squared, the units for K_n are $\text{kg/m}^{n-1} \cdot \text{sec}^2$. Thus if units are changed, say to the English system, the coefficients K_n will change in the manner prescribed by their respective units. (It is sometimes said that a power series cannot be applied to a dimensional quantity, because the terms will have different dimensions and cannot be added. This is true only of power series such as exp or cos, in which the coefficients are nondimensional quantities.)

At this point the equation of our spring–mass system has the form $M d^2Y/dT^2 = -R(Y)$, or $M d^2Y/dT^2 + K_1 Y + K_3 Y^3 = 0$. Sometimes we will want to add a linear damping term $\Delta dY/dT$ or a periodic forcing term $F \cos \Omega T$, or both; the most general form is then

$$M \frac{d^2Y}{dT^2} + \Delta \frac{dY}{dT} + K_1 Y + K_3 Y^3 = F \cos \Omega T,$$

$$Y(0) = A, \qquad \frac{dY}{dT}(0) = B. \tag{2.2.1}$$

The forcing term can be thought of as due to an electromagnet placed on the floor under the spring–mass system, creating a field which varies periodically with frequency Ω, alternately attracting and repelling the mass. The linear damping term is a commonly used expression representing frictional forces such as air resistance which are actually very difficult to model accurately. The following is a list of the natural parameters in this problem and their units:

M	kg
Δ	kg/sec
K_1	kg/sec^2
K_3	kg/m$^2 \cdot$ sec^2
F	kg $\cdot$ m/sec^2
Ω	1/sec
A	m
B	m/sec.

These are the units necessary to make each term have the units of force, and to make ΩT dimensionless (so that, as explained above, $\cos \Omega T$ will make sense).

As it stands, (2.2.1) is not a perturbation problem. In order to create perturbation problems based upon (2.2.1), it is necessary to find solvable special cases which can serve as reduced problems, and then to introduce a perturbation parameter. Since (2.2.1) can be solved by elementary methods if K_3 is zero, one possibility is to study (2.2.1) under the assumption that K_3 is small. But we will begin with an easier problem: The solution of (2.2.1) is very simple if both K_3 and F are zero, so we will take this as the reduced problem, and then attempt to find approximate solutions by perturbation theory when K_3 and F are small. An even easier reduced problem is obtained if Δ is also zero; this reduced problem can be used if Δ, K_3, and F are all small. In any case, a convenient first step in fitting (2.2.1) into the form of a perturbation problem is to move the terms which are to be considered small to the right hand side. In the two cases we will consider, this results in either

$$M \frac{d^2Y}{dT^2} + \Delta \frac{dY}{dT} + K_1 Y = -K_3 Y^3 + F \cos \Omega T \qquad (2.2.2)$$

or else,

$$M \frac{d^2Y}{dT^2} + K_1 Y = -\Delta \frac{dY}{dT} - K_3 Y^3 + F \cos \Omega T. \qquad (2.2.3)$$

We have not yet introduced a perturbation parameter, but by putting the "small" terms on the right hand side we have made it easy to visualize the reduced problem in each case: it is the problem obtained when the right hand side is set equal to zero. In the case of (2.2.2), the reduced problem has damped solutions, whereas in the case of (2.2.3) they are steady-state oscillations. It would seem therefore that (2.2.3) should be used to describe systems in which the damping is small enough that many oscillations occur before the amplitude decreases appreciably, because in these cases a steady-state oscillation is a reasonable "zeroth order" approximation. When the amplitude decreases more rapidly, (2.2.2) is probably better. The perturbation theories resulting from (2.2.2) and (2.2.3) are called, respectively, theories of *strongly damped* and *weakly damped* oscillations. These terms have already been introduced briefly in Section 2.1. The point to be emphasized here is that the formulation of a perturbation problem can be influenced by experience with the applications. With practice the individual terms in a differential equation should become associated with specific effects, so that experimental results can suggest that certain effects are less important than others in a given case and can be omitted from the reduced problem ("moved to the right hand side").

Of course, it is still necessary to introduce a perturbation parameter into (2.2.2) or (2.2.3). The simplest way to do this is to write each natural parameter that is considered small as a multiple of ε. Thus in (2.2.3) we would set $\Delta = \varepsilon\tilde{\Delta}$, $K_3 = \varepsilon\tilde{K}_3$, and $F = \varepsilon\tilde{F}$ to obtain

$$M\frac{d^2Y}{dT^2} + K_1Y = \varepsilon\left\{-\tilde{\Delta}\frac{dY}{dT} - \tilde{K}_3Y^3 + \tilde{F}\cos\Omega T\right\}. \qquad (2.2.4)$$

The perturbation parameter ε that is introduced in this way is dimensionless (that is, has no units), since we take each of the quantities $\tilde{\Delta}$, $\tilde{K}_3$, and $\tilde{F}$ to have the same units as the original natural parameter. Equation (2.2.4) is now in a form suitable for the use of perturbation methods to be introduced in Section 2.4 and in Chapters 3, 5, and 6. However, there are great advantages to be gained by putting (2.2.3) into what is called *nondimensional form* before introducing a perturbation parameter. This subject will be taken up in Section 2.3 (Example 2.3.1).

Example 2.2.2. Diffusion-Reaction Problem

Suppose that a long pipe is filled with a liquid, into which a gas is diffusing from the ends. Suppose further that the gas reacts chemically with the liquid to produce a third chemical (called the product). The entire process is assumed to be in a steady state, that is, the concentration of gas in the liquid at each point inside the pipe is independent of time; whatever gas disappears in the formation of the product is exactly replaced by diffusion. This is a reasonable assumption if the gas has been diffusing into the liquid for a long time and if there is some mechanism for removing the product and replacing the used liquid without stirring the contents of the pipe. (One might imagine that the product is a solid and settles out through a screen at the bottom of the pipe, and that the top of the pipe is a moist sponge from which more liquid enters the pipe when needed.) The problem to be solved is as follows: The concentration of gas in the liquid is measured at two points inside the pipe (not the endpoints where the liquid is in contact with the gas, but two points inside the liquid); the concentration at each point between these two points is to be determined.

Let X denote distance in centimeters measured from one of the measuring points (which is then $X = 0$) toward the other (at $X = N > 0$). Let the concentration of gas at each point X be $Y(X)$. A suitable unit for measuring concentration might be moles per liter (a "mole" is a certain number of molecules of gas), but we will take advantage of the fact that the pipe has constant cross-section and measure the concentration in moles per centimeter. (That is, the number of moles of gas between X and $X + \Delta X$ is approximately $Y(X) \cdot \Delta X$ if ΔX is small.)

In order to derive an equation for $Y(X)$, consider a small section of the pipe from X to $X + \Delta X$. According to the law of conservation of matter and the steady-state assumption, the total number of moles of gas entering the section through its end by diffusion must equal the number of moles leaving the section in the form of product. According to Fick's law of diffusion, the amount of gas (in moles/sec) entering the section at X will equal $-DY'(X)$ for some positive constant D; the minus sign expresses the fact that gas diffuses from regions of higher concentration to lower; if $Y'(X) > 0$, then the concentration is greater inside the section than outside, and gas is leaving the section at that end rather than entering. Similarly, the amount of gas (in moles/sec) entering the section through the end at $X + \Delta X$ is $DY'(X + \Delta X)$. Here D must have units of cm^2/sec, since Y' will be in moles/cm^2. Now the amount of gas (in moles/sec) disappearing from the section in the form of product will be the integral from X to $X + \Delta X$ of some function $H(X)$ expressing the rate of disappearance (in moles/sec $\cdot$ cm) at the point X. (We will discuss the form of this function below.) Therefore the steady-state hypothesis takes the form

$$DY'(X + \Delta X) - DY'(X) = \int_{X}^{X+\Delta X} H(S)dS. \qquad (2.2.5)$$

Dividing (2.2.5) by ΔX, the left hand side becomes a difference quotient and the right hand side becomes the average of $H(X)$ over the interval. Therefore letting $\Delta X \to 0$ implies

$$DY''(X) = H(X). \qquad (2.2.6)$$

It is reasonable to assume that the "reaction rate" $H(X)$ depends only on the concentration of gas at X, since other factors that might influence the rate (such as temperature) are assumed equal at all points. (This would not be true if the reaction itself gave off or consumed heat.) In fact, in most treatments of this problem it is assumed that the dependence is linear:

$$H(X) = KY(X),$$ (2.2.7)

where K is a constant with the units of 1/sec. But this can only be true for a certain range of concentrations. (I was once given the following advice: To write an instant paper on any area of science, chose any two variables and claim that the dependence of one on the other is approximately linear over a suitable range. This is of course nothing more than claiming that the relation between the variables is a differentiable function.) For our purposes, we will assume that for the range of concentrations in which we are interested, the function H is not quite linear, but is adequately represented by a *quadratic* function

$$H(X) = KY(X) + LY^2(X).$$ (2.2.8)

Of course, K will still have the units of 1/sec, and L will be cm/mole·sec. Putting (2.2.6) and (2.2.8) together, the problem now becomes

$$DY'' = KY + LY^2,$$
$$Y(0) = A,$$ (2.2.9)
$$Y(N) = B,$$

where A and B are the measured concentrations at the endpoints. This is a boundary value problem. Since it is easily solvable for $L = 0$, it gives a natural perturbation problem for small L. In Example 2.3.2 it will be reformulated in a nondimensional manner.

Example 2.2.3. The Elastica

Whereas the last two examples were developed from physical principles, the present example is more complicated, and the differential equation will merely be given without derivation. The problem concerns a flexible rod with pressure applied at both ends. When the pressure is zero, the rod assumes the shape of a straight line segment. As the pressure is increased, a critical value is reached at which the straight solution becomes unstable, and the rod assumes the shape of a bow. At still higher pressures there are stable solutions with one or more inflection points; a solution with one inflection point would have an S-shaped appearance. This is a typical problem in bifurcation theory, with the pressure being the bifurcation parameter and the number of solutions changing at a sequence of critical values.

The equilibrium position of the rod is governed by the following boundary value problem:

$$u'' + \lambda \sin u = 0,$$
$$u'(0) = 0, \qquad\qquad (2.2.10)$$
$$u'(1) = 0.$$

Here (although this does not matter for our purposes) the independent variable x is arc length along the rod, and $u(x)$ is the angle between the tangent to the rod and the horizontal at the point x (assuming that the rod is horizontal when the pressure λ is zero). The variables in (8.2.10) are already nondimensional.

Without further attention to the physics, our aim is to see how (8.2.10) leads to a perturbation problem of the type (2.1.12). There is clearly no perturbation parameter in (2.2.10), and the problem is strongly nonlinear. However, in studying (2.2.10) an important role is played by the solutions for which u is small. (These are the nearly horizontal configurations of the rod.) To this end, it is convenient to introduce the scaling

$$u = \varepsilon y,$$

under which the differential equation in (2.2.10) becomes

$$y'' + \lambda \frac{\sin \varepsilon y}{\varepsilon}.$$

Since εy is small, the sine can be approximated by a few terms of its Taylor series, say $\sin \varepsilon y \cong \varepsilon y - \varepsilon^3 y^3 / 6$. This leads at once to the problem

$$y'' + \lambda y = \varepsilon^2 \lambda y^3 / 6,$$
$$y'(0) = 0,$$
$$y'(1) = 0,$$

which, although not identical to (2.1.12), belongs to the same class of problems.

Exercises 2.2

1. Introduce a perturbation parameter into (2.2.2) in the same way that we did for (2.2.3), and compare the result with (2.2.4).

2. Solve the reduced problem for (2.2.4) and also for the perturbation family which you created in Exercise 2.2.1.

2.3. DIMENSIONAL ANALYSIS

Applied mathematics is concerned with making connections between the "real world" of physical objects (or of more complex chemical, biological, or social systems) and the abstract world of mathematical equations. The crucial step in making such a connection is *measurement*: the assigning of numbers to objects or phenomena in the real world by carrying out procedures of various kinds. (An example of such a "procedure" is stretching a tape measure across a room. Of course, in science the procedures become quite complex, and the rules for carrying out an acceptable measurement must be spelled out in great detail.) The result of a measurement process is a number, together with a *unit*. A unit is a short name designating the meaning of the number.

Suppose there were a single, universally accepted object used as a standard of length, and suppose that it were called the "standard inch-rod." Then to say "John's height is 70" would mean that seventy copies of the standard inch-rod would equal John's height. There would be no need to say John's height was 70 *inches*, because everyone would understand that the number 70 was the ratio of John's height to the standard inch. The need to name units comes about because there are many, potentially infinitely many, possible standard lengths that can be used for measuring other lengths, for instance the "standard centimeter-rod." In terms of the standard centimeter-rod, John's height comes out to be 177.8, and the standard inch-rod comes out to have a length of 2.54.

When more than one system of measurement is in use, it is useful to tag these numbers with unit names in order to keep track of which system is being used, and to enable familiar calculations such as

$$70 \, \text{in} = (70 \, \text{in}) \left(\frac{2.54 \, \text{cm}}{1 \, \text{in}} \right) = 177.8 \, \text{cm}, \tag{2.3.1}$$

in which "units" are "cancelled" just like algebraic symbols. This cancellation does not imply that a unit is a "thing" of the same sort as a number, or that the presence of the unit makes the accompanying number into something other than an ordinary number. A unit is simply a description of how a number is obtained.

Suppose now that a new method of measuring heights is proposed: Each person's height will be measured using the length of the last knuckle of his or her right index finger. Suppose John's knuckle is 1.1 inches (2.794 cm) long. Then his height, in the new system, works out to

$$\frac{70 \, \text{in}}{1.1 \, \text{in}} = \frac{177.8 \, \text{cm}}{2.794 \, \text{cm}} = 63.63. \tag{2.3.2}$$

The inches cancel; the centimeters cancel; the number 63.63 seems to have no units and is sometimes called a "pure number" or "nondimensional length." There is something a little strange about this; one must admit that the units do cancel, and yet the nagging feeling remains that there is in fact a unit involved in the number 63.63: John's height is not 63.63 in some absolute sense; it is, it seems, "63.63 knuckles." This feeling grows stronger if someone suggests using the width of the palm in place of knuckles. John's palm is (let us say) 3.5 inches wide, and his height measured by his palm is

$$\frac{70 \text{ in}}{3.5 \text{ in}} = \frac{177.8 \text{ cm}}{8.89 \text{ cm}} = 20. \qquad (2.3.3)$$

Once again, there seem to be no units; yet there are now two distinct ways of arriving at a "nondimensional length." After determining that the length of John's palm in terms of his knuckle is 3.18, the numbers representing John's height turn out to be related by the following calculation, which has exactly the same form as any change of units such as (2.3.1):

$$63.63 \text{ knuckles} = (63.63 \text{ knuckles}) \left(\frac{1 \text{ palm}}{3.17 \text{ knuckles}} \right) = 20 \text{ palms.} \quad (2.3.4)$$

How are we to understand this? Are the numbers 63.63 and 20 occurring in (2.3.4) actually dimensionless, or do they have units of knuckles and palms?

To answer this question, remember that we said a "unit" was simply a brief description of how a number is obtained. Since the numbers 63.63 and 20 are obtained from the real world by carrying out a measurement procedure, of course they have units. The reason that the units cancel in (2.3.2) and (2.3.3) is that not enough information is included in these equations to tell what the new unit should be called. For instance, in looking at (2.3.2), there is nothing there to tell that the "1.1 in" is the length of John's knuckle. In this case one must fall back on one's additional knowledge of the measurement processes involved in order to create a unit name appropriate to the situation.

But there is still something very important to say about the situation we have been discussing, something far more important than the mere question of whether or not a "nondimensional length" really has units. There is a fundamental difference between units such as inches and centimeters, on the one hand, and units such as knuckles and palms, on the other. Namely, the former type of unit is based on a standard length that is *external* to the object being measured (in this case, John), while the latter is *internal* to the object. When height is measured in inches or in centimeters, John's height is compared to an external standard which is the same standard used to measure everyone else. When height is measured in knuckles or in palms, John's height is measured using his knuckle or palm and Eva's height is measured using hers. In fact, it is clear that what is measured by a person's "height in palms" is not the person's *size*, but the person's *shape* (or one aspect of it). Two people of different size but the same bodily proportions will have the same "height" in palms, whereas two people with the same height (in the usual sense) but with hands of different sizes will have different "height" in palms.

Now we are in a better position to explain why a "nondimensional length" is usually considered not to have units. We have explained that unit names are necessary when more than one system of measurement is in use at the same time. To be more precise, unit names are necessary when there is more than one way of measuring *a specific feature of an object*. (If length were measured in either inches or centimeters, but weight were always measured in pounds, then the unit of length must be specified but there would be no need to mention the unit "pounds.") We have also pointed out that what is measured in knuckles or palms is a *different feature* than what is measured in inches or centimeters (shape rather than size). Suppose, then, that many systems of *external* units are in use (such as inches and centimeters) but that only one system of *internal* units is used (such as palms). Then there is no need to mention palms, and no harm will result from leaving the number 20 in (2.3.3) as a "pure number" without units. This number will be called a *nondimensional length*, but it should be remembered that it does not actually measure length. In fact, *the real advantage gained by defining the nondimensional length is that a new significant feature of the object in question has been discovered and a way to measure it has been defined.* Very often the quantities obtained by "nondimensionalizing" turn out to be far more significant for the problem being investigated than the quantities that were originally measured using the standard concepts of length, mass, time, and so forth.

Generalizing from the preceding discussion, the process of "nondimensionalizing" can be described as consisting of three steps. First, a class of objects must be specified as the objects to be studied (in our example, people). Second, a systematic way of choosing a unit (of length or whatever) *internal to the object* is prescribed (for instance, for each person, choose that person's palm). Third, all quantities of the same type (say, all lengths) within each object are to be measured using the prescribed standard unit for that object (in our example, the person's palm), and no other choice of internal standard (such as knuckles) is to be allowed. Once these conditions are met, the measurements made in this way may be regarded as nondimensional with respect to all *external* unit systems such as inches and centimeters. When calculated using these external units, these nondimensional quantities will always appear as ratios in which the (external) units cancel, and are independent of the system of (external) units used. The reason for excluding any other internal standard than the one assigned in step 2 is to avoid having to introduce a complete system of internal units (palms, knuckles, etc.) after the external units have been eliminated.

Let us give a brief example of this process, which will emphasize the point that nondimensional quantities tend to acquire great significance. Step 1: The objects to be studied are right triangles. Step 2: The internal length unit, or as it is commonly called, the *characteristic length*, for each right triangle, is its hypotenuse. Step 3: The legs of each right triangle are measured in terms of its hypotenuse. The resulting "nondimensional length" of each leg is in fact just the cosine of the angle which that leg makes with the hypotenuse. Of course, two right triangles which are "similar" in the geometric sense, that is, two right triangles with the same shape but possibly different sizes, will come out having legs of equal nondimensional length. If we did not already know the concept of cosine, we would have discovered it merely by the process of nondimensionalizing. In fact a great many concepts as important as the cosine in their own fields have been discovered in exactly this way; the tradition is to name such concepts by combining the name of their discoverer with the word "number," as in Reynolds number, Rayleigh number, and so forth. It is in the creation of such concepts that the real value of nondimensionalizing lies, not in the mere fact (although it is convenient) that one obtains numbers "without units."

Example 2.3.1. The Nonlinear Spring (Example 2.2.1 continued)

For simplicity we will first consider the nonlinear spring without forcing or damping. In this case (2.2.1) reduces to

$$M\frac{d^2Y}{dT^2} + K_1Y + K_3Y^3 = 0,$$
$$Y(0) = A, \tag{2.3.5}$$
$$\frac{dY}{dT}(0) = B.$$

Our aim is to find a nondimensional version of this problem, and in particular to find a suitable nondimensional small parameter ε for the formulation of a perturbation problem in the case of small nonlinearity K_3. According to the list on page 91, the dimensions of the natural parameters occurring in (2.3.5) are as follows:

$$\begin{array}{ll} M & \text{kg} \\ K_1 & \text{kg/sec}^2 \\ K_3 & \text{kg/m}^2 \cdot \text{sec}^2 \\ A & \text{m} \\ B & \text{m/sec.} \end{array} \tag{2.3.6}$$

Since units of length, mass, and time all appear in this list, the first step is to determine a characteristic length, a characteristic mass, and a characteristic time that can be used to nondimensionalize all of these parameters as well as the coordinates Y and T. One therefore looks for combinations of the natural parameters which have the dimensions of length, mass, or time. Two obvious choices suggest themselves: M for the characteristic mass and A for the characteristic length. (Of course, A cannot be used if it is zero. This case will be considered below. The mass M cannot be zero since this would not make physical sense.) There is no such instant choice for a characteristic time; three possibilities suggest themselves, $\sqrt{M/K_1}$, $\sqrt{M/K_3 A^2}$, and A/B. (To find these, look for the parameters that involve time, namely K_1, K_3, and B. Then eliminate mass and length by multiplying or dividing by the already determined characteristic mass and length.) Of these three, the second possibility can be eliminated, since our purpose is to study problems for which K_3 is near zero, and therefore we must include the case $K_3 = 0$ as the reduced problem; but $\sqrt{M/K_3 A^2}$ is undefined in this case. (Such false choices can be avoided by first nondimensionalizing the reduced problem and then going on to the full problem. This will be illustrated below.) To decide between the remaining two possible choices for the characteristic time, it is helpful to think about the physical meaning of the respective quantities. The first possibility, $\sqrt{M/K_1}$, is closely related to the physical properties of the spring-mass system itself, whereas the second, A/B, is related only to the initial conditions and seems much more "accidental" to physical reality. This argues strongly in favor of the first choice. If further evidence is needed, consider the fact that the exact solution of the reduced problem involves the sine and cosine of $\sqrt{K_1/M} \cdot T$, which is precisely the nondimensionalized time obtained using $\sqrt{M/K_1}$ as characteristic time. This is certainly the correct characteristic time for this problem. (Whenever any doubt remains, it is always possible to try different possibilities. We will see later that some problems require the use of more than one characteristic length or time.)

Having settled on the characteristic length, mass, and time, the coordinates and parameters can be nondimensionalized by multiplying or dividing by these "internal units" raised to appropriate powers. In this way we define

$$y := Y/A,$$
$$t := T\sqrt{K_1/M},$$
$$m := M/M = 1,$$
$$k_1 := K_1(M/K_1)(1/M) = 1, \qquad (2.3.7)$$
$$k_3 := K_3 A^2/K_1,$$
$$a := A/A = 1,$$
$$b := (B/A)\sqrt{M/K_1}.$$

Notice that three of the new parameters turn out to be 1; therefore nondimensionalizing has reduced the number of independent parameters by three. To obtain the nondimensionalized equations, it is only necessary to write (2.3.5) using small letters, that is,

$$m\frac{d^2y}{dt^2} + k_1y + k_3y^3 = 0,$$

$$y(0) = a,$$

$$\frac{dy}{dt}(0) = b,$$

and then eliminate those which are equal to 1, obtaining

$$\ddot{y} + y + k_3y^3 = 0,$$

$$y(0) = 1, \qquad\qquad (2.3.8)$$

$$\dot{y}(0) = b.$$

This procedure for obtaining (2.3.8) is valid because the equations (2.2.1) take the same form under any system of units; changing (2.3.5) to (2.3.8) is exactly like changing (2.3.5) from metric to English units (except that our new "units" are internal and hence appear not to be units at all, according to the previous discussion). For a more careful approach, see Exercise 2.3.2. Equation (2.3.8) reveals that in nondimensional form, the nonlinearity is represented by k_3. Therefore, this should be the small parameter; that is, we define

$$\varepsilon := k_3 := \frac{K_3A^2}{K_1}. \qquad\qquad (2.3.9)$$

This illustrates the idea that dimensional analysis "automatically" finds the important and meaningful combinations of variables in a problem, for it now becomes clear that "small nonlinearity" in our problem does not necessarily mean that K_3 is small; it could equally well mean that A is small, or that K_1 is large, or some combination of these. Our perturbation results will apply equally well to all of these cases.

There is one respect in which (2.3.8) is not yet entirely satisfactory. In choosing A as the characteristic length, we have excluded the possibility that $A = 0$, and there seems to be no reason why this case should not be handled on an equal footing with any other case. There is another possibility for a characteristic length, namely $B\sqrt{M/K_1}$, but this has a similar difficulty since it is not valid when $B = 0$. The fact is that problem (2.3.5) simply *does not have* a characteristic length that is defined in terms of the natural parameters and is suitable for all choices of the initial conditions. In such cases it is sometimes useful to introduce a characteristic length arbitrarily. (One author, speaking of the insoluble problem of finding a characteristic length for an infinite flat plate, suggests painting a red stripe of some arbitrary length on the plate. Then the stripe becomes part of the system, and its length can be used as the characteristic length!) If we select an arbitrary length L (measured in meters) as the characteristic length, still using M and $\sqrt{M/K_1}$ as the characteristic mass and time, the nondimensionalized variables become

$$
\begin{aligned}
y &:= Y/L, \\
t &:= T\sqrt{K_1/M}, \\
m &:= M/M = 1, \\
k_1 &:= K_1(M/K_1)(1/M) = 1, \\
\varepsilon &:= K_3 L^2/K_1, \\
\alpha &:= A/L, \\
\beta &:= (B/L)\sqrt{M/K_1},
\end{aligned}
\tag{2.3.10}
$$

and the equations become

$$
\begin{aligned}
&\ddot{y} + y + \varepsilon y^3 = 0, \\
&y(0) = \alpha, \\
&\dot{y}(0) = \beta.
\end{aligned}
\tag{2.3.11}
$$

These equations are usable for all initial conditions. Although (2.3.11) has the advantage of being valid for all initial conditions, it also has a disadvantage: the perturbation parameter ε defined in (2.3.10) does not involve A and so does not become small when A is small, as in (2.3.9). Therefore this new ε is not as sensitive a measure of the "true" nonlinearity of the problem as is the ε of (2.3.9). On the other hand, it is easy to recover (2.3.9) from (2.3.10) by setting $L = A$ and $\alpha = 1$. Therefore we consider (2.3.11) to be a more flexible version of the nondimensionalized equations than (2.3.8), and it is this form that will usually be used in this book. In fact we will usually add even a little more flexibility by allowing k_1 to remain in the equation, even though we have seen that (when the nondimensionalizing is done as above) it is equal to 1. Renaming k_1 as k^2 to emphasize that it is positive, the unforced Duffing equation takes the form

$$
\begin{aligned}
&\ddot{y} + k^2 y + \varepsilon y^3 = 0, \\
&y(0) = \alpha, \\
&\dot{y}(0) = \beta.
\end{aligned}
\tag{2.3.12}
$$

This form includes both (2.3.8) and (2.3.11) as well as certain alternative ways of nondimensionalizing the system that might be needed in certain problems. For instance, if a Duffing oscillator occurs as part of a larger system (perhaps a system of several coupled oscillators), a dimensional analysis of the entire system may require a value of k^2 unequal to 1 in (2.3.12); it would not be possible to make the frequencies of all of the oscillators equal to 1 if those frequencies are different.

Having thoroughly discussed the case of (2.2.1) without forcing or damping, we now briefly treat the full equation,

$$M\frac{d^2Y}{dT^2} + \Delta\frac{dY}{dT} + K_1Y + K_3Y^3 = F\cos\Omega T,$$

$$Y(0) = A, \qquad \frac{dY}{dT}(0) = B. \qquad (2.3.13)$$

Suppose that it is desired to study this equation when all three of the quantities Δ, K_1, and F are small. Then the reduced problem is

$$M\frac{d^2Y}{dT^2} + K_1Y = 0,$$

$$Y(0) = A, \qquad \frac{dY}{dT}(0) = B.$$

As explained above, it is simplest to begin by nondimensionalizing the reduced problem, because the parameters that do not appear in the reduced problem must be allowed to equal zero and so cannot be used in defining characteristic units. Now the characteristic units for this reduced problem have already been found; they are M (for mass), $\sqrt{M/K_1}$ (for time), and L for length, where L can either be A, $B\sqrt{M/K_1}$, or an arbitrary fixed length, according to the application in mind. Applying these characteristic units to introduce nondimensional coordinates and parameters, one finds as in (2.3.7) that m and k_1 are equal to 1, and the same is possibly true of a or b (depending on the choice of characteristic length). Using δ, λ, γ, and ω, respectively, for the nondimensionalized values of Δ, K_3, F, and Ω, the nondimensionalized equations become

$$\ddot{y} + \delta\dot{y} + y + \lambda y^3 = \gamma\cos\omega t,$$

$$y(0) = \alpha, \qquad (2.3.14)$$

$$\dot{y}(0) = \beta.$$

See Exercise 2.3.3 for the expressions for δ, λ, and γ in terms of the original coordinates; it is these expressions that show what must "really" be small in order to say that the target equation is close to the reduced equation. As discussed in connection with (2.3.12), it is sometimes useful to include a factor k^2 in the y term of (2.3.14), even though this can be taken to equal 1.

Notice that the dimensional analysis of (2.3.13) does not lead to a final selection of a perturbation parameter ε, as it did in the simpler case of (2.3.5). That is because there are three small quantities (δ, λ, and γ) in (2.3.14), which ideally should be treated as independent small parameters. Unfortunately the mathematical techniques needed for handling a problem with several independent small parameters are considerably more advanced than those dealt with in this book, and it is necessary somehow to reduce the three parameters to functions of a single small parameter ε. The simplest way to do this is to set

$$\delta = \varepsilon\delta_1,$$
$$\gamma = \varepsilon\gamma_1,$$
$$\lambda = \varepsilon\lambda_1,$$

obtaining

$$\ddot{y} + \varepsilon\delta_1\dot{y} + y + \varepsilon\lambda_1 y^3 = \varepsilon\gamma_1 \cos \omega t,$$
$$y(0) = \alpha, \qquad (2.3.15)$$
$$\dot{y}(0) = \beta.$$

This amounts to choosing a straight-line path beginning at the origin in the parameter space (δ, γ, λ), with ε as parameter along the path. (The subscript 1 in δ_1, etc., is meant to indicate that these are coefficients of $\varepsilon = \varepsilon^1$.) A more general approach is taken in Section 4.7.

Example 2.3.2. A Diffusion-Reaction Problem (Example 2.2.2 continued)

Recall that the equations derived in Example 2.2.2 were

$$D\frac{d^2Y}{dX^2} = KY + LY^2,$$
$$Y(0) = A, \qquad (2.3.16)$$
$$Y(N) = B,$$

with variables and constants having the following dimensions:

X	cm
Y	moles/cm
D	cm^2/sec
K	1/sec
L	cm/mole $\cdot$ sec
A	moles/cm
B	moles/cm
N	cm.

Notice that moles never occurs as a unit by itself, but only in the combination moles/cm (even in L, where this combination occurs inverted). To nondimensionalize the variables X and Y, then, it is necessary to find, in terms of the constants, a characteristic length (in cm) and a characteristic concentration (in moles/cm); there is no need (or possibility) of finding a "characteristic quantity of gas" to replace the unit of moles. The most immediate choice for characteristic length and concentration are N and A (or B); see Exercise 2.3.4. Another possible choice is $\sqrt{D/K}$ and K/L, but this choice does not lead to a suitable perturbation problem; see Exercise 2.3.5. The choice that we will make here is $\sqrt{D/K}$ for characteristic length and A for characteristic concentration; this seems to lead to the most "enlightening" perturbation family for this problem.

Substituting

$$x := X/\sqrt{D/K},$$
$$y := Y/A \tag{2.3.17}$$

into (2.3.16), remembering that $d/dX = (dx/dX) \cdot d/dx$, leads immediately to

$$y'' - y = \varepsilon y^2,$$
$$y(0) = 1, \tag{2.3.18}$$
$$y(n) = \gamma,$$

where

$$\varepsilon := LA/K,$$
$$n := N/\sqrt{D/K}, \tag{2.3.19}$$
$$\gamma := B/A$$

are nondimensional parameters. Notice that in this problem there are no choices to make with regard to nondimensional constants: once the characteristic quantities for the variables are chosen, the choice of nondimensional constants is forced. Equation (2.3.18) is a nonoscillatory boundary value problem that is solvable when $\varepsilon = 0$, so it is well suited for study by perturbation methods. (An almost identical problem, given in (2.5.1), will be solved explicitly later in this chapter.) Equation (2.3.19) shows that such methods will be suitable whenever LA/K is small. For fixed A, this means that the ratio of the nonlinear coefficient L to the linear coefficient K is small; this is a natural interpretation of what it means to be "nearly linear." On the other hand, without the dimensional analysis it would not be clear that having a low concentration at one end could also make the problem amenable to perturbation analysis.

Exercises 2.3

1. A *sector* is a pie-shaped slice of a circle. Explain why the radian measure of the angle of a sector is a nondimensional quantity. (Follow the three-step process by which a nondimensional quantity is defined.) Explain why a radian nevertheless functions like a unit when changing from radians to degrees.

2. Apply (2.3.7) as a change of variables in (2.3.5) to obtain (2.3.8). For instance, this involves computing d^2y/dt^2 in terms of d^2Y/dT^2 using the chain rule.

3. Carry out the details of nondimensionalizing (2.3.13) to obtain (2.3.14). In the process, obtain the expressions for δ, λ, and γ in terms of the natural parameters.

4. Nondimensionalize (2.3.16) using characteristic quantities N and A. Your differential equation should have the form $y'' = ky + ly^2$.

5. Nondimensionalize (2.3.16) using characteristic quantities $\sqrt{D/K}$ and K/L. Your differential equation will have the form $y'' = y + y^2$. Since there is no coefficient of y^2 which can be taken to be small, this does not provide a perturbation family having a solvable reduced problem.

2.4. INITIAL VALUE PROBLEMS

Consider first the specific initial value problem

$$\ddot{y} - y = \varepsilon y^2,$$
$$y(0) = \alpha, \qquad\qquad (2.4.1)$$
$$\dot{y}(0) = \beta$$

as an example of (2.1.3). To seek a formal second order approximation, write

$$y = y_0(t) + \varepsilon y_1(t) + \varepsilon^2 y_2(t) + \mathcal{O}_F(\varepsilon^3) \qquad\qquad (2.4.2)$$

and substitute this into (2.4.1):

$$(\ddot{y}_0 - y_0) + \varepsilon(\ddot{y}_1 - y_1) + \varepsilon^2(\ddot{y}_2 - y_2) + \mathcal{O}_F(\varepsilon^3)$$
$$= 0 + \varepsilon y_0^2 + \varepsilon^2(2y_0 y_1) + \mathcal{O}_F(\varepsilon^3),$$
$$y_0(0) + \varepsilon y_1(0) + \varepsilon^2 y_2(0) + \mathcal{O}_F(\varepsilon^3) = \alpha,$$
$$\dot{y}_0(0) + \varepsilon \dot{y}_1(0) + \varepsilon^2 \dot{y}_2(0) + \mathcal{O}_F(\varepsilon^3) = \beta.$$

Equating powers of ε, and putting the initial conditions for each y_i with the differential equation for the same y_i, leads to the following sequence of problems:

$$\ddot{y}_0 - y_0 = 0, \qquad y_0(0) = \alpha, \qquad \dot{y}_0(0) = \beta; \qquad (2.4.3a)$$

$$\ddot{y}_1 - y_1 = y_0^2, \qquad y_1(0) = 0, \qquad \dot{y}_1(0) = 0; \qquad (2.4.3b)$$

$$\ddot{y}_2 - y_2 = 2y_0 y_1, \qquad y_2(0) = 0, \qquad \dot{y}_2(0) = 0. \qquad (2.4.3c)$$

The solution of (2.4.3a) is (see Appendix C)

$$y_0 = \left(\frac{\alpha + \beta}{2}\right) e^t + \left(\frac{\alpha - \beta}{2}\right) e^{-t}. \qquad (2.4.4a)$$

Having found y_0, it is possible to write (2.4.3b) explicitly as

$$\ddot{y}_1 - y_1 = \frac{(\alpha + \beta)^2}{4} e^{2t} + \frac{\alpha^2 - \beta^2}{2} + \frac{(\alpha - \beta)^2}{4} e^{-2t}.$$

A particular solution for y_1 may be found by the method of undetermined coefficients. Attempting a solution in the form dictated by the rules of that method, namely $y_1 = Ae^{2t} + B + Ce^{-2t}$, it is found by computation that the coefficients must have the values $A = (\alpha + \beta)^2/12$, $B = -(\alpha^2 - \beta^2)/2$, and $C = (\alpha - \beta)^2/12$. The general solution for y_1 is then $c_1 e^t + c_2 e^{-t}$ plus the particular solution just found. Evaluating c_1 and c_2 from the initial conditions in (2.4.3b) yields

$$y_1 = \left(\frac{\alpha^2 - 2\alpha\beta - 2\beta^2}{6}\right) e^t + \left(\frac{\alpha^2 + 2\alpha\beta - 2\beta^2}{6}\right) e^{-t}$$

$$+ \frac{(\alpha + \beta)^2}{12} e^{2t} \quad - \quad \frac{(\alpha^2 - \beta^2)}{2} \quad + \quad \frac{(\alpha - \beta)^2}{12} e^{-2t}. \quad (2.4.4b)$$

To calculate y_2 requires first finding $2y_0 y_1$ from (2.4.4a,b) and putting this into (2.4.3c). The right hand side, and also the solution, will involve e to the powers $\pm t$, $\pm 2t$, and $\pm 3t$. The calculation is straightforward but tedious, and is omitted. The calculations which we have just carried out using exponential functions can also be done using hyperbolic functions.

The results, equivalent to (2.4.4), are

$$y_0 = \alpha \cosh t + \beta \sinh t, \tag{2.4.5a}$$

$$
\begin{aligned}
y_1 = {}& \left(\frac{\alpha^2 - 2\beta^2}{3} \right) \cosh t - \left(\frac{2\alpha\beta}{3} \right) \sinh t \\
& + \left(\frac{\alpha^2 + \beta^2}{6} \right) \cosh 2t \; + \; \left(\frac{\alpha\beta}{3} \right) \sinh 2t \; + \; \frac{\beta^2 - \alpha^2}{2} . \tag{2.4.5b}
\end{aligned}
$$

(If you try this calculation, you will need to use some identities for the hyperbolic functions in order to evaluate y_0^2.) The solution of (2.4.1) has been presented in a formal manner, but it is not difficult to justify it, for a finite interval of time $0 \le t \le T$. Remember that to "justify" an asymptotic solution means to show that the formal big-oh symbol $\mathcal{O}_F$ in (2.4.1), signifying that the omitted terms involve ε to the third power or higher, can be replaced by the asymptotic big-oh symbol $\mathcal{O}$ indicating that the error is bounded by a constant times ε^3.

The essential principles for showing this have been set down in Sections 1.8 and 2.1. The function $f(t, y, \dot{y}, \varepsilon) = y^2$ occurring in (2.4.1) is infinitely differentiable. Therefore according to Theorem 2.1.1 and the remarks which follow it, there is a unique solution for $|\varepsilon|$ less than some ε_0, having an asymptotic expansion of the form

$$y = y_0 + \varepsilon y_1 + \varepsilon^2 y_2 + \mathcal{O}(\varepsilon^3) \tag{2.4.6}$$

uniformly for α, β, and t in compact intervals. This already looks the same as (2.4.2) with $\mathcal{O}_F$ replaced by $\mathcal{O}$, but we are not yet finished. It still remains to show that the coefficients y_n in (2.4.6) are the same as those in (2.4.2). This is done by showing that the formal operations which were performed on (2.4.2) to arrive at (2.4.4) or (2.4.5) are permissible operations on the asymptotic series (2.4.6). Differentiating (2.4.6) twice with respect to time is justified by Theorem 1.8.5. (Notice that this requires the existence and continuity of two derivatives beyond those needed for (2.4.6) itself, which is no problem here.) The computation $y^2 = y_0^2 + \varepsilon(2y_0 y_1) + \mathcal{O}(\varepsilon^2)$ is justified by Theorem 1.8.4c, since $y_0 = y_0(t; \alpha, \beta)$ is known to be bounded (because it is continuous and t, α, and β are confined to compact subsets). Equating terms of each power of ε, to arrive at (2.4.3), is justified by the uniqueness of ordinary asymptotic expansions (Theorem 1.8.3). Finally, the solutions of the initial value problems given in (2.4.3) are unique. (Why is this important?) These arguments remove any doubt that the coefficients in (2.4.6) are exactly those given by (2.4.4) or (2.4.5).

The argument given above is typical of regular perturbation theory, in which the validity of an approximation often follows from general principles. The same result can be established in another way, by directly estimating the error in the approximation (2.4.2); this is more difficult, but sometimes makes it possible to find the constants c and ε_1. (A similar situation was encountered in Section 1.4.) Direct error estimation for this type of problem will be presented in Section 3.2.

The same reasoning used to solve (2.4.1) can be used to solve the general initial value problem (2.1.3), that is,

$$
\begin{aligned}
a\ddot{y} + b\dot{y} + cy &= \varepsilon f(t, y, \dot{y}, \varepsilon), \\
y(0) &= \alpha, \\
\dot{y}(0) &= \beta.
\end{aligned}
\tag{2.4.7}
$$

There are two steps: showing that the formal computation of the approximate solutions is possible, and showing that the desired asymptotic error estimates hold. These steps will be carried out as separate theorems (Theorems 2.4.1 and 2.4.2). Both of these theorems are easiest to prove when the perturbation f in (2.4.7) is infinitely differentiable, and this case covers most applications. But something interesting happens when f has only finitely many (say r) derivatives: The formal perturbation series can be calculated up to order r, but in general the last term of this series is not asymptotically valid. This provides a valuable illustration of the fact that an operation which is formally possible is not always justifiable.

Theorem 2.4.1. *If f has continuous partial derivatives of all orders $\leq r$, then the coefficients of the formal regular perturbation series for (2.1.3) are unique and can be determined recursively up to order r. (Warning: The last term is not asymptotically valid.)*

Proof. Assume first that f is infinitely differentiable; in this case the theorem simply states that the perturbation series can be computed to any order. For simplicity assume also that f depends only on y and $\dot{y}$: $f = f(y, \dot{y})$. Substituting $y = y_0 + \varepsilon y_1 + \varepsilon^2 y_2 + \cdots$ and $\dot{y} = \dot{y}_0 + \varepsilon \dot{y}_1 + \varepsilon^2 \dot{y}_2 + \cdots$ into (2.4.7) yields

$$
\sum_{n=0}^{\infty} \varepsilon^n (a\ddot{y}_n + b\dot{y}_n + cy_n) = \varepsilon f \left(\sum_{i=0}^{\infty} \varepsilon^i y_i, \sum_{j=0}^{\infty} \varepsilon^j \dot{y}_j \right).
\tag{2.4.8}
$$

When the right hand side is expanded in powers of ε, using a formal Taylor expansion, the first few terms are

$$
\varepsilon f(y_0, \dot{y}_0) + \varepsilon^2 \{ f_y(y_0, \dot{y}_0) y_1 + f_{\dot{y}}(y_0, \dot{y}_0) \dot{y}_1 \} + \cdots .
\tag{2.4.9}
$$

The fundamental fact about this expansion is that the coefficient of ε^n

depends only upon f and its derivatives and $y_0, \dot{y}_0, \ldots, y_{n-1}, \dot{y}_{n-1}$. (See Exercise 2.4.1.) If (2.4.9) is written as

$$\varepsilon K_1(y_0, \dot{y}_0) + \varepsilon^2 K_2(y_0, \dot{y}_0, y_1, \dot{y}_1) + \varepsilon^3 K_3(y_0, \dot{y}_0, y_1, \dot{y}_1, y_2, \dot{y}_2) + \cdots,$$

$$(2.4.10)$$

then the result of equating coefficients of ε^n is

$$
\begin{aligned}
& a\ddot{y}_0 + b\dot{y}_0 + cy_0 = 0, \\
& a\ddot{y}_n + b\dot{y}_n + cy_n = K_n(y_0, \dot{y}_0, \ldots, y_{n-1}, \dot{y}_{n-1}), \qquad n \geq 1.
\end{aligned}
$$

$$(2.4.11)$$

The initial conditions of (2.4.7) become

$$
\begin{aligned}
& y_0(0) = \alpha, \qquad \dot{y}_0(0) = \beta, \\
& y_n(0) = 0, \qquad \dot{y}_n(0) = 0, \qquad n \geq 1.
\end{aligned}
$$

$$(2.4.12)$$

This sequence of initial value problems can be solved recursively, since the equation for y_n requires knowledge only of previously calculated terms $y_0, \ldots, y_{n-1}$ and their derivatives (and, of course, f). Therefore the computation can be carried to any order.

If f has continuous derivatives only up to order r, a finite perturbation series $y = y_0 + \varepsilon y_1 + \cdots + \varepsilon^k y_k$ can be found for any $k \leq r$: Since K_n requires derivatives of f only up to order $n - 1$ (see Exercise 2.4.1), it follows that K_n is continuously differentiable for $n \leq r$ and the equations (2.4.11) are solvable. (The touchiest case is K_r, which has only one continuous derivative, but this is sufficient.) If f depends on t and ε, these arguments can be carried out in a similar way. (See Exercise 2.4.2.) ∎

The next theorem addresses the asymptotic validity of the series we have just constructed. When f is infinitely differentiable, the result is as expected: The series are asymptotically valid to any order. But as we have already remarked, when f has continuous derivatives only up to order r, the solutions constructed above are not asymptotically valid up to order r, but only at most up to order $r - 1$. This is not surprising; the same thing happens with any Taylor series of a finitely differentiable function. What is more surprising is that by the methods available to us now it is only easy to prove validity up to order $r - 3$. With some extra work we can attain $r - 2$ in the general case and $r - 1$ (the best that is to be expected) if f does not depend upon $\dot{y}$. The method of error estimation given in Section 3.2 does not have these limitations.

Theorem 2.4.2. *Let f satisfy the hypotheses of Theorem 2.1.1, so that the exact solution $\varphi(t; \alpha, \beta; \varepsilon)$ of (2.4.7) exists and is smooth of order r for $0 \leq t \leq T$ and $|\varepsilon| < \varepsilon_0$. Then for any $k \leq r - 2$, the approximate solution*

$$\tilde{\varphi}(t; \alpha, \beta; \varepsilon) := y_0(t) + \varepsilon y_1(t) + \cdots + \varepsilon^k y_k(t)$$

computed according to Theorem 2.4.1 satisfies

$$\varphi(t; \alpha, \beta; \varepsilon) = \tilde{\varphi}(t; \alpha, \beta; \varepsilon) + \mathcal{O}(\varepsilon^{k+1}) \qquad (2.4.13)$$

uniformly for $0 \le t \le T$ and for α, β in compact subsets. That is, given $0 < \varepsilon_1 < \varepsilon_0$ and compact intervals A, B, there exists a constant $c > 0$ such that

$$|\varphi(t; \alpha, \beta; \varepsilon) - \tilde{\varphi}(t; \alpha, \beta; \varepsilon)| \le c\varepsilon^{k+1}$$

for $0 \le t \le T$, $\alpha \in A$, $\beta \in B$, and $|\varepsilon| \le \varepsilon_1$. If f does not depend on $\dot{y}$, then the same statements are true for all $k \le r - 1$.

Proof. As in the proof of Theorem 2.4.1, we will first assume that f is infinitely differentiable and that f depends only on y and $\dot{y}$. In this case the argument is almost the same as for the special case of equation (2.4.1) discussed above. Namely, the solution φ exists and is infinitely differentiable by Theorem 2.1.1. Therefore φ has a Taylor expansion to any order k which satisfies (2.4.13). Knowing this, the formal steps carried out in the proof of Theorem 2.4.1 can be justified using the theorems of Section 1.8. The only point in which the general case differs from the example is that the expansion (2.4.9) must be shown to be a valid asymptotic expansion of the right hand side of (2.4.8) to all orders. This is a consequence of Exercise 1.8.7, suitably modified to allow f to be a function of two variables. The details are left to the reader. (This question did not arise in the discussion of (2.4.1) because f was simply y^2 which could be expanded using the multiplication theorem for asymptotic power series, Theorem 1.8.4. When the function f is not specified in such a concrete way, the only way to expand the right hand side of (2.4.8) is by a Taylor series.) We also leave to the reader the case in which f depends on t and ε. This involves no new difficulties.

> In the case of finitely many derivatives, the argument is more technical and may be omitted if desired. The solution φ is known from Theorem 2.1.1 to be smooth of order r, and therefore it has a Taylor series up to order $r - 1$; the last derivative cannot be used in the series because it is needed (see the proof of Theorem A.1) in order to obtain the $\mathcal{O}(\varepsilon^r)$ error estimate. So there is no hope of justifying the last term in the formal series computed in Theorem 2.4.1. But there is an additional difficulty. We know that the series $y \sim y_0 + \varepsilon y_1 + \cdots + \varepsilon^{r-1} y_{r-1}$ exists. But in order to substitute it into the differential equation to obtain (2.4.8) (which this time is only expected to hold up to order $r - 1$), it is necessary to differentiate it twice with respect to t. This "uses up" two more derivatives and only allows us to justify the series through order $r - 3$. (See Theorem 1.8.5, with x being the scalar variable t. According to that theorem, taking one derivative is possible if the series is restricted to order $r - 2$; applying the theorem twice restricts to $r - 3$.)

There is a way to get around part of this difficulty and justify the solution through order $r - 2$ if f depends upon $\dot{y}$, and all the way through order

$r - 1$ if it does not. The trick is to replace the differential equation (2.4.7), with its initial conditions, by the integral equation

$$y(t, \varepsilon) = \alpha u(t) + \beta v(t) + \varepsilon \int_0^t K(t, \tau) f(\tau, y(\tau, \varepsilon), \dot{y}(\tau, \varepsilon), \varepsilon) \, d\tau. \quad (2.4.14)$$

Here u and v denote the two solutions of $a\ddot{y} + b\dot{y} + cy = 0$ such that $u(0) = 1$, $\dot{u}(0) = 0$, $v(0) = 0$, and $\dot{v}(0) = 1$. The integral kernel K is defined by $K(t, \tau) := (u(\tau)v(t) - u(t)v(\tau))/a$. This integral equation is derived in Appendix C, where it appears in slightly different notation as (C.30). For our purposes the advantage of the integral equation is that only the first derivative of y appears in (2.4.14). The series $y \sim y_0 + \cdots + \varepsilon^{r-2} y_{r-2}$ may be differentiated once with respect to t and substituted into (2.4.15); or if f does not depend on $\dot{y}$, there is no need to differentiate at all, and the series $y \sim y_0 + \cdots + \varepsilon^{r-1} y_{r-1}$ may be substituted into (2.4.14). The integrand may be expanded in powers of ε through order $r - 2$, and then integrated term by term using Exercise 1.8.8. (It is easier to justify integrating an asymptotic series than differentiating one.) The coefficients y_n can then be found as usual by equating terms of equal degree using the uniqueness theorem for ordinary asymptotic expansions (Theorem 1.8.3); these coefficients are now fully justified. Finally, it can be shown that the coefficients computed in this way are the same as those computed from the differential equation according to the method of Theorem 2.4.1. Rather than present the details, we give a computational example in Exercise 2.4.5 which should be convincing. ∎

It is sometimes necessary to solve an initial value problem in which the initial conditions depend upon ε:

$$\begin{aligned} a\ddot{y} + b\dot{y} + cy &= \varepsilon f(t, y, \dot{y}, \varepsilon), \\ y(0) &= \alpha(\varepsilon), \\ y'(0) &= \beta(\varepsilon). \end{aligned} \quad (2.4.15)$$

There are two ways to do this. The first is based on the following lemma:

Lemma 2.4.3. *If $\varphi(t; \alpha, \beta; \varepsilon)$ is the solution of (2.4.7), then $\varphi(t; \alpha(\varepsilon), \beta(\varepsilon), \varepsilon)$ is the solution of (2.4.15).*

In words, this lemma states that if the solution is known for arbitrary constant initial conditions, then the solution for initial conditions that depend upon ε is obtained merely by allowing these constants to become functions of ε.

Proof. To say that φ is the solution of (2.4.7) is to say that

$$\begin{aligned} a\varphi_{tt}(t; \alpha, \beta; \varepsilon) + b\varphi_t(t; \alpha, \beta; \varepsilon) + c\varphi(t; \alpha, \beta; \varepsilon) \\ = \varepsilon f(t, \varphi(t; \alpha, \beta; \varepsilon), \varphi_t(t; \alpha, \beta, \varepsilon), \varepsilon). \end{aligned}$$

Since this is an identity in all of the variables, it remains true if the constants α and β are replaced by functions $\alpha(\varepsilon)$ and $\beta(\varepsilon)$. ∎

The practical consequence of Lemma 2.4.3 is that if the asymptotic power series solution, say,

$$\varphi(t; \alpha, \beta; \varepsilon) = \varphi_0(t; \alpha, \beta) + \varepsilon\varphi_1(t; \alpha, \beta) + \mathcal{O}(\varepsilon^2), \qquad (2.4.16)$$

has been found for (2.4.7), it is only necessary to substitute $\alpha(\varepsilon)$ and $\beta(\varepsilon)$ into (2.4.16) and expand the resulting functions $\varphi_n(t; \alpha(\varepsilon), \beta(\varepsilon))$ in ε by Taylor's theorem and regroup in powers of ε to obtain the asymptotic power series solution of (2.4.15).

The second method of solving (2.4.15) is best if the asymptotic solution of (2.4.7) is not available in advance. Expand $\alpha(\varepsilon) = \alpha_0 + \varepsilon\alpha_1 + \cdots$, $\beta(\varepsilon) = \beta_0 + \varepsilon\beta_1 + \cdots$, and solve (2.4.11) recursively for $y_0, y_1, \ldots$ using the initial conditions

$$y_n(0) = \alpha_n, \qquad \dot{y}_n(0) = \beta_n, \qquad n = 0, 1, \ldots \qquad (2.4.17)$$

in place of (2.4.12). These initial conditions follow immediately from

$$y_0(0) + \varepsilon y_1(0) + \cdots = \alpha_0 + \varepsilon\alpha_1 + \cdots,$$
$$\dot{y}_0(0) + \varepsilon\dot{y}_1(0) + \cdots = \beta_0 + \varepsilon\beta_1 + \cdots,$$

obtained by putting $y = y_0 + \varepsilon y_1 + \cdots$ into (2.4.15).

Exercises 2.4

1. (a) Comparison of (2.4.9) and (2.4.10) shows that $K_1 = f(y_0, \dot{y}_0)$ and $K_2 = f_y(y_0, \dot{y}_0)y_1 + f_{\dot{y}}(y_0, \dot{y}_0)\dot{y}_1$. Continuing in this way, calculate K_3 explicitly and show that it depends only on y_j and $\dot{y}_j$ for $j < 3$, as indicated in (2.4.10). Check that K_3 involves derivatives of order f only up to the second order.

    (b) For those familiar with mathematical induction: Prove that K_n depends only on y_j and $\dot{y}_j$ for $j < n$.

2. Carry out the proof of Theorem 2.4.1 in the case when $f = f(t, y, \dot{y}, \varepsilon)$. Hint: The crucial point is to find the expansions corresponding to (2.4.9) and (2.4.10). Check your results to third order (as in Exercise 2.4.1a) or to all orders (as in Exercise 2.4.1b).

3. (a) Solve the problem $\ddot{y} + y = \varepsilon y^2$, $y(0) = \alpha$, $\dot{y}(0) = 0$ by an asymptotic solution of first order. (That is, carry the solution out to and including the term in ε.) Hint: Express y_0^2 in Fourier form using Table E.1 before solving for y_1.

    (b) State precisely the extent to which this solution is known to be valid. What kind of error estimate holds, for what kind of intervals of t, α, and β? State this error estimate without using big-oh notation. (Given ... these exist ... such that ... .)

(c) Your solution should be periodic of period 2π in t. Does this prove that the exact solution is periodic?

(d) Suppose it were true that the exact solution were periodic of period 2π. Show (with almost no work) that in this case the error estimate which you gave in part (b) would actually be good for all time. (This idea will appear again in Section 3.2.)

(e) Compute the next term $(y_2\varepsilon^2)$ in the expansion of y. (Hint: Eliminate all powers and products of trig functions by using Table E.1 before solving the differential equation for y_2.) Do you believe now that the exact solution is periodic? (Be prepared to be surprised again. This problem is continued in Exercises 3.2.4 and 4.3.3.)

4. Solve the following oscillator problems by ordinary asymptotic power series to first order. These are problems whose solutions will be improved greatly in later chapters, using generalized asymptotic series. They are regular perturbation problems when considered on a finite interval $0 \le t \le T$, but singular when considered on unbounded intervals. The first is a *nonlinear spring* or *Duffing equation* without forcing or damping (see Example 2.2.1). The second is an unforced *Van der Pol equation*.

(a)　$\ddot{y} + y + \varepsilon y^3 = 0,$
$y(0) = \alpha,$
$\dot{y}(0) = 0;$

(b)　$\ddot{y} + \varepsilon(y^2 - 1)\dot{y} + y = 0,$
$y(0) = \alpha,$
$\dot{y}(0) = 0.$

5. This exercise concerns the initial value problem $\ddot{y} + y = \varepsilon y^2|y|$, $y(0) = 1$, $\dot{y}(0) = 0$.

(a) What is the order of smoothness of f in this equation? What order of asymptotic expansion can you expect to find for the solution? (See Theorem 2.4.2.)

(b) Compute a formal perturbation series solution to first order on the interval $0 \le t \le \pi$ by the method of Theorem 2.4.1. Hint: In computing y_1 you will need to take note of when y_0 changes sign.

(c) What formal operation(s) have you performed in part (b) that are not rigorously justifiable?

(d) Compute a perturbation series solution to first order on the interval $0 \leq t \leq \pi$ by substituting the series into the integral equation
(2.4.14) after computing u, v, and K for this problem. Does your
solution agree with your solution to part (b)? Have you performed
any unjustifiable steps?

2.5. BOUNDARY VALUE PROBLEMS

Boundary value problems for second order ordinary differential equations
are as easy to solve as initial value problems in the nondegenerate case,
that is, when the reduced problem has a unique solution. As pointed out
in Section 2.1, nonoscillatory boundary value problems are always nondegenerate, and we concentrate on this case here (although the methods presented work in all nondegenerate cases). The regular perturbation method
for these problems will first be illustrated by an example. Then the theory
will be presented, making use of the shooting method, which consists in
replacing the boundary value problem by an initial value problem plus a
"shooting problem." In this way it is possible to prove the existence of
solutions to the boundary value problem and at the same time justify the
regular perturbation method. But it is important to emphasize that the
shooting method is not intended as a practical computational procedure.
The simple way to solve these problems is the direct method used in the
example of equation (2.5.1) below; the shooting method is merely a round-
about way to justify the solution. (There is another form of the shooting
method, used in numerical analysis to solve problems of the same type on
a digital computer. The basic idea is the same: Solve initial value problems,
varying the initial conditions until you hit the right boundary value at the
far end. But in that setting, the shooting method is a practical method and
not only a theoretical tool.)

Consider the boundary value problem

$$y'' - y = \varepsilon y^2,$$
$$y(0) = \alpha, \qquad\qquad (2.5.1)$$
$$y(1) = \gamma.$$

This problem involves the same differential equation as (2.4.1) and is
nonoscillatory because of the minus sign before the y. The first boundary condition is the same as the first initial condition of (2.4.1), but the
second boundary condition is (of course) imposed at the other end of the
interval on which the solution is sought, which in this case is $0 \leq x \leq 1$.
Notice that we use the letter β to denote the second initial condition in
(2.4.1) and γ for the second boundary condition in (2.5.1); this notation
will help to keep things clear in the shooting method, where initial and

boundary values appear at the same time. Trying a formal perturbation series solution $y = y_0 + \varepsilon y_1 + \mathcal{O}_F(\varepsilon^2)$, it follows by substituting into (2.5.1) that

$$y_0'' - y_0 = 0, \qquad y_0(0) = \alpha, \qquad y_0(1) = \gamma \qquad (2.5.2a)$$

and

$$y_1'' - y_1 = y_0^2, \qquad y_1(0) = 0, \qquad y_1(1) = 0. \qquad (2.5.2b)$$

As usual in regular perturbation problems, the equations which must be solved to find the leading term y_0 are the same as those of the reduced problem obtained by setting $\varepsilon = 0$ in (2.5.1). The solution (see Exercise 2.1.2 and Appendix C) is

$$y_0 = \left(\frac{\alpha - \gamma e}{1 - e^2} \right) e^x - e \left(\frac{\alpha e - \gamma}{1 - e^2} \right) e^{-x}. \qquad (2.5.3)$$

Then (2.5.2b) becomes

$$y_1'' - y_1 = ae^{2x} + b + ce^{-2x},$$

with

$$a = \left(\frac{\alpha - \gamma e}{1 - e^2} \right)^2, \qquad b = \frac{-2e(\alpha - \gamma e)(\alpha e - \gamma)}{(1 - e^2)^2},$$

$$c = e^2 \left(\frac{\alpha e - \gamma}{1 - e^2} \right)^2.$$

The general solution for y_1, ignoring for a moment the boundary conditions in (2.5.2b), can be found by the method of undetermined coefficients; it is

$$y_1 = c_1 e^x + c_2 e^{-x} + \tfrac{1}{3} ae^{2x} - b + \tfrac{1}{3} ce^{-2x}.$$

Next the boundary conditions are used to determine c_1 and c_2; it is not hard to see that these must satisfy the simultaneous linear equations

$$c_1 + c_2 = -\tfrac{1}{3}a + b - \tfrac{1}{3}c,$$
$$c_1 e + c_2 e^{-1} = -\tfrac{1}{3}ae^2 + b - \tfrac{1}{3}ce^{-2}.$$

It is possible (but very messy) to solve these equations for c_1 and c_2 in terms of a, b, and c, which in turn depend on α and γ. When all of this is substituted back into y_1 the result can be arranged in the form

$$y_1 = A + Be^x + Ce^{-x} + De^{2x} + Ee^{-2x}, \qquad (2.5.4)$$

where A, B, C, D, E are complicated but explicitly calculable functions of α and γ. Since these formulas become so awkward, it is probably best to put numerical values for α and γ into (2.5.3) and proceed from there by the same steps as above, but with numerical coefficients at each stage. As the simplest illustration, take the case $\alpha = 1$, $\gamma = e$. Then (2.5.3) becomes $y_0 = e^x$, and (2.5.4) is $y_1 = -0.996e^x + 0.662e^{-x} + 0.333e^{2x}$, giving for the complete solution to first order

$$y = e^x + \varepsilon(-0.996e^x + 0.662e^{-x} + 0.333e^{2x}) + \mathcal{O}_F(\varepsilon^2). \qquad (2.5.5)$$

In order to justify this result, that is, to replace the $\mathcal{O}_F(\varepsilon^2)$ in (2.5.5) with "$\mathcal{O}(\varepsilon^2)$ uniformly for $0 \leq x \leq 1$," all that is needed is to show that the solution to (2.5.1) exists and depends smoothly on ε. Then the existence of an asymptotic expansions for the solution, and the validity of the procedure we have just used to find it, follows from Taylor's theorem and the theorems of Section 1.8, exactly as for initial value problems. This program will now be carried out for the general problem (2.5.7) stated below, which includes (2.5.1). The same argument works for the still more general problem (2.1.8).

In order to motivate the shooting method, it is convenient to look at it first "in reverse." Suppose for the moment that there exists a solution

$$y = \psi(x; \alpha, \gamma; \varepsilon) \qquad (2.5.6)$$

to the boundary value problem

$$\begin{aligned}
y'' - k^2 y &= \varepsilon f(y, y'), \\
y(0) &= \alpha, \\
y(1) &= \gamma.
\end{aligned} \qquad (2.5.7)$$

Although this is a boundary value problem, its solution (2.5.6) of course has initial values; that is, $y(0)$ and $y'(0)$ exist. Specifically, $y(0) = \alpha$ and $y'(0) = \beta(\varepsilon) := \psi_x(0, \alpha, \gamma; \varepsilon)$. Therefore (2.5.6) is also the solution of the initial value problem

$$\begin{aligned}
y'' - k^2 y &= \varepsilon f(y, y'), \\
y(0) &= \alpha, \\
y'(0) &= \beta(\varepsilon).
\end{aligned} \qquad (2.5.8)$$

Notice that $y'(0)$ in this problem generally depends upon ε (for fixed α and γ). The initial value problem defined by (2.5.8) is equivalent to the boundary value problem (2.5.7), in the sense that both problems have the same solution. If the correct function $\beta(\varepsilon)$ were known (for a given α and γ), the boundary value problem could be replaced by the initial value problem. But things are not this simple; $\beta(\varepsilon)$ is not known in advance; the expression for $\beta(\varepsilon)$ given above depends upon the solution of the boundary value problem.

The idea of the shooting method is to reverse this argument. That is, consider the initial value problem

$$y'' - k^2 y = \varepsilon f(y, y'),$$
$$y(0) = \alpha, \qquad\qquad (2.5.9)$$
$$y'(0) = \beta$$

(in which β is a constant) and denote its solution by

$$y = \varphi(x; \alpha, \beta; \varepsilon). \qquad\qquad (2.5.10)$$

By Theorems 2.1.1 and 2.4.2 the solution (2.5.10) exists and has valid asymptotic expansions for $0 \le x \le 1$ and for sufficiently small ε. (A little more care must be taken on this point in the proof of Lemma 2.5.1 below.) According to Lemma 2.4.3, the solution function φ for (2.5.9), in which β is constant, can also be used to represent the solution of an initial value problem of the form (2.5.8) in which β depends upon ε; namely,

$$y = \varphi(x; \alpha, \beta(\varepsilon); \varepsilon). \qquad\qquad (2.5.11)$$

(We are not assuming here that the "correct" $\beta(\varepsilon)$ is known. We are just saying that (2.5.11) gives the solution to (2.5.8) whatever $\beta(\varepsilon)$ might be.) In particular, the solution of (2.5.8) exists for $0 \le x \le 1$; it has asymptotic expansions, provided that $\beta(\varepsilon)$ is smooth, so that (2.5.11) can be expanded in a Taylor series. If it is possible to choose $\beta(\varepsilon)$ so that

$$\varphi(1; \alpha, \beta(\varepsilon); \varepsilon) = \gamma, \qquad\qquad (2.5.12)$$

then (2.5.11) will solve not only the initial value problem (2.5.8) but also the boundary value problem (2.5.7). In other words if a smooth function $\beta(\varepsilon)$ satisfying (2.5.12) exists, then

$$\psi(x; \alpha, \gamma; \varepsilon) = \varphi(x; \alpha, \beta(\varepsilon); \varepsilon), \qquad\qquad (2.5.13)$$

proving both existence and smoothness of ψ. As we have remarked, this suffices to justify calculations such as (2.5.5).

Our attention is now focused upon (2.5.12), the so-called *shooting equation* whose graphical interpretation was given in Fig. 2.1.1.

Lemma 2.5.1. *Given α and γ, there exists an $\varepsilon_1 > 0$ and a unique smooth function $\beta(\varepsilon)$ defined for $|\varepsilon| < \varepsilon_1$ satisfying (2.5.12).*

Proof. We apply the implicit function theorem to the equation

$$\Phi(\beta, \varepsilon) := \varphi(1; \alpha, \beta; \varepsilon) - \gamma = 0 \qquad\qquad (2.5.14)$$

for a fixed α and γ. First consider the reduced problem $\Phi(\beta_0, 0) = 0$. Since $\varphi(x; \alpha, \beta; 0)$ is the solution of $y'' - k^2 y = 0$, $y(0) = \alpha$, $y'(0) = \beta$, it can be computed explicitly:

$$\varphi(x; \alpha, \beta; 0) = \alpha \cosh kx + \frac{\beta}{k} \sinh kx. \qquad (2.5.15)$$

Then $\Phi(\beta_0, 0) = 0$ is just the equation

$$\alpha \cosh k + \frac{\beta_0}{k} \sinh k = \gamma$$

with unique solution

$$\beta_0 = k \left(\frac{\gamma - \alpha \cosh k}{\sinh k} \right).$$

Next we show that $\varphi_\beta(1; \alpha, \beta_0; 0) \neq 0$. In fact, (2.5.15) implies $\varphi_\beta(1; \alpha, \beta; 0) = \frac{1}{k} \sinh k$ which is never zero (and so is $\neq 0$ when $\beta = \beta_0$). We have shown that

$$\Phi(\beta_0, 0) = 0 \qquad \text{and} \qquad \Phi_\beta(\beta_0, 0) \neq 0,$$

which seems to be enough to call upon the implicit function theorem. However, we also need to check that Φ is defined and smooth in a neighborhood of $(\beta_0, 0)$. Here Theorem 2.1.1 must be used. Choose a compact interval B around β_0. (Since α is fixed, the interval A in Theorem 2.1.1 can be taken to be the point α.) Then there exists $\varepsilon_0 > 0$ such that $\varphi(1; \alpha, \beta; \varepsilon)$, and hence $\Phi(\beta, \varepsilon)$, is defined for all β in B and $|\varepsilon| < \varepsilon_0$. This provides a neighborhood of $(\beta_0, 0)$, and Theorem B.1 gives the existence of ε_1 (with $(0 < \varepsilon_1 \leq \varepsilon_0)$ and of a smooth $\beta(\varepsilon)$. Uniqueness of $\beta(\varepsilon)$ requires a further comment. The implicit function theorem guarantees that $\beta(\varepsilon)$ is the unique continuous function satisfying (2.5.12) *and* $\beta(0) = \beta_0$. But in this case the latter condition is implied by the former, since β_0 is the only solution of the reduced equation. Therefore $\beta(\varepsilon)$ is the unique continuous solution of (2.5.12) defined for $|\varepsilon| < \varepsilon_1$. This does not rule out the possibility of other solutions $\beta(\varepsilon)$ of (2.5.12) that are not defined when $\varepsilon = 0$, such as solutions that approach $\pm\infty$ as $\varepsilon \to 0$. ∎

Lemma 2.5.1 shows that $\psi(x; \alpha, \gamma; \varepsilon)$, defined by (2.5.13), is a smooth function of x and ε for fixed α and γ. This implies that perturbation series solutions of (2.5.7) are asymptotically valid, uniformly for $0 \leq x \leq 1$. In order to show uniformity with respect to α and γ in compact subsets, it is necessary to show that φ is also smooth in α and γ. Only slight modification of the proof of Lemma 2.5.1 is necessary. (See Exercise 2.5.1.)

As a summary of this section, we state the following theorem covering the most general boundary value problem of the class that we are considering. This theorem corresponds to Theorems 2.1.1, 2.4.1, and 2.4.2 for initial value problems.

Theorem 2.5.2. *Let $f(x, y, y', \varepsilon)$ be defined for $0 \le x \le L$, for all y and y', and for all ε near zero. Let f have continuous partial derivatives of all orders $\le r$. Let compact intervals A and C be specified for α and γ. Then there exists $\varepsilon_1 > 0$ such that for $|\varepsilon| < \varepsilon_1$, there is an exact solution $\psi(x; \alpha, \gamma; \varepsilon)$ of*

$$ay'' + by' + cy = \varepsilon f(x, y, y', \varepsilon),$$
$$y(0) = \alpha, \qquad\qquad\qquad (2.5.16)$$
$$y(L) = \gamma,$$

provided this problem is nondegenerate; in particular, this will be true if the problem is nonoscillatory, that is, if $b^2 - 4ac > 0$. This solution ψ is defined for $0 \le x \le L$, $\alpha \in A$, $\gamma \in C$, and $|\varepsilon| < \varepsilon_1$, and is the only continuous family of solutions of (2.5.16) defined over this range of parameters. (There could be other solutions with α, γ and ε in this range, but not forming a continuous family valid for $\varepsilon = 0$.) For any $k \le r - 2$, it is possible to compute an approximate solution

$$\tilde{\psi}(x; \alpha, \gamma; \varepsilon) = y_0(x) + \varepsilon y_1(x) + \cdots + \varepsilon^k y_k(x) \qquad (2.5.17)$$

by substituting (2.5.17) into (2.5.16) and carrying out the obvious formal operations. This approximate solution satisfies

$$\psi(x; \alpha, \gamma; \varepsilon) = \tilde{\psi}(x; \alpha, \gamma; \varepsilon) + \mathcal{O}(\varepsilon^{k+1}) \qquad (2.5.18)$$

uniformly for $0 \le x \le L$, $x \in A$, and $\gamma \in C$. That is, for $0 < \varepsilon_2 < \varepsilon_1$, there exists a constant $c > 0$ such that

$$|\psi(x; \alpha, \gamma; \varepsilon) - \tilde{\psi}(x; \alpha, \gamma; \varepsilon)| \le c\varepsilon^{k+1}$$

for $0 \le x \le L$, $\alpha \in A$, $\gamma \in C$, and $|\varepsilon| \le \varepsilon_2$.

Exercises 2.5

1. Prove that the solution (2.5.6) of (2.5.7) is smooth in α and γ as well as in ε. Hint: Show, using the implicit function theorem in the form of Theorem B.2, that the shooting problem (2.5.14) can be solved to give a smooth function $\beta(\alpha, \gamma; \varepsilon)$.

2. Solve the following boundary value problems by perturbation series to first order, or explain why the problem cannot be solved.

(a) $y'' - 25y = \varepsilon y^3$, $y(0) = 1$, $y(1) = 0$.

(b) $y'' - 9y = \varepsilon y\dot{y}$, $y(0) = 0$, $y(1) = 1$.

(c) $y'' + \dot{y} + y = \varepsilon y^2$, $y(0) = 1$, $y(1) = 0$.

(d) $y'' + \dot{y} + y = \varepsilon y^2$, $y(0) = 1$, $y(2\pi/\sqrt{3}) = 0$.

2.6* NONLINEAR EIGENVALUE PROBLEMS

In Section 2.1, second order differential equations were classified as degenerate and nondegenerate, and also as oscillatory and nonoscillatory. Nondegenerate boundary value problems were treated in Section 2.5. In this section we will consider a particular type of degenerate boundary value problem (which must therefore be oscillatory), namely

$$y'' + \lambda y = \alpha f(y),$$
$$y(0) = 0, \tag{2.6.1}$$
$$y(1) = 0,$$

where f is a nonlinear function of y such that $f(0) = 0$, λ and α are nondimensional natural parameters, and $\lambda > 0$. This problem can be handled by regular perturbation theory, but introduces certain features which will occur again with singular perturbations, in particular with the Lindstedt method in Chapter 4.

Equations (2.6.1) do not as yet contain a specified perturbation parameter ε. The first step, then, is to select a suitable reduced problem (which must be solvable); the second is to introduce a perturbation family (or path through the parameter space) connecting the reduced problem with unsolvable problems close to it. In order to be solvable, the reduced problem must have $\alpha = 0$. In this case (2.6.1) is the linear eigenvalue problem

$$y'' + \lambda y = 0,$$
$$y(0) = 0, \tag{2.6.2}$$
$$y(1) = 0.$$

The general solution of $y'' + \lambda y = 0$ is $y = A \sin \sqrt{\lambda} x + B \cos \sqrt{\lambda} x$, and the first boundary condition $y(0) = 0$ implies $B = 0$. The second boundary condition is then $A \sin \sqrt{\lambda} = 0$, which holds if $A = 0$ or if $\lambda = n^2\pi^2$ for an integer $n \geq 1$. Thus (2.6.2) has the solution $y = A \sin \sqrt{\lambda} x$, provided either $A = 0$ or $\lambda = n^2\pi^2$, that is, provided pair (λ, A) lies on the graph shown in Fig. 2.6.1. So (2.6.2) has the unique solution $y = 0$ when $\lambda \neq n^2\pi^2$ and infinitely many solutions $y = A \sin n\pi x$ when $\lambda = n^2\pi^2$. The values

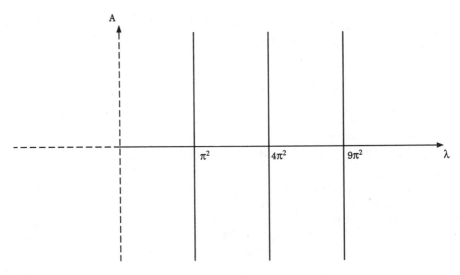

FIG. 2.6.1. The set of values of (λ, A) corresponding to solutions of (2.6.2). Equivalently, the set of values of (λ, A) making (2.6.9) solvable when $\varepsilon = 0$; each vertical line corresponds to a value of $n = 1, 2, 3, \ldots$.

$\lambda = n^2\pi^2$ are called *eigenvalues* of the problem (2.6.2), and the solutions $A \sin n\pi x$ are called *eigenfunctions*.

Our aim in this section is to show that for $\alpha \neq 0$, the vertical straight lines in Fig. 2.6.1 become bent into curves as shown in Fig. 2.6.2. What this means will not be clear until the end of the section, but this much should be emphasized at once: Although the values of λ at points on the curves in Fig. 2.6.2 are sometimes called *nonlinear eigenvalues*, they are not eigenvalues at all in the usual sense of linear mathematics.

In linear algebra an *eigenvalue* of a linear transformation T is defined as a number λ such that $Tv = \lambda v$ for some vector v called an *eigenvector*. When the vector space is a space of functions, eigenvectors are called *eigenfunctions*. Applying these ideas to (2.6.2), the functions $y(x)$ satisfying $y(0) = 0$ and $y(1) = 0$ form a vector space, and $y'' + \lambda y = 0$ can be thought of as $Ty = \lambda y$, with $T = -d^2/dx^2$. It follows from the linearity of T that any multiple of an eigenvector is also an eigenvector; this fundamental property of eigenvectors fails in any attempt to extend the concept of eigenvector to a nonlinear transformation or to a nonlinear differential equation such as (2.6.1) with $\alpha \neq 0$.

With a clear understanding of the solutions when $\alpha = 0$, it would seem to be a simple matter to choose a reduced problem and a perturbation family. However, it turns out that the "obvious" choices in this problem do not work, and the "correct" perturbation problem does not seem at all reasonable at first sight. It is this which makes the problem interesting.

Since we have assumed $f(0) = 0$, it follows that the identically zero function $y(x) = 0$ is a solution of (2.6.1) for any λ and α, called the *trivial*

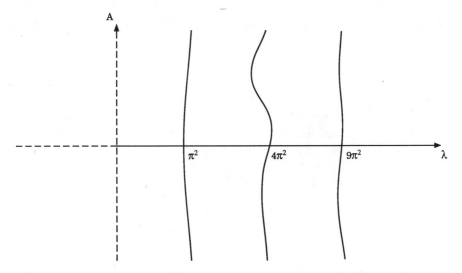

FIG. 2.6.2. A typical graph of the set of values of (λ, A) making (2.6.9) solvable for a fixed small value of ε.

solution. Therefore we will focus on those solutions which are not trivial. One might guess, wrongly, that the way to find such solutions is to set $\lambda = n^2\pi^2$ (for integer values of n), take α as the small parameter (that is, set $\varepsilon := \alpha$), and look for solutions of

$$y'' + n^2\pi^2 y = \varepsilon f(y),$$
$$y(0) = 0, \tag{2.6.3}$$
$$y(1) = 0$$

in the form

$$y \sim A \sin n\pi x + \varepsilon y_1(x) + \varepsilon^2 y_2(x) + \cdots. \tag{2.6.4}$$

In (2.6.4) the leading term $y_0(x)$ has been taken to be a solution of the reduced problem obtained by setting $\varepsilon = 0$ in (2.6.3). Since A is arbitrary in this reduced solution, it seems reasonable to choose A arbitrarily in (2.6.4) and then substitute (2.6.4) into (2.6.3) to obtain equations for y_1, y_2, and so forth. Unfortunately, when this is done, it turns out that the boundary value problems that should be satisfied by the y_i have no solutions (for most functions $f(y)$). See Exercise 2.6.1 for an example.

What went wrong? The answer is that we were too hasty in concluding that λ should be set equal to $n^2\pi^2$. Let us go back to the original problem (2.6.1) in natural parameters and consider more carefully how to introduce ε. "Introducing a perturbation parameter" means making the natural parameters λ and α into functions of ε in such a way that the solution is known when $\varepsilon = 0$. The only known solutions are the eigenfunctions of (2.6.2); this implies that the functions $\alpha(\varepsilon)$ and $\lambda(\varepsilon)$ must be taken to satisfy

$$\alpha(0) = 0$$
$$\lambda(0) = n^2\pi^2$$

for some n. It *does not follow* that $\lambda(\varepsilon)$ should be taken to be *identically* equal to $n^2\pi^2$. In fact, the key to finding a correct formulation is to allow λ to vary with ε; only when the function $\lambda(\varepsilon)$ is chosen correctly do solutions to the problem exist. One feature of our first (incorrect) attempt will be retained: No difficulty results from taking α to be the perturbation parameter, so once again $\alpha(\varepsilon) := \varepsilon$. At this point, then, it would appear that the problem should look something like this:

$$y'' + \lambda(\varepsilon)y = \varepsilon f(y),$$
$$y(0) = 0, \qquad\qquad\qquad (2.6.5)$$
$$y(1) = 0.$$

It is not yet clear what $\lambda(\varepsilon)$ should be, except that $\lambda(0) = n^2\pi^2$. *From here on we will assume that a value of n has been chosen and will remain fixed.* That is, we are looking for the solutions of the perturbed problem lying close to eigenfunctions of the unperturbed problem having a specified eigenvalue. Of course, solutions will exist for every choice of n.

There is one other issue which must be addressed before attempting a final formulation of this problem. We have already seen that the solution $A \sin n\pi x$ of the reduced problem (for a given n) is not unique, and that to make it unique one must specify the amplitude A. In our first attempt at a perturbation problem, namely (2.6.3) and (2.6.4), we chose the value of A in the leading term arbitrarily and hoped to find the higher order terms y_i, but solutions for y_i turned out not to exist. When the problem is changed to (2.6.5) and $\lambda(\varepsilon)$ is chosen correctly, it turns out that the opposite difficulty arises: All of the coefficients y_i in the solution (2.6.4) exist, but *none* of them are unique. It is not enough to specify the amplitude A of the leading term; some stronger condition must be imposed in order to yield a unique solution for each coefficient. Such a condition can be stated in terms of the Fourier sine series for the solution.

Any solution of (2.6.1) is a function $y(x)$ which is defined for $0 \leq x \leq 1$ and satisfies $y(0) = y(1) = 0$. Such a function has a Fourier sine series representation (see Appendix E)

$$y(x) = \sum_{m=1}^{\infty} b_m \sin m\pi x, \qquad\qquad (2.6.6a)$$

with

$$b_m = 2 \int_0^1 y(x) \sin m\pi x \, dx. \qquad\qquad (2.6.6b)$$

The eigenfunctions of the reduced problem are merely $y = A \sin n\pi x$; that is, for these known solutions, each Fourier coefficient b_m with $m \neq n$ vanishes, and the nth coefficient $b_n = A$ is arbitrary. That is, *for the reduced problem, specifying A means specifying the amplitude of the nth harmonic, and doing so makes the solution unique.* We will now simply carry this idea over to the perturbed problem and specify in advance the amplitude of the nth harmonic of the solution that we seek. In other words, fix A (and recall that we have already fixed n) and add to (2.6.5) the requirement that

$$2 \int_0^1 y(x) \sin n\pi x \, dx = A. \tag{2.6.7}$$

This condition turns out to make the solution unique.

It may seem paradoxical that a third condition (2.6.7) can be added to (2.6.5), since a second order differential equation usually admits only two side conditions. The explanation is that $\lambda(\varepsilon)$ has not yet been specified. We will see below that three conditions are exactly what are needed to completely determine both $y(x, \varepsilon)$ and $\lambda(\varepsilon)$.

Let us now assemble the pieces from the above discussion and formulate the precise perturbation problem to be solved. Choose values of n and A, which will remain fixed. This is the same as choosing a point $(n^2\pi^2, A)$ on one of the vertical lines in Fig. 2.6.1, corresponding to a solution $y_0 = A \sin n\pi x$ of (2.6.2). We then seek a function

$$\lambda = \lambda(\varepsilon) = \lambda^{(n)}(A, \varepsilon) \tag{2.6.8}$$

such that $\lambda(0) = n^2\pi^2$ and such that the problem

$$\begin{aligned}
&y'' + \lambda(\varepsilon)y = \varepsilon f(y) \\
&y(0) = 0 \\
&y(1) = 0 \\
&2 \int_0^1 y(x) \sin n\pi x \, dx = A
\end{aligned} \tag{2.6.9}$$

has a unique solution

$$y = y(x, \varepsilon) = \psi(x; A, n; \varepsilon). \tag{2.6.10}$$

To solve this problem by perturbation methods requires finding two functions of ε simultaneously:

$$\begin{aligned}
&\lambda(\varepsilon) = \lambda_0 + \varepsilon\lambda_1 + \mathcal{O}_F(\varepsilon^2), \\
&y(x, \varepsilon) = y_0(x) + \varepsilon y_1(x) + \mathcal{O}_F(\varepsilon^2).
\end{aligned} \tag{2.6.11}$$

(Notice the difference between subscripts and superscripts on λ: $\lambda(\varepsilon) = \lambda^{(n)}(A, \varepsilon) = \lambda_0 + \varepsilon\lambda_1 + \cdots$, so that actually λ_k should be denoted $\lambda_k^{(n)}$.) Substituting these into (2.6.9) yields the following sequence of problems:

$$\begin{aligned}
&y_0''(x) + \lambda_0 y_0(x) = 0 \\
&y_0(0) = 0 \\
&y_0(1) = 0 \\
&2 \int_0^1 y_0(x) \sin n\pi x \, dx = A;
\end{aligned} \tag{2.6.12a}$$

$$y_1''(x) + \lambda_0 y_1(x) = f(y_0(x)) - \lambda_1 y_0(x)$$

$$y_1(0) = 0$$

$$y_1(1) = 0 \tag{2.6.12b}$$

$$2 \int_0^1 y_1(x) \sin n\pi x \, dx = 0.$$

The value of $\lambda_0 = \lambda(0) = \lambda^{(n)}(A, 0)$ has already been selected: $\lambda_0 = n^2 \pi^2$. Then the first three equations of (2.6.12a) imply $y_0 = c \sin n\pi x$ for arbitrary c, and the last equation fixes $c = A$, so that

$$y_0(x) = A \sin n\pi x \tag{2.6.13}$$

as expected.

Proceeding on to (2.6.12b), the first equation can now be written as

$$y_1'' + n^2 \pi^2 y_1 = f(A \sin n\pi x) - \lambda_1 A \sin n\pi x. \tag{2.6.14}$$

Because of the hypothesis that $f(0) = 0$, which was stated initially along with (2.6.1) but has not been used until now, the composite function $f(A \sin n\pi x)$ vanishes at $x = 0$ and $x = 1$ and can be expanded in a Fourier sine series

$$f(A \sin n\pi x) = \sum_{m=1}^{\infty} a_m \sin m\pi x. \tag{2.6.15}$$

For certain functions f, such as $f(y) = y^3$, this expansion can be calculated by trig identities and the series (2.6.15) is finite, while for others, such as $f(y) = y^2$, an infinite series is needed. (See Exercises 2.6.1 and 2.6.2.) Putting (2.6.15) into (2.6.14) yields

$$y_1'' + n^2 \pi^2 y_1 = (a_n - \lambda_1 A) \sin n\pi x + \sum_{m \neq n} a_m \sin m\pi x. \tag{2.6.16}$$

At this point we pause to discuss the inhomogeneous linear boundary value problem

$$y'' + n^2 \pi^2 y = a \sin k\pi x,$$

$$y(0) = 0, \tag{2.6.17}$$

$$y(1) = 0.$$

If $k \neq n$ the general solution of the first equation (obtained by undetermined coefficients) is $y = c_1 \sin n\pi x + c_2 \cos n\pi x + (a/(n^2 - k^2)\pi^2) \sin k\pi x$. The condition $y(0) = 0$ implies $c_2 = 0$, and then $y(1) = 0$ is satisfied automatically, so c_1 is arbitrary. On the other hand, if $k = n$ the general solution is $y = c_1 \sin n\pi x + c_2 \cos n\pi x - (a/2\pi n)x \cos n\pi x$. Again $y(0) = 0$ implies $c_2 = 0$, but this time $y(1) = 0$ gives $a/2\pi n = 0$, in other words, if $k = n$, there are no solutions to (2.6.17) unless $a = 0$.

Since (2.6.16) is linear, the superposition principle applies, and the solution can be formed as a sum of solutions treating the terms on the right hand side independently. There will be no solution which satisfies the boundary conditions of (2.6.12b) unless $a_n - \lambda_1 A = 0$; it is this which determines the value of $\lambda_1 = \lambda_1^{(n)}$:

$$\lambda_1 = a_n/A. \tag{2.6.18}$$

Then the solution of (2.6.16) with $y(0) = y(1) = 0$ is

$$y_1(x) = c_1 \sin n\pi x + \sum_{n \neq m} \frac{a_m}{(n^2 - m^2)\pi^2} \sin m\pi x, \tag{2.6.19}$$

with c_1 arbitrary. Finally, the last condition of (2.6.12b) implies $c_1 = 0$, and

$$y_1(x) = \sum_{m \neq n} \frac{a_m}{(n^2 - m^2)\pi^2} \sin m\pi x. \tag{2.6.20}$$

Putting our results together, the solution of (2.6.9) is

$$\lambda = n^2\pi^2 + \frac{\varepsilon a_n}{A + \mathcal{O}_F(\varepsilon^2)},$$
$$y = A \sin n\pi x + \varepsilon \sum_{m \neq n} \frac{a_m}{(n^2 - m^2)\pi^2} \sin m\pi x + \mathcal{O}_F(\varepsilon^2). \tag{2.6.21}$$

It should now be clear how to continue the solution to higher orders (see Exercise 2.6.1). Notice that λ is only known approximately (to whatever order it is calculated). Since we never know exactly what λ is, we never know exactly what differential equation it is that we are solving!

This completes the formal part of the solution of our nonlinear eigenvalue problem. In the remainder of this section we will do two things: (a) Discuss the meaning of this solution and how it leads to Fig. 2.6.2, and (b) indicate how to make the formal solution rigorous. Both of these steps should help to eliminate any remaining confusion about why the problem should be formulated as in (2.6.8)–(2.6.11) rather than as in (2.6.3)–(2.6.4).

In order to understand Fig. 2.6.2, it is necessary first of all to have a clear conception of the meaning of A in the perturbation problem (2.6.9). Recall that n and A are fixed and that when $\varepsilon = 0$ the solution is $A \sin n\pi x$. Since the perturbation problem is intended to find solutions close to this solution for small nonzero ε, these solutions will *a priori* take the form of the Fourier sine series (2.6.15) with all of the coefficients b_m being small except for $m = n$; the nth coefficient b_n will be close to A. This much is to be expected by the mere idea of perturbation, without making use of the last condition in (2.6.9). According to this condition, A is the amplitude of the nth harmonic in the solution y; in other words, b_n is not merely *close* to A but is exactly *equal* to A.

These expectations are confirmed by (2.6.21): The leading term of the perturbation solution is $A \sin n\pi x$, and the first order correction introduces small amounts of the other harmonics with $m \neq n$ but does not in any way change the nth harmonic. Carrying the computations to higher order continues this pattern. Therefore specifying A means specifying the amplitude of the dominant harmonic in the Fourier sine series for the solution, and the approach represented by (2.6.8)–(2.6.11) is one which obtains the dominant harmonic *exactly* in the first step, with the higher order terms being confined to the other harmonics.

As we have seen, it is necessary to add a third condition to (2.6.5) in order to formulate a well-posed problem (that is, one for which the solution exists and is unique). Our approach has been to specify the amplitude of the dominant harmonic, but this is not the only approach. There are several concepts of the "amplitude" of a Fourier sine series; one is the amplitude of the dominant harmonic, another is the maximum absolute value of the function, and a third is the L^2-norm (which equals the square root of the infinite sum of the squares of the b_m). Any one of these "amplitudes" could be specified. Or instead of an amplitude condition, an initial condition could be used: One could add $y'(0) = \beta$ to (2.6.5), where β is specified instead of A. This approach is carried out in Exercise 2.6.1. We have specified the amplitude of the dominant harmonic because that seems to be the most meaningful (one is not likely to care about the initial condition for a boundary value problem) and also leads to the simplest computations (since in any other approach, each step of the calculation modifies all of the harmonics). The only drawback to our method is that a condition like (2.6.7) is less familiar than an initial or boundary condition, since it is a condition on y that is "spread over" the entire interval $0 \leq x \leq 1$.

Next, consider equation (2.6.8). This equation states that for fixed n and A, changing ε will change the value of λ for which (2.6.9) has a solution. Now for $\varepsilon = 0$, fixing n and A means picking a point on one of the vertical lines in Fig. 2.6.1; notice that when $\varepsilon = 0$, fixing n automatically fixes $\lambda = n^2 \pi^2$. Then subsequently varying ε away from zero causes λ to change while A remains fixed. Thus each point on a vertical line in Fig. 2.6.1 moves horizontally either to the left or to the right, producing (for fixed ε) a picture such as Fig. 2.6.2. That is, the curves in Fig. 2.6.2 represent all points (λ, A) that are obtained from (2.6.8) when ε is fixed (at a small value). For each point (λ, A) on the curve (for fixed ε) passing through $(n^2\pi^2, 0)$, there exists a unique function $y(x)$ defined for $0 \leq x \leq 1$ which satisfies $y'' + \lambda y = \varepsilon f(x)$, $y(0) = 0$, $y(1) = 0$, and has $A \sin n\pi x$ as its dominant harmonic. (Or more precisely: This conjecture is suggested by our formal perturbation analysis. It will be proved rigorously, with minor technical refinements, below.) Notice that the dominant harmonic is $A \sin n\pi x$ and not $A \sin \sqrt{\lambda} x$; these are the same for the unperturbed problem, but not for the perturbed problem (because λ moves).

Figure 2.6.2 shows that if λ is chosen arbitrarily, (2.6.1) may have one solution (the trivial solution) or more than one. Furthermore if a nontrivial solution exists for a given λ and ε, it will usually cease to exist if ε is changed (since the curves in Fig. 2.6.2 will move). In other words, there do not exist families of solutions depending continuously on ε for fixed λ. But if A is fixed rather than λ, there is always a unique solution. This explains graphically why (2.6.3) is incorrectly formulated.

Figure 2.6.2 can be thought of as a bifurcation diagram for (2.6.1), because it shows how the number of solutions of (2.6.1) varies as λ is varied. Several types of bifurcations are illustrated in the figure; $(\pi^2, 0)$ is a pitchfork bifurcation point, $(4\pi^2, 0)$ is a transcritical bifurcation point, and the branch through $(9\pi^2, 0)$ contains a reverse pitchfork and two pair bifurcations. The *bifurcation parameter* in this diagram is λ, not ε as in Chapter 1.

The first step in a rigorous treatment of this problem is to consider the related problem

$$y'' + \lambda y = \varepsilon f(y),$$
$$y(0) = 0, \tag{2.6.22}$$
$$2 \int_0^1 y(x) \sin n\pi x \, dx = A,$$

in which λ is an independent control parameter, not a function of ε, and the second boundary condition is dropped. This problem is not an eigenvalue problem and has the expected number of side conditions for a second order equation; it has a unique solution

$$y = \varphi(x; \lambda, A; \varepsilon). \tag{2.6.23}$$

(This can be proved by a contraction mapping argument similar to the existence proof for initial value problems, but we will accept it without proof.)

For certain combinations of λ, A, and ε, the solution (2.6.23) will satisfy the boundary condition $y(1) = 0$; this happens if and only if

$$\varphi(1; \lambda, A; \varepsilon) = 0. \tag{2.6.24}$$

When $\varepsilon = 0$, this equation reduces to

$$\varphi(1; \lambda, A; 0) = A \sin \sqrt{\lambda} = 0, \tag{2.6.25}$$

since $\varphi(x; \lambda, A; 0) = A \sin \sqrt{\lambda} x$. A solution of (2.6.25) is a point $(\lambda, A) = (n^2\pi^2, A)$ on one of the vertical lines in Fig. 2.6.1. Taking partial derivatives reveals that $\varphi_A(1; n^2\pi^2, A; 0) = 0$ but $\varphi_\lambda(1; n^2\pi^2, A; 0) \neq 0$ as long as $A \neq 0$. By the implicit function theorem, then, it is not possible (in general) to solve (2.6.24) for A as a function of ε for fixed λ, but it is possible to solve for λ as a function of ε with fixed $A \neq 0$. (This is, once again, the reason that our first attempt at a perturbation problem was incorrectly posed.) That is, there exist functions

$$\lambda = \lambda^{(n)}(A, \varepsilon) \tag{2.6.26}$$

such that $\varphi(1; \lambda^{(n)}(A, \varepsilon), A; \varepsilon) = 0$. It follows that (2.6.26) and

$$\psi^{(n)}(x; A; \varepsilon) := \varphi(x; \lambda^{(n)}(A, \varepsilon), A; \varepsilon) \tag{2.6.27}$$

satisfy (2.6.9); this proves the existence of the solutions that we have already computed formally. Since these solutions are smooth, they have asymptotic power series expansions, and it follows in the usual way (by various theorems in Section 1.8) that the procedure in (2.6.12) for calculating these expansions is valid. Since ψ_n satisfies the amplitude condition in (2.6.22), the nth harmonic has exact amplitude A, independently of ε. Therefore the correction terms to all orders will be free of the nth harmonic.

One technical point has been overlooked in this discussion. What has been shown is that each point at a nonzero height A on one of the vertical lines in Fig. 2.6.1 moves when ε is varied, and corresponds to a solution of (2.6.1) for some range of values $|\varepsilon| \leq \varepsilon_0$; however, ε_0 may depend on A and n, and in particular, ε_0 may approach zero either as A approaches the "bad" value $A = 0$ where the implicit function theorem fails, or as A approaches infinity. Therefore there is no guarantee that the entire picture in Fig. 2.6.2 holds for a particular ε. But the uniform implicit function theorem shows that for any compact segment of a vertical line in Fig. 2.6.1 not intersecting the λ-axis there is a uniform range of validity in ε. A separate analysis can be carried out in a neighborhood of the intersection point. This analysis confirms the general picture presented in Fig. 2.6.2 and also determines what type of bifurcation point (pitchfork or transcritical) occurs at the intersection. See Exercise 2.6.4.

Exercises 2.6

1. Letting $f(y) = y^3$, investigate the correct and incorrect perturbation problems as follows.

    (a) Attempt to solve (2.6.3)–(2.6.4) for y_1. Obtain the differential equation for y_1, find its general solution, and show that the constants of integration cannot be chosen to satisfy the boundary conditions.

    (b) Compute the solution of (2.6.9) to second order, that is, $y \cong y_0 + \varepsilon y_1 + \varepsilon^2 y_2$. Hint: Carry out the steps indicated in (2.6.11)–(2.6.21). You will need to find the next set of equations "(2.6.12c)" for y_2.

Use Table E.1 in Appendix E to find the expansion (2.6.15); for the next order, use Table E.1 and equations (E.5).

2. Letting $f(y) = y^2$, compute the solution of (2.6.9) to first order. Use the Fourier coefficient formulas (E.4) to compute (2.6.15). Why can't you use Table E.1?

3. Apply the shooting method to (2.6.1) by carrying out the following steps.

   (a) The initial value problem

   $$y'' + \lambda y = \varepsilon f(y),$$
   $$y(0) = 0,$$
   $$y'(0) = \beta$$

   has a unique solution $y = \varphi(x; \lambda, \beta; \varepsilon)$. Show that for fixed β and n there exists a function $\lambda = \lambda(\varepsilon) = \lambda^{(n)}(\beta, \varepsilon)$ such that $\varphi(1; \lambda^{(n)}(\beta, \varepsilon), \beta; \varepsilon) = 0$. Use this to define a function $\psi^{(n)}(x; \beta; \varepsilon)$ that satisfies (2.6.1). This solution differs from (2.6.25) in that the initial slope, rather than the amplitude of the dominant harmonic, is held constant as ε varies.

   (b) Show how to compute the solution found in part (a) to first order (i.e., $y_0 + \varepsilon y_1$). The principal difference concerns the evaluation of the constant c_1 in (2.6.19). Note: Although the solutions of (2.6.1) found in this problem seem different from those found in the text, they are actually the same solutions, only organized according to their initial slopes β rather than the amplitude of their dominant harmonic. The advantage of the shooting method is that the initial value problem is simpler than (2.6.21). The advantage of the method in the text is that A is a more significant quantity than β.

4. There are three types of points on the graph in Fig. 2.6.1: points where $A \neq 0$; points where $A = 0$ and $\lambda \neq n^2\pi^2$; and points $(\pi^2 n^2, 0)$. In the text we have dealt only with the first type. Discuss the fate of the other points for small ε by carrying out the following steps.

   (a) Show that (2.6.1) admits the solution $y \equiv 0$ for all λ and ε.

   (b) Show that when the starting solution for (2.6.24) is taken to be $(\lambda, A) = (\lambda_0, 0)$ with $\lambda_0 \neq n^2\pi^2$, there is a unique solution $A(\varepsilon)$ for λ constant, and that this leads to the trivial solution found in part (a). Therefore the λ axis remains on the graph in Fig. 2.6.2 and has no bifurcation points other than $(\pi^2 n^2, 0)$. Note that for those points one solves for $A(\varepsilon)$ with fixed λ rather than $\lambda(\varepsilon)$ for fixed A.

   (c) For starting solutions $(n^2\pi^2, 0)$, show that the implicit function theorem cannot be applied to (2.6.24).

   (d) For the sample problem with $f(y) = y^3$, compute the function $\varphi(x; \lambda, A; \varepsilon)$ to first order in ε. Then write out the shooting equation (2.6.24) to first order and apply the methods of Section 1.6 to rescale both λ and A near $(\pi^2, 0)$ so that the implicit function theorem applies. Determine the type of bifurcation point which exists at $(\pi^2, 0)$: Is it a pitchfork (one solution for λ on one side of π^2 and three on the other) or transcritical (two on each side) bifurcation?

2.7* PARTIAL DIFFERENTIAL EQUATIONS

The method of regular perturbations may be applied in partial differential equations when it is known (or suspected) that the solution depends smoothly on the small parameter ε. When the smooth dependence is known, the asymptotic validity of the solution follows in the usual manner.

We will illustrate the technique briefly with the following example of a perturbed Laplace equation:

$$u_{xx} + u_{yy} = \varepsilon u^2. \tag{2.7.1}$$

Here $u = u(x, y)$ is a function of two spatial variables. It seems reasonable to assign the same side conditions to this equation, as would be appropriate when $\varepsilon = 0$. Equation (2.7.1), then, might be posed on a region such as the unit disk in (x, y) space, with a boundary condition such as

$$u(x, y) = f(x, y) \qquad \text{for} \quad x^2 + y^2 = 1, \tag{2.7.2}$$

where f is a smooth function (that is, the restriction to the unit circle of a smooth function defined in a neighborhood of the circle). Regular perturbation theory then proceeds on the assumption that there exists a smooth solution $u(x, y, \varepsilon)$ of (2.7.1) and (2.7.2) which can be expanded in the form

$$u(x, y, \varepsilon) \sim u_0(x, y) + \varepsilon u_1(x, y) + \cdots. \tag{2.7.3}$$

This assumption can in fact be proved for this problem, and the steps to follow are all justifiable on this basis together with the theorems of Section 1.8. Substituting (2.7.3) into (2.7.1) leads to the sequence of problems

$$u_{0xx} + u_{0yy} = 0,$$
$$u_0(x, y) = f(x, y) \qquad \text{for} \quad x^2 + y^2 = 1; \tag{2.7.4a}$$

$$u_{1xx} + u_{1yy} = u_0^2,$$
$$u_1(x, y) = 0 \qquad \text{for} \quad x^2 + y^2 = 1, \tag{2.7.4b}$$

$$\vdots$$

Notice that (2.7.4a) is a homogeneous Laplace equation with nonzero boundary data, whereas (2.7.4b) is an inhomogeneous Laplace equation with known right hand side (after solving (2.7.4a)) and zero boundary data; both of these are linear (as opposed to the original problem, (2.7.1)) and are solvable by elementary means, for instance by separation of variables or by Poisson's kernel (that is, a Green's function). Since the details would take us far from the topic at hand, they are left to the reader (see Exercise 2.7.1).

Exercises 2.7

1. Solve (2.7.4a) and (2.7.4b) for the case in which f is identically 1, by any method from elementary partial differential equations.

2.8. NOTES AND REFERENCES

The subject of regular perturbation theory is so old and so basic that it is seldom written about at any length in the current literature. Therefore many of the details in this chapter were developed "from scratch" with no particular sources in mind. Much of the original work on regular perturbations appears in the guise of finding *convergent* series for solutions of *analytic* differential equations. In fact, throughout this chapter if the function f appearing in the differential equations is analytic, then the solution is analytic (both in the coordinates and the parameters), and it follows that the perturbation series (if carried to infinitely many terms) are convergent.

Regular perturbations appear in the books by Nayfeh under the heading of "straightforward expansions," primarily with the purpose of pointing out when they do *not* work (because the regular expansion is not uniformly valid on the intended domain and the problem calls for one of the methods from Parts II and III of this book). But as this chapter has shown, there are many problems for which the regular expansion is quite satisfactory.

Regular perturbation methods for partial differential equations, with a number of examples, are covered in Section 9.2 of the book by Zauderer (listed in the Preface).

CHAPTER 3

DIRECT ERROR ESTIMATION

3.1. BOUNDARY VALUE PROBLEMS FOR SECOND ORDER DIFFERENTIAL EQUATIONS

Up to this point, the principal method for proving asymptotic validity of perturbation series has been to appeal to Taylor's theorem. Even when the solutions did not take the form of (ordinary) power series, they were reducible to power series by a change of variables, such as $x = x_0 + y\mu$, $\mu = \varepsilon^\nu$ in Section 1.6. This reliance upon Taylor's theorem will continue in Chapter 4, but will not suffice for later chapters. The present chapter is intended as a transition from "elementary" perturbation theory, in which Taylor's theorem plays the central role, to "advanced" perturbation theory, in which error estimates—when they are possible at all—must be established by direct arguments, involving specific features of the problem at hand. This distinction between "elementary" and "advanced" theory roughly corresponds to the distinction between regular and singular perturbations (see Section 1.8). (An exception is the Lindstedt method, treated in Chapter 4, which belongs to singular theory because it produces generalized asymptotic series, but which is nevertheless "elementary" because the error estimates follow from Taylor's theorem.) The purpose of this chapter is to review some of the error estimates obtained in Chapter 2 and derive them again by direct methods in order to develop techniques which will be useful in later chapters. In so doing, some of the previous results will be strengthened. In particular it is sometimes possible to evaluate the constants (c and ε_0) occurring in asymptotic error estimates, and for certain damped systems it is possible to show that estimates established previously for finite time intervals actually hold for all time.

Direct error estimation begins with a given problem and a given (conjectural) approximate solution whose error is to be estimated. The first step is to write the exact solution as the given approximation plus an unknown remainder. By substituting this expression into the original problem, an equation for the remainder can be derived. One does not attempt to solve this equation; if it were possible to solve it, then one would have solved the original equation as well; and if this were possible, there would be no need for perturbation theory. Instead one attempts to derive a bound for the remainder from the equation which it satisfies. This technique has already been encountered in Section 1.4: equation (1.3.8) is the equation satisfied by the remainder, and the estimate derived from this is (1.3.10). The major difference between this chapter and Section 1.4 is that the equation for the remainder here will be a differential or integral equation, and obtaining bounds from this equation will involve differential and integral inequalities.

The first problem to be addressed is the nonoscillatory boundary value problem

$$
\begin{aligned}
y'' - y &= \varepsilon f(y), \\
y(0) &= \alpha, \\
y(1) &= \gamma
\end{aligned}
\tag{3.1.1}
$$

considered in Section 2.5. According to regular perturbation theory, this problem can be solved approximately in the form

$$
y(x, \varepsilon) \cong y_0(x) + \varepsilon y_1(x),
\tag{3.1.2}
$$

where y_0 and y_1 are solutions of

$$
\begin{array}{lll}
y_0'' - y_0 = 0, & y_0(0) = \alpha, & y_0(1) = \gamma, \\
y_1'' - y_1 = f(y_0), & y_1(0) = 0, & y_1(1) = 0.
\end{array}
\tag{3.1.3}
$$

We will prove that (3.1.2) gives an $\mathcal{O}(\varepsilon^2)$ approximation to the solution of (3.1.1), using nothing besides (3.1.1), (3.1.2), and (3.1.3); that is, we take (3.1.2) together with (3.1.3) as a conjectured approximate solution to (3.1.1), and proceed to estimate the error of this approximate solution without making any reference to the way we arrived at (3.1.3) in Chapter 2. To state it more picturesquely, it would not matter if someone fell asleep while struggling with problem (3.1.1) and dreamed of the solution defined by (3.1.2) and (3.1.3); the reasoning carried out below would still provide a complete justification of this solution. It is helpful to remember this when faced with a "sloppy" heuristic derivation of an approximate solution: There is no need to worry about the validity of the derivation if the result can be checked afterwards by direct error estimation. (On the other hand, *good* heuristics often provides a clue as to how the direct error estimation can be accomplished.)

In Chapter 2, it was proved by the shooting method that there exists a solution to (3.1.1). In order to keep the present discussion entirely independent of Chapter 2, we will not assume this, but instead sketch an alternative proof which belongs naturally with the ideas considered here. *If* there is an exact solution to (3.1.1), it can be written in the form

$$y(x, \varepsilon) = y_0(x) + \varepsilon y_1(x) + R(x, \varepsilon), \qquad (3.1.4)$$

where y_0 and y_1 are the solutions of (3.1.3), and where $R(x)$ denotes the remainder or error; y_0 and y_1 can be considered as known, since they can be found easily once f is specified. Substituting (3.1.4) into (3.1.1) leads to the following differential equation for $R(x)$:

$$R'' - R = \varepsilon\{f(y_0 + \varepsilon y_1 + R) - f(y_0)\}. \qquad (3.1.5)$$

The boundary conditions for $R(x)$ are zero because the approximation (3.1.2) satisfies exactly the boundary conditions imposed in (3.1.1). Now (3.1.5) is entirely equivalent to (3.1.1); that is, any solution $y(x)$ of (3.1.1) corresponds to a solution $R(x)$ of (3.1.5) and vice versa. In fact, equation (3.1.4) can be interpreted as a change of variable from y to R, so that (3.1.5) is nothing but (3.1.1) in the new variable. So if we prove that a solution of (3.1.5) exists, we will have shown the existence of a solution to (3.1.1); and if we find a bound for the solution of (3.1.5), we will have found an error bound for the approximation (3.1.2). This is the strategy which will be followed in the remainder of this section.

It is easiest to work with (3.1.5) after converting it to an integral equation. First, observe that the homogeneous problem $R'' - R = 0$ associated with (3.1.5) has two solutions, R_1 and R_2, satisfying $R_1(0) = 1$, $R_1(1) = 0$, $R_2(0) = 0$, and $R_2(1) = 1$; these solutions exist because the problem is nonoscillatory. According to Appendix C, this is just the condition under which a Green's function exists for the boundary value problem (3.1.5); see equations (C.31), (C.32), (C.39), and (C.40). In the present notation, there is a Green's function $G(x, \xi)$ such that any solution $R(x, \varepsilon)$ of (3.1.5) is also a solution of the integral equation

$$R(x, \varepsilon) = \varepsilon \int_0^1 G(x, \xi)\{f(y_0(\xi) + \varepsilon y_1(\xi) + R(\xi, \varepsilon)) - f(y_0(\xi))\}d\xi. \quad (3.1.6)$$

The reader should check that according to (C.38), the Green's function for the present problem is given by

$$G(x, \xi) = \begin{cases} (e^{2-x} - e^x)(e^\xi - e^{-\xi})/2(e^2 - 1) & \text{for } 0 \le \xi \le x \\ (e^{2-\xi} - e^\xi)(e^x - e^{-x})/2(e^2 - 1) & \text{for } x \le \xi \le 1, \end{cases}$$

$$(3.1.7)$$

and that this function is bounded for $0 \le x \le 1, 0 \le \xi \le 1$.

The existence of a unique solution to equation (3.1.6) can be proved by the method of iteration. For the most part, fundamental existence theorems (such as the implicit function theorem or the existence theorem for differential equations) are not proved in this book, but merely quoted from other courses. Since many students may not encounter nonlinear integral equations in other courses, a proof will be sketched here; the details can be filled in by anyone familiar with the iteration technique (or, equivalently, the method of contraction mappings). For a starting approximation, choose the function $R_0(x) = 0$, which satisfies (3.1.6) when $\varepsilon = 0$. Then define the functions $R_n(x, \varepsilon)$ recursively by

$$R_{n+1}(x, \varepsilon) = \varepsilon \int_0^1 G(x, \xi)\{f(y_0 + \varepsilon y_1 + R_n) - f(y_0)\} d\xi, \qquad (3.1.8)$$

where the functions y_0, y_1, and R_n are evaluated at ξ. It can be proved that for small enough ε this sequence of functions converges uniformly to a continuous function $R(x, \varepsilon)$ which satisfies (3.1.6). (The precise restriction on ε is $|\varepsilon| < 1/LA$, where L and A are constants defined below.) Then by differentiating (3.1.6) twice under the integral sign and using the definition (3.1.7) of $G(x, \xi)$ it can be shown that $R(x, \varepsilon)$ is twice differentiable and satisfies (3.1.5). Finally, it has already been remarked that a solution of (3.1.5) implies a solution of (3.1.1).

Knowing that there is a solution to (3.1.6), we proceed to the calculation of an upper bound. Letting A denote an upper bound for $|G(x, \xi)|$, (3.1.6) implies that

$$|R(x, \varepsilon)| \leq \varepsilon A \int_0^1 |f(y_0 + \varepsilon y_1 + R) - f(y_0)| d\xi. \qquad (3.1.9)$$

It is important to realize that (3.1.9) is not already an upper bound for $|R|$, since R occurs on both sides of the inequality sign. Instead, (3.1.9) is a *nonlinear integral inequality* satisfied by the unknown function R. The integrand of (3.1.9) can be bounded using a Lipschitz constant for f, that is, a constant L such that $|f(a) - f(b)| \leq L|a - b|$. According to Appendix F, such a constant exists provided a and b are confined to a compact convex set. In the present instance, both the exact solution $y = y_0 + \varepsilon y_1 + R$ and the leading order approximate solution y_0 are bounded as x ranges over $0 \leq x \leq 1$ and as ε ranges over a fixed closed interval; therefore they remain in some compact convex set, and there exists a Lipschitz constant L such that

$$|f(y_0 + \varepsilon y_1 + R) - f(y_0)| \leq L(\varepsilon|y_1| + |R|). \qquad (3.1.10)$$

Let M be such that

$$L|y_1(x)| \leq M \qquad (3.1.11)$$

for $0 \le x \le 1$. Putting (3.1.9)–(3.1.11) together gives

$$|R(x, \varepsilon)| \le A \int_0^1 \{\varepsilon L |R(\xi, \varepsilon)| + \varepsilon^2 M\} d\xi. \qquad (3.1.12)$$

This is still only an integral inequality for $|R|$, but now it is a *linear integral inequality* because R no longer occurs inside the (presumably nonlinear) function f. The following procedure allows (3.1.12) to be converted into an inequality in which $|R|$ occurs only on the left hand side. Define the function

$$S(\varepsilon) := A \int_0^1 \{\varepsilon L |R(\xi, \varepsilon)| + \varepsilon^2 M\} d\xi. \qquad (3.1.13)$$

Then (3.1.12) says

$$|R(x, \varepsilon)| \le S(\varepsilon) \qquad (3.1.14)$$

for $0 \le x \le 1$. Feeding this back into (3.1.13) in a way which seems almost circular but is not, one finds

$$S(\varepsilon) \le A \int_0^1 \{\varepsilon L S(\varepsilon) + \varepsilon^2 M\} d\xi = \varepsilon ALS(\varepsilon) + \varepsilon^2 AM; \qquad (3.1.15)$$

the integral here can at last be calculated, since the integrand is no longer a function of ξ. It follows that $S(\varepsilon) \le \varepsilon^2 AM/(1 - \varepsilon AL)$, provided $1 - \varepsilon AL > 0$, i.e., $0 \le \varepsilon < \varepsilon_0 := 1/AL$. (This is the same interval of ε for which the iteration argument (3.1.8) is valid.) Therefore, using (3.1.14) one more time,

$$|R(x, \varepsilon)| \le \varepsilon^2 AM/(1 - \varepsilon AL). \qquad (3.1.16)$$

This has the appearance of an $\mathcal{O}(\varepsilon^2)$ estimate, but is not quite, because of the ε in the denominator. This is easy to remedy by taking a slightly smaller interval of ε. Namely, for any ε_1 in the interval $0 < \varepsilon_1 < 1/AL$, let $c := AM/(1 - \varepsilon_1 AL)$; then one has the following theorem.

Theorem 3.1.1. *The error R in the approximate solution (3.1.2) of the boundary value problem (3.1.1) satisfies*

$$|R(x, \varepsilon)| \le c\varepsilon^2 \quad \text{for } 0 \le x \le 1 \text{ and } 0 \le \varepsilon \le \varepsilon_1. \qquad (3.1.17)$$

That is, the error is $\mathcal{O}(\varepsilon^2)$ uniformly for $0 \le x \le 1$. In addition to the big-oh estimate, we have explicit values for c and ε_1, which can be found as actual numbers, provided that A and L can be found. (See Exercise 3.1.2.)

Exercises 3.1

1. Discuss the initial value problem for (3.1.1). Replace the independent variable x by t and the second boundary condition by the initial condition $\dot{y}(0) = \beta$. Assume that the exact solution y (and hence also the remainder R) exists for the time interval $0 \leq t \leq T$. Find the appropriate replacement for (3.1.3) and carry out an analysis parallel to (3.1.4)–(3.1.7) and (3.1.9)–(3.1.17). (Since you are assuming existence, you do not need (3.1.8).) The biggest difference lies in the fact that (C.30) is used in place of (C.40) and that the integral in (C.30) is from 0 to t rather than from 0 to 1. Use the fact that the integral of a positive quantity from 0 to t is less than or equal to the integral from 0 to T for $0 \leq t \leq T$. (Another way of handling these integrals is given in the next section.)

2. Find a numerical upper bound A for $|G|$ using (3.1.7). Assuming $f(y) = y^2$, find M satisfying (3.1.12). From these values, find a suitable pair of numerical constants c and ε_1 for (3.1.17).

3.2. INITIAL VALUE PROBLEMS FOR SECOND ORDER DIFFERENTIAL EQUATIONS

This section concerns the nearly linear second order initial value problem

$$\begin{aligned}
a\ddot{y} + b\dot{y} + cy &= \varepsilon f(y), \\
y(0) &= \alpha, \\
\dot{y}(0) &= \beta
\end{aligned} \tag{3.2.1}$$

for arbitrary constant coefficients $a \neq 0$, b, and c. The right hand side is restricted to $f(y)$ rather than $f(y, \dot{y}, t)$ merely to keep the equations simple; the general case can be handled in the same way. Equation (3.2.1) may be oscillatory or nonoscillatory. The results obtained in this section will repeat those in Section 2.4 and will improve upon them in the case of strongly damped oscillatory problems. (It may be appropriate to remind the reader that for most oscillatory initial value problems better results will be obtained by the methods of Chapters 4 through 6, which make use of generalized asymptotic expansions. At this point we are still estimating the error of the regular perturbation solution, and although we sometimes improve the *error estimate* over what was obtained in Chapter 2, we make no attempt here to improve the *solution* itself.) As in the last section, we will consider the first order approximation

$$y(t, \varepsilon) \cong y_0(t) + \varepsilon y_1(t), \tag{3.2.2}$$

where y_0 and y_1 are solutions of

$$a\ddot{y}_0 + b\dot{y}_0 + cy_0 = 0, \qquad y_0(0) = \alpha, \qquad \dot{y}_0(0) = \beta,$$
$$a\ddot{y}_1 + b\dot{y}_1 + cy_1 = f(y_0), \qquad y_1(0) = 0, \qquad \dot{y}_1(0) = 0. \tag{3.2.3}$$

The solutions $y_0(t)$ and $y_1(t)$ of (3.2.3) exist for all time, since these equations are linear. The exact solution $y(t, \varepsilon)$ of (3.2.1) may not exist for all future time $0 \le t < \infty$, but if not, then it will exist for some interval $0 \le t < T(\varepsilon)$ which depends on ε in such a way that $T(\varepsilon) \to \infty$ as $\varepsilon \to 0$. (See Exercise 3.2.1.) For the same range of t the remainder or error R of the approximation can be defined by

$$y(t, \varepsilon) = y_0(t) + \varepsilon y_1(t) + R(t, \varepsilon). \tag{3.2.4}$$

Just as in the last section, differentiating (3.2.4) and substituting it into (3.2.1) leads to a second order differential equation satisfied by $R(x, e)$:

$$a\ddot{R} + b\dot{R} + cR = \varepsilon\{f(y_0 + \varepsilon y_1 + R) - f(y_0)\},$$
$$R(0) = 0, \tag{3.2.5}$$
$$\dot{R}(0) = 0.$$

Equation (C.30) allows this differential equation to be converted to the integral equation

$$R(t, \varepsilon) = \varepsilon \int_0^t K(t, \tau) \{f(y_0(\tau) + \varepsilon y_1(\tau) + R(\tau, \varepsilon)) - f(y_0(\tau))\} d\tau, \tag{3.2.6}$$

where $K(t, \tau)$ is the integral kernel constructed from the two fundamental solutions of the homogeneous linear equation $a\ddot{y} + b\dot{y} + cy = 0$; see (C.25), (C.26), and Exercise 3.2.1. In particular, K does not depend on ε or on the perturbation f.

In order to change (3.2.6) into an integral inequality, it is necessary to have some bounds on K, y_0, and y_1. Three possibilities arise:

1. K, y_0, and y_1 bounded on a finite interval $0 \le t \le T$;
2. K, y_0, and y_1 bounded for all $t \ge 0$;
3. K exponentially decreasing and y_0 and y_1 bounded for all $t \ge 0$.

Case 1 always holds for any T, since K, y_0, and y_1 are always continuous for $t \ge 0$. Cases 2 and 3 are successively stronger. (It is always possible to check whether case 2 or 3 holds in a given problem, since K, y_0, and y_1 are computable. See Exercises 3.2.2 and 3.2.3.) For the present we will consider only case 1, and therefore assume that

$$|K(t, \tau)| \le A \tag{3.2.7}$$

for $0 \le t, \tau \le T$. On the basis of (3.2.7), then, we can replace (3.2.6) by

$$|R(t, \varepsilon)| \le \varepsilon A \int_0^t |f(y_0 + \varepsilon y_1 + R) - f(y_0)| d\tau \qquad (3.2.8)$$

for $0 \le t \le T$, where the functions inside the integral are all evaluated at τ.

The next step is to write

$$|f(y_0 + \varepsilon y_1 + R) - f(y_0)| \le L|R| + \varepsilon M, \qquad (3.2.9)$$

where L is a Lipschitz constant and M is an upper bound for $L|y_1(t)|$. As long as we confine our attention to case 1 (so that $0 \le t \le T$), there is no difficulty about the existence of L and M. (For the existence of L, the argument is the same as in the last section: Given T, there is some ε_0 such that the exact solution y exists and is continuous for $0 \le t \le T$. Then both y_0 and $y_0 + \varepsilon y_1 + R$ remain bounded for $0 \le t \le T$, so they remain in a compact convex set on which, according to Appendix F, f has a Lipschitz constant.) It follows from (3.2.8) and (3.2.9) that

$$|R(t, \varepsilon)| \le \varepsilon A \int_0^t \{L|R(\tau, \varepsilon)| + \varepsilon M\} d\tau. \qquad (3.2.10)$$

The argument which is about to be applied to (3.2.10) is important and will occur frequently in the rest of this book. The first step is to define a function equal to the right hand side of (3.2.10):

$$S(t, \varepsilon) := \varepsilon A \int_0^t \{L|R(\tau, \varepsilon)| + \varepsilon M\} d\tau. \qquad (3.2.11)$$

Then, of course,

$$|R(t, \varepsilon)| \le S(t, \varepsilon). \qquad (3.2.12)$$

Differentiating (3.2.11) by the fundamental theorem of calculus yields

$$\frac{dS}{dt} = \varepsilon A\{L|R(t, \varepsilon)| + \varepsilon M\}.$$

Using (3.2.12), this becomes a differential inequality:

$$\frac{dS}{dt} \le \varepsilon ALS + \varepsilon^2 AM. \qquad (3.2.13)$$

This first order linear differential inequality can be "solved" (some authors prefer to say "resolved," since the result is not a "solution" but another inequality) in the same way that a first order linear differential equation is solved. That is, subtract εALS from both sides and multiply by the "integrating factor" $\exp(-\varepsilon ALt)$, which is positive and therefore preserves the inequality:

$$\left(\frac{dS}{dt} - \varepsilon ALS\right) e^{-\varepsilon ALt} \leq \varepsilon^2 AMe^{-\varepsilon ALt}.$$

Express the left hand side as a derivative:

$$\frac{d}{dt}(Se^{-\varepsilon ALt}) \leq \varepsilon^2 AMe^{-\varepsilon ALt}.$$

Integrate both sides from 0 to t using the initial condition $S(0) = 0$, which follows immediately from (3.2.11); note that definite integration preserves inequalities (whereas differentiation, for instance, does not),

$$S(t)e^{-\varepsilon ALt} \leq -\frac{\varepsilon M}{L}(e^{-\varepsilon ALt} - 1).$$

Finally, express this result as

$$S(t) \leq \frac{\varepsilon M}{L}(e^{\varepsilon ALt} - 1)$$

and combine it with another use of (3.2.12) to obtain

$$|R(t, \varepsilon)| \leq \varepsilon M(e^{\varepsilon ALt} - 1)/L. \tag{3.2.14}$$

This result is one form of what is called *Gronwall's inequality*, which appears in a great number of variations in different contexts.

In order to interpret (3.2.14), notice first that subtracting 1 from the power series for the exponential function leaves a series whose dominant term is εALt. Therefore the right hand side of (3.2.14) is of order $\mathcal{O}(\varepsilon^2)$ uniformly in the finite interval $0 \leq t \leq T$. We have proved the following theorem.

Theorem 3.2.1. *Given T, there exist c and ε_0 such that the remainder R in the solution (3.2.4) of (3.2.1) satisfies*

$$|R(t, \varepsilon)| \leq c\varepsilon^2 \quad \text{for} \quad 0 \leq t \leq T \quad \text{and} \quad 0 \leq \varepsilon \leq \varepsilon_1. \tag{3.2.15}$$

This theorem has been proved twice before: in Chapter 2, and in Exercise 3.1.1 (which is easily extended to the general case). The method of Exercise 3.1.1 is simpler than the present method, since it does not make any use of Gronwall's inequality. But it is limited to finite intervals, whereas the present method can be extended in some cases. We will now investigate this.

Suppose first that case 2 holds; specifically, that (3.2.7) holds for all $t \geq 0$ and that y_0 and y_1 are bounded for all time. Then most of the steps leading to (3.2.14) can be carried out without restricting to a finite interval. There is one difficulty: The exact solution of (3.2.1) may not remain bounded and may not even exist for all time. In many problems, it can be shown that this difficulty does not arise. (For instance, if $b = 0$ and f depends only on y, not on $\dot{y}$ or t, then it follows from Section 4.3 that all solutions with fixed α and β are bounded, in fact periodic, for sufficiently small ε.) Assuming boundedness of the exact solution, a Lipschitz constant L exists such that (3.2.9) holds for all time, and there are no further obstacles to proving that (3.2.14) holds for all time. However, even in this case the conclusion of Theorem 3.2.1 does not hold for all time. The reason is that the right hand side of (3.2.14) is not of order $\mathcal{O}(\varepsilon^2)$ uniformly for all time; in fact it approaches infinity exponentially as $t \to \infty$. The best that can be said beyond a finite time interval is that since $e^{\varepsilon A L t}$ is bounded for $0 \leq t \leq 1/\varepsilon$, one has

$$|R(t, \varepsilon)| \leq c\varepsilon \qquad \text{for} \quad 0 \leq t \leq 1/\varepsilon \quad \text{and} \quad 0 \leq \varepsilon \leq \varepsilon_1. \qquad (3.2.16)$$

In other words, in case 2 (with an additional boundedness assumption on the exact solution) the first order approximation has the expected order of error uniformly on bounded intervals (according to Theorem 3.2.1) and has an error which is "one order larger" on an *expanding interval* of length $1/\varepsilon$. (This is called an expanding interval because it increases in length as ε decreases, and is therefore "larger than any finite interval" without being infinite. Chapters 4 through 6 will expand upon this subject, although the length of the chapters will remain finite.) Exercise 3.2.4 gives one instance in which (3.2.16) holds, and shows that in general this result cannot be strengthened. Much better methods for solving oscillatory problems on expanding intervals will be given in Chapters 4 through 6.

In case 3, it is possible to prove that the error is $\mathcal{O}(\varepsilon^2)$ uniformly for all $t \geq 0$. Case 3 is characterized by exponential decay of K on the interval $t \geq 0$, together with boundedness of y_0 and y_1 on the same interval. Exponential decay means that there exist positive constants B and h such that

$$|K(t, \tau)| \leq Be^{h(\tau - t)} \qquad \text{for} \quad 0 \leq \tau \leq t. \qquad (3.2.17)$$

For instance, according to Exercise 3.2.2 these conditions hold if $a = 1$, $b = 2h > 0$, and $c = k^2$. In this case (3.2.1) is called a *strongly damped oscillator*. (See Section 2.1.)

The first step in deriving an error estimate for Case 3 is to find suitable constants L and M such that

$$|f(y_0 + \varepsilon y_1 + R) - f(y_0)| \leq L|R| + \varepsilon M. \qquad (3.2.18)$$

The difficulty (as in case 2) is that we do not know *a priori* that the exact solution (and hence also R) is bounded for all time, so there is no obvious compact set on which to base a Lipschitz constant. In case 2 we solved this problem (in a manner of speaking) by assuming the boundedness, but in case 3 it is not necessary to make any additional assumptions. The technique for avoiding a boundedness assumption is called *bootstrapping*, and proceeds as follows. It is known that $R(0, \varepsilon) = 0$; therefore there is an interval of time (beginning at $t = 0$ and possibly depending on ε) over which

$$|R| \leq 1. \tag{3.2.19}$$

We will prove that various statements hold *for as long as* (3.2.19) holds. At the end of the argument it will become clear that in fact (3.2.19) holds for all $t \geq 0$, and therefore all of the statements that depend upon (3.2.19) also hold for all (future) time. The argument superficially appears to be circular ("picking yourself up by your own bootstraps"), but in fact is not.

Fix $\varepsilon_0 > 0$, and let K be the compact interval $|y| \leq D_0 + \varepsilon_0 D_1 + 1$, where D_0 and D_1 are upper bounds for $y_0(t)$ and $y_1(t)$, respectively, for $t \geq 0$. Then as long as (3.2.19) holds, y_0 and $y_0 + \varepsilon y_1 + R$ belong to K. (Precisely: For any fixed ε in the range $0 \leq \varepsilon \leq \varepsilon_0$, $y_0(t)$ and $y_0(t) + \varepsilon y_1(t) + R(t, \varepsilon)$ belong to K for t in any interval $0 \leq t \leq t_0$ on which $|R(t, \varepsilon)| \leq 1$.) Let L be a Lipschitz constant for f on K and let $M = LD_1$. Then

$$|f(y_0 + \varepsilon y_1 + R) - f(y_0)| \leq L|R| + \varepsilon M \tag{3.2.20}$$

for as long as (3.2.19) holds. From (3.2.6), (3.2.17), and (3.2.20) it follows that

$$|R(t, \varepsilon)| \leq \varepsilon \int_0^t Be^{h(\tau - t)} \left(L|R(\tau, \varepsilon)| + \varepsilon M \right) d\tau \tag{3.2.21}$$

for as long as (3.2.19) holds. Using (3.2.19) inside the integral, it follows (since the integral of $e^h(\tau - t)$ is less than 1) that

$$|R(t, \varepsilon)| \leq c_0 \varepsilon \tag{3.2.22}$$

as long as (3.2.19) holds, where $c_0 := B(L + \varepsilon_0 M)$. Now choose ε_1 to be the minimum of ε_0 and $c_0/2$. Then (3.2.22) implies that

$$|R(t, \varepsilon)| \leq \tfrac{1}{2} \tag{3.2.23}$$

for as long as (3.2.19) holds, provided $0 \leq \varepsilon \leq \varepsilon_1$. This is the point at which the "bootstrapping" takes place. Since R begins at zero, if (3.2.23) ever fails, it must fail *before* (3.2.19) fails, since $1/2$ is less than 1. And yet this is exactly what cannot happen, since we have shown that as long as (3.2.19) holds, (3.2.23) cannot fail. Therefore both (3.2.19) and (3.2.23), and all the equations in between, must hold for all $t \geq 0$, provided $0 \leq \varepsilon \leq \varepsilon_1$.

The sharpest bound on R obtained in the last paragraph is the $\mathcal{O}(\varepsilon)$ bound given in (3.2.22), whereas we have claimed a bound of $\mathcal{O}(\varepsilon^2)$. It is not difficult to prove this bound, knowing that (3.2.19)–(3.2.23) hold for all positive time. It is only necessary to use (3.2.22) as a bound for R inside the integral in (3.2.21) and integrate one more time. (This procedure of "jacking up" the order of an error estimate by successive integrations may be called "recycling" and should not be confused with "bootstrapping," which refers to the time-extension argument in the last paragraph.) It follows that

$$|R(t, \varepsilon)| \leq c_1 \varepsilon^2 \qquad (3.2.24)$$

for $t \geq 0$ and $0 \leq \varepsilon \leq \varepsilon_1$ for a suitable c_1. (What is c_1 in terms of the constants already defined?) Notice that the presence of the M term in (3.2.21) prevents us from improving this result any further (by recycling again, that is, substituting (3.2.24) back into (3.2.21) and integrating a third time). We have proved the following theorem.

Theorem 3.2.2. *Suppose it is found (after computation) that the functions $K(t, \tau)$, $y_0(t)$, and $y_1(t)$ associated with the initial value problem (3.2.1) are bounded for all time and that K is exponentially decreasing in the sense that it satisfies (3.2.17). Then the solution of (3.2.1) exists and is bounded for all $t \geq 0$, and there exist c and ε_1 such that the remainder R in the solution (3.2.4) satisfies*

$$|R(t, \varepsilon)| \leq c\varepsilon^2 \qquad \text{for} \quad t \geq 0 \quad \text{and} \quad 0 \leq \varepsilon \leq \varepsilon_1.$$

These results can, of course, be extended to higher order approximations. We will conclude this section with some remarks concerning error estimation, residuals, uniformity, and uniform ordering.

Given an equation expressed in such a way that the right hand side is equal to zero, and given a proposed approximate solution to that equation, the *residual* is the actual value (or function) obtained (instead of zero) when the approximate solution is substituted into the left hand side of the equation. (This concept has already been discussed in Example 1.3.3 and Exercise 1.4.3.) The approximate solution of (3.2.1) defined by (3.2.2) and (3.2.3) has been constructed in such a way that it "satisfies (3.2.1) up to order ε"; that is merely to say that the residual of this solution is formally of order ε^2, or $\mathcal{O}_F(\varepsilon^2)$. It is not difficult to show that in fact the residual is pointwise $\mathcal{O}(\varepsilon^2)$ for $t \geq 0$ and is uniformly of this order on finite intervals. In case 3, the residual is uniformly $\mathcal{O}(\varepsilon^2)$ for $t \geq 0$. These facts may suggest that the order of the residual is the same as the order of the remainder (or error), but this is not always true; it fails in case 2, as shown in Exercise 3.2.4. This question is pursued further in Exercise 3.2.5.

According to Theorem 1.8.1(a), the two-term perturbation series (3.2.2) is uniformly ordered if and only if y_1 is bounded. Therefore in case 1, the approximate solution is uniformly ordered on finite intervals, and in cases 2 and 3 it is uniformly ordered on the entire real line. Therefore case 2 provides examples in which the approximate solution is uniformly ordered but not uniformly asymptotically valid; once again, the example in Exercise 3.2.4 is a specific instance. (Such an example, coming from a realistic problem and not merely an academic curiosity, was promised in Section 1.8 in connection with Theorem 1.8.1.) This serves to emphasize the fact that uniform validity concerns the estimate of the error, whereas uniform ordering concerns only the relative sizes of the terms that are retained in the approximation and makes no reference to the error at all. Despite the fact that some authors treat uniform validity as synonymous with uniform ordering (which they therefore do not define as a separate concept), this is not the case.

Exercises 3.2

1. Show that if the solution of (4.2.1) does not exist for all time $t \geq 0$, then it must exist for an interval $0 \leq t < T(\varepsilon)$, where $T(\varepsilon) \to \infty$ as $\varepsilon \to 0$. Hint: Use Theorem 2.1.1 and the definition of "$\to \infty$." (Recall that one says $f(x) \to \infty$ as $x \to a$ if for every positive real number r, there is a positive real number δ such that $f(x) > r$, provided $|x - a| < \delta$.)

2. In each of the following cases, find an explicit formula for $K(t, \tau)$: $a = 1$, $b = 0$, $c = k^2$; $a = 1$, $b = 0$, $c = -k^2$; $a = 1$, $b = 2h$, $c = k^2$ with $0 < h < k$. Show that in the first case K is bounded for all t and τ; in the second, K is bounded only for finite intervals; and in the third, K is not only bounded but exponentially decaying, in the sense that

$$|K(t, \tau)| \leq Be^{h(\tau-t)}$$

for a suitable constant B. Hint: Find the solutions of the homogeneous linear equation satisfying (C.25) in Appendix C, and use the definition of K immediately following (C.26). The first two cases are easy and simplify to a single term using identities for trig functions and hyperbolic trig functions respectively. The third is a little messier; use the notations suggested in (C.7).

3. Show that when $a = 1$, $b = 2h > 0$, and $c = k^2$, the solutions y_0 and y_1 of (3.2.3) are bounded for all time. Hint: Solve for y_0 explicitly. Since y_0 is bounded, it follows that $f(y_0)$ is bounded. Write the solution for y_1 as an integral using K, use the known bounds for the integrand, and integrate.

4. Consider the problem $\ddot{y} + y = \varepsilon y^2$, $y(0) = \alpha$, $\dot{y}(0) = 0$. The formal solution (3.2.2) for this problem has been computed in Exercise 2.4.3a.

Compute the residual of this approximation and show that it is uniformly $\mathcal{O}(\varepsilon^2)$ for all time. It will be shown later (in Exercise 4.3.3(b)) that the exact solution is periodic but has period different from 2π. Assuming this, show that the remainder defined by (3.2.4) is $\mathcal{O}(1)$ on $t \geq 0$ and that no sharper estimate valid for all time is possible. (Hint: This does not involve any error estimation techniques, merely the comparison of two periodic functions of different periods. Your argument that no sharper estimate is possible should be intuitive, not highly rigorous; a rigorous proof requires the notion of topological density.) Verify that this problem shows that the order of the residual (even a rigorous uniform estimate) is not always the same as the order of the remainder. It also shows that Theorem 3.2.2 cannot be extended to case 2.

5. Suppose that a proposed approximate solution $\tilde{y}$ of (3.2.1) has a residual $\rho(t, \varepsilon) = a d^2\tilde{y}/dt^2 + b d\tilde{y}/dt + c\tilde{y} - \varepsilon f(\tilde{y})$ which is uniformly $\mathcal{O}(\varepsilon^n)$ for all $t \geq 0$. Suppose also that $\tilde{y}$ satisfies the initial conditions exactly.

   (a) Show that the remainder is $\mathcal{O}(\varepsilon^n)$ uniformly on any finite interval $0 \leq t \leq T$.

   (b) Show that the remainder is $\mathcal{O}(\varepsilon^n)$ uniformly for $t \geq 0$ if (3.2.17) holds. (For simplicity you may assume the exact solution exists and is bounded for all time. For a more challenging problem, prove these things at the same time as you prove the error estimate.)

6. Give another proof of Theorem 3.2.1 using the "recycling" idea (not the bootstrapping idea) contained in the proof of Theorem 3.2.2. Hint: It is known that R is bounded by a constant on $0 \leq t \leq T$. Use this in (3.2.10) to obtain first an $\mathcal{O}(\varepsilon)$ bound and then an $\mathcal{O}(\varepsilon^2)$ bound.

3.3. INITIAL VALUE PROBLEMS FOR NEARLY LINEAR SYSTEMS

The method given in the last section is limited to initial value problems for second order scalar differential equations. It is possible to extend it to higher order equations and to systems of equations, but this requires the construction of a suitable integral kernel K for each case. The method to be presented in this section avoids the use of integral equations altogether, making use instead of some elementary inequalities involving vector and matrix norms. The only disadvantage of this method is that it does not apply to boundary value problems, such as those considered in Section 3.1; that is where integral equations, based on Green's functions, find their greatest usefulness.

The second order initial value problem (assuming $a = 1$ for convenience)

$$\ddot{y} + b\dot{y} + cy = \varepsilon f(y, \dot{y}, t, \varepsilon),$$
$$y(0) = \alpha,$$
$$\dot{y}(0) = \beta$$
$$(3.3.1)$$

can be written as a system of two differential equations by introducing the variables $u = y$, $v = \dot{y}$:

$$\dot{u} = v,$$
$$\dot{v} = -cu - bv + \varepsilon f(u, v, t, \varepsilon),$$
$$u(0) = \alpha,$$
$$v(0) = \beta.$$
$$(3.3.2)$$

This is a special case of the general nearly linear system of differential equations

$$\dot{\mathbf{x}} = A\mathbf{x} + \varepsilon \mathbf{f}(\mathbf{x}, t, \varepsilon),$$
$$\mathbf{x}(0) = \alpha,$$
$$(3.3.3)$$

where $\mathbf{x} = (x_1, x_2, \ldots, x_N)$ is an N-dimensional vector, A is an $N \times N$ constant matrix, and $\mathbf{f}$ is an N-vector valued function. (Although the components of $\mathbf{x}$ are often written as an N-tuple separated by commas for typographical convenience, it is important to think of $\mathbf{x}$ as a column vector in (3.3.3) in order for the matrix multiplication to make sense. See Appendix D for a review of systems of differential equations.) Most of this section will be devoted to the general system (3.3.3); at the end, the results will be applied to (3.3.1 and 3.3.2).

First it is necessary to outline the regular perturbation method for (3.3.3), since systems of differential equations were not treated in Chapter 2. One seeks a solution in the form

$$\mathbf{x}(t, \varepsilon) \sim \mathbf{x}_0(t) + \varepsilon \mathbf{x}_1(t) + \cdots$$
$$(3.3.4)$$

and finds, by substituting into (3.3.3) and equating coefficients of powers of ε, that

$$\dot{\mathbf{x}}_0 = A\mathbf{x}_0,$$
$$\mathbf{x}_0(0) = \alpha;$$
$$(3.3.5a)$$

$$\dot{\mathbf{x}}_1 = A\mathbf{x}_1 + \mathbf{f}(\mathbf{x}_0),$$
$$\mathbf{x}_1(0) = \mathbf{0}.$$
$$(3.3.5b)$$

The equation for x_0 is a homogeneous linear system with constant coefficients; the equation for x_1 is inhomogeneous but still linear (since the nonlinear function $\mathbf{f}$ is only evaluated at $x_0(t)$ and so gives a function of t which is known at the time (3.3.5b) is to be solved.) These equations are solvable by elementary procedures, and the solutions can be written (although we will not need these expressions) as

$$\mathbf{x}_0(t) = e^{At}\alpha,$$

$$\mathbf{x}_1(t) = \int_0^t e^{A(t-s)}\mathbf{f}(e^{As}\alpha)ds. \tag{3.3.6}$$

See Exercise 3.3.1 for the derivation of the equation for x_1, and Exercise 3.3.2 for a practical method of solution when A is diagonalizable. Exercise 3.3.3 compares the present method for (3.3.2) with the method of Chapter 2 for (3.3.1). When (3.3.4) is truncated at a given order, the resulting approximate solution has an error which is of the order of the first omitted term uniformly on finite intervals of time. This can be shown in the same manner as in Chapter 2. (That is, the solution of (3.3.3) is smooth, so it has a Taylor series with the desired error estimates. It is only necessary to show that the terms of this Taylor series can be legitimately calculated by the steps just outlined, and this follows from the theorems in Section 1.8.) However, in the spirit of the current chapter we will justify the approximation (to first order) by direct error estimation. Therefore we write the exact solution as

$$\mathbf{x}(t, \varepsilon) = \mathbf{x}_0(t) + \varepsilon \mathbf{x}_1(t) + \mathbf{R}(t, \varepsilon) \tag{3.3.7}$$

and deduce from this an estimate for the Euclidean norm $\|\mathbf{R}(t, \varepsilon)\|$. (The Euclidean norm is defined in equation (F.2).)

As before, the first step is to find a differential equation satisfied by $\mathbf{R}$. Substituting (3.3.7) into (3.3.3) and using (3.3.5), it is seen that

$$\dot{\mathbf{R}} = A\mathbf{R} + \varepsilon\{\mathbf{f}(\mathbf{x}_0 + \varepsilon\mathbf{x}_1 + \mathbf{R}(t, \varepsilon)t, \varepsilon) - \mathbf{f}(\mathbf{x}_0, t, \varepsilon)\},$$
$$\mathbf{R}(0) = \mathbf{0}. \tag{3.3.8}$$

To convert (3.3.8) into a differential inequality satisfied by $\|\mathbf{R}\|$ requires two ideas: the "running away inequality" and the notion of matrix norm.

Suppose that an angry lion is about to be released from the origin of a coordinate system in the plane. Suppose that your position in this coordinate system at time t is $\mathbf{r}(t)$. Your distance from the origin at any time is $\|\mathbf{r}(t)\|$. It is clear that the best way to increase your distance from the origin is to run directly away from the origin, in which case the rate at which your distance from the origin increases is equal to the speed at which you run; if you run in any other direction, your distance from the origin will increase at a rate less than your speed. In symbols,

$$\frac{d}{dt}\|\mathbf{r}(t)\| \le \left\|\frac{d}{dt}\mathbf{r}(t)\right\|, \tag{3.3.9}$$

or "you can't *run away* faster than you can *run.*" The same principle holds in any number of dimensions. One technical proviso is necessary: The real-valued function $\|\mathbf{r}(t)\|$ may not be differentiable at times t for which $\mathbf{r}(t) = 0$, since the function can have a corner at such points; the derivative on the left side of (3.3.9) should be interpreted as a right hand derivative in this case. A rigorous proof of this result will now be sketched; some of the details are left to Exercise 3.3.4.

Lemma 3.3.1 ("Running Away Inequality"). *Let $\mathbf{r}(t)$ be a differentiable vector-valued function of a real variable. Then $\|\mathbf{r}(t)\|$ is differentiable at each point where $\mathbf{r}(t) \ne 0$, and right-differentiable everywhere. Furthermore (3.3.9) holds, with the right hand derivative being used when necessary.*

Proof. The first step is to prove that the right hand derivative of $\|\mathbf{r}(t)\|$ exists. It follows from the triangle inequality (with a little care) that for fixed vectors $\mathbf{r}$ and $\mathbf{s}$,

$$\frac{\|\mathbf{r} + h\mathbf{s}\| - \|\mathbf{r}\|}{h}$$

is a nondecreasing function of h, bounded below by $-\|\mathbf{s}\|$. Therefore the limit of this expression, as h approaches zero from above, exists. It follows that for any t,

$$\lim_{h \to 0^+} \frac{\|\mathbf{r}(t) + h\dot{\mathbf{r}}(t)\| - \|\mathbf{r}(t)\|}{h}$$

exists. The right hand derivative of $\|\mathbf{r}(t)\|$ is by definition

$$\lim_{h \to 0^+} \frac{\|\mathbf{r}(t + h)\| - \|\mathbf{r}(t)\|}{h}.$$

To show that this limit exists (and in fact equals the previous limit), it is enough to show that

$$\lim_{h \to 0^+} \frac{\|\mathbf{r}(t + h)\| - \|\mathbf{r}(t) + h\dot{\mathbf{r}}(t)\|}{h} = 0;$$

but this follows from the triangle inequality and the Taylor expansion of $\mathbf{r}(t + h)$ in h.

Knowing that the right hand derivative exists, the rest of the proof is very simple. From the triangle inequality again,

$$\frac{\|\mathbf{r}(t+h)\| - \|\mathbf{r}(t)\|}{h} \le \left\|\frac{\mathbf{r}(t+h) - \mathbf{r}(t)}{h}\right\|.$$

The result follows by letting h approach zero from above. ∎

For any $N \times N$ matrix A, there exists a nonnegative constant k such that $\|A\mathbf{r}\| \le k\|\mathbf{r}\|$ for all $\mathbf{r}$. Of course, if k is such a constant, any larger constant k' will also work in the inequality. The smallest such constant is called the *operator norm* of A. (There is a smallest k, since the set of k is closed.) The operator norm is not computable by any simple formula from the components of A, but it is not difficult to show that the computable quantity $\sqrt{N}\sum_{ij}|a_{ij}|$ is a suitable value for k, in general larger than the operator norm. (See Exercise 3.3.5.) We will use the notation $\|A\|$ for any such "matrix norm"; the only property that will be used is

$$\|A\mathbf{r}\| \le \|A\|\|\mathbf{r}\|, \tag{3.3.10}$$

although both the operator norm and the computable norm (and other matrix norms) also satisfy $\|A + B\| \le \|A\| + \|B\|$ and $\|cA\| = |c|\|A\|$.

Returning now to (3.3.8), an application of (3.3.8), (3.3.9), and the triangle inequality yield

$$\frac{d}{dt}\|\mathbf{R}\| \le \|A\|\|\mathbf{R}\| + \varepsilon\|\mathbf{f}(\mathbf{x}_0 + \varepsilon\mathbf{x}_1 + \mathbf{R}, t, \varepsilon) - \mathbf{f}(\mathbf{x}_0, t, \varepsilon)\|. \tag{3.3.11}$$

For a finite interval of t, all of the functions involved are bounded, and the rightmost term of (3.3.11) can be estimated as in (3.2.9), so that

$$\frac{d}{dt}\|\mathbf{R}\| \le (\|A\| + \varepsilon L)\|\mathbf{R}\| + \varepsilon^2 M, \tag{3.3.12}$$
$$\mathbf{R}(0) = 0.$$

This can be resolved by a Gronwall argument (see Exercise 3.3.6, and notice that there is no need to introduce a function S as in the last section) to give

$$\|\mathbf{R}(t)\| \le \frac{\varepsilon^2 M}{\|A\| + \varepsilon L}\left(e^{(\|A\| + \varepsilon L)t} - 1\right). \tag{3.3.13}$$

The right hand side is $\mathcal{O}(\varepsilon^2)$ for bounded t (Exercise 3.3.7), and we have justified our regular perturbation method for finite time intervals. If the eigenvalues of A all have negative real parts, and if $\mathbf{f}$ either does not depend on t or its partial derivatives are all bounded in time (so that a Lipschitz constant exists for all time), then the estimate can be proved for all time. (The eigenvalue condition is the equivalent of the damping condition in Section 3.2.)

In applying this result to (3.3.2), we notice that we have gained an unexpected bonus over what was proved in Section 3.2. R is now a two-dimensional vector whose first component is the error in the approximation to u (which equals y) and whose second component is the error in the approximation to v (which equals $\dot{y}$). It is shown in Exercise 3.3.3 that the asymptotic series solution for u given by the first component of (3.3.4) is the same as the asymptotic series solution for y given by (3.2.2), and that the solution for v given by the second component of (3.3.4) is the termwise derivative of the solution for y. Therefore our estimate (3.3.13) not only shows that (3.2.2) is an $\mathcal{O}(\varepsilon^2)$ approximation to y, as we proved in Section 3.2, but also that the derivative of (3.2.2) is an $\mathcal{O}(\varepsilon^2)$ approximation to the derivative of y. This can be proved by the methods of Section 3.2, but only by introducing yet another integral equation satisfied by the error in the approximation of the derivative. It is much more natural to prove the result in the present manner.

Exercises 3.3

1. Derive formula (3.3.6) for x_1. Hint: Multiply both sides of (3.3.5b) by e^{-At} and proceed as if you were solving a first order scalar linear differential equation. Make sure that the steps are valid in the matrix context; this involves writing matrices on the correct side of column vectors, and not "dividing by a matrix."

2. Use the regular perturbation method described by equations (3.3.3)–(3.3.5) to solve the problem

$$
\frac{d}{dt}\begin{bmatrix} x \\ y \\ z \end{bmatrix} = \begin{bmatrix} 1 & -1 & -1 \\ 0 & 1 & 3 \\ 0 & 3 & 1 \end{bmatrix}\begin{bmatrix} x \\ y \\ z \end{bmatrix} + \varepsilon \begin{bmatrix} yz \\ x^2 \\ x \end{bmatrix}.
$$

Hint: Instead of using (3.3.6) to solve (3.3.5), find a matrix S such that $S^{-1}AS$ is diagonal and use this matrix to split each differential equation in (3.3.5) into three scalar equations by a change of coordinates. Be sure to express your solution in the original variables.

3. Prove that if $(u, v) = (u_0 + \varepsilon u_1, v_0 + \varepsilon v_1)$ is the two-term perturbation solution of (3.3.2), and if $y_0 + \varepsilon y_1$ is the two-term perturbation solution of (3.3.1), then $u_0 + \varepsilon u_1 = y_0 + \varepsilon y_1$ and $v_0 + \varepsilon v_1 = \dot{y}_0 + \varepsilon \dot{y}_1$. Hint: Find the differential equations and initial values satisfied by these functions and use the uniqueness of solutions to initial value problems. Do not attempt to find the solutions.

4. Complete the details of the proof of Lemma 3.3.1.

5. Let A be an $N \times N$ matrix and let $\|A\|_{\text{sum}} = \sum_{ij} |a_{ij}|$ and $\|\mathbf{r}\|_{\text{sum}} = \sum_k |r_k|$. Show that $\|A\mathbf{r}\|_{\text{sum}} \leq \|A\|_{\text{sum}} \cdot \|\mathbf{r}\|_{\text{sum}}$. Then let $\|A\| = \sqrt{N}\|A\|_{\text{sum}}$ and $\|\mathbf{r}\| = \sqrt{r_1^2 + \cdots + r_N^2}$ and show that $\|A\mathbf{r}\| \leq \|A\|\|\mathbf{r}\|$. Also show that $\|AB\|_{\text{sum}} \leq \|A\|_{\text{sum}}\|B\|_{\text{sum}}$ and $\|AB\| \leq \|A\|\|B\|$.

6. Carry out the argument leading from (3.3.12) to (3.3.13).

7. Show that the right hand side of (3.3.13) is $\mathcal{O}(\varepsilon^2)$ uniformly for t in compact intervals, provided $\|A\| \neq 0$. If $\|A\| = 0$ (which implies $A = 0$), determine the order of the right hand side on expanding intervals of length $\mathcal{O}(1/\varepsilon)$.

3.4. NOTES AND REFERENCES

The basic techniques of this chapter will be found in any graduate textbook on ordinary differential equations in various disguises, for instance, under the heading of "continuous (or smooth) dependence on parameters" or in conjunction with certain kinds of existence proofs. The use of these techniques (and others) as methods of error estimation is a main theme of Smith's *Singular Perturbation Theory* (listed in the Preface), but the focus there is on error estimation for the perturbation methods developed in Chapters 5 through 8 below. I do not know of a reference that uses these techniques explicitly to do error estimates for regular perturbations as done in this chapter, since regular perturbations can be justified without this (as in Chapter 2). Theorem 3.2.2, giving an error estimate for all future time with strong damping, may possibly be new, although similar theorems for the averaging method with weak damping (having much harder proofs) are known.

PART II

OSCILLATORY PHENOMENA

CHAPTER 4

PERIODIC SOLUTIONS AND LINDSTEDT SERIES

4.1. SECULAR TERMS

This chapter continues the study of nearly linear second order differential equations, focusing on periodic solutions of oscillatory equations of the form

$$\ddot{y} + k^2 y = \varepsilon f(t, y, \dot{y}, \varepsilon) \qquad (4.1.1)$$

where f is either periodic in t or independent of t. Some equations of this form have only periodic solutions, but most have one or more isolated periodic solutions intermixed with nonperiodic solutions which may approach periodic solutions as $t \to \infty$. The method developed in this chapter, known as the Lindstedt method, applies only to the periodic solutions; more general methods given in Chapters 5 and 6 (the methods of multiple scales and averaging) can handle the transient solutions as well. But for the periodic solutions, the Lindstedt method has a distinct advantage over these other methods, both in simplicity and in accuracy. In addition the Lindstedt method provides a good introduction to the more powerful methods. In fact it occupies a middle ground between regular and singular perturbation theory, making this chapter a transitional one in this book. The perturbation series given by the Lindstedt method are generalized asymptotic expansions in the sense of (1.8.16), which places them within singular perturbation theory. On the other hand they are reducible to ordinary asymptotic expansions by a change of variables and can therefore be justified by Taylor's theorem in much the same way as the expansions in Chapter 2.

Historically they were the first generalized asymptotic expansions to be discovered, and this historical role continues to make good expository sense. Periodic solutions are special but important, and continue to occupy a prominent place in current research, for instance in the form of Hopf bifurcations and periodic stages in transition to turbulence in fluid flow.

In Section 2.4 it was shown that initial value problems for (4.1.1) have regular perturbation solutions uniformly valid for finite intervals of time. In order to understand the need for methods such as the Lindstedt method, the method of multiple scales, or the method of averaging, it is necessary to understand why the solutions of initial value problems obtained by the regular perturbation method are not uniformly valid for all time. The clue lies in the usual presence of *secular terms* such as $t \sin t$ or $t^2 \sin t$ in these regular expansions. (Such terms are found in Exercises 2.4.3 and 2.4.4, although they are absent in Exercise 2.4.3(a).) A secular term is a term that becomes unbounded as $t \to \infty$. The word *secular* means "pertaining to an age, a long period of time" and is used in this connection because in the study of planetary motions the time scale is so long that a secular term might not be noticeable over hundreds of years. But eventually any secular term will become unbounded.

It is important to understand in exactly what sense a secular term is "bad." First, if one is interested in approximate solutions over short intervals of time, secular terms do no harm whatever.

Second, there is nothing bad about an approximate solution becoming unbounded, provided that the actual solution becomes unbounded also. Exercise 4.1.1 gives a trivial example in which a secular term is necessary in order to represent the true unbounded nature of the solution. On the other hand if the exact solution is bounded, for instance if it is periodic, any approximation which is unbounded must deviate from the actual solution with unbounded error and hence cannot be uniformly valid for all time. A more general statement about series with secular terms follows from Theorem 1.8.1, parts (a) and (b'). Namely, if a series contains secular terms, its coefficients y_n are not bounded on $0 < t < \infty$; therefore the series is not uniformly ordered on this interval; therefore it is not uniformly valid on this interval. (This applies, somewhat surprisingly, even to the solution $y = y_0 + \varepsilon y_1$ obtained in Exercise 4.1.1. Although this solution has zero error and is thus the best solution possible, it nevertheless is not a uniformly valid asymptotic expansion, because the approximation $y \cong y_0$ does not satisfy $y = y_0 + o(1)$ uniformly for $0 < t < \infty$; recall that in an asymptotic expansion, *each truncation* must satisfy an error estimate. *The solution of this problem simply does not have a uniformly valid asymptotic expansion*, because the exact solution has secular terms.)

To gain insight into the meaning of secular terms, consider the problem

$$\ddot{y} + (1 + \varepsilon)^2 y = 0,$$
$$y(0) = \alpha, \qquad\qquad\qquad (4.1.2)$$
$$\dot{y}(0) = 0.$$

This problem is linear and has the explicit solution

$$y = \alpha \cos(1 + \varepsilon)t. \tag{4.1.3}$$

This solution is periodic with period $T(\varepsilon) = 2\pi/(1 + \varepsilon)$, frequency $\nu(\varepsilon) = 1 + \varepsilon$. If (4.1.3) is expanded in a Taylor series in ε, the result is

$$y = \alpha \cos t - \varepsilon(\alpha t \sin t) - \tfrac{1}{2}\varepsilon^2(\alpha t^2 \cos t) + \mathcal{O}(\varepsilon^3), \tag{4.1.4}$$

with the error estimate $\mathcal{O}(\varepsilon^3)$ holding uniformly on finite intervals $0 \leq t \leq t_1$. The same result is obtained if $y = y_0 + \varepsilon y_1 + \cdots$ is substituted into (4.1.2) to obtain the regular perturbation series. Now (4.1.4) contains secular terms at each stage. Any finite partial sum of (4.1.4) becomes unbounded as $t \rightarrow \infty$, whereas the actual solution (4.1.3) remains bounded. It follows that the error committed by using a finite number of terms of (4.1.4) becomes unbounded as $t \rightarrow \infty$.

The secular terms present in (4.1.4) are called *spurious* or *false secular terms* because the actual solution (4.1.3) does not exhibit secular (unbounded) behavior, being in fact periodic. But of course (4.1.4), the series with secular terms, is nothing but the power series expansion of the periodic solution (4.1.3). This suggests that when a solution is periodic with period depending upon ε, it is not always wise to expand the solution in powers of ε, because the periodicity will be lost. The Lindstedt method, the method of multiple scales, and the method of averaging do not completely expand solutions in powers of ε. These methods single out what part of the dependence on ε is best expressed by expanding in powers and what part is best left unexpanded. This, of course, results in generalized, rather than ordinary, asymptotic expansions.

Section 4.2 describes the idea of Lindstedt expansions of periodic functions. Section 4.3 addresses problems of the form (4.1.1) in which f does not contain t or $\dot{y}$. These problems are conservative, in the sense that they have an "energy" which is constant on solutions. For a certain range of initial conditions, all solutions of these systems are periodic with periods which, in general, depend upon ε, and we show how to construct Lindstedt approximations for these solutions. In Section 4.4, f is allowed to depend upon $\dot{y}$, but still not upon t. These systems are not conservative and may have isolated periodic solutions. We give criteria for the existence of such solutions and construct their Lindstedt expansions. The periodic solutions discussed so far are called *free oscillations*. In Sections 4.5 and 4.6, we consider *forced oscillations*. These are solutions of differential equations that involve time explicitly by way of a periodic forcing term. Again we give criteria for the existence of periodic solutions and a method for expanding the solutions.

A major difference between free and forced oscillations is that the period of a free oscillation is unknown in advance, whereas the period of a forced oscillation must equal the period of the forcing (or an integral multiple of that period).

Exercises 4.1

1. Solve the initial value problem $\ddot{y} + y = \varepsilon \sin t$, $y(0) = 1$, $\dot{y}(0) = 0$ by the regular perturbation method to first order $(y_0 + \varepsilon y_1)$. Show that this in fact gives the exact solution to the problem and thus has zero error, in spite of the fact that y_1 contains a secular term. (An approximation without a secular term could not correctly represent the behavior of the solution.)

4.2. LINDSTEDT EXPANSIONS

Let $\varphi(t, \varepsilon)$ be a smooth function of t and ε defined for all t and for $|\varepsilon| < \varepsilon_0$, and let its kth Taylor polynomial in ε be

$$p_k(t, \varepsilon) = \varphi_0(t) + \varepsilon \varphi_1(t) + \cdots + \varepsilon^k \varphi_k(t). \tag{4.2.1}$$

Theorem A.2 shows that in general such a series satisfies asymptotic estimates uniformly for t in compact intervals, but not for all t. In other words given $[t_0, t_1]$ and a constant ε_1 with $0 < \varepsilon_1 < \varepsilon_0$ there exists $c > 0$ such that

$$|\varphi(t, \varepsilon) - p_k(t, \varepsilon)| \leq c\varepsilon^{k+1} \tag{4.2.2}$$

for $t_0 \leq t \leq t_1$ and $|\varepsilon| \leq \varepsilon_1$, but there is usually no constant c such that (4.2.2) holds for all t. One case in which such a constant does exist occurs if φ is periodic in t with period T independent of ε:

Theorem 4.2.1. *If $\varphi(t + T, \varepsilon) = \varphi(t, \varepsilon)$ for all t, then given $0 < \varepsilon_1 < \varepsilon_0$ there exists $c > 0$ such that (4.2.2) holds for all time t.*

Proof. Each Taylor coefficient

$$\varphi_n(t) = \frac{1}{n!} \frac{\partial^n}{\partial \varepsilon^n} \varphi(t, \varepsilon) \Big|_{\varepsilon=0}$$

is periodic in t with period T, and therefore $p_k(t, \varepsilon)$ is T-periodic as well. There is a constant c such that (4.2.2) holds for $0 \leq t \leq T$ and $|\varepsilon| \leq \varepsilon_1$. But any time t can be expressed in the form $t = t_1 + mT$ for $0 \leq t_1 \leq T$ and some integer m. Then $|\varphi(t, \varepsilon) - p_k(t, \varepsilon)| = |\varphi(t_1 + mT, \varepsilon) - p_k(t_1 + mT, \varepsilon)| = |\varphi(t_1, \varepsilon) - p_k(t_1, \varepsilon)| \leq c\varepsilon^{k+1}$. ∎

The result just obtained applies only when $\varphi(t, \varepsilon)$ has a period which is *independent of ε*. Suppose that the period is a function $T(\varepsilon)$, that is, $\varphi(t + T(\varepsilon), \varepsilon) = \varphi(t, \varepsilon)$ for all t and for $|\varepsilon| < \varepsilon_0$. Associated with the period $T(\varepsilon)$ there is a frequency $\nu(\varepsilon) = 2\pi/T(\varepsilon)$. The idea of the Lindstedt method is to introduce a new time variable such that the given function works out to have a period independent of ε when expressed in the new variable; usually the new constant period is taken to be 2π. Of course, in order to bring this about, the new time variable must depend upon ε. This new variable is called the *strained time* and is denoted

$$t^+ := \nu(\varepsilon)t. \tag{4.2.3a}$$

The function φ expressed in strained time now becomes

$$\psi(t^+, \varepsilon) := \varphi(t^+/\nu(\varepsilon), \varepsilon). \tag{4.2.3b}$$

Another form of the last equation is

$$\varphi(t, \varepsilon) = \psi(\nu(\varepsilon)t, \varepsilon). \tag{4.2.3c}$$

From (4.2.3b) and the periodicity of φ it follows, as expected, that $\psi(t^+ + 2\pi, \varepsilon) = \psi(t^+, \varepsilon)$, and so by Theorem 4.2.1, $\psi(t^+, \varepsilon) = \psi_0(t^+) + \cdots + \psi_k(t^+)\varepsilon^k + \mathcal{O}(\varepsilon^{k+1})$ uniformly for all t^+. Therefore

$$\varphi(t, \varepsilon) = \psi_0(\nu(\varepsilon)t) + \cdots + \psi_k(\nu(\varepsilon)t)\varepsilon^k + \mathcal{O}(\varepsilon^{k+1}) \tag{4.2.4}$$

uniformly for all t. This expansion of φ is not a Taylor expansion, since ε appears not only in the powers but also in the coefficient of each power; it is *incompletely expanded* in ε. Such an expansion is called a *generalized asymptotic power series*, and is a special case of a generalized asymptotic expansion (1.8.16). A generalized power series such as (4.2.4), in which the coefficient of each ε^n is a periodic function with ε entering only through the frequency, is called a *Lindstedt expansion*. In practice these equations are used in the following way: strained time is introduced into a problem by (4.2.3a), allowing the problem to be posed in terms of ψ rather than φ using (4.2.3b). After the problem is solved, (4.2.3c) or (4.2.4) is used to return to the original function and time variable. The expression *strained time* originated with other applications of the same idea; for instance, in elasticity theory *strain* refers to the stretching of a solid, and a coordinate axis embedded in the solid and stretched with it defined a *strained coordinate*. Strained time has the property that when $\varepsilon = 0$ the variable t^+ remains well defined. This contrasts with time variables such as *slow time* $\tau = \varepsilon t$, used in the method of multiple scales (Chapter 5), which are only meaningful for $\varepsilon > 0$.

Two situations arise when trying to find Lindstedt expansions for periodic solutions of differential equations. Sometimes the frequency of the solution is known in advance, and at other times the frequency must be discovered as part of the solution process. Both of these situations will appear in later sections of this chapter. When the frequency $\nu(\varepsilon)$ is known, recursive calculations yield $\psi_0, \psi_1, \ldots, \psi_k$ up to the desired order, and (4.2.4) provides an approximation to $\varphi(t, \varepsilon)$ which is uniformly valid for all time.

When the frequency is not known, the solution procedure calls for finding $\nu(\varepsilon)$ in the form of an expansion $\nu(\varepsilon) = \nu_0 + \varepsilon\nu_1 + \dots$. The coefficients of ν and ψ will be found alternately in the following order: $\nu_0, \psi_0, \nu_1, \psi_1, \nu_2, \dots$. Finding the ψ_n is considerably more difficult than finding the ν_n. Therefore one usually stops calculating after finding $\nu_0, \nu_1, \dots, \nu_{k+1}$ and $\psi_0, \psi_1, \dots, \psi_k$ for some k; that is, one more coefficient is calculated for the frequency than for the solution. The best approximation that can be constructed with this information is

$$\varphi(t, \varepsilon) \cong \tilde{\psi}(\tilde{\nu}(\varepsilon)t, \varepsilon), \tag{4.2.5}$$

where

$$\begin{aligned} \tilde{\psi}(t^+, \varepsilon) &:= \psi_0(t^+) + \varepsilon\psi_1(t^+) + \cdots + \varepsilon^k\psi_k(t^+), \\ \tilde{\nu}(\varepsilon) &:= \nu_0 + \varepsilon\nu_1 + \cdots + \varepsilon^{k+1}\nu_{k+1}. \end{aligned} \tag{4.2.6}$$

This approximation (4.2.5) differs from (4.2.4), containing the same number of terms, in that the frequency $\nu(\varepsilon)$ is replaced by its approximation $\tilde{\nu}(\varepsilon)$. Thus the period of the right hand side of (4.2.5) does not exactly equal that of the left, and the two will drift apart over a long period of time. So we cannot expect (4.2.5) to retain the accuracy $\mathcal{O}(\varepsilon^{k+1})$ for all time as (4.2.4) does.

To analyze this situation more carefully, given $0 < \varepsilon_1 < \varepsilon_0$ there exists $c_0 > 0$ such that

$$|\nu(\varepsilon) - \tilde{\nu}(\varepsilon)| \leq c_0\varepsilon^{k+2}. \tag{4.2.7}$$

Let

$$\ell = \max\left\{\left|\frac{\partial\tilde{\psi}}{\partial t^+}(t^+, \varepsilon)\right| : 0 \leq t^+ \leq 2\pi, \quad |\varepsilon| \leq \varepsilon_1\right\}.$$

This maximum exists since the intervals are compact, and in fact, since $\tilde{\psi}$ is 2π-periodic, ℓ is the maximum over all t^+. It follows from the mean value theorem that

$$\left|\tilde{\psi}(t_1^+, \varepsilon) - \tilde{\psi}(t_2^+, \varepsilon)\right| \leq \ell|t_1^+ - t_2^+| \tag{4.2.8}$$

for all t_1^+ and t_2^+, $|\varepsilon| \leq \varepsilon_1$. (Thus ℓ is a global Lipschitz constant for $\tilde{\psi}$ with respect to t^+.) From (4.2.7) and (4.2.8) we have

$$\left|\tilde{\psi}(\nu(\varepsilon)t, \varepsilon) - \tilde{\psi}(\tilde{\nu}(\varepsilon)t, \varepsilon)\right| \leq c_0\ell\varepsilon^{k+2}|t|. \tag{4.2.9}$$

From (4.2.4) there exists $c_1 > 0$ such that

$$\left|\varphi(t, \varepsilon) - \tilde{\psi}(\nu(\varepsilon)t, \varepsilon)\right| \leq c_1\varepsilon^{k+1} \tag{4.2.10}$$

for all t. From (4.2.9), (4.2.10), and the triangle inequality, we conclude

$$\left|\varphi(t,\varepsilon) - \check{\psi}\big(\check{\nu}(\varepsilon)t,\varepsilon\big)\right| \le c_1\varepsilon^{k+1} + c_0\ell\varepsilon^{k+2}|t|. \tag{4.2.11}$$

This is the error estimate for the best approximation to φ that can be found from a knowledge of $\nu_0,\dots,\nu_{k+1}$ and $\psi_0,\dots,\psi_k$. This error bound grows as $t \to \pm\infty$. (Actually, unlike an approximation having secular terms, the error never really approaches infinity, since there is another bound for the error that "takes over" from (4.2.11) when $|t|$ is large. Namely, since both $\varphi(t,\varepsilon)$ and $\check{\psi}\big(\check{\nu}(\varepsilon)t,\varepsilon\big)$ are periodic, the difference between them can never become greater than the sum of their amplitudes. But this bound is so large as to be useless. There is no bound valid for all time that is small with ε.)

In order to grasp the meaning of (4.2.11), we ask the question: how long does the error remain of order ε^{k+1}? The answer is that if L is a positive constant, then

$$\left|\varphi(t,\varepsilon) - \check{\psi}\big(\check{\nu}(\varepsilon)t,\varepsilon\big)\right| \le (c_1 + c_0\ell L)\varepsilon^{k+1} \qquad \text{for} \quad |\varepsilon| \le \varepsilon_1 \quad \text{and} \quad |t| \le L/\varepsilon. \tag{4.2.12}$$

This is an immediate consequence of (4.2.11). Thus the approximation is uniformly of order ε^{k+1} for $-L/\varepsilon \le t \le L/\varepsilon$. This time interval is called an *expanding interval of order* $\mathcal{O}(1/\varepsilon)$, since as ε decreases, the length of the interval increases. In fact as ε decreases, the estimate (4.2.12) improves in two ways: The bound for the error decreases, and the length of time for which it is valid increases. Of course the complicated constant $c_1 + c_0\ell L$ in (4.2.12) is of no concern to us, since the constants in asymptotic estimates are usually not computable anyway. What matters from the asymptotic point of view is that there is such a constant, and that it depends upon ε_1 and L, but not upon ε.

Suppose $L > 0$ has been chosen. Then the estimate (4.2.12) remains valid until $t = L/\varepsilon$. But the estimate (4.2.11) from which it was derived continues to give information beyond this time. For instance, up to $t = L/\varepsilon^2$ (which is later than L/ε if $\varepsilon < 1$), (4.2.11) gives an error $\le c_1\varepsilon^{k+1} + c_0\ell L\varepsilon^k \le (c_1\varepsilon_0 + c_0\ell L)\varepsilon^k$. Thus the approximation is of order $\mathcal{O}(\varepsilon^k)$ uniformly on the expanding interval $|t| \le L/\varepsilon^2$ of order $\mathcal{O}(1/\varepsilon^2)$. This "trade-off" of accuracy for length of validity continues to longer intervals $L/\varepsilon^3,\dots,L/\varepsilon^{k+1}$. (For length L/ε^{k+2}, the error in (4.2.11) is $\mathcal{O}(1)$, which says nothing since we have already seen that the error is $\mathcal{O}(1)$ for all time merely because φ and its approximation are both periodic.) These results can be summarized in the following theorem.

Theorem 4.2.2. *If $\varphi(t,\varepsilon)$ is smooth, $\nu(\varepsilon)$ is smooth, and*

$$\varphi\left(t + \frac{2\pi}{\nu(\varepsilon)},\varepsilon\right) = \varphi(t,\varepsilon)$$

for all t and ε, and if ψ, ψ̃, and ν̃ are defined by (4.2.3) and (4.2.6), then

$$\varphi(t,\varepsilon) = \tilde{\psi}\big(\nu(\varepsilon)t, \varepsilon\big) + \mathcal{O}(\varepsilon^{k+1}) \qquad \text{uniformly for all } t, \qquad (4.2.13)$$

and

$$\varphi(t,\varepsilon) = \tilde{\psi}\big(\tilde{\nu}(\varepsilon)t, \varepsilon\big) + \mathcal{O}(\varepsilon^{k+1}) \qquad \begin{array}{l}\textit{uniformly on expand-}\\ \textit{ing intervals of length}\\ \mathcal{O}(1/\varepsilon).\end{array} \qquad (4.2.14)$$

Furthermore for $1 \le j \le k + 1$

$$\varphi(t,\varepsilon) = \tilde{\psi}\big(\tilde{\nu}(\varepsilon)t, \varepsilon\big) + \mathcal{O}(\varepsilon^{k+2-j}) \qquad \begin{array}{l}\textit{uniformly on expand-}\\ \textit{ing intervals of length}\\ \mathcal{O}(1/\varepsilon^{j}).\end{array} \qquad (4.2.15)$$

In summary, the essential idea of the Lindstedt method is to approximate a periodic function by another periodic function of the same period or as close to the same period as possible. If the periods are exactly equal, the approximation is uniform for all time. If the periods are slightly different, the approximation is only uniformly valid on an "expanding interval."

4.3. FREE CONSERVATIVE OSCILLATIONS

A differential equation of the form

$$\ddot{y} + F(y) = 0 \qquad (4.3.1)$$

is called *conservative* because the quantity

$$E := E(y, \dot{y}) = \frac{1}{2}\dot{y}^2 + V(y), \qquad \text{where} \quad V(y) := \int_0^y F(\eta)d\eta, \qquad (4.3.2)$$

is conserved (that is, remains constant) along solutions. To see this, let $y(t)$ be a solution of (4.3.1) and check that

$$\frac{d}{dt}E\big(y(t), \dot{y}(t)\big) = \Big(\ddot{y}(t) + F\big(y(t)\big)\Big)\dot{y}(t) = 0. \qquad (4.3.3)$$

The roots of this concept lie in the physical principle of conservation of energy: a particle of unit mass moving on a line with position y in a field with potential $V(y)$ satisfies (4.3.1), and E is the sum of its kinetic and potential energy. The mathematical fact (4.3.3) is not dependent on this interpretation, but the quantity E defined by (4.3.2) will nevertheless be referred to as the *energy*, or the *energy integral*. (A conserved quantity is frequently called an *integral* or *first integral* of the differential equation, because it is usually obtained by integrating and is a first step in the solution of the problem.)

The fact that E is constant along solutions makes it possible, in principle, to solve the initial value problem

$$\ddot{y} + F(y) = 0$$
$$y(0) = \alpha \qquad (4.3.4)$$
$$\dot{y}(0) = \beta$$

in the following way. First calculate the energy $E(\alpha, \beta)$ at the initial point. Then put this value of E, which remains constant for all time, into (4.3.2) to obtain the first order initial value problem

$$\dot{y} = \pm\sqrt{2(E - V(y))},$$
$$y(0) = \alpha. \qquad (4.3.5)$$

This problem is separable and has the solution

$$t = \int_\alpha^y \frac{d\eta}{\pm\sqrt{2(E - V(\eta))}}. \qquad (4.3.6)$$

Care must be taken with the ambiguous sign in these equations; according to (4.3.5), the sign to be chosen must equal the sign of $\dot{y}$. If the initial value β of $\dot{y}$ is positive, the positive sign should be used at first. If the solution reaches a point where $\dot{y} = 0$, it will cross over to $\dot{y} < 0$, and after this (until the next crossing) the negative sign must be used. At such a crossing point $V(y) = E$ and the integral (4.3.6) becomes improper; in addition, the value of y begins to decrease, and (4.3.6) must be understood as a sum of two improper integrals,

$$t = \int_\alpha^{y_1} \frac{d\eta}{+\sqrt{2(E - V(\eta))}} + \int_{y_1}^y \frac{d\eta}{-\sqrt{2(E - V(\eta))}},$$

where y_1 is the crossing point. Here $y < y_1$, so the second integral is positive in spite of the negative sign. To obtain the solution explicitly requires performing the integration and then inverting the result, which has the form $t = t(y)$, to obtain $y = y(t)$. These steps are frequently difficult or impossible. (In one of the simplest cases, the pendulum $\ddot{y} + \sin y = 0$, the integral is an elliptic integral, and the inverse involves the Jacobian elliptic function sn.) At least, the improper integrals will in general be convergent. If the leading term in the expansion of $E - V(\eta)$ at a singularity is of order η, then the integral is like $\int d\eta/\sqrt{\eta}$, which is convergent (for improper integrals of the second kind). See Exercise 4.3.7 for an example.

Since the energy method does not lead to sufficiently simple solutions, it is useful to construct approximate solutions in cases when (4.3.1) is close to a solvable linear system. If $F(y) = k^2 y - \varepsilon f(y)$, then (4.3.1) becomes

$$\ddot{y} + k^2 y = \varepsilon f(y), \tag{4.3.7}$$

which is the special case of (4.1.1) in which f does not involve $\dot{y}$ or t. Although we will not use (4.3.6) to solve these systems, the energy function will still play a role: It will be used to show that there exist many periodic solutions. This in turn suggests that the Lindstedt method is appropriate for these problems.

In order to see how energy can be used to find periodic solutions, we return to the general conservative second order equation (4.3.1) and consider some additional consequences of the energy integral (4.3.2). It is useful to introduce the *phase plane* (u, v) with $u = y$, $v = \dot{y}$, so that (4.3.1) becomes the system of first order equations

$$\begin{aligned} \dot{u} &= v, \\ \dot{v} &= -F(u), \end{aligned} \tag{4.3.8}$$

and $E = E(u, v) = v^2/2 + V(u)$. Since (4.3.8) is autonomous (see Appendix D), the phase plane is filled with curves called *orbits* that do not intersect; each solution $u(t)$, $v(t)$ moves along an orbit, and any two solutions lying on the same orbit differ only by a shift of time. The first equation of (4.3.8) implies that solutions move to the right along orbits in the upper half plane, to the left in the lower. *Rest points* (see Appendix D), also called *equilibrium points* or *singular points*, occur where the right hand side of (4.3.8) vanishes; therefore they occur only on the u axis, at points where $F(u) = 0$. For a conservative system, the orbits are just the *level curves* of E, that is, the curves on which $E(u, v)$ is constant. The curve $E = E_0$ is given by

$$v = \pm\sqrt{2(E_0 - V(u))}. \tag{4.3.9}$$

This is just equation (4.3.5) in new variables. To determine the shape of this curve, draw a horizontal line at E_0 on a graph of $V(u)$ and determine the interval (or intervals) of u on which $V(u) < E_0$. On each such interval (4.3.9) is defined and is symmetrical about the u axis. If the interval is bounded, its endpoints satisfy $V(u) = E_0$; and at such points $v = 0$. In this case the parts of the curve above and below the u axis join at the endpoints to make a single closed curve. If neither endpoint is a rest point, solutions on these closed orbits circulate clockwise and hence are periodic. Figure 4.3.1 shows a typical situation with energy curves drawn for several values of E_0. It is clear that a dip in the graph of $V(u)$, called a *potential well* in physics because of its shape, leads to a region of nested periodic

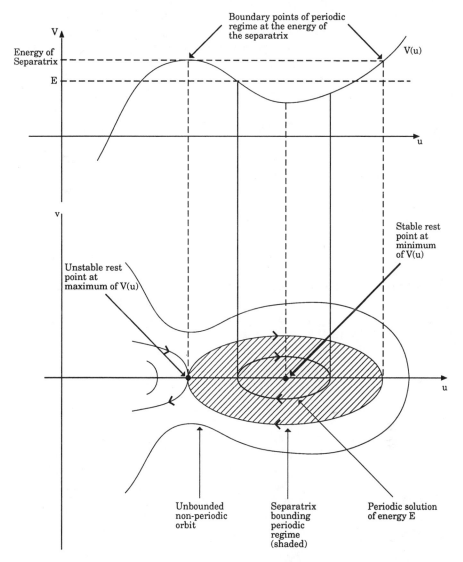

FIG. 4.3.1. A typical phase plane portrait of a conservative system, showing how the rest points, periodic regimes, and separatrices are derived from the graph of the potential energy function $V(u)$.

solutions in the phase plane. The points at which solutions cross the u axis (so that $v = 0$, that is, $\dot{y} = 0$) are just the points at which a sign change must be made in (4.3.5) and the integral in (4.3.6) must be split into a sum of integrals. For periodic solutions these switches occur repeatedly. Notice that the rest points, indicated by dots in Fig. 4.3.1, occur at the critical points of V (points where V has a horizontal tangent): We have already noted that rest points occur on the u axis at points where $F(u) = 0$; but $F = V'$, so these are just the critical points of V. The dark curve in Fig. 4.3.1 is a *separatrix*. It is not a periodic solution, but actually consists of two orbits, one being the rest point at the left end, the other being an orbit which approaches the rest point from below as $t \to +\infty$ and from above as $t \to -\infty$. The separatrix separates the *periodic regime* (shaded) from a regime in which solutions are unbounded. It is also possible for a periodic regime to be bounded by four orbits: two rest points on the u axis, and two orbits connecting them. (See Fig. 4.3.3.)

Every periodic orbit of a conservative equation must intersect the u axis. Let α be such an intersection point. Then the solution (of the original second order equation) with initial conditions $y(0) = \alpha$, $\dot{y}(0) = 0$ will be a periodic solution, and the point $(y(t), \dot{y}(t))$ will trace out the periodic orbit in the phase plane. Since a conservative equation is autonomous and time can be shifted along solutions, all other periodic solutions which trace out the same orbit can be found from this one in the form $y(t - t_0)$, where t_0 is one of the times at which the desired solution crosses through $(\alpha, 0)$. For this reason there is no need, when looking for periodic solutions, to solve initial value problems with initial velocities other than zero. For the periodic regime surrounding the origin in a nearly linear conservative equation, we may further limit ourselves to $\alpha > 0$ since every solution crosses the positive u axis. This simplifies the arguments in the remainder of this section.

Bringing these techniques to bear on the unforced Duffing equation

$$\ddot{y} + y + \varepsilon y^3 = 0 \tag{4.3.10}$$

with potential $V(y) = \frac{1}{2}y^2 + \frac{1}{4}\varepsilon y^4$, there are two cases according to whether $\varepsilon > 0$ or $\varepsilon < 0$. These are illustrated in Figs. 4.3.2 and 4.3.3. If $\varepsilon > 0$, all solutions are periodic except for the rest point at the origin. If $\varepsilon < 0$, there is a large periodic regime around a rest point at the origin. There are two additional rest points at $u = \pm 1/\sqrt{-\varepsilon}$, and four regimes of unbounded solutions. The points $\pm 1/\sqrt{-\varepsilon}$ retreat to $\pm\infty$ as $\varepsilon \to 0$, leaving a larger and larger region of periodic solutions.

Theorem 4.3.1. *For any $\alpha > 0$, the solution $\varphi(t, \alpha, \varepsilon)$ of*

$$\ddot{y} + y + \varepsilon y^3 = 0,$$
$$y(0) = \alpha, \tag{4.3.11}$$
$$\dot{y}(0) = 0$$

is periodic for ε in the interval $-1/\alpha^2 < \varepsilon < \infty$. The period $T(\alpha, \varepsilon)$ is a smooth function of α and ε, and therefore there exist Lindstedt expansions

$$\varphi(t, \alpha, \varepsilon) = \psi_0(\nu(\alpha, \varepsilon)t, \alpha) + \varepsilon\psi_1(\nu(\alpha, \varepsilon)t, \alpha) + \cdots \tag{4.3.12}$$

satisfying error estimates of the type described in Theorem 4.2.2.

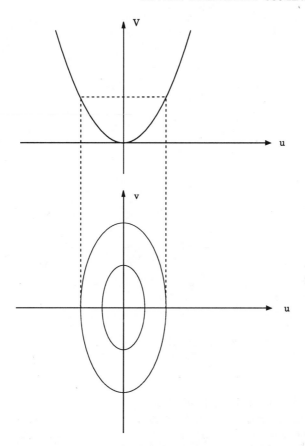

FIG. 4.3.2. The potential energy and phase portrait for Duffing's equation (4.3.10) with $\varepsilon > 0$, called a *hard spring*. The periodic regime is the entire phase plane.

Remark. To clarify the intended error estimates, observe that the Lindstedt expansion (4.3.12) is written using the exact frequency, so it satisfies the error estimate (4.2.13). In fact, however, the exact frequency $\nu(\alpha, \varepsilon)$ is unknown in advance for this problem, and for computational purposes it is necessary to expand ν in powers of ε; the procedure will be developed below. When this is done, (4.2.14) and (4.2.15) will be the appropriate types of error estimates.

Proof. By the graphical arguments above, the solution is periodic for all $\varepsilon \geq 0$, and if $\varepsilon < 0$ it is periodic provided α lies between the rest points at $\pm 1/\sqrt{-\varepsilon}$. That is, the solution is periodic provided $-1/\alpha^2 < \varepsilon < \infty$. To show the smooth dependence of the period on α and ε, we will establish the formula

$$T(\alpha, \varepsilon) = 2 \int_{-\alpha}^{\alpha} \frac{d\eta}{+\sqrt{(\alpha^2 - \eta^2) + \varepsilon(\alpha^4 - \eta^4)/2}}. \tag{4.3.13}$$

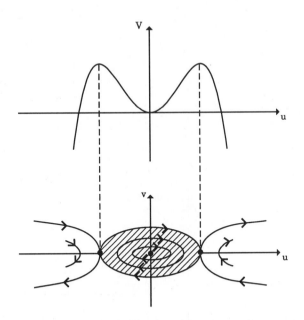

FIG. 4.3.3. The potential energy and phase portrait for Duffing's equation (4.3.10) with $\varepsilon < 0$, called a *soft spring*. The periodic regime is bounded by a separatrix terminating in rest points at $\pm 1/\sqrt{-\varepsilon}$.

To prove (4.3.13), observe from Figs. 4.3.2 and 4.3.3 that the periodic solution through $(\alpha, 0)$ also passes through $(-\alpha, 0)$, and that the period is twice the time of motion from $(-\alpha, 0)$ to $(\alpha, 0)$ in the upper half plane. (The symmetry of the periodic solution about the u axis holds for all conservative systems (4.3.1), but the symmetry about the v axis results from the oddness of y^3 in (4.3.11). See Exercise 4.3.3(c).) The energy at $(\alpha, 0)$ is $\alpha^2/2 + \varepsilon\alpha^4/4$, and (4.3.13) then follows from (4.3.6). The existence of the Lindstedt expansions follows from Theorem 4.2.2. ∎

Theorem 4.3.1 can be generalized to an arbitrary equation of the form (4.3.7), although it is often better to treat each instance separately. The most that can be said in the general case is the following theorem, which will be proved in Exercise 4.4.5d (at the end of the next section).

Theorem 4.3.2. *For sufficiently small ε, (4.3.7) has a rest point at a point $u = u(\varepsilon)$, $v = 0$, near $(0, 0)$; there are no other rest points near the origin. If $f(0) = 0$, this rest point is exactly $(0, 0)$. Around this rest point is a regime of periodic orbits. For any $\alpha > 0$ there exists $\varepsilon_0 > 0$ such that for $|\varepsilon| < \varepsilon_0$ the solution through $(\alpha, 0)$ is periodic with period $T(\alpha, \varepsilon)$ depending smoothly on α and ε, and this solution has a Lindstedt expansion.*

We now turn to the procedures for calculation of the Lindstedt expansion of the solution through $(\alpha, 0)$, which exists according to Theorem 4.3.2. In addition to finding the solution y itself, it is necessary to find the frequency $\nu(\varepsilon) = \nu_0 + \nu_1\varepsilon + \cdots$. For the Duffing equation, one way to find the frequency is to expand the integrand of (4.3.13) in a Taylor series in ε, integrate term by term to find $T = T_0 + \varepsilon T_1 + \cdots$, and use $\nu = 2\pi/T$. (See Exercises 4.3.2 and 4.3.3c.) A similar integral can be written for any conservative problem, and the expanded integrand can often be integrated even though the original integral cannot. However, a more efficient technique is to evaluate $\nu_0, \nu_1, \ldots$ in the course of finding $\psi_0, \psi_1, \ldots$. The procedure resembles the determination of $\lambda_0, \lambda_1, \ldots$ in Section 2.6.

Recall that the Lindstedt expansion (4.2.4) of a periodic function $\varphi(t, \varepsilon)$ becomes an ordinary Taylor expansion of a function $\psi(t^+, \varepsilon)$ with fixed period 2π when the variable is changed to $t^+ = \nu(\varepsilon)t$ as in (4.2.3). Therefore the first step in computing a Lindstedt expansion is to make this change of variable in

$$\frac{d^2y}{dt^2} + k^2 y = \varepsilon f(y),$$
$$y(0) = \alpha, \qquad\qquad (4.3.14)$$
$$\frac{dy}{dt}(0) = 0$$

to obtain

$$\nu(\varepsilon)^2 \frac{d^2y}{(dt^+)^2} + k^2 y = \varepsilon f(y),$$
$$y(0) = \alpha, \qquad\qquad (4.3.15)$$
$$\frac{dy}{dt^+}(0) = 0.$$

Substitute

$$y = y_0 + \varepsilon y_1 + \cdots,$$
$$\nu = \nu_0 + \varepsilon \nu_1 + \cdots \qquad\qquad (4.3.16)$$

into (4.3.15), expanding $f(y)$ in a power series in ε and equating coefficients of equal powers to obtain (letting $' = d/dt^+$)

$$y_0'' + y_0 = 0,$$
$$y_1'' + y_1 = f(y_0)/k^2 - 2\nu_1 y_0''/k,$$
$$y_2'' + y_2 = \{f'(y_0)y_1 - \nu_1^2 y_0'' - 2k\nu_1 y_1''\}/k^2 - 2\nu_2 y_0''/k, \qquad (4.3.17)$$
$$\vdots$$
$$y_n'' + y_n = K_n(t^+) - 2\nu_n y_0''/k,$$

where $K_n(t^+)$ becomes a known function of t^+ as soon as $y_0, \dots, y_{n-1}$ are determined. These equations are to be solved successively, with initial conditions

$$y_0(0) = \alpha y_0'(0) = 0,$$
$$y_1(0) = 0 y_1'(0) = 0,$$
$$\vdots \qquad \qquad \vdots \qquad (4.3.18)$$
$$y_n(0) = 0 y_n'(0) = 0.$$

In addition each $y_n(t^+)$ is required to be a 2π-periodic function of t^+. It is this condition which will be used to determine $\nu_1, \nu_2, \dots$. Just as in Section 2.6, the unexpected number of conditions imposed at each order in ε (three, that is, two initial conditions plus periodicity) balances the number of constants that need to be determined (two constants in the general solution of the differential equation and one coefficient in the expansion of ν). Such a balance is not a guarantee that the equations are solvable at each stage, since it is still possible that the conditions might prove incompatible; the details which follow are needed in order to be sure that everything works. But the balance between the number of conditions imposed and the number of undetermined constants is an indication that we are on the right track.

Since the solution frequency $\nu(\varepsilon)$ should reduce to the frequency k of the unperturbed solutions when $\varepsilon = 0$, ν_0 should be taken to equal k. Then $y_0 = \alpha \cos t^+$, whereupon the equation for y_1 becomes

$$y_1'' + y_1 = f(\alpha \cos t^+)/k^2 + (2\alpha\nu_1/k) \cos t^+. \qquad (4.3.19)$$

The quantity ν_1 on the right is as yet undetermined. However, y_1 must be 2π-periodic, which is impossible if the Fourier series of the right hand side contains the first harmonic ($\sin t^+$ or $\cos t^+$): such a term is resonant and gives rise to a secular term ($t^+ \cos t^+$ or $t^+ \sin t^+$) in the solution. (See Appendix C.) Since the right hand side of (4.3.19) is an even function, there will be no sine terms, and ν_1 must be chosen so as to eliminate the term in $\cos t^+$. That is to say, using the formula for Fourier coefficients,

$$\nu_1 = -\frac{1}{2\pi\alpha k} \int_0^{2\pi} f(\alpha \cos t^+) \cos t^+ \, dt^+. \qquad (4.3.20)$$

(In practice, for simple f the Fourier expansion of $f(\alpha \cos t^+)$ is found by trigonometric identities rather than by the Fourier formula. See Exercise 4.3.1.) With this choice of ν_1, the unique solution for y_1 satisfying the initial conditions will be 2π-periodic and may be inserted into the right hand side of the equation for y_2. The pattern of solution is now clear; it must only be proved that $K_n(t^+)$ will be an even function for every n, so that the equation

$$y_n'' + y_n = K_n(t^+) + (2\alpha\nu_n/k)\cos t^+$$

will have periodic solutions for the proper choice of ν_n. But the exact solution $y = \varphi(t, \alpha, \varepsilon)$ of (4.3.14) is an even function of t, because of the symmetry of the orbits in the $(y, \dot{y})$ phase plane with respect to the y axis. (If $y(t)$ is the solution of (4.3.14), then $y(-t)$ also satisfies (4.3.14), and by the uniqueness of solutions to initial value problems, both solutions coincide.) Therefore the exact solution $y = \psi(t^+, \alpha, \varepsilon)$ of (4.3.15) is also even in t^+, and so is every term in the expansion of $f(y) = f(\psi(t^+, \alpha, \varepsilon))$ in powers of ε.

In summary, the recursive procedure described above has been shown to yield asymptotic expansions to any order for $\varphi(t, \alpha, \varepsilon)$ for any α. For sufficiently small ε these correspond to actual periodic solutions which we have previously shown to exist. In certain cases, such as the Duffing equation for $\varepsilon < 0$, the solution through $(\alpha, 0)$ ceases to be periodic for ε beyond a certain threshold value. For such ε, then, the asymptotic solution bears no relation to the true solution. This underlines the danger in any attempt to argue from the existence of an asymptotic solution with certain properties (such as periodicity) to conclude that the actual solution has those properties.

We have already discussed the error estimates for Lindstedt expansions. Usually, in speaking of the error in a solution such as $\varphi(t, \alpha, \varepsilon)$, what is meant is the difference between the exact and approximate solution for a given value of t. But the phase plane concepts introduced in this section suggest another notion of error, the *orbital error*. Suppose that $y(t, \varepsilon)$ is a family of periodic solutions and that $\tilde{y}(t, \varepsilon)$ is an approximation to this family. The orbit of the exact solution (for a given ε) is the closed curve in the phase plane traced out by $(y(t, \varepsilon), \dot{y}(t, \varepsilon))$. The approximation gives rise to an approximate orbit, the curve traced by $(\tilde{y}, d\tilde{y}/dt)$. The orbital error is a measure of the distance between these curves. (Several ways might be used to define this "distance," and we will not give a precise definition.) Since the orbit contains a record of the positions and velocities of the solution, but not of the time at which these are taken on, the orbital error ignores the phase drift due to the use of $\tilde{\nu}$ in place of ν in a Lindstedt series. Thus the kth order Lindstedt method gives the orbit accurately to order $\mathcal{O}(\varepsilon^{k+1})$ for all time. If we watch $(y, dy/dt)$ and $(\tilde{y}, d\tilde{y}/dt)$ as functions of time, they will trace out their respective orbits for all time, but will gradually drift out of phase with one another. This drift is called *in-track error*, since it is error that occurs "within the track" of the orbit. Thus the total error is a combination of the orbital error and the in-track error. We have not attempted to define these notions exactly, but they are helpful in visualizing the way that a Lindstedt approximation works.

Although we have said that it is sufficient to consider initial conditions with $\dot{y}(0) = 0$ for these problems, the reader may wonder whether the Lindstedt method can be applied directly to $y(0) = \alpha$, $\dot{y}(0) = \beta$. The answer of course is yes, but the solutions are no longer even functions of time, and their Fourier series will in general have both sine and cosine terms. It may then seen surprising that the choice of a single quantity ν_n eliminates both the resonant sine and cosine terms from the equation for y_n. This can be shown to be true (Exercise 4.3.5), and the deeper reason is that although y is no longer even, it still has a "hidden symmetry": The sine and cosine terms are not independent.

Exercises 4.3

1. Compute ν_0, y_0, ν_1, y_1, and ν_2 in the Lindstedt expansion for the solution of Duffing's equation $\ddot{y} + y + \varepsilon y^3 = 0$, $y(0) = \alpha$, $\dot{y}(0) = 0$. Use Table E.1 in place of (4.3.10).

2. Compute the period of the solution of Duffing's equation from the formula (4.3.13) by expanding the integrand in powers of ε and integrating term by term. Then use $T = 2\pi/\nu$ to compare your answers to Problems 1 and 2.

3. Consider the problem $\ddot{y} + y = \varepsilon y^2$, $y(0) = \alpha$, $\dot{y}(0) = 0$ begun in Exercise 2.4.3.

   (a) Draw energy and phase plane portraits as in Fig. 4.3.2 and 4.3.3. Determine conditions on α and ε under which the solutions are periodic.

   (b) Carry out a Lindstedt expansion far enough to show that the solutions do not have period 2π, as you might have conjectured after solving Exercise 2.4.3(c). Compare your solution with the solution of Exercise 2.4.3(e), and observe how the presence of a secular term in y_2 is connected with the fact that y_1 has the wrong period.

   (c) Does there exist a formula like (4.3.13) for the period of the solution through $(\alpha, 0)$? Hints: The orbit is not symmetric about the v axis. The upper and lower limits of integration are two of the three roots of a cubic polynomial and must be expanded in ε by methods from Chapter 1.

(d) Compare part (b) with the solution of Exercise 2.6.2 which involves a similar differential equation but different side conditions. Why does $f(y_0)$ give a finite Fourier series in this problem and an infinite one in Exercise 2.6.2?

4. Draw energy and phase plane portraits for the pendulum equation $\ddot{y} + \sin y = 0$. This is not a perturbation problem, but is an example of (4.3.1). Carry out the solution by energy methods as far as finding the explicit form of (4.3.6). The integral you get cannot be integrated using elementary functions, but is an example of an *elliptic integral* for which an extensive theory exists.

5. Work through the steps from (4.3.14) to (4.3.20) in the case $y(0) = \alpha$, $\dot{y}(0) = \beta$. All of these equations must be modified to include sine terms.

6. The equation $\ddot{y} + \varepsilon y\dot{y} + y = 0$ is not conservative. Nevertheless, all solutions are periodic. To see this, show that when written as a vector field in the phase plane, the vector at $(-u, v)$ is the mirror image (in the v axis) of the vector at (u, v). Therefore any solution starting on the positive v axis reaches the same point on the negative v axis both forward and backward in time, forming a closed orbit. Knowing that all solutions are periodic, use the Lindstedt method to find a leading order approximation to the solution with initial conditions $y(0) = 1$, $\dot{y}(0) = 0$. (Continued in Exercise 4.4.6.)

7. Show that the improper integral (4.3.13) is convergent at both endpoints. Hint: Use the fact that $\int_0^1 dx/\sqrt{x}$ is convergent at 0, and use one of the tests for convergence of improper integrals of the second kind (which may be found in any advanced calculus text).

4.4. FREE SELF-SUSTAINED OSCILLATIONS

The equation

$$\ddot{y} + k^2 y = \varepsilon f(y, \dot{y}) \tag{4.4.1}$$

studied in this section is like that of the last section except that f depends on $\dot{y}$ as well as on y. At first sight, then, this section would seem to include the previous section as a special case. However, the theorems in this section require that f actually depend on $\dot{y}$, that is, that $\partial f/\partial \dot{y} \neq 0$, which excludes the conservative case.

To gain some insight into systems depending on $\dot{y}$, recall that the linear system $\ddot{y} + 2h\dot{y} + k^2 y = 0$ gives damped oscillations if $h > 0$, and oscillations growing in amplitude if $h < 0$. One of the classic equations of the form

(4.4.1), which arises in the theory of electronic circuits, is the *Van der Pol equation*

$$\ddot{y} + \varepsilon(y^2 - 1)\dot{y} + y = 0. \tag{4.4.2}$$

For this equation the perturbation term is a nonlinear expression consisting of a function of y times $\dot{y}$; since this somewhat resembles a damping term, it is commonly placed in the "damping position" in (4.4.2) rather than on the right hand side. Since the reduced problem is a linear oscillator, the solutions ought to be oscillatory. The coefficient of $\dot{y}$ is negative for $|y| < 1$, so one might expect (by analogy with the linear case) that for y in this range the oscillations should be amplified; when $|y| > 1$, they should be damped. Suppose that y is oscillating between, say, approximately ± 5. Then while $-5 \leq y < -1$, and again while $1 < y \leq 5$, the system should experience damping, but during the time when $-1 < y < 1$, it should undergo amplification. The net result—whether the oscillations grow or decline— should depend upon the balance between these opposing influences over the course of an oscillation. Oscillations of large amplitude should diminish, as they spend most of their time outside the interval $-1 < y < 1$, while oscillations spending only a short time, or none at all, outside this interval should grow. In fact for this system there is one orbit for which damping and excitation exactly balance, and which is therefore periodic. Orbits with smaller amplitude grow toward this one, orbits with larger amplitude decay toward it. Such an orbit is called a *limit cycle*. Figure 4.4.1 shows the limit cycle of the Van der Pol equation in the $y, \dot{y}$ phase plane, and Fig. 4.4.2 shows the same periodic solution graphed against time. In these figures ε is fixed at some value which is reasonably small and yet large enough that the limit cycle differs visibly from a circle (to which it of course reduces when $\varepsilon = 0$). The existence of a limit cycle for (4.4.2) for all real values of ε (not necessarily small) can be proved using geometrical arguments that are rather specialized and will not be given here. We will, however, prove the existence of a limit cycle for sufficiently small ε, using perturbation methods, and show how to approximate it by a Lindstedt series. The arguments will be carried out for the general equation (4.4.1).

Oscillations of the type just illustrated arise very frequently and are called *self-excited* or *self-sustained* oscillations because if the system is brought to rest it will return to its limit cycle oscillation spontaneously. The reason is that the rest point at the origin, or equilibrium solution, is unstable, and any deviation from the origin will be quickly amplified. Often an equation of the form (4.4.1) contains an additional parameter λ with the property that for λ less than some critical value λ_0 the origin is stable and there is no limit cycle, but for $\lambda > \lambda_0$ the origin is unstable and is surrounded by a stable limit cycle; this limit cycle is small when λ is close to λ_0, and collapses onto the rest point at the origin as λ approaches λ_0 from above. This situation is called *supercritical Hopf bifurcation* and is illustrated in Exercise 4.4.3. A typical physical example is an ice-coated telephone wire hanging between two telephone poles in a crosswind. For

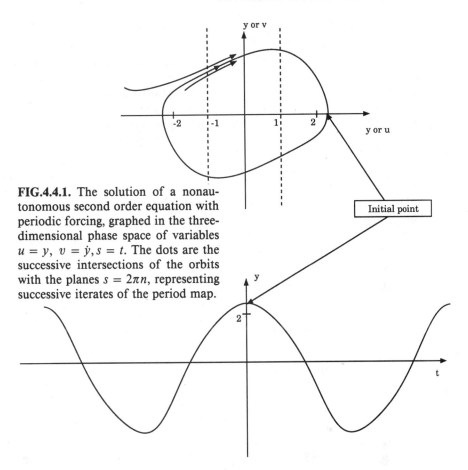

FIG.4.4.1. The solution of a nonautonomous second order equation with
periodic forcing, graphed in the three-
dimensional phase space of variables
$u = y$, $v = \dot{y}$, $s = t$. The dots are the
successive intersections of the orbits
with the planes $s = 2\pi n$, representing
successive iterates of the period map.

FIG. 4.4.2. The same limit cycle as in Fig. 4.4.1, graphed as a *solution* (that is, y
as a function of t) rather than as a phase plane orbit. It is distorted from a simple
harmonic motion of amplitude 2.

low wind velocity λ the normal hanging equilibrium position of the wire
will be stable, but at a certain wind speed λ_0 it loses its stability, and
from there on the stable solution is an oscillation (a limit cycle) about
the equilibrium, with an amplitude which increases with the wind velocity. Sophisticated methods exist to study Hopf bifurcation in complex
systems. Our approach is simply to study (4.4.1) looking for limit cycles. If a specific example contains control parameters, it may happen
that there are limit cycles arising through Hopf bifurcation under variation of these parameters; that fact will appear automatically after our
analysis has been carried out. (Hopf bifurcation does not exactly fit the
definition of bifurcation given in Chapter 1, which was "a change in the
number of solutions of an equation as a parameter is varied, usually by
a splitting or joining of solutions." In Hopf bifurcation, a stable rest

point is thought of as "splitting" not into several rest points but into an unstable rest point and a stable limit cycle.)

Since the solutions of (4.4.1) in general undergo both damping and amplification, it is clear that energy is not conserved, and the method of finding the phase plane portrait given in Section 4.3 cannot be used. We will briefly sketch an approach which enables the main features to be established. The first step is to introduce phase plane coordinates. If these are defined by $u = y$, $v = \dot{y}$ as in Section 4.3, it is easy to see that the orbits of the reduced problem $\varepsilon = 0$ are ellipses; this did not matter in Section 4.3, but for our present purposes it is more convenient for the reduced orbits to be circles. This can be achieved by modifying the definition of phase plane coordinates to $u = y$, $v = \dot{y}/k$. This leads immediately to the phase plane representation

$$\dot{u} = kv,$$
$$\dot{v} = -ku + \frac{\varepsilon}{k} f(u, kv). \tag{4.4.3}$$

Next, introduce polar coordinates (r, θ) into the phase plane by setting $u = r \cos \theta$, $v = r \sin \theta$, giving

$$\dot{r} = \frac{\varepsilon}{k} f(r \cos \theta, kr \sin \theta) \sin \theta,$$
$$\dot{\theta} = -k + \frac{\varepsilon}{kr} f(r \cos \theta, kr \sin \theta) \cos \theta. \tag{4.4.4}$$

Finally, since t does not appear (except in the form dt in $\dot{r}$ and $\dot{\theta}$), these equations can be divided to obtain a single first order equation for the orbits in polar form $r = r(\theta)$:

$$\frac{dr}{d\theta} = \frac{\varepsilon f(r \cos \theta, kr \sin \theta) \sin \theta}{-k^2 + \varepsilon f(r \cos \theta, kr \sin \theta)(\cos \theta)/r}$$
$$= -\frac{\varepsilon}{k^2} f(r \cos \theta, kr \sin \theta) \sin \theta + \mathcal{O}(\varepsilon^2). \tag{4.4.5}$$

Equation (4.4.5) is not equivalent to (4.4.4); the solutions of (4.4.5) do not give the *solutions* of (4.4.4) but only the *orbits*. Even so, (4.4.5) is still too difficult to solve directly and must be attacked by perturbation methods.

It is clear from (4.4.3) that there is a rest point at the origin when $\varepsilon = 0$. The implicit function theorem implies that there is a rest point near the origin for small ε, which always lies on the u axis. (Exercise 4.4.5a.) From (4.4.4), it is seen that $\dot{\theta} < 0$ when $\varepsilon = 0$, so θ is a decreasing function of t and solutions wind clockwise around the origin. This continues to hold for small ε; a precise statement is that given a point $(u, v) = (\alpha, 0)$ with $\alpha > 0$, there exists ε_0 such that for $|\varepsilon| < \varepsilon_0$, the solution of (4.4.4) starting at $r = \alpha$, $\theta = 0$ winds at least once around the rest point near the origin and intersects the positive u axis at some point $r = \hat{\alpha}$, $\theta = -2\pi$, where $\hat{\alpha} = \Phi(\alpha, \varepsilon)$ depends on α and ε. (Exercise 4.4.5b.) Of course $\hat{\alpha}(\alpha, 0) = \alpha$, since when $\varepsilon = 0$ the orbits are circles. In order to compute $\hat{\alpha}$ approximately for small ε, it is sufficient to solve (4.4.5) by the regular perturbation method, which will be valid over the finite interval $-2\pi \leq \theta \leq 0$. (Notice that the equation is to be solved "backwards" from $\theta = 0$ because θ decreases with time.) Substituting $r(\theta) = r_0(\theta) + \varepsilon r_1(\theta) + \mathcal{O}(\varepsilon^2)$ into (4.4.5), with initial conditions $r(0) = \alpha$, gives $r_0(\theta) \equiv \alpha$ and

$$
\frac{dr_1}{d\theta} = \frac{-1}{k^2} f(\alpha \cos \theta, k\alpha \sin \theta) \sin \theta,
$$

$$
r_1(0) = 0.
$$

(4.4.6)

Because the right hand side of (4.4.6) does not depend upon r_1, it can be solved simply by integrating with respect to θ:

$$
\begin{aligned}
r_1(-2\pi) &= -\frac{1}{k^2} \int_0^{-2\pi} f(\alpha \cos \theta, k\alpha \sin \theta) \sin \theta \, d\theta \\
&= +\frac{1}{k^2} \int_{-2\pi}^0 f(\alpha \cos \theta, k\alpha \sin \theta) \sin \theta \, d\theta \\
&= +\frac{1}{k^2} \int_0^{2\pi} f(\alpha \cos \theta, k\alpha \sin \theta) \sin \theta \, d\theta,
\end{aligned}
$$

using the periodicity of the integrand. The next intersection with the positive u axis then occurs at $\hat{\alpha} = r_0(-2\pi) + \varepsilon r_1(-2\pi) + \mathcal{O}(\varepsilon^2)$, that is, at

$$
\hat{\alpha} = \Phi(\alpha, \varepsilon) = \alpha + \frac{\varepsilon}{k^2} \int_0^{2\pi} f(\alpha \cos \theta, k\alpha \sin \theta) \sin \theta \, d\theta + \mathcal{O}(\varepsilon^2). \quad (4.4.7)
$$

The function $\hat{\alpha} = \Phi(\alpha, \varepsilon)$ that we have just computed is called the *first return map* or *Poincaré map* of the orbits with respect to the positive u axis, and we will see that this map contains all of the information needed to determine the existence and stability of limit cycles.

To understand what (4.4.7) says, let

$$
\varphi(\alpha) := \int_0^{2\pi} f(\alpha \cos \theta, k\alpha \sin \theta) \sin \theta \, d\theta. \quad (4.4.8)
$$

If $\varphi(\alpha) < 0$, then it follows from (4.4.7) that $\hat{\alpha} < \alpha$ for sufficiently small positive ε, and the orbit through $(\alpha, 0)$ is being damped (on the average, over the course of an oscillation), while if $\varphi(\alpha) > 0$, it is being amplified. If $\varphi(\alpha_0) = 0$, then the orbit through $(\alpha_0, 0)$ is nearly periodic, since the return point $\hat{\alpha}$ differs from α_0 only by $\mathcal{O}(\varepsilon^2)$. If in addition $\varphi'(\alpha_0) \neq 0$, that is, if α_0 is a simple zero of φ, then the implicit function theorem can be used to show that there is a point very near to α_0 which does return exactly to itself, and therefore gives a limit cycle. (Exercise 4.4.5(c).) But this is not all; for simple zeroes, the Poincaré map not only gives the existence of a limit cycle but also determines its stability. Suppose that $\varphi(\alpha_0) = 0$ and $\varphi'(\alpha_0) < 0$. Then $\varphi(\alpha)$ is negative for α slightly larger than α_0 and positive for α slightly smaller; therefore orbits outside the limit cycle are damped, and orbits inside are amplified. This means that the limit cycle is stable. Conversely, if $\varphi'(\alpha_0) > 0$, the limit cycle is unstable. The arguments given above are only heuristic, since they are based on "ignoring ε^2" in (4.4.7), but Exercise 4.4.5 gives rigorous versions of the existence arguments, and a rigorous proof of the stability criterion will be given in Section 6.3 using the method of averaging. (See the discussion surrounding equation (6.3.21).) The final result is the following theorem.

Theorem 4.4.1. *For every positive simple root $\alpha_0 > 0$ of the determining equation (4.4.8), there exists a function $\alpha(\varepsilon)$ such that $\alpha(0) = \alpha_0$ and for sufficiently small ε, the orbit of (4.4.3) through the point $u = \alpha(\varepsilon)$, $v = 0$ is periodic. Solutions lying on this orbit have a period $T(\varepsilon)$ depending smoothly on ε, and hence have Lindstedt expansions. Furthermore the orbit through $\alpha(\varepsilon)$ is stable for small positive ε if $\varphi'(\alpha_0) < 0$, and unstable if $\varphi'(\alpha_0) > 0$; the opposite is true for negative ε near zero.*

The equation $\varphi(\alpha) = 0$ is known as the *determining equation* for limit cycles. It is often possible to compute the function φ explicitly by integrating (4.4.8). For later use, note that the derivative $\varphi'(\alpha)$ can be expressed as an integral by differentiating (4.4.8) with respect to α under the integral sign.

In the opening paragraph of this section, it was said that the theorems in this section require $\partial f/\partial \dot{y} \neq 0$. This hypothesis does not appear explicitly in Theorem 4.4.1. But if $\partial f/\partial \dot{y} = 0$, the integrand in (4.4.8) reduces to $f(\alpha \cos \theta) \sin \theta$, and the integral vanishes for all α because the integrand is an odd periodic function. Therefore it is impossible for the determining equation to have simple roots in the conservative case. (This is obvious from the geometry of the orbits discussed in Section 4.3, but deserves to be noted analytically as well.)

According to Theorem 4.4.1, periodic solutions of (4.4.1) generally occur in isolation, and the initial condition $(\alpha(\varepsilon), 0)$ for a periodic solution depends upon ε. This is in contrast with the conservative case, where (in a periodic regime) every solution is periodic and the initial condition can be taken as fixed. This fact strongly influences the way in which a Lindstedt expansion is to be sought: In the present case both the frequency $\nu(\varepsilon)$ and the initial condition $y(0) = \alpha(\varepsilon)$ must be expanded in powers of ε with coefficients which are to be determined recursively along with the coefficients of the solution. Since (4.4.1) is autonomous, it is still sufficient to treat only the case $\dot{y}(0) = 0$; all other periodic solutions lying on the same orbit can be found by a shift of time. In constructing Lindstedt expansions for (4.4.1) we will take from Theorem 4.4.1 only the fundamental idea that the initial condition for a periodic solution must depend upon ε, and we will find that the Lindstedt procedure forces us in a natural way to rediscover the determining equation and the requirement that it have simple roots.

Introducing $t^+ = \nu(\varepsilon)t$ into (4.4.1), with $' = d/dt^+$, and taking arbitrary initial conditions having zero velocity, leads to

$$\nu^2 y'' + k^2 y = \varepsilon f\left(y, \nu y'\right),$$
$$y(0) = \alpha, \tag{4.4.9}$$
$$y'(0) = 0. \cdot$$

Into this, substitute

$$y = y_0 + \varepsilon y_1 + \cdots,$$
$$\nu = k + \varepsilon \nu_1 + \cdots, \tag{4.4.10}$$
$$\alpha = \alpha_0 + \varepsilon \alpha_1 + \cdots,$$

taking $\nu_0 = k$ because that is the unperturbed frequency. The result of the substitution is the following sequence of differential equations:

$$k^2(y_0'' + y_0) = 0,$$
$$k^2(y_1'' + y_1) = f(y_0, ky_0') - 2k\nu_1 y_0'',$$
$$k^2(y_2'' + y_2) = \{f_y(y_0, ky_0')y_1 + f_{\dot{y}}(y_0, ky_0')ky_1' - 2k\nu_1 y_1'' - 2k\nu_2 y_0''\}$$
$$\qquad + \{f_{\dot{y}}(y_0, ky_0')\nu_1 y_0' - \nu_1^2 y_0''\}, \tag{4.4.11}$$

$$\vdots$$

$$k^2(y_n'' + y_n) = \{f_y(y_0, ky_0')y_{n-1} + f_{\dot{y}}(y_0, ky_0')ky_{n-1}' - 2k\nu_1 y_{n-1}''$$
$$\qquad -2k\nu_n y_0''\} + h_n,$$

where h_n denotes terms known when $y_1, \ldots, y_{n-1}$ and $\nu_1, \ldots, \nu_{n-1}$ are known. Equations (4.4.11) are to be solved with initial conditions

$$y_n(0) = \alpha_n, \qquad y_n'(0) = 0. \tag{4.4.12}$$

The solutions must also be 2π-periodic in t^+, i.e., $y_n(t^+ + 2\pi) = y_n(t^+)$, in order for the result to be a Lindstedt series. The quantities α_n and ν_n are to be determined in such a way that the solutions of (4.4.11) with initial conditions (4.4.12) are indeed 2π-periodic.

Theorem 4.4.2. *The formal construction of the three series in (4.4.10) is possible in a unique way, provided that α_0 is taken to be a simple zero* *of the determining equation (4.4.8). Thus for each such root (if any exist) there is a Lindstedt series, which corresponds to a periodic solution given by Theorem 4.4.1.*

Proof. The solution of the first equation in (4.4.11) is $y_0 = \alpha_0 \cos t^+$. Inserting this into the equation for y_1 yields

$$k^2(y_1'' + y_1) = f(\alpha_0 \cos t^+, -k\alpha_0 \sin t^+) + 2k\nu_1\alpha_0 \cos t^+. \qquad (4.4.13)$$

Remember that α_0 has not yet been fixed and also ν_1 is unknown. Now (4.4.13) has 2π-periodic solutions only if the right hand side has no first harmonics (resonant terms) when expanded as a Fourier series in t^+. Using the formulas for the coefficients of $\sin t^+$ and $\cos t^+$ in a Fourier series, the requirement of no first harmonics is

$$\int_0^{2\pi} f(\alpha_0 \cos t^+, -k\alpha_0 \sin t^+) \sin t^+ \, dt^+ = 0,$$

$$2k\nu_1\alpha_0 + \frac{1}{\pi} \int_0^{2\pi} f(\alpha_0 \cos t^+, -k\alpha_0 \sin t^+) \cos t^+ \, dt^+ = 0. \qquad (4.4.14)$$

The first equation in (4.4.14) is equivalent to the determining equation (4.4.8), as can be seen by the substitution $\theta = -t^+$. Therefore α_0 must be a root of this equation. (The fact that it must be a *simple* root appears later.) The second equation in (4.4.14) is solvable for ν_1, provided $\alpha_0 \neq 0$. At this point we have found α_0, y_0, and ν_1. (Note that y_0 is not fully determined until α_0 is fixed. If there are several simple zeroes α_0, a separate Lindstedt series must be calculated for each one.)

Equation (4.4.13) is ready to be solved for y_1, now that its resonant terms have been removed. This equation can be written $y_1'' + y_1 = \psi_1(t^+)$, where $\psi_1(t^+)$ is a known function of t^+ lacking the first harmonic. Since this equation is linear, the solution for y_1 with initial conditions (4.4.12) can be written as the sum of a "homogeneous solution" with correct initial conditions and an "inhomogeneous solution" with zero initial conditions; thus:

$$y_1(t^+) = \alpha_1 \cos t^+ + g_1(t^+), \qquad (4.4.15)$$

where $g_1(t^+)$ is the solution of $z'' + z = \psi_1(t^+)$, $z(0) = 0$, $z'(0) = 0$. Of course in any specific problem, with f given, ψ_1 and g_1 can be computed explicitly. (See Exercises 4.4.1–4.4.4.) Working with an arbitrary f, we can only proceed as though $g_1(t^+)$ is a known 2π-periodic function. (No secular terms arise because there are no resonant terms in $\psi_1(t^+)$.) So y_1 is known except for the value of α_1.

At this point $y_1(t^+)$ is substituted into the right hand side of the equation for y_2 in (4.4.11). Instead of discussing this case, we discuss the general case since all steps from here on follow the same pattern. Suppose, then, that the following are known: $\alpha_0, y_0, \nu_1, \alpha_1, y_1, \nu_2, \ldots, \alpha_{n-2}, y_{n-2}, \nu_{n-1}$. Suppose it is also known that

$$y_{n-1} = \alpha_{n-1} \cos t^+ + g_{n-1}(t^+), \tag{4.4.16}$$

where $g_{n-1}(t^+)$ is 2π-periodic and known, but α_{n-1} remains undetermined. Substitute (4.4.16) and all of the lower-order known functions and constants into the right hand side of the equation for y_n in (4.4.11). Observe that $h_n(t^+)$ is completely determined in this process since it depends only on fully known data. The result of the substitution may be written

$$k^2(y_n'' + y_n) = \{ f_y \cdot \alpha_{n-1} \cos t^+ - f_{\dot y} \cdot k\alpha_{n-1} \sin t^+ + 2k\nu_1\alpha_{n-1} \cos t^+ $$
$$+ 2k\nu_n\alpha_0 \cos t^+ \} + H_n(t^+), \tag{4.4.17}$$

where $f_y = f_y(\alpha_0 \cos t^+, -k\alpha_0 \sin t^+)$, $f_{\dot y} = f_{\dot y}(\alpha_0 \cos t^+, -k\alpha_0 \sin t^+)$, and where $H_n(t^+)$ is a new known function consisting of $h_n(t^+)$ plus terms involving $g_{n-1}(t^+)$. In order to show that the nth step in the recursive process is solvable, we must show that α_{n-1} and ν_n can be determined so that the right hand side of (4.4.17) has no first harmonic. This imposes two conditions:

$$\alpha_{n-1} \int_0^{2\pi} \{ f_y \cdot \cos t^+ - k f_{\dot y} \cdot \sin t^+ \} \sin t^+ \, dt^+ + \int_0^{2\pi} H_n(t^+) \sin t^+ \, dt^+ = 0 \tag{4.4.18a}$$

and

$$\nu_n(2k\alpha_0) + \int_0^{2\pi} [\ldots] \cos t^+ \, dt^+ = 0, \tag{4.4.18b}$$

where $[\ldots]$ stands for various terms whose details are unimportant, except that they do not involve ν_n. Equation (4.4.18a) can be solved for α_{n-1}, provided

$$\int_0^{2\pi} \{ f_y(\alpha_0 \cos t^+, -k\alpha_0 \sin t^+) \cos t^+ $$
$$- k f_{\dot y}(\alpha_0 \cos t^+, -k\alpha_0 \sin t^+) \sin t^+ \} \sin t^+ \, dt^+ \neq 0. \tag{4.4.19}$$

We will return to condition (4.4.19) in a moment. Equation (4.4.18b) can be solved for ν_n since $\alpha_0 \neq 0$. With these choices, (4.4.17) becomes

$$y_n'' + y_n = \psi_n(t^+) \tag{4.4.20}$$

with no resonant terms, having the solution

$$y_n(t^+) = \alpha_n \cos t^+ + g_n(t^+), \tag{4.4.21}$$

without secular terms. We have found α_{n-1}, y_{n-1}, and ν_n (as before, y_{n-1} is not completely determined until α_{n-1} is found), and have completed the nth stage in the recursion.

The first two stages in this process are quite different from the general stage, which occurs from $n = 2$ onward. The solution for y_0 required no hypotheses. The solution for y_1 required that α_0 satisfy the determining equation. The solution for y_2 and all further stages requires that α_0 also satisfy (4.4.19). In fact, condition (4.4.19) is nothing other than the requirement that α_0 be a *simple* zero of the determining equation. To see this, put $\theta = -t^+$ in (4.4.8) and differentiate with respect to α_0. Thus the complete set of hypotheses necessary to construct the Lindstedt expansion coincides with the set of hypotheses in Theorem 4.4.1, as expected. ∎

The technique of finding the Lindstedt expansion for a limit cycle will only become clear by actually following through the steps for one or two specific equations of the form (4.4.1). Of course, in doing so it is important to work through each example according to the pattern of the general case, rather than merely substituting into the equations derived in the above discussion. The necessary experience will be provided by Exercises 4.4.1–4.4.4, provided these are used thoughtfully as learning tools and not merely as rote exercises.

The techniques of proof and computation used in this section can be compared with the shooting method as used in Sections 2.5 and 2.6. Periodicity with period T, for a second order autonomous differential equation, is equivalent to the boundary conditions $y(0) = y(T)$ and $y'(0) = y'(T)$, so the search for periodic solutions may be regarded as a kind of boundary value problem. One difference is that the value of T can depend on ε; this problem is handled either by eliminating t as an independent variable and replacing it by θ as in (4.4.5), or else by expanding $\nu = 2\pi/T$ in powers of ε as in (4.4.10). Once this is settled, the idea is to "shoot" from given initial conditions $y(0) = \alpha$, $\dot{y}(0) = 0$ (or $r = \alpha$, $\theta = 0$), trying to hit the same point at the next intersection with the positive u (that is, y) axis. It is here that the greatest difference between this problem and those of Sections 2.5 arises. In Section 2.5 the boundary value problem had a unique solution for $\varepsilon = 0$, and the implicit function theorem could be used to extend this solution to small ε. In the present section, every starting point α returns to itself when $\varepsilon = 0$, and the reduced problem gives no clue as to which initial conditions give rise to periodic solutions which continue for $\varepsilon = 0$. The implicit function theorem cannot be applied to the reduced problem, because it is degenerate, in fact degenerate in the extreme sense that the derivative of the Poincaré map with respect to α is zero everywhere. In order to find suitable candidates for continuable periodic solutions, it is necessary to begin not with the solution of the reduced problem (the zeroth approximation), but with the first order perturbation solution (see (4.4.6) and (4.4.7)). Once a periodic solution has "survived" the first approximation, it has the possibility of being "nondegenerate at the next level" (that is, of being a simple root of the determining equation); in this case it will survive as a periodic solution of the exact system. Notice that the starting value α_0 for the initial condition for such a solution depends upon the function f, as can be seen from the determining equation (4.4.8). This is already apparent from our heuristic discussion of damping and amplification in the Van der Pol equation, which clearly depends on the form of the Van der Pol term. In Section 2.5, on the contrary, the starting value β_0 of the (second) initial condition for a solution of the boundary value problem is determined by (2.5.14), which does not involve the function f.

In the terminology to be developed in the next section, the boundary value problems of Section 2.5 are *noncritical*, whereas those of the present section are *critical*. For a noncritical problem the starting solutions do not depend on the perturbation term, and they lead to a nondegenerate implicit function problem at once. For a critical problem it is necessary to complete one or more stages of the perturbation calculation before it is possible to set up a nondegenerate implicit function problem, and therefore the starting solutions depend upon the specific perturbation terms present.

Exercises 4.4

1. Use the Lindstedt method to find the limit cycle of the Van der Pol equation (4.4.2) for small positive ε. Find and solve the determining equation, and find α_0, y_0, ν_1, α_1, y_1, and ν_2. What is the approximate solution resulting from these calculations? Give the error estimate in big-oh notation, and the intervals on which it is uniformly valid. Then interpret this statement carefully using constants and inequalities. Show that the limit cycle is stable (for positive ε).

2. Find an approximation to the limit cycle of $\ddot{y} + \varepsilon(y - 1)(y + 2)\dot{y} + y = 0$ having error $\mathcal{O}(\varepsilon)$ on expanding intervals of the form $0 \leq t \leq 1/\varepsilon$.

3. Consider the modified Van der Pol equation $\ddot{y} + \varepsilon(y^2 - \lambda)\dot{y} + y = 0$. Show that the limit cycle results from a Hopf bifurcation at a critical value of λ.

4. Look for limit cycles in the equation $\ddot{y} + \varepsilon(y^2 - 1)(y^2 - \lambda^2)\dot{y} + y = 0$ and test them for stability. Sketch the phase plane portrait for small ε. This problem exhibits the phenomenon of *pair bifurcation of limit cycles*, in which two limit cycles merge and disappear as the parameter λ is varied. What are the critical values of λ at which such bifurcations occur? Do the theorems of this chapter apply at the critical values of λ?

5. This exercise concerns the phase plane portrait of equations (4.4.3), or equivalently (4.4.4) in polar coordinates. These results apply also to Section 4.3, in which f does not depend on v; in this case, parts (a) and (d) give the proof of Theorem 4.4.2.

   (a) Observe from (4.4.3) that any rest point must lie on the u axis. Use the implicit function theorem to show that there exists a unique $u(\varepsilon)$

such that $u(0) = 0$ and such that $(u, v) = (u(\varepsilon), 0)$ is a rest point of (4.4.3), for small ε. In other words, the rest point at $(0,0)$ for the unperturbed system ($\varepsilon = 0$) continues to exist for small ε, although it may move slightly from the origin. Under what condition on f does the rest point remain at the origin?

(b) Consider the solution of (4.4.4) beginning at the point $r = \alpha > 0$, $\theta = 0$ on the positive u axis. Show that for sufficiently small ε, this solution winds clockwise around the origin at least once, intersecting the positive u axis at some point $r = \beta > 0$, $\theta = 2\pi$. Hint: Consider the annulus $\alpha/2 < r < 3\alpha/2$, $0 \leq \theta \leq 2\pi$. Let M be the maximum of $|f|$ on this annulus, and show that for sufficiently small ε, $\dot{\theta} < -k/2$ as long as the solution remains in the annulus. This implies that the solution spirals clockwise at least until it leaves the annulus. Then use the first equation of (4.4.4) to show that for sufficiently small ε, the solution remains in the annulus until it has made at least one complete turn. (How small must ε be for all of this to hold? What is the longest time that one revolution could take?)

(c) Prove the first sentence of Theorem 4.4.1 by applying the implicit function theorem.

(d) Suppose now that f does not depend on v, so that energy is conserved. Use part (b) together with (4.4.9) to prove that for any $\alpha > 0$, the solution beginning at $r = \alpha$, $\theta = 0$ is periodic for sufficiently small ε. Hint: Use the symmetry of the graph of (4.4.9).

6. Show that the method of this section fails for $\ddot{y} + \varepsilon y \dot{y} + y = 0$, and explain why. (See Exercise 4.3.6.)

4.5. FORCED OSCILLATIONS: GENERAL PRINCIPLES

This section and the next are devoted to equations such as the forced Duffing equation

$$\ddot{y} + \varepsilon \delta_1 \dot{y} + y + \varepsilon \lambda_1 y^3 = \varepsilon \gamma_1 \cos \omega t, \tag{4.5.1}$$

the forced Van der Pol equation

$$\ddot{y} + \varepsilon \lambda_1 (y^2 - 1)\dot{y} + y = \varepsilon \gamma_1 \cos \omega t, \tag{4.5.2}$$

and the Mathieu equation

$$\ddot{y} + (1 + \varepsilon \cos \omega t)y = 0. \tag{4.5.3}$$

These equations have the general form

$$\ddot{y} + k^2 y = \varepsilon f(y, \dot{y}, \omega t), \tag{4.5.4}$$

where $f(u, v, s)$ is 2π-periodic in s, so that $f(y, \dot{y}, \omega t)$ has period $2\pi/\omega$ in t. Equation (4.5.4) represents an oscillator placed in some periodically time-dependent environment, for instance an iron mass hanging from a nearly linear spring in a weak, periodically varying magnetic field. The oscillator is considered to be *forced* or *driven* by its environment. Therefore the control parameter ω is called the *forcing frequency*; k is called the *free frequency* because it is the frequency of the free (unforced) oscillation occurring when $\varepsilon = 0$. The search for periodic solutions of such equations is quite a complicated matter. In this section we consider what types of periodic solutions can exist and how they are classified according to *entrainment, resonance*, and *criticality*. The next section shows how to construct asymptotic approximation to the periodic solutions in the most important special case, that of *harmonic resonance*.

Since (4.5.4) is nonautonomous, the phase plane $(u, v) = (y, \dot{y}/k)$ is no longer suitable for graphical analysis of the solutions. This is because the solution passing through a given point (α, β) at time t_1 need not follow the same path as the solution passing through the same point at a different time t_2. In other words the *solutions* do not lie on *orbits* in the plane as they do in the autonomous case. The phase plane must be replaced by a three-dimensional graph having axes $(y, \dot{y}/k, t)$ in order to have a picture in which solutions fill the space with nonintersecting curves. One way to do this is to set $u = y$, $v = \dot{y}/k$, $s = \omega t$, and write (4.5.4) as a three-dimensional system

$$\dot{u} = kv,$$
$$\dot{v} = -ku + \frac{\varepsilon}{k} f(u, kv, s), \qquad (4.5.5)$$
$$\dot{s} = \omega.$$

This system is autonomous (since t no longer appears on the right hand side) and has orbits which fill the three-dimensional space with axes u, v, s. For small ε these spiral around the s axis, with s increasing steadily at rate ω. See Fig. 4.5.1. As with any autonomous system, each orbit carries infinitely many solutions differing only by a time shift. The last equation of (4.5.5) implies that $s = \omega t + s_0$; only those solutions for which $s_0 = 0$ correspond to solutions of (4.5.4). The others are solutions of the phase-shifted equation

$$\ddot{y} + k^2 y = \varepsilon f(y, \dot{y}, \omega t + s_0). \qquad (4.5.6)$$

Since $s_0 = 0$ implies that $s = 0$ when $t = 0$, the solutions of (4.5.5) which correspond to solutions of (4.5.4) are those whose initial values lie in the uv plane. This condition picks out one solution on each orbit.

The right hand side of (4.5.5) defines a vector field in (u, v, s) space, and orbits are tangent to this vector field at every point. (See Appendix D.) The vector field is 2π-periodic in s, so that the field in the infinite "slab" $0 \le s < 2\pi$ is repeated in $2\pi \le s < 4\pi$, and so on. Let $(\alpha, \beta, 0)$

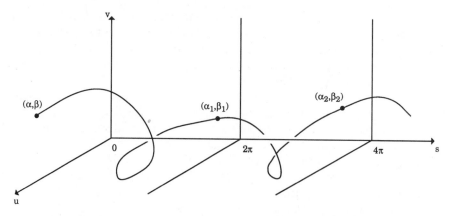

FIG. 4.5.1. The solution of a nonautonomous second order equation with periodic forcing, graphed in the three-dimensional phase space of variables $u = y$, $v = \dot{y}$, $s = t$. The dots are the successive intersections of the orbits with the planes $s = 2\pi n$, representing successive iterates of the period map.

be any initial point in the plane $s = 0$. The orbit through this point intersects the plane $s = 2\pi$ in a point $(\alpha_1, \beta_1, 2\pi)$. The map Φ which assigns to each (α, β) its associated (α_1, β_1) is called the *Poincaré map* or the *period map* of (4.5.4); see Appendix D for a discussion of period maps. If it happens that for some $(\alpha, \beta, 0)$ the orbit hits the plane $s = 2\pi$ in the point $(\alpha, \beta, 2\pi)$—that is, if (α, β) is a fixed point of the period map Φ—then the orbit will repeat, for $2\pi \leq s < 4\pi$, the same path that it followed for $0 \leq s < 2\pi$, since the starting values (α, β) at the left side of each slab are the same, and the vector fields within the slabs are the same. Such an orbit would be periodic in s with period 2π, and hence periodic in t with period $2\pi/\omega$. This is one way for periodic solutions to exist. Another way is for the orbit through $(\alpha, \beta, 0)$ to reach some different value $(\alpha_1, \beta_2, 2\pi)$ at $s = 2\pi$ but then to return to $(\alpha, \beta, 4\pi)$ at $s = 4\pi$. Such a solution repeats itself after every two slabs and so has period 4π in s or $4\pi/\omega$ in t. If the orbit through $(\alpha, \beta, 0)$ passes successively through $(\alpha_1, \beta_1, 2\pi)$, $(\alpha_2, \beta_2, 4\pi), \ldots, (\alpha_{\ell-1}, \beta_{\ell-1}, 2\pi(\ell-1))$, and finally returns to $(\alpha, \beta, 2\pi\ell)$ after passing through ℓ slabs, the orbit will repeat itself thereafter in blocks ℓ slabs long, and the corresponding solution of (4.5.4) will have period $2\pi\ell/\omega$. The frequency of such an orbit will be $\nu = \omega/\ell$. It would seem difficult for periodic solutions of any other period (not a multiple of $2\pi/\omega$) to exist, since they would have to repeat themselves forever with a period unrelated to that of the environment through which they are passing. In fact in many cases it can be shown that such solutions do not exist. We will prove a result of this sort which covers equations such as (4.5.1) and (4.5.2), in which the perturbation splits into the sum of two terms, one involving only y and $\dot{y}$, the other only t. The statement of this theorem uses the concept of *least period*, which is the smallest (positive) number that is a period of a given function. For example, $\sin t$ has least period 2π, although 4π, 6π, ... are also periods of this function. When two functions of a given period are combined, the resulting function will have the same period, but its *least* period may be shorter; for instance, $2 \sin t \cos t$ is a combination of functions having least period 2π, but since it equals $\sin 2t$ its least period is π.

Theorem 4.5.1. *Let $h(s)$ be periodic with least period 2π. Then any periodic solution of*

$$\ddot{y} + k^2 y = \varepsilon g(y, \dot{y}) + \varepsilon h(\omega t) \qquad (4.5.7)$$

for $\varepsilon \neq 0$ must have least period equal to $2\pi \ell / \omega$ for some integer ℓ.

Proof. Let $y(t)$ be a periodic solution of (4.5.7) for some value of $\varepsilon \neq 0$, and let its least period be T. Then $\ddot{y}(t) + k^2 y(t) - \varepsilon g(y(t), \dot{y}(t)) = \varepsilon h(\omega t)$. The left hand side has period T, but this is not necessarily its least period; its least period must be of the form T/ℓ for some integer ℓ. But the right hand side has least period $2\pi/\omega$. Since these are equal, $T = 2\pi \ell / \omega$. (Notice that this argument does not require ε to be small. In fact it is not a perturbation result at all, and could be stated without using ε, but we have used (4.5.7) in order to fit the form of our applications. ∎

The phenomenon we are discussing—that the period of a forced oscillation is an integer multiple of the period of the forcing—goes under the name of *frequency entrainment*: the frequency ν of the solution is said to be *entrained* by the frequency ω of the forcing, so that

$$\nu = \omega / \ell \qquad (4.5.8)$$

for some positive integer ℓ called the *entrainment index*.

The generalization of Theorem 4.5.1 to equation (4.5.4) states that any periodic solution has a period of the form $2\pi \ell / \omega$, provided that f_t is not identically zero along the solution, but does not guarantee that the *least* period is of the form $2\pi \ell / \omega$. The condition that f_t does not vanish is to exclude cases in which f does not actually depend upon t in certain regions. In such a case the solutions in those regions are actually solutions of an autonomous equation and are not entrained. See Exercise 4.5.1 for an example. See Section 4.10 for references to the general theorem.

Theorem 4.5.1 (and its generalization) are of great importance for the correct formulation of perturbation problems for forced equations. In the first place, it tells us that there is no use looking for solutions of (4.5.4) having any other period than $2\pi \ell / \omega$ for some integer ℓ. Therefore for each ℓ we should pose the problem: Are there periodic solutions of period $2\pi \ell / \omega$? In attacking this problem, the expected period is known exactly, in advance, so there is no need to set up an unknown solution frequency $\nu(\varepsilon)$ and solve for unknown coefficients ν_n as in the last two sections. Furthermore, according to Theorem 4.2.2 and equation (4.2.13), the Lindstedt expansion will be uniformly valid for all time and not merely on expanding intervals. On the other hand, it is no longer possible to deal only with solutions beginning on the u axis. To look for solutions with entrainment index ℓ it is necessary to consider arbitrary initial conditions (α, β) in the plane $s = 0$ (in Fig. 4.5.1) and try to find functions $\alpha(\varepsilon)$ and $\beta(\varepsilon)$ so that the solution through $(\alpha(\varepsilon), \beta(\varepsilon), 0)$ arrives at $(\alpha(\varepsilon), \beta(\varepsilon), 2\pi\ell)$ after passing through ℓ slabs. Therefore in using the Lindstedt method, both $\alpha(\varepsilon)$ and $\beta(\varepsilon)$ must be expanded in powers of ε with unknown coefficients. There are the same number of undetermined coefficients at each stage as in Section 4.4, but they occur in a different place: Two initial conditions are unknown, rather than one initial condition and the solution frequency.

There is a second consequence of Theorem 4.5.1 that is also fundamental. To understand it, consider (4.5.4) when $\varepsilon = 0$. There are two types of solutions: the origin, which is a rest point but can be considered as a periodic solution of any period; and the other solutions, which all have least period $2\pi/k$. Any family $y(t; \varepsilon)$ of periodic solutions must reduce to one or the other of these types when $\varepsilon = 0$. This fact, together with Theorem 4.5.1 (which applies when $\varepsilon \neq 0$), imposes further restrictions on the possible periodic solution families. Recall that a real number is called *rational* if it can be written as a ratio of integers (what is called in elementary school a *common fraction*). Most real numbers (such as $\sqrt{2}$ and π) cannot be written as a ratio of integers but only as an infinitely long nonrepeating decimal fraction; these are called *irrational*. ("Ir-*ratio*-nal" means "not a *ratio* (of integers)." It does not mean "unreasonable.") A rational number is said to be expressed *in lowest terms* if it is written as a ratio of integers which have no common factors (for instance, $1/2$ rather than $2/4$).

Corollary 4.5.2. *If ω/k is irrational (the nonresonant case), then any periodic solution family $y(t; \varepsilon)$ for (4.5.7) must reduce to the zero solution at $\varepsilon = 0$ (and therefore must have small amplitude when ε is small).*

Proof. Let $y(t; \varepsilon)$ be a periodic solution family for which $y(t; 0) \neq 0$. Since $y(t; \varepsilon)$ must have least period of the form $2\pi\ell/\omega$ when $\varepsilon \neq 0$, it follows by letting $\varepsilon \to 0$ that $y(t; 0)$ must have period $2\pi\ell/\omega$. (It does not follow that this is the *least* period of $y(t; 0)$, but only that is it *a* period. The function $\varepsilon \sin t + \sin 2t$ is an example of a function having least period 2π for $\varepsilon \neq 0$ and π for $\varepsilon = 0$; the least period undergoes a jump, but there is still a common period 2π valid for all ε.) On the other hand, it is known that $y(t; 0)$ has least period $2\pi/k$. Therefore $2\pi\ell/\omega$ must be an integer multiple of $2\pi/k$, that is, $2\pi\ell/\omega = 2\pi m/k$ for some m, that is, $\omega/k = \ell/m$. Therefore ω/k must be rational. We have proved that solutions not reducing to zero when $\varepsilon = 0$ are possible only in the resonant case. ∎

Corollary 4.5.2 is valid for equation (4.5.4) under the additional assumption that f_t does not vanish identically along any solution, but since we have only proved Theorem 4.5.1 for (4.5.7) our version of Corollary 4.5.2 must be similarly restricted.

In connection with Corollary 4.5.2 we introduce the following definition. The forcing frequency ω is said to be *in $p : q$ resonance* with the free frequency k if

$$\omega/k = p/q \qquad \text{in lowest terms,} \qquad (4.5.9)$$

where p and q are integers (with no common factors). There are three natural frequencies associated with any periodic solution of (4.5.4): the solution frequency ν, the forcing frequency ω, and the free frequency k. (The last is not actually present; it is the frequency that the solution would have in the absence of forcing.) Entrainment, defined by equation (4.5.8), is a relation between the forcing frequency and the *solution* frequency. Resonance, defined in (4.5.9), is a relation between the forcing frequency and the *free* frequency.

In *linear* forced oscillation problems (see Appendix C) only the 1 : 1 resonance plays a role, and the effect of this resonance is to cause solutions to grow with time (so that no solutions are periodic). For nonlinear problems resonance has much more subtle effects. Resonances of higher orders $p : q$ must be taken into account, and the effect of a resonance is not usually to cause unbounded growth, but rather to create new periodic solutions. A rough explanation of these effects is as follows. Nonlinear periodic processes are not simple harmonic oscillations. Instead they are (in general) represented by Fourier series with many harmonics. If a nonlinear periodic motion has frequency ω_1, it will resolve into a sum of simple harmonic motions with frequencies $n\omega_1$ for $n = 1, 2, \ldots$. A second such motion with frequency ω_2 will have components $m\omega_2$. These two processes will interact resonantly (in the original sense of the word) if a component of the first motion has the same frequency as a component of the second motion, that is, if $n\omega_1 = m\omega_2$ for some n and m. But this is the same as saying $\omega_1/\omega_2 = m/n$. So a "higher order resonance" is just a 1 : 1 resonance between higher harmonics of the motion. Now suppose that a periodic forcing is applied to a nonlinear oscillator, such as the Duffing oscillator represented in Fig. 4.3.2. The periodic solutions shown in this figure do not all have the same period; the period depends on the amplitude. Therefore a given forcing term can only be in resonance with one periodic orbit. The resonance effect will start to cause this solution to grow, but as soon as it grows, its frequency will change and it will begin to move out of resonance. Therefore its amplitude will not continue to grow, but instead it will settle into an entrained periodic motion slightly larger than otherwise expected. This explanation of how nonlinear resonance produces periodic solutions is not useful for perturbation analysis, because it is based on the application of forcing to an already nonlinear system; it is as if we considered equation (4.5.7) for fixed ε, first without and then with the h-term. The perturbation theory for (4.5.7) instead considers first $\varepsilon = 0$

and then $\varepsilon \neq 0$. Nevertheless this discussion gives some useful insights into the meaning of nonlinear resonance.

These preliminaries suggest a strategy for finding periodic solutions of (4.5.4) which will now be outlined. (The details will be given later.) The first step (which is based on Theorem 4.5.1) is to select an integer ℓ which will be the entrainment index of the periodic solutions being sought. The second step is to set up a shooting problem, shooting from the plane $s = 0$ to $s = 2\pi\ell$ and looking for fixed points $\alpha(\varepsilon)$ and $\beta(\varepsilon)$ using the implicit function theorem. (This is similar to "shooting" from $\theta = 0$ to $\theta = 2\pi$ in Section 4.4.) This step will go quite differently in the resonant and nonresonant cases. In the nonresonant case it is known from Corollary 4.5.2 that any such fixed point must originate (for $\varepsilon = 0$) at the point $(0, 0)$, that is, must satisfy $\alpha(0) = 0$ and $\beta(0) = 0$. It turns out that the implicit function problem for the continuation of this fixed point (for $\varepsilon \neq 0$) is always nondegenerate, so a unique fixed point always exists. Furthermore, there is no need (in the nonresonant case) to consider entrainment indices other than $\ell = 1$. Indeed, we have just pointed out that a unique fixed point exists (for each ℓ); but certainly the fixed point for $\ell = 1$, which returns to itself after moving through one "slab," also returns to itself after moving through any number of slabs. Therefore there is only one fixed point which is the same for all ℓ, and there is only one periodic solution, whose least period is equal to the period of the forcing. In the resonant case, the problem is more difficult since the values of $\alpha(0)$ and $\beta(0)$ cannot be known in advance. Having selected ℓ one again sets up a shooting problem, which this time is degenerate. A determining equation is needed in order to locate suitable starting values for α and β; each simple root (α, β) of the determining equations (if there are any) will give rise to a family of periodic solutions. Once again it turns out that all periodic solutions may be found using a single value of ℓ, in this case $\ell = p$, where p is specified by the resonance as in (4.5.9).

The strategy outlined in the last paragraph is essentially correct. However, if we were to carry out these steps immediately, the resulting theory would be almost useless in practice. For we made a serious blunder at the very beginning of this section, which must now be corrected. This blunder occurred at the point when (4.5.4) was chosen as the general equation for a forced oscillator. This equation is mathematically quite reasonable, but it is usually a mistake to write down an equation without examining the experimental data which the equation is supposed to explain. We will now pause to look at this data and see that the observed phenomena can never be explained by equation (4.5.4) as it stands. One small but crucial change must be made: It is necessary to allow the forcing frequency ω to depend upon ε. It is once again a question of finding the proper way to introduce a perturbation parameter into a problem that is initially given in natural parameters.

Consider, then, a physical system having two control parameters μ and ω and described by

$$\ddot{y} + y = \mu f(y, \dot{y}, \omega t). \tag{4.5.10}$$

Here μ controls the strength of the forcing and ω the frequency, and these can be varied independently by the experimentalist. (The reader with access to an analog computer can easily simulate such experiments by wiring the computer so that values of μ and ω can be set by potentiometers.) For any values of μ and ω, the initial conditions can be varied until a periodic solution is located. Actually only the *stable* periodic solutions can be observed experimentally, and these are usually easy to find because the initial conditions need not be located exactly; as long as the initial conditions are in the "basin of attraction" of the periodic solution, it is only necessary to let the system run until the transients have died out and the system falls into the periodic solution automatically. Typically the results of such an experiment are as follows: For most choices of μ and ω there is only one periodic solution. This solution has the same period as the forcing, and it has a fairly small amplitude; in fact it is the nearest thing to an equilibrium solution that is possible in the presence of the forcing. On the other hand, when ω is close to certain rational numbers such as 1, 1/2, 1/3, or 2/3, there exist other periodic solutions besides this basic one. (Note that in (4.5.10) we have taken $k = 1$, so resonance is defined by $\omega = p/q$.) What is noteworthy here is that additional periodic solutions appear for ω *nearly equal to* a rational number, not only *exactly equal*. Furthermore not every rational number seems to produce this effect, but only rational numbers that are ratios of small integers. (What is "small" depends on the situation; occasionally resonance effects can be observed up to p or q equal to 15 or 20.) The periodic solutions that appear when ω is close to p/q have entrainment index $\ell = p$; thus, near 1, 1/2, or 1/3 the new solutions have the same period as the forcing, but near 2/3 they have twice that period. If the values of ω and μ that give additional periodic solutions are plotted in the (ω, μ) plane, they are found to form cusp-shaped regions with the vertex at the point $(p/q, 0)$; see Fig. 4.5.2. These regions are called *resonance horns, resonance tongues*, or *entrainment domains*.

Incidentally, it is fortunate that the existence of resonance does not require that ω take on a specific rational value exactly, because otherwise resonance effects would be unobservable: No physical quantity can be set or measured exactly in a laboratory. It is also fortunate that only finitely many resonances actually have an effect in most systems. There are some systems (those possessing essentially no friction, such as planetary motions) for which every rational number can (at least in principle) produce a resonance effect, and these are very difficult to analyze mathematically.

It is now possible to see why (4.5.4) is not an adequate formulation of a perturbation problem for (4.5.10). A perturbation problem for (4.5.10)

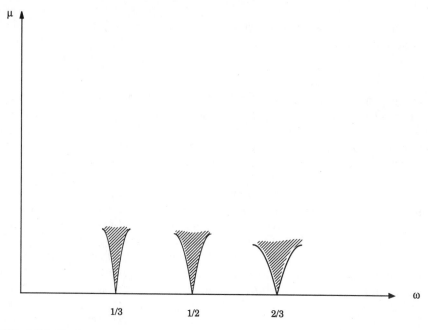

FIG. 4.5.2 Typical resonance horns, resonance tongues, or entrainment domains.

can be viewed as a path $(\omega(\varepsilon), \mu(\varepsilon))$ in the (ω, μ) plane, parameterized by ε. The path must begin on the ω axis, because the reduced problem must have $\mu = 0$ in order to be solvable. Three such paths are shown in Fig. 4.5.3, together with a resonance horn. The path labeled (1) is defined by $\omega(\varepsilon) = p/q$, $\mu(\varepsilon) = \varepsilon$. This path has constant ω and thus corresponds to (4.5.4). It represents a resonant case, but it does not lie in the $p : q$ resonance horn. Therefore no resonance effects will appear in the perturbation problem defined by this path. (Some resonance horns, such as those drawn in Fig. 4.5.2, contain the vertical straight-line path beginning at the vertex, and others do not.) The path labeled (2) is a nonresonant path with constant ω. Although this path enters the resonance horn, perturbation results obtained from this problem will not be valid once the horn is reached. This is because perturbation methods only detect families of periodic solutions which exist along a segment of the path beginning at the reduced problem. Families which come into existence at a distance from the reduced problem are not detected. Thus it is clear that the only way to find the periodic solutions that exist in the resonance horn in Fig 4.5.3 is to use a path such as the one labeled (3), on which ω becomes a function of ε and not merely a constant. (For the resonance horns in Fig. 4.5.2, the periodic solutions in *exact* resonance can be found with constant ω, but not those in *near*-resonance.

In Chapter 1 it was found (see Example 1.3.3 and Section 1.7) that a nearly degenerate root-finding problem was best reformulated as a degenerate one. The degenerate problem was obtained by making a control

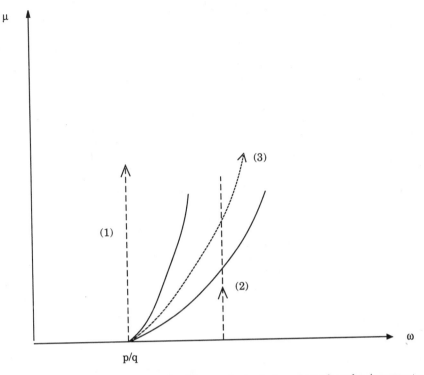

FIG. 4.5.3. Three perturbation families, represented as dotted paths (parameterized by ε) in a space of natural parameters; the starting point $\varepsilon = 0$ is at the bottom of each path. The solid lines bound a resonance horn. Path (1) misses the resonance altogether. Path (2) does not show resonance because results along this path are valid only near the starting point. Only paths like (3), on which the forcing frequency ω varies with ε, can properly detect the resonance behavior.

parameter (in that case τ) into a function of ε. The issue there was simply one of numerical accuracy. The present example illustrates the same principle, although the question goes deeper than mere accuracy. A near-resonant target problem (if it lies in the resonance horn) must not be embedded in a nonresonant perturbation family, but in a resonant one. The only way to do this is to allow the control parameter ω to become a function of ε.

These considerations force us to generalize equation (4.5.4) to the form

$$\ddot{y} + k^2 y = \varepsilon f(y, \dot{y}, \omega(\varepsilon)t) \qquad (4.5.11)$$

and to redefine $p : q$ resonance by the condition

$$\omega(0)/k = p/q. \qquad (4.5.12)$$

Theorem 4.5.1 is easily generalized to show that any periodic solution family $y(t, \varepsilon)$ of (4.5.11) must have a frequency

$$\nu(\varepsilon) = \omega(\varepsilon)/\ell \tag{4.5.13}$$

for some integer entrainment index ℓ. (Our proof, of course, only applies if $f = g(y, \dot{y}) + h(\omega(\varepsilon)t)$.) Note that the entrainment condition will hold (exactly) for all ε for which the periodic solution exists, whereas the resonance condition is imposed only at $\varepsilon = 0$ (and for $\varepsilon \neq 0$ holds only approximately).

In the engineering literature the word *detuning* is often used in connection with approximate resonance. To understand this term, first notice that (4.5.4) can be written as $\ddot{y} + \omega^2 y = (\omega^2 - k^2)y + \varepsilon f(y, \dot{y}, \omega t)$. This form suggests that if $\omega^2 - k^2$ is small (that is, in the case of near $1:1$ resonance) it can be considered part of the perturbation; then the problem can be considered as exactly resonant, with a new free frequency of ω. To carry out this idea, define a new parameter by

$$\Delta := \omega^2 - k^2 \tag{4.5.14}$$

so that (4.5.4) becomes

$$\ddot{y} + \omega^2 y = \Delta y + \varepsilon f(y, \dot{y}, \omega t). \tag{4.5.15}$$

Δ is called the *detuning*, because it measures the extent to which the forcing frequency ω is "out of tune" with the free frequency k. The next step is to introduce a perturbation parameter ε into (4.5.15). Since k has been eliminated, it is now natural to regard ω (rather than k) as constant, and make Δ a function of ε with $\Delta(0) = 0$. Such a function can be expanded as $\Delta(\varepsilon) = \varepsilon \Delta_1 + \varepsilon^2 \Delta_2 + \cdots$, and then (4.5.15) becomes

$$\ddot{y} + \omega^2 y = \varepsilon[\Delta_1 y + f(y, \dot{y}, \omega t)] + \cdots. \tag{4.5.16}$$

Now (4.5.15) is an exactly resonant perturbation problem in which the free frequency is ω. Of course, this is a slightly different perturbation family than (4.5.11), but either way the same ground is covered: all combinations of ω and k that are sufficiently close compared to ε will be accessible by perturbation families of either type. We will work with (4.5.11) rather than (4.5.15) because it seems more flexible. To handle resonances of type $p:q$ by the detuning approach requires a definition like $\Delta := q^2 \omega^2 - p^2 k^2$; that is, it is necessary to define a different detuning parameter for each resonance. But (4.5.11) can be used in the same form in all cases.

Finally the perturbation problem (4.5.11) is formulated correctly. The solution strategy, outlined previously, now takes the following form. Assume that the forcing frequency $\omega(\varepsilon)$ is specified, either as a constant or as a function of ε. Select an integer ℓ, define $\nu(\varepsilon)$ by (4.5.13), and ask the question: Do there exist any periodic solutions $y(t, \varepsilon)$, defined for ε in an interval about zero, having frequency $\nu(\varepsilon)$? This question can be attacked by the shooting method: "Shoot" from $(\alpha, \beta, 0)$ in Fig. 4.5.1 to the plane $s = 2\pi\ell$, trying to hit $(\alpha, \beta, 2\pi\ell)$.

Consider first the case of $\ell = 1$. Let Φ be the period map from $s = 0$ to $s = 2\pi$ in Fig. 4.5.1. It is easy to see (Exercise 4.5.2) that when $\varepsilon = 0$, Φ is a rotation through angle $2\pi k/\omega(0)$. If this angle is 2π, or a multiple of 2π, then Φ is the identity map when $\varepsilon = 0$, and all points are fixed points. This is the *critical case*: A determining equation is needed to find which of these fixed points, if any, continue as fixed points (and hence as periodic solutions) for $\varepsilon \neq 0$. On the other hand if $2\pi k/\omega(0)$ is not a multiple of 2π, then the origin is the only fixed point of Φ for $\varepsilon = 0$, and an implicit function theorem argument shows that this fixed point continues to a fixed point near the origin for $\varepsilon \neq 0$ and that there are no other fixed points. (Exercise 4.5.3.) This is the *noncritical case*. In the noncritical case there is a unique periodic solution for $\varepsilon \neq 0$ and it has small amplitude. In the critical case there is a possibility of one or more periodic solutions of significant amplitude, but whether they exist depends upon f (through the determining equation). These remarks also apply to entrainment indices $\ell > 1$, except that here one is shooting from $s = 0$ to $s = 2\pi\ell$ by the map Φ^ℓ, which for $\varepsilon = 0$ is a rotation through $2\pi\ell k/\omega(0)$, and the case will be critical if this is a multiple of 2π.

The significance of resonance arises from the fact that $2\pi\ell k/\omega(0)$ becomes a multiple of 2π for certain ℓ if and only if $k/\omega(0)$ is rational. We will now go through the various cases of resonance and nonresonance, showing for each case which values of ℓ are significant and whether they are critical or noncritical. The results are summarized in Table 4.5.1.

If $\omega(0)/k$ is irrational (the first line of the table), then $2\pi\ell k/\omega(0)$ is never a multiple of 2π for any ℓ. Every choice of ℓ gives a noncritical problem having a unique periodic solution $y(t, \varepsilon)$, which reduces to $y(t, 0) \equiv 0$ when $\varepsilon = 0$. But the solution for $\ell = 1$, which has the period $2\pi/\omega(\varepsilon)$, also has period $2\pi\ell/\omega(\varepsilon)$ for any ℓ, and so must be the unique solution with that period. Therefore there is no need to consider any ℓ other than 1.

TABLE 4.5.1

Name	Resonance or Nonresonance	Useful Values of ℓ
nonresonance	$\omega(0)/k$ irrational	$\ell = 1$ (noncritical)
harmonic resonance	$\omega(0) = k$	$\ell = 1$ (critical)
subharmonic resonance	$\omega(0)/k = p$ $\quad(p \neq 1)$	$\ell = 1$ (noncritical) and $\ell = p$ (critical)
superharmonic resonance	$\omega(0)/k = 1/q$ $\quad(q \neq 1)$	$\ell = 1$ (critical)
supersubharmonic resonance	$\omega(0)/k = p/q$ $\quad(p \neq 1, q \neq 1)$	$\ell = 1$ (noncritical) and $\ell = p$ (critical)

If $\omega(0)/k = 1$ (the *harmonic* or 1 : 1 resonance), the map from $s = 0$ to $s = 2\pi$ is a complete rotation (the identity map). This is a critical problem. The determining equation will be found in the next section by the Lindstedt method, and again in Chapters 5 and 6 by other methods.

If $\omega(0)/k = p$ *(subharmonic resonance)*, the free period $2\pi/k$ is approximately p times the forcing period $2\pi/\omega(\varepsilon)$. In this case the forcing is capable of exciting two types of oscillations: a small oscillation of period exactly equal to the forcing period (i.e., $\ell = 1$), and possibly also other oscillations have a period $2\pi p/\omega(\varepsilon)$ close to the free period ($\ell = p$). The former type of oscillation always exists, since the map from $s = 0$ to $s = 2\pi$ is not a complete rotation, and the problem is noncritical. The map from $s = 0$ to $s = 2\pi p$ is a complete rotation, and it requires a determining equation to find whether any such "subharmonic oscillations" exist. No additional periodic solutions can be found by using any other value of ℓ.

In the superharmonic case the forcing frequency is slow compared to the free frequency: One period of the forcing equals approximately q free periods. The map from $s = 0$ to $s = 2\pi$ is q complete rotations when $\varepsilon = 0$, so the $\ell = 1$ problem is critical. No additional solutions can be found for larger ℓ. We leave it to the reader to check that in the supersubharmonic case, $\ell = 1$ is noncritical and $\ell = p$ is the least critical value of ℓ. Again nothing new is found for larger ℓ.

Exercises 4.5

1. Verify that $y = \cos t$ is a solution of $\ddot{y} + y = \varepsilon(1 - y^2 - \dot{y}^2) \sin \sqrt{2} t$. Show that this solution is not entrained (it does not have a period which is a multiple of the forcing period). Why is this not a violation of the generalization of Theorem 4.5.1 (stated in the paragraph following that theorem)?

2. Find the explicit solution of (4.5.5) for $\varepsilon = 0$. Use this to show that when $\varepsilon = 0$, the Poincaré map Φ is a rotation of the plane through angle $2\pi k/\omega(0)$. For what values of $k/\omega(0)$ is this map the identity? For what values of $k/\omega(0)$ is an iterate of this map equal to the identity? What types of resonance do these cases represent (in terms of Table 4.5.1)?

3. Prove that in the noncritical case, there exists a unique periodic solution with entrainment index 1 having initial conditions near the origin. Hint: Apply the vector implicit function theorem to the Poincaré map, using the known form of that map when $\varepsilon = 0$.

4. If k is constant and $\omega = k + \varepsilon\omega_1 + \varepsilon^2\omega_2 + \cdots$, compute Δ_1 and Δ_2 in the expansion of (4.5.14). Conversely, if $\Delta(\varepsilon) = \varepsilon\Delta_1 + \varepsilon^2\Delta_2 + \cdots$, compute the expansions of $\omega(\varepsilon)$ if k is considered constant, and of $k(\varepsilon)$ if ω is constant. These formulas are useful in comparing results obtained from (4.5.11) and (4.5.16).

4.6. HARMONIC RESONANCE

The harmonic resonance problem is the following: to find periodic solutions of period $2\pi/\omega(\varepsilon)$ for the equation

$$\ddot{y} + k^2 y = \varepsilon f(y, \dot{y}, \omega(\varepsilon)t) \qquad (4.6.1)$$

when

$$\omega(0) = k. \qquad (4.6.2)$$

According to the general theory of the last section, this is a critical problem. Therefore the existence of periodic solutions depends upon a determining equation; simple zeroes of the determining equation will pick out starting values of the initial conditions for periodic solutions. Our goal in this section is to find the determining equations for the harmonic resonance problem. This will be done by attempting to construct periodic solutions using the Lindstedt method.

Recall that in Section 4.4 the determining equations for a self-excited problem were obtained in two ways. In Theorem 4.4.1 the existence of periodic solutions was studied by a shooting method/implicit function argument, whereas in Theorem 4.4.2 the same determining equations were found by the Lindstedt method. The same two procedures are possible here. In this section the Lindstedt method will be used. As in Section 4.4, this method does not result in a rigorous proof that the *periodic solutions* exist; it only shows that for each simple root of the determining equations a *formal Lindstedt series solution* exists. The shooting method for these problems does prove the existence of the periodic solutions; the details are left to the interested reader, with hints given in Exercises 4.6.5 and 4.6.6. A much more general theorem about periodic solutions, which includes all of the existence theorems in this chapter, will be given in Section 6.3 using the method of averaging.

According to the discussion in the last section, the forcing frequency $\omega(\varepsilon)$ is to be treated as an arbitrary smooth function, subject only to the harmonic resonance condition (4.6.2). This function will be considered as fixed throughout the derivation of the Lindstedt series. It may be expressed as an asymptotic power series

$$\omega \sim k + \varepsilon\omega_1 + \varepsilon^2\omega_2 + \cdots \qquad (4.6.3)$$

in which the coefficients ω_n are arbitrary, but fixed in advance; they are not unknowns to be determined in the course of the solution, as were the coefficients ν_n of the solution frequency in Section 4.4. The initial conditions for the periodic solutions of (4.6.1), on the other hand, are not arbitrary, but depend upon ε in a way that must be determined. Therefore they are expanded in asymptotic power series

$$y(0) \sim \alpha_0 + \varepsilon\alpha_1 + \varepsilon^2\alpha_2 + \cdots,$$
$$\dot{y}(0) \sim \beta_0 + \varepsilon\beta_1 + \varepsilon^2\beta_2 + \cdots \qquad (4.6.4)$$

in which the coefficients are treated as unknowns to be found. Our experience with Lindstedt expansions suggests that the periodic solutions (if there are any) should be expressed in the form

$$y \sim y_0(t^+) + \varepsilon y_1(t^+) + \varepsilon^2 y_2(t^+) + \cdots, \tag{4.6.5}$$

where

$$t^+ = \omega(\varepsilon)t. \tag{4.6.6}$$

In order to substitute (4.6.5) into the differential equation, it should first be expressed in terms of the variable t^+:

$$\omega^2 y'' + k^2 y = \varepsilon f(y, \omega y', t^+), \tag{4.6.7}$$

where $' = d/dt^+$. Squaring (4.6.3) gives

$$\omega^2 \sim k^2 + \varepsilon(2k\omega_1) + \cdots. \tag{4.6.8}$$

Substituting everything into (4.6.7) gives

$$\{k^2 + \varepsilon(2k\omega_1) + \cdots\}\{y_0'' + \varepsilon y_1'' + \cdots\} + k^2\{y_0 + \varepsilon y_1 + \cdots\}$$
$$\sim \varepsilon f(y_0, ky_0', t^+) + \cdots,$$

which separates into

$$y_0'' + y_0 = 0,$$
$$y_1'' + y_1 = -2\omega_1 y_0''/k + f(y_0, ky_0', t^+)/k^2, \tag{4.6.9}$$
$$\vdots$$

These equations are to be solved with initial conditions determined from (4.6.4). The second equation of (4.6.4) must be changed to the new time variable t^+:

$$y'(0) = \frac{\dot{y}(0)}{\omega} \sim \frac{\beta_0 + \varepsilon\beta_1 + \cdots}{k + \varepsilon\omega_1 + \cdots} \sim \frac{\beta_0}{k} + \varepsilon\frac{k\beta_1 - \omega_1\beta_0}{k^2} + \cdots. \tag{4.6.10}$$

Therefore the initial conditions for (4.6.9) are

$$y_0(0) = \alpha_0, \quad y_0'(0) = \beta_0/k, \quad y_1(0) = \alpha_1, \quad y_1'(0) = (k\beta_1 - \omega_1\beta_0)/k^2. \tag{4.6.11}$$

The solution for $y_0(t^+)$ is

$$y_0(t^+) = \alpha_0 \cos t^+ + \frac{\beta_0}{k} \sin t^+, \tag{4.6.12}$$

after which y_1 is to be determined from

$$y_1'' + y_1 = \frac{2\omega_1\alpha_0}{k} \cos t^+ + \frac{2\omega_1\beta_0}{k^2} \sin t^+$$
$$+ \frac{1}{k^2} f(\alpha_0 \cos t^+ + \frac{\beta_0}{k} \sin t^+, -\alpha_0 k \sin t^+ + \beta_0 \cos t^+, t^+).$$

(4.6.13)

In order for the solution $y_1(t^+)$ to give an acceptable coefficient for a Lindstedt series, it must be periodic in t^+ with period 2π. In order for such solutions to exist, the right hand side of (4.6.13) must be free of the first harmonic when expanded in a Fourier series in t^+; any first harmonic terms (terms in $\sin t^+$ or $\cos t^+$) will be resonant (in the sense of linear resonance) and will give rise to secular terms in the solution. Thus the Fourier coefficients of $\sin t^+$ and $\cos t^+$ of the right hand side of (4.6.13) must be set equal to zero, giving two equations

$$\varphi_1(\alpha_0, \beta_0) = \frac{2\omega_1\alpha_0}{k} + \frac{1}{\pi} \int_0^{2\pi} \hat{f}(\alpha_0, \beta_0, t^+) \cos t^+ \, dt^+ = 0,$$
$$\varphi_2(\alpha_0, \beta_0) = \frac{2\omega_1\beta_0}{k^2} + \frac{1}{\pi} \int_0^{2\pi} \hat{f}(\alpha_0, \beta_0, t^+) \sin t^+ \, dt^+ = 0$$

(4.6.14)

to be solved for α_0 and β_0; here $\hat{f}(\alpha_0, \beta_0, t^+)$ denotes the last term of (4.6.13), that is,

$$\hat{f}(\alpha_0, \beta_0, t^+) := \frac{1}{k^2} f(\alpha_0 \cos t^+ + \frac{\beta_0}{k} \sin t^+, -\alpha_0 k \sin t^+ + \beta_0 \cos t^+, t^+).$$

There may exist one solution (α_0, β_0) of these equations, or none, or more than one. If there are none, it is impossible to continue, and there are no periodic solution families for (4.6.1). If there are solutions for (α_0, β_0), then each of them must be followed up separately. The procedure is to solve (4.6.13) for $y_1(t^+)$, which will be periodic since (α_0, β_0) has been chosen properly. Then $y_0(t^+)$ and $y_1(t^+)$ are substituted into the right hand side of an equation for $y_2(t^+)$ derived by carrying (4.6.7) to one more term. The equation for $y_2(t^+)$ has a periodic solution if and only if α_1 and β_1 are chosen properly. It turns out to be possible to solve for α_1 and β_1 uniquely, provided that (α_0, β_0) is a simple zero of the equations (4.6.14); see Exercise 4.6.4 for details. In this setting, a "simple zero" is one for which the Jacobian determinant does not vanish:

$$\left. \frac{\partial(\varphi_1, \varphi_2)}{\partial(\alpha, \beta)} \right|_{(\alpha_0, \beta_0)} \neq 0.$$

(4.6.15)

The further stages in the calculation do not impose any additional restrictions on the starting values (α_0, β_0) of the initial conditions for the periodic solutions.

At this point it is reasonable to make the following conjecture.

Theorem 4.6.1. *The determining equations for the harmonic resonance problem (4.6.1,2) are given by (4.6.14). That is, for any simple zero (α_0, β_0) of (4.6.14) there exist functions $\alpha(\varepsilon)$ and $\beta(\varepsilon)$, defined for ε in some interval $|\varepsilon| < \varepsilon_0$, such that the solution of (4.6.1) with initial conditions $y(0; \varepsilon) = \alpha(\varepsilon)$, $\dot{y}(0; \varepsilon) = \beta(\varepsilon)$ is periodic with period $2\pi/\omega(\varepsilon)$. Furthermore these solutions have Lindstedt expansions, constructed by the procedure given above, which are uniformly valid for all time.*

Of course, the approximations will only be valid for all time if the exact forcing frequency $\omega(\varepsilon)$ is used in (4.6.6) when writing the final solution, and not a truncation of its asymptotic expansion (4.6.3). (See Theorem 4.2.2.) But since the forcing frequency is *given* in these problems rather than *computed*, it is possible (in theory) to use the exact frequency.

Exercises 4.6

1. Show that the determining equations for the Duffing equation (4.5.1) under the harmonic resonance assumption (4.6.2) are

$$2\omega_1\alpha_0 - \delta_1\beta_0 - \tfrac{3}{4}\lambda_1\alpha_0^3 - \tfrac{3}{4}\lambda_1\alpha_0\beta_0^2 + \gamma_1 = 0,$$
$$2\omega_1\beta_0 + \delta_1\alpha_0 - \tfrac{3}{4}\lambda_1\alpha_0^2\beta_0 - \tfrac{3}{4}\lambda_1\beta_0^3 = 0.$$

Do not do this by applying equations (4.6.14), but instead by following the steps of the Lindstedt method. (Hint: Instead of evaluating any integrals, the Fourier expansions can be found using Table E.1 in Appendix E.) Then introduce polar coordinates by $\alpha_0 = r\cos\theta$, $\beta_0 = r\sin\theta$ and obtain the determining equations in the form

$$2\omega_1 r - \tfrac{3}{4}\lambda_1 r^3 + \gamma_1 \cos\theta = 0$$
$$\delta_1 r - \gamma_1 \sin\theta = 0.$$

These equations will be studied in the next section.

2. Find the determining equations for the forced Van der Pol equation (4.5.2) in the harmonic resonance case, both in rectangular and polar coordinates.

3. Find the determining equations for the Mathieu equation (4.5.3) in the harmonic resonance case.

4. Find the equation for y_2 which comes next in (4.6.9). Show that if (α_0, β_0) is a simple zero of the determining equations then there exist

unique α_1 and β_1 such that the equation for y_2 has a periodic solution. Do this *without* attempting to find the solution for y_2. Hint: Compare the discussion of (4.4.19).

5. Prove Theorem 4.6.1. Hint: Solve equation (4.6.7) by the regular perturbation method (Chapter 2) over the finite interval $0 \le t^+ \le 2\pi$, with arbitrary initial conditions α, β/k. (At this point you should treat α and β as constants, not functions of ε.) Write down the shooting equations stating that the solution returns to its initial conditions at $t^+ = 2\pi$. (This is the same as shooting from the plane $s = 0$ to $s = 2\pi$ in Fig. 4.5.1.) Apply the vector implicit function theorem (Appendix B) to show that the shooting equations can be solved for α and β as functions of ε with $\alpha(0) = \alpha_0$ and $\beta(0) = \beta_0$, provided that (α_0, β_0) is a simple zero of (4.6.14).

6. Theorem 4.6.1 is stated without specifying the smoothness requirements. Analyze your proof (from Exercise 4.6.5) to determine what smoothness is needed for existence of the periodic solutions and what smoothness is needed for the Lindstedt expansions to be asymptotic to order k.

4.7. DUFFING'S EQUATION

Duffing's equation has been introduced in Examples 2.2.1 and 2.4.1. The unforced conservative version has been solved by the regular perturbation method in Exercise 2.4.4(a), and by the Lindstedt method in Theorem 4.4.1 and Exercises 4.4.1 and 4.4.2. In this section we will study some of the periodic solutions of the forced Duffing equation, with and without damping. This equation will continue to provide interesting examples in Chapters 5 and 6. It exhibits such a large number of nonlinear phenomena in their simplest form that it is still a subject of current research.

The general form of Duffing's equation (see (2.3.14)) is

$$\ddot{y} + \delta\dot{y} + y + \lambda y^3 = \gamma \cos \omega t,$$
$$y(0) = \alpha, \qquad\qquad (4.7.1)$$
$$\dot{y}(0) = \beta.$$

This system has six nondimensional natural parameters, α, β, γ, δ, λ, and ω. In order to create perturbation problems for this equation, the natural parameters must be made into functions of a small parameter ε in such a way that the problem for $\varepsilon = 0$ is solvable. A perturbation problem, then, can be viewed as a path in the six-dimensional space of the natural parameters. (See the discussion associated with equations (2.3.14) and (2.3.15) for an elementary way to introduce ε into this problem.)

The simplest solvable problem that is a special case of (4.7.1) is the unforced and undamped linear oscillator $\ddot{y} + y = 0$, which is solvable for any initial conditions α and β. The simplest perturbation problems for Duffing's equation are those that reduce to this problem when $\varepsilon = 0$. The most general such perturbation problem can be found by letting α, β, and ω be completely arbitrary functions of ε, while δ, λ, and γ are arbitrary except that they must vanish at $\varepsilon = 0$. Assuming smoothness, such functions have expansions

$$\alpha(\varepsilon) = \alpha_0 + \varepsilon\alpha_1 + \cdots ,$$
$$\beta(\varepsilon) = \beta_0 + \varepsilon\beta_1 + \cdots ,$$
$$\gamma(\varepsilon) = \varepsilon\gamma_1 + \cdots ,$$
$$\delta(\varepsilon) = \varepsilon\delta_1 + \cdots \tag{4.7.2}$$
$$\lambda(\varepsilon) = \varepsilon\lambda_1 + \cdots ,$$
$$\omega(\varepsilon) = \omega_0 + \varepsilon\omega_1 + \cdots .$$

In order to establish the determining equations and study the existence of periodic solutions, the terms of these expansions are needed only through the first order. The last four of the functions in (4.7.2) are commonly taken to be exactly given by $\gamma(\varepsilon) = \varepsilon\gamma_1$, $\delta(\varepsilon) = \varepsilon\delta_1$, $\lambda(\varepsilon) = \varepsilon\lambda_1$, and $\omega(\varepsilon) = \omega_0 + \varepsilon\omega_1$, without higher order terms. This is of course permissible, but is more restrictive than necessary. (Incidentally, this explains the notation in equation (4.5.1).) On the other hand, the expansions for α and β cannot be truncated because they are not fixed in advance, but must be determined in the course of the analysis because we are looking for periodic solutions.

The choice of ω_0 is crucial because it determines the type of resonance (or nonresonance) being studied. For harmonic resonance, which we will consider first, $\omega_0 = 1$. Ignoring second order terms in the control parameters at our disposal (or setting them equal to zero), we can specify the harmonic resonance problem for Duffing's equation

$$\ddot{y} + \varepsilon\delta_1\dot{y} + y + \varepsilon\lambda_1 y^3 = \varepsilon\gamma_1 \cos(1 + \varepsilon\omega_1)t. \tag{4.7.3}$$

The determining equations for periodic solutions of this problem have been found in Exercise 4.6.1; in polar coordinates ($\alpha_0 = r\cos\theta$, $\beta_0 = r\sin\theta$) they are

$$2\omega_1 r - \tfrac{3}{4}\lambda_1 r^3 + \gamma_1 \cos\theta = 0,$$
$$\delta_1 r - \gamma_1 \sin\theta = 0. \tag{4.7.4}$$

Our primary goal here is to study the solutions of these equations to discover how many periodic solutions exist for the harmonic resonance problem, and how they bifurcate as the control parameters are varied. Then we will carefully interpret what the results say about the original Duffing equation in natural parameters.

Before doing these things, however, we will run through the derivation of the determining equations in order to point out some useful tricks that simplify the calculations, which you probably didn't notice (and weren't intended to notice) in working out Exercise 4.6.1. So, if you are reading this before doing Exercise 4.6.1, you might want to stop and work that exercise before looking at the rest of this paragraph. After making the change of independent variable from t to t^+ according to (4.6.6) and expanding to obtain (4.6.9), the equation for y_1 turns out to be

$$y_1'' + y_1 = -2\omega_1\{-\alpha_0 \cos t^+ - \beta_0 \sin t^+\} - \delta_1\{-\alpha_0 \sin t^+ + \beta_0 \cos t^+\}$$
$$- \lambda_1\{\alpha_0 \cos t^+ + \beta_0 \sin t^+\}^3 + \gamma_1 \cos t^+. \tag{4.7.5}$$

At this point, rather than cubing the expression in parentheses, it is convenient to introduce polar coordinates immediately. One finds, using trigonometric identities, that

$$\alpha_0 \cos t^+ + \beta_0 \sin t^+ = r\cos(t^+ - \theta),$$
$$\alpha_0 \sin t^+ - \beta_0 \cos t^+ = r\sin(t^+ - \theta).$$

The cubed term can now be computed by one reference to Table E.1 rather than four. The only remaining difficulty is the term $\gamma_1 \cos t^+$, which must be expressed in terms of $t^+ - \theta$ in order to be comparable with the other terms. Write

$$\cos t^+ = \cos[(t^+ - \theta) + \theta] = \cos\theta\cos(t^+ - \theta) - \sin\theta\sin(t^+ - \theta).$$

Now the equation for y_1 is

$$y_1'' + y_1 = \left\{2\omega_1 r - \frac{3}{4}\lambda_1 r^3 + \gamma_1\cos\theta\right\}\cos(t^+ - \theta)$$
$$+ \{\delta_1 r - \gamma_1\sin\theta\}\sin(t^+ - \theta) - \frac{1}{4}\lambda_1\cos 3(t^+ - \theta).$$

The terms in braces are resonant (in the sense of linear resonance) and will give rise to secular terms in y_1; setting them equal to zero gives the determining equations directly in the form (4.7.4). (To be precise, one should point out that $\cos(t^+ - \theta)$ and $\sin(t^+ - \theta)$ are linearly independent functions of t^+ spanning the same two-dimensional subspace as $\cos t^+$ and $\sin t^+$; therefore they serve to identify the resonant terms.)

We will begin the analysis of the determining equations (4.7.4) in the special case of no damping (or at least, of no damping to first order, $\delta_1 = 0$). In this case the second equation of (4.7.4) reduces to $\sin\theta = 0$, so that $\theta = 0$ or π. It is easy to see that if (r, π) is a solution of (4.7.4) then so is $(-r, 0)$, and these two solutions correspond to the same point in the (α_0, β_0) plane. Therefore it is not necessary to find all of the solutions of the determining equation; it is enough *either* to find all the solutions with $\theta = 0$, *or else* all the solutions having $r > 0$. Choosing the former convention ($\theta = 0$), the first equation of (4.7.4) becomes $2\omega_1 r - 3\lambda_1 r^3/4 + \gamma_1 = 0$. This is a cubic equation for r and cannot easily be solved. However, it can be solved for ω_1:

$$\omega_1 = \frac{3}{8}\lambda_1 r^2 - \frac{\gamma_1}{2r}. \tag{4.7.6}$$

Now it is not difficult to graph ω_1 as a function of r for various fixed values of λ_1 and γ_1. First, if $\gamma_1 = 0$, the curve is merely the parabola $\omega_1 = 3\lambda_1 r^2/8$. If $\gamma_1 \neq 0$, then it follows from (4.7.6) that $\omega_1 \to \pm\infty$ as $r \to 0$ and $\omega_1 \to 3\lambda_1 r^2/8$ as $r \to \pm\infty$. Thus the curve is asymptotic to the ω_1 axis and to the parabola found above in the case $\gamma_1 = 0$. It is easy to show that there is a single vertical tangent. A typical curve with these properties is shown in Fig. 4.7.1a in the case $\lambda_1 > 0$, $\gamma_1 > 0$. The same curve is shown in Fig. 4.7.1b under the convention that $r > 0$ rather than $\theta = 0$; the solutions from Fig. 4.7.1a having negative r are reflected into the upper half-plane and now must be labeled as having $\theta = \pi$. The vertical lines shown in these figures correspond to a fixed value of ω_1 (as well as fixed values of λ_1 and γ_1) for which the cubic equation (4.7.6) has three solutions for r; the location of the corresponding three points in the (α_0, β_0) plane is shown in Fig. 4.7.1c. Notice that as ω_1 is decreased, the two points on the negative side of the α_0 axis move together and annihilate one another at the vertical tangent point; this is a pair bifurcation point. Except at this bifurcation point, the zeroes of the determining equation are simple and therefore actually correspond to periodic solutions of Duffing's equation. (At the bifurcation point the zero is not simple, and strictly speaking there is no conclusion about the behavior of the system. But in practice, nothing unusual happens at this point. It may be assumed that any actual physical value of ω_1 will not exactly equal the critical value but will lie slightly on one side or the other.) Remember that the three points in Fig. 4.7.1b are the limiting positions as $\varepsilon \to 0$ of the initial conditions for these periodic solutions and that the zero-order approximations to these solutions are the simple harmonic oscillations with the same initial conditions, namely, $y \cong \alpha_0 \cos t$; the phase plane graphs of these approximate solutions are the three concentric circles about the origin passing through the three points in Fig. 4.7.1c, shown in Fig. 4.7.1d. The polar coordinate r at a zero of the determining equations is the approximate amplitude of the corresponding periodic solution. Therefore a graph such as Fig. 4.7.1b gives the approximate amplitudes of periodic solutions when a nonlinear spring is forced with a frequency $1 + \varepsilon\omega_1$ slightly out of exact harmonic resonance.

If the damping coefficient δ_1 is not equal to 0, it is not quite as simple to find the graphical solution to (4.7.4), but our experience with the undamped case is a good guide. The two equations together are unchanged if

(r, θ) is replaced by $(-r, \theta + \pi)$, so it is sufficient to look for solutions with $r > 0$. (This is the second convention in the undamped case, the convention of Fig. 4.7.1b. The convention $\theta = 0$ does not work this time, since θ can take any value.) The second equation gives $\sin \theta = \delta_1 r / \gamma_1$, which implies $\cos^2 \theta = 1 - \delta_1^2 r^2 / \gamma_1^2$. The first equation then implies

$$\left(\omega_1 - \frac{3}{8} \lambda_1 r^2 \right)^2 = \frac{\gamma_1^2}{4r^2} - \frac{\delta_1^2}{4}, \tag{4.7.7a}$$

which can also be written

$$\left(2\omega_1 r - \frac{3}{4} \lambda_1 r^3 \right)^2 + \delta_1^2 r^2 = \gamma_1^2. \tag{4.7.7b}$$

Figure 4.7.2 shows the shape of these curves in the (ω_1, r) plane for fixed δ_1 and λ_1. The plane is filled with curves having different values of γ_1; two of these are shown, together with the loci of their horizontal and vertical tangents. (See Exercise 4.7.1.) Notice that there are two different types of curves, depending on the magnitude of the forcing parameter γ_1 in relation to the damping δ_1 and nonlinearity λ_1. When $|\gamma_1| \leq \sqrt{8\delta_1 / 3\lambda_1}$, then the curves are not folded over, and there is only one periodic solution for each ω_1. Otherwise the curve is folded, and there is an interval $\omega_1^- < \omega_1 < \omega_1^+$ in which three periodic solutions exist for each ω_1. For small ω_1 there is one periodic solution; as ω_1 is increased across the first critical value ω_1^-, a new pair of periodic solutions is created; these move apart, and at the second critical value the larger meets the original periodic solution and annihilates it, leaving one solution (but not the "same" one that existed at first). As before, the roots are simple except at the bifurcation points, where the theorems of the last section do not (strictly speaking) apply.

It is worth emphasizing that the quantities involved in the determining equations are not the original parameters in Duffing's equation in the form (4.7.1). To choose values of these quantities is not to select a particular Duffing equation; it is to select a particular perturbation family of Duffing equations. The graphical solution of the determining equations then tells whether there are one or three families of periodic solutions for that perturbation family, for sufficiently small ε. It would be much more convenient, in practice, to have equations and graphs expressed in natural parameters that would decide the number and approximate amplitude of periodic solutions for any Duffing equation. To this end it is customary to substitute $\omega = 1 + \varepsilon\omega_1$, $\lambda = \varepsilon\lambda_1$, $\delta = \varepsilon\delta_1$, and $\gamma = \varepsilon\gamma_1$ into (4.7.7). It happens that ε cancels out, leaving the equation

$$\left(\omega - 1 - \frac{3}{8} \lambda r^2 \right)^2 = \frac{\gamma}{4r^2} - \frac{\delta^2}{4}, \tag{4.7.8}$$

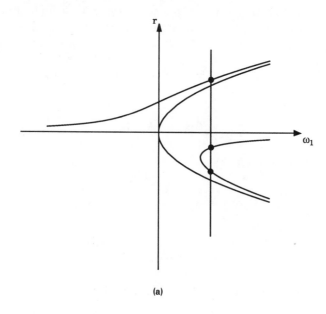

(a)

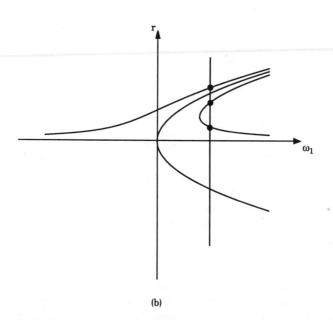

(b)

FIG. 4.7.1. Response curves for the harmonically forced Duffing equation without damping. The solid dots in each figure refer to the same three periodic solutions. (a) The response curves under the convention $\theta = 0$, showing three periodic solutions for a given forcing frequency. (b) The same response curves under the convention $r > 0$. (c) The limiting position, as $\varepsilon \to 0$, of the initial conditions for the three periodic solutions. (d) The limiting position, as $\varepsilon \to 0$, of the three periodic orbits in the phase plane.

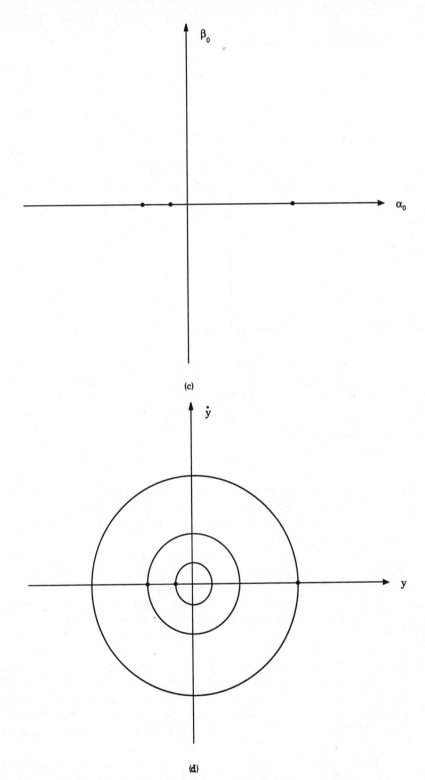

(c)

(d)

FIG. 4.7.1 *(Continued)*

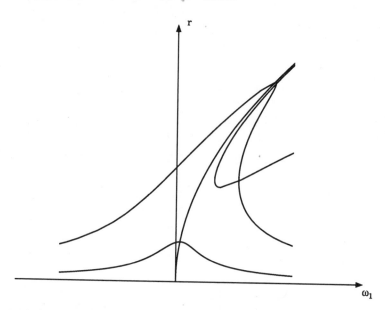

FIG. 4.7.2. Response curves for the harmonically forced Duffing equation with damping, under the convention $r > 0$; this should be compared to Fig. 4.7.1b. The curve through the origin is the locus of horizontal tangents to the response curves. The fingerlike curve is the locus of vertical tangents. Each point on this response curve is correct and is useful for sufficiently small ε, but the range of ε may be different at different points (nonuniformity). Contrast this with Fig. 4.7.3.

called the *frequency response equation* for Duffing's equation. It is intended to give the approximate amplitude r of the periodic solutions (one or three) of Duffing's equation forced at a given frequency ω. Typical graphs of (4.7.8) are shown in Fig. 4.7.3a.

Although the mathematical operations leading to (4.7.8) seem simple, it is necessary to take this equation with a grain of salt. The reason is that conclusions drawn from (4.7.7) only apply *for sufficiently small ε*. Considering how the natural parameters are related to the corresponding constants with subscript 1, this means that (4.7.8) applies only when ω is "close" to 1 and when λ, δ, and γ are all "sufficiently" small. Now there is a precise meaning to "sufficiently small" when this phrase is applied to ε: It means that given ω_1, δ_1, γ_1, and λ_1, there exists ε_0 such that for all ε with $|\varepsilon| < \varepsilon_0$ the periodic solutions exist and are approximated by their Lindstedt series. It is not so easy to say what "sufficiently small" means when applied to several independent parameters at the same time, because the allowable range for one of the parameters can depend upon the values of the others.

One way out of this difficulty is simply to draw the curves given by (4.7.8), announce that they are predictions of the amplitudes of periodic solutions and that these predictions are most likely to be true when the

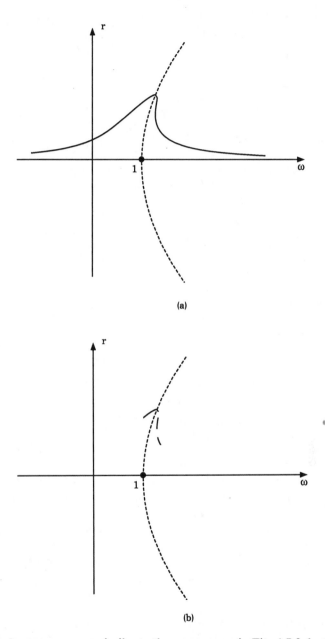

FIG. 4.7.3. A response curve similar to the upper curve in Fig. 4.7.2, but expressed in natural parameters for a fixed value of ε. Notice that the horizontal axis is ω and not ω_1. (a) shows the entire response curve, but not all of this curve is reliable because of the nonuniformity mentioned in Fig. 4.7.2. (b) shows the portions of the curve that are guaranteed to be reliable.

quantities mentioned are small, and then leave it to experiment (or computer simulation) to determine the range of validity. Ultimately this is what must be done, because as usual in perturbation theory we do not know the value of ε_0.

Nevertheless, there is insight to be gained by working out exactly what can be said on a theoretical basis about equation (4.7.8). To this end, observe that until now the four basic control parameters with subscript 1 have been regarded as fixed; therefore ε_0 depends upon these parameters. In other words, different points on the curve in Fig. 4.7.2 may apply for different ranges of ε. In order to pass from Fig. 4.7.2 to Fig. 4.7.3a it is necessary to choose a value of ε; but since this ε may not be valid for all points in Fig. 4.7.2, it may be that only a portion of Fig. 4.7.3a is valid. To clarify the situation, let us consider only the three parameters λ_1, δ_1, and γ_1 as fixed (this singles out a particular curve for the determining equation), and let us confine ω_1 to a compact interval $a \leq \omega_1 \leq b$ not containing the points ω_1^- and ω_1^+ at which the vertical tangents occur in Fig. 4.7.2. Then an application of the implicit function theorem in the form of Theorem B.3 implies that there is a single value of ε_0 which can be used throughout the interval $[a, b]$. Now choose an ε smaller than ε_0. This fixes the values of $\lambda = \varepsilon\lambda_1$, $\delta = \varepsilon\delta_1$, and $\gamma = \varepsilon\gamma_1$, thus singling out a particular frequency-response curve. Notice that this curve will not be very greatly bent over from the vertical, because the amount of bending is determined by small quantities. Choosing ε also fixes the interval of ω for which the response curve is known to be valid; namely, $1 + \varepsilon a \leq \omega \leq 1 + \varepsilon b$. This is a small interval on the ω axis. Of course, what we have just done for a single interval can be done for any compact set; Fig. 4.7.3b shows the result for a compact set consisting of three closed intervals on the ω_1 axis, coming close to (but omitting) the vertical tangent points which must be omitted because of Theorem B.4.

Our next objective is to explain the existence of a resonance horn for the 1 : 1 resonance in Duffing's equation, using the frequency-response equation. The discussion of resonance horns in Section 4.5 was based on equation (4.5.10), which contains only two control parameters, μ and ω. In order to bring (4.7.1) into this form, we will assume that λ, δ, and $\gamma/2$ are equal, and call their common value μ. (Of course other cases can be studied in the same way.) Then Duffing's equation becomes

$$\ddot{y} + y = \mu\{-\dot{y} - y^3 + 2\cos\omega t\}, \tag{4.7.9}$$

and the frequency-response equation (4.7.8) becomes

$$\left(\omega - 1 - \frac{3}{8}\mu r^2\right)^2 = \frac{\mu}{2r^2} - \frac{\mu^2}{4}. \tag{4.7.10}$$

Now for each value of μ there is a frequency-response curve having vertical tangents at two points $\omega^-(\mu) < \omega^+(\mu)$. (The reason we took $\gamma = 2\mu$ is that this is a large enough γ to produce vertical tangents in the response curve when $\lambda = \delta = \mu$.) The curves $\omega^-(\mu)$ and $\omega^+(\mu)$ in the (ω, μ) plane define the 1 : 1 resonance horn; within the horn there are three periodic solutions with period equal to the forcing period, and outside of it there is only one. (Of course, near the boundary of the horn the validity of the response curve is doubtful.) Figure 4.7.4 shows the resonance horn and its derivation from the vertical tangent points of the response curves drawn in three dimensions.

There are other interesting ways to make three-dimensional graphs of the frequency-response equation. Figure 4.7.5 shows the graph obtained when λ and δ are held constant and r is considered as a (sometimes multiple-valued) function of γ and ω. Each "slice" of this graph for constant γ gives one of the previously drawn response curves; the three-dimensional graph shows how the "pleat" in the surface giving rise to triple solutions is "pulled out" when the forcing term becomes too small. Any surface having this general shape is known as a *cusp catastrophe surface*. The word *cusp* refers to the cusp-shaped projection of the vertical tangent points into the base plane. A *catastrophe* is an abrupt change in the state of a system.

To see how Fig. 4.7.5 predicts such a change, suppose that a Duffing oscillator is executing a periodic motion corresponding to a point on the

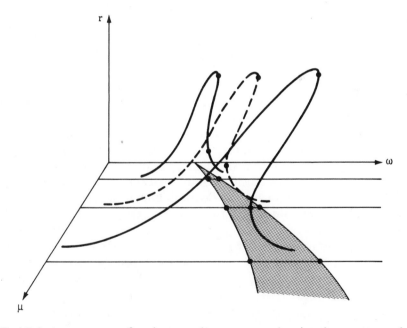

FIG. 4.7.4. A response surface in natural parameters, showing three cross-sections for various μ. Each cross-section is a response curve of the type shown in Fig. 4.7.3. The solid dots on the response surface are the vertical tangents; these project downward onto the solid dots in the (μ, ω) plane, which trace out the boundary of the resonance horn.

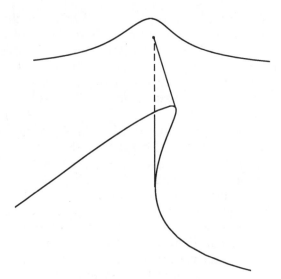

FIG. 4.7.5 A cusp catastrophe surface.

upper sheet of the surface. Suppose that the control parameters γ and ω of the oscillator are changed slowly enough that the oscillator continues to execute a slowly modulated motion close to the periodic motion predicted by Fig. 4.7.5. This is possible provided that the parameters do not cross one of the cusp lines in the base plane. If such a crossing takes place, the periodic solution corresponding to the upper sheet of the surface ceases to exist. The theory which we have developed in this chapter is not sufficient to predict what happens next, because we have not spoken about the nonperiodic solutions of Duffing's equation or about the stability of the periodic solutions. In Chapter 6 it will be seen that whenever a damped Duffing equation has only one periodic solution, it is stable and all other solutions tend toward that one. Therefore when the "upper sheet" solution disappears, the system undergoes an abrupt (although not discontinuous) change to (a solution close to) the "lower sheet" solution. This transition is the so-called "catastrophe."

It is possible to analyze other $p : q$ resonances for Duffing's equation in a similar way. This will be addressed in the exercises and also in Chapter 6.

Exercises 4.7

1. Show that the graph of (4.7.7) is correctly represented in Fig. 4.7.2, by carrying out the following steps. Differentiate (4.7.7b) implicitly with respect to ω_1 and set $dr/d\omega_1 = 0$; show from this that the curve has a horizontal tangent where it intersects the parabola $\omega_1 = 3\lambda_1 r^2/8$.

Similarly, show that the curve has a vertical tangent where it intersects the curve

$$\left(2\omega_1 - \tfrac{3}{4}\lambda_1 r^2\right)\left(2\omega_1 - \tfrac{9}{4}\lambda_1 r^2\right) + \delta_1^2 = 0. \qquad (4.7.11)$$

To show that this locus of vertical tangents has the shape indicated in Fig. 4.7.2, show first that the graph of (4.7.11) in the (ω_1, r) plane is contained between the parabolas $\omega_1 = 3\lambda_1 r^2/8$ and $\omega_1 = 9\lambda_1 r^2/8$. (These parabolas are where the two parentheses in (4.7.11) vanish.) Next, use implicit differentiation again to show that the curve defined by (4.7.11) has a horizontal tangent at the point $(\omega_1, r) = (\delta_1, \sqrt{4\delta_1/3\lambda_1})$. This is the minimum point on the locus of vertical tangents of the response curves. Finally, use (4.7.7b) to show that the response curve passing through this minimum point is the one for which $\gamma_1^2 = 8\delta_1^3/3\lambda_1$. This establishes that response curves with smaller γ_1^2 have no vertical tangents, and response curves with greater γ_1^2 have two. (Warning: If you differentiate (4.7.7a) implicitly, instead of (4.7.7b), you will arrive at a different equation for the vertical tangents than (4.7.11) since δ_1 will disappear upon differentiation and γ_1 will be retained. It is possible to recover (4.7.11) by using (4.7.7a) once again to eliminate γ_1. The reason that (4.7.11) is the desired form is that in Fig. 4.7.2, λ_1 and δ_1 are fixed, whereas γ_1 varies for the different response curves in the figure.)

2. For the undamped Duffing equation, assume that α_0 and β_0 are simple roots of the determining equation and solve equation (4.7.5) for y_1. (Remember that this equation is considerably simplified because the determining equation holds and there is no damping.) Set up the equation for y_2 and find the equations for α_1 and β_1. Sketch the shape of a typical periodic solution. (The determining equation only fixes y_0, giving a simple harmonic oscillation as the leading order approximation. Finding the next order, y_1, reveals what terms in the Fourier series are most important after the fundamental, and therefore shows the most important part of the deviation from simple harmonic motion introduced by the nonlinearity.)

3. Investigate the subharmonic resonance case for Duffing's equation with small forcing. According to Table 4.5.1, since $k = 1$, the subharmonic case is $\omega(0) = p$ with p an integer greater than 1. Therefore the equation to be studied is the same as (4.7.3) with the right hand side replaced by $\varepsilon\gamma_1 \cos(p + \varepsilon\omega_1)t$. Again according to Table 4.5.1, the critical problem for this equation concerns the existence of solutions having period exactly p times the period of the forcing; that is, the solution frequency is $\nu(\varepsilon) = \omega(\varepsilon)/p = (p + \varepsilon\omega_1)/p$. Find the determining equations for such solutions by looking for Lindstedt series for solutions of this frequency. From your determining equations, show: (a) that there are no subharmonic solutions to this problem when there is damping, and (b) when there is no damping, the determining equations are degenerate (that is,

the zeroes are not simple) so that no conclusion about the existence of subharmonic solutions can be drawn by the present method. (Remark: The problem being discussed here is not the one that is most often posed with regard to subharmonics of Duffing's equation. If the forcing is not small, that is, if no ε multiplies the right hand side, then subharmonic solutions do exist for $p = 3$; this problem will be addressed in Chapter 6.)

4. Consider the equation

$$\ddot{y} + \varepsilon\delta_1\dot{y} + y + \varepsilon\lambda_1 y^3 = \varepsilon\left(\gamma_1\cos\omega t + \mu_1\cos\frac{\omega t}{2}\right).$$

Letting $\omega = p + \varepsilon\omega_1$, show that the determining equations are nondegenerate in both the harmonic case $(p = 1)$ and the subharmonic case of order 2 $(p = 2)$.

4.8. MULTIPLE SCALE APPROXIMATIONS DERIVED FROM LINDSTEDT SERIES

One of the three or four most frequently used asymptotic methods is the *method of multiple scales* which will be developed in Chapter 5. The method of multiple scales is able to handle all of the problems which have been considered in this chapter, and in addition can be used for nonperiodic solutions. However, in the periodic case the approximate solutions given by the multiple scale method are not exactly the same as those found by the Lindstedt method. From one point of view they are simpler than the Lindstedt expansions, but this simplicity is gained at a cost: The multiple scale approximations lack the "trade-off" property of Lindstedt series. That is, a multiple scale approximation valid for an expanding time interval of order $1/\varepsilon$ does not necessarily satisfy any error estimate on an interval of order $1/\varepsilon^2$. In fact, the multiple scale approximations of a periodic solution are not even periodic. (If it seems surprizing that there can be several different asymptotic series for the same solution, remember that both Lindstedt series and multiple scale series are *generalized* asymptotic expansions, and these do not have a uniqueness theorem. In fact the work done in this section illustrates why Theorem 1.8.3 fails for generalized expansions.)

In order to motivate the method of multiple scales, we will show in this section how to obtain approximations of the multiple scale type from Lindstedt expansions. It should be emphasized that the method used here to obtain multiple scale solutions is not the actual "method of multiple scales"; that method does not begin with a Lindstedt expansion, but constructs the solutions from the beginning according to its own principles.

Consider a typical Lindstedt series:

$$y \sim y_0(t^+) + \varepsilon y_1(t^+) + \varepsilon^2 y_2(t^+) + \cdots,$$
$$t^+ = \nu t, \tag{4.8.1}$$
$$\nu \sim \nu_0 + \varepsilon \nu_1 + \varepsilon^2 \nu_2 + \cdots.$$

An essential feature of the Lindstedt method is that the coefficients $y_n(t^+)$ in this expansion are periodic in t^+. Suppose that it is desired to create an approximation having error $\mathcal{O}(\varepsilon^2)$ on an expanding interval of length $\mathcal{O}(1/\varepsilon)$. The usual Lindstedt prescription for such an approximation is to take two terms for y and three terms for ν in (4.8.1):

$$y \cong y_0(\nu_0 t + \nu_1 \varepsilon t + \nu_2 \varepsilon^2 t) + \varepsilon y_1(\nu_0 t + \nu_1 \varepsilon t + \nu_2 \varepsilon^2 t). \tag{4.8.2}$$

The approximation (4.8.2) satisfies not only the desired error estimate but also a trade-off estimate of $\mathcal{O}(\varepsilon)$ on expanding intervals of length $\mathcal{O}(1/\varepsilon^2)$. Introduce the notations

$$T_0 = t, \quad T_1 = \varepsilon t, \quad T_2 = \varepsilon^2 t, \quad \ldots, \quad T_n = \varepsilon^n t. \tag{4.8.3}$$

These quantities are called *time scales*. Observe that the time scale T_1 is bounded on expanding intervals of order $1/\varepsilon$; that is, if $0 < t < L/\varepsilon$, then $0 < T_1 < L$. Write (4.8.2) in the following form using (for the present) only the two time scales T_0 and T_1:

$$y \cong y_0(\nu_0 T_0 + \nu_1 T_1 + \nu_2 \varepsilon T_1) + \varepsilon y_1(\nu_0 T_0 + \nu_1 T_1 + \nu_2 \varepsilon T_1). \tag{4.8.4}$$

Now expand (4.8.4) in a formal Taylor series in ε, holding T_0 and T_1 constant (as though T_1 were an independent variable and not a function of ε), and delete terms of formal order ε^2:

$$y \cong y_0(\nu_0 T_0 + \nu_1 T_1) + \varepsilon\{y_{0s}(\nu_0 T_0 + \nu_1 T_1)\nu_2 T_1 + y_1(\nu_0 T_0 + \nu_1 T_1)\}. \tag{4.8.5}$$

This *two-time-scale approximation* is a special case of the following general form:

$$y \cong y_0(T_0, T_1) + \varepsilon y_1(T_0, T_1) = y_0(t, \varepsilon t) + \varepsilon y_1(t, \varepsilon t). \tag{4.8.6}$$

We will show in the next paragraph that the approximation (4.8.5) has an error of the same order as (4.8.2) on intervals of order $1/\varepsilon$, but does not satisfy any useful estimate on intervals of order $1/\varepsilon^2$. Observe that (4.8.5) is not periodic; the second term is a periodic function times T_1 and becomes unbounded as $t \to \infty$. However, this term (which is called a T_1-*secular term*) is bounded on intervals of order $1/\varepsilon$, which partially explains why the approximation is successful on such intervals.

In order to analyze the error committed in passing from (4.8.2) to (4.8.5), notice that since y_0 is periodic in t^+, Theorem 4.2.1 implies that

$$y_0(t^+ + h) = y_0(t^+) + y_0'(t^+)h + \mathcal{O}(h^2) \qquad (4.8.7)$$

uniformly for all t^+. Applying this with $t^+ = \nu_0 T_0 + \nu_1 T_1$ and $h = \nu_2 \varepsilon T_1$, the error that results from the y_0 terms of (4.8.5) is less than a constant times $(\varepsilon T_1)^2$. Since T_1 is bounded on intervals of order $1/\varepsilon$, this error is of order ε^2, the same order as the error already present in (4.8.2). A similar analysis shows that the error committed in replacing the y_1 term of (4.8.2) by the y_1 term of (4.8.5) is of the same order. On the other hand, $(\varepsilon T_1)^2$ is unbounded on intervals of order $1/\varepsilon^2$, so no error estimate can be deduced for (4.8.5) on these intervals.

It is possible to carry out the process just indicated, using only the two time scales T_0 and T_1, to any order in ε. (See Exercise 4.8.1.) This produces two-time-scale expansions having error of any desired order on intervals of order $1/\varepsilon$. None of these approximations satisfy any trade-off estimates on longer intervals. In order to obtain approximations valid on expanding intervals of order $1/\varepsilon^2$ it is necessary to make use of three time scales T_0, T_1, and T_2. (At least, it is necessary in the present case of periodic solutions. It will be seen in Chapter 5 that approximations to damped oscillations can be uniformly valid for all time with just two time scales.) For a three-time approximation with error $\mathcal{O}(\varepsilon)$ for time $1/\varepsilon^2$ one can simply use (4.8.2) itself, written in the form

$$y \cong y_0(\nu_0 T_0 + \nu_1 T_1 + \nu_2 T_2) + \varepsilon y_1(\nu_0 T_0 + \nu_1 T_1 + \nu_2 T_2). \qquad (4.8.8)$$

(Warning: The error in approximations like (4.8.8) is not of the order of the first omitted term, when the interval in question is longer than $\mathcal{O}(1/\varepsilon)$.) To improve the accuracy of (4.8.8) it is necessary to return to (4.8.1) and retain at least one more term in both y and ν. (See Exercise 4.8.2.)

In discussing the various kinds of multiple scale approximations, it is useful to define *secular terms* of various kinds. If (4.8.2) is expanded in powers of ε, the result contains secular terms proportional to t. (The series obtained is, of course, the regular perturbation series which would be obtained by the methods of Chapter 2 and is uniformly valid only for finite intervals.) Since $t = T_0$, such terms are called T_0-*secular terms*. We have already mentioned that (4.8.5) contains a T_1-*secular term* and that this is permissible in a solution that is not intended to last beyond time $1/\varepsilon$. A solution that is to last up to time $1/\varepsilon^2$ cannot contain T_1-secular terms (at least if the actual solution is periodic or even just bounded); to eliminate them, it is necessary to bring in the time scale T_2. This strategy of "buying" more time by bringing in more time scales and eliminating secular terms of successively higher order is always successful with periodic solutions. For nonperiodic solutions, the use of more time scales does not always "buy" accuracy for longer times; there are difficult unsolved problems here, and the question will be discussed in more detail in Section 5.4.

In conclusion, it must be said that no one would actually wish to discard a Lindstedt expansion in favor of a multiple scale expansion if the Lindstedt expansion is already known. To do so is to discard both the periodicity of the approximation and the trade-off accuracy on long intervals of time. Our purpose in this discussion was to show that nonperiodic approximations of periodic solutions are possible on expanding intervals and to show that the concept of multiple time scales, to be developed in Chapter 5, is already implicitly present in Lindstedt approximations.

Exercises 4.8

1. Beginning with (4.8.1), write a Lindstedt approximation having error $\mathcal{O}(\varepsilon^3)$ on expanding intervals of order $\mathcal{O}(1/\varepsilon)$. Convert this to a two-time-scale approximation with the same order of error.

2. The Lindstedt approximation that you used in Exercise 4.8.1 also has error $\mathcal{O}(\varepsilon^2)$ on intervals of order $\mathcal{O}(1/\varepsilon^2)$. Convert this to a three-time-scale approximation with the same order of error. What type of secular terms are present in your result?

4.9* PARTIAL DIFFERENTIAL EQUATIONS

As an illustration of the Lindstedt method for partial differential equations, consider the problem

$$
\begin{aligned}
u_{tt} + u_{xx} + u &= \varepsilon(1 - u^2)u_t, \\
u(0, t) &= 0, \\
u(\pi, t) &= 0
\end{aligned}
\tag{4.9.1}
$$

for a function $u(x, t)$ defined on the infinite strip $0 \leq x \leq \pi$, $-\infty < t < \infty$. The first two terms of (4.9.1), that is, the highest derivative terms, are the same as for the wave equation; therefore this equation is called a nonlinear hyperbolic equation and can be expected (or hoped) to have solutions satisfying the same kind of initial and boundary conditions as are usually imposed on the wave equation. In fact the equation can be thought of as describing the vibrations of some unusual sort of string subject to energy inputs and losses which we will not attempt to describe physically. The boundary conditions in (4.9.1), then, should not be sufficient to determine u completely, and one should also impose initial conditions on $u(x, 0)$ and $u_t(x, 0)$.

But there is another aspect to (4.9.1) that suggests a different approach: The perturbation resembles a self-sustaining term (as in Section 4.4) and therefore suggests the possible existence of periodic solutions. In fact, equation (4.9.1) without the boundary conditions has nontrivial periodic solutions that are independent of x. To see this, note that if u is independent of x then the u_{xx} term drops out from (4.9.1), leaving an ordinary differential equation which is in fact just the Van der Pol equation (Section 4.4). Thus we might think of each point x on the segment (or "string") $0 \le x \le \pi$ as a self-excited oscillator coupled to the oscillators on either side of it; the boundary conditions mean that the oscillators at the ends are pinned and prevented from oscillating. (Of course, on a continuous string it is not correct to think of each point as having a point "on either side of it," but this picture of coupled oscillators does describe the coupled system of ordinary differential equations obtained when (4.9.1) is discretized with respect to x.) Motivated by this analogy, we will look for periodic solution families $u(x, t, \varepsilon)$ of (4.9.1) having unknown initial conditions and an unknown frequency $\nu(\varepsilon)$; because of our experience with self-excited oscillators, we expect both the initial conditions and the frequency to depend upon ε. The periodicity condition for such a solution family is

$$u\left(x, t + \frac{2\pi}{\nu(\varepsilon)}, \varepsilon\right) = u(x, t, \varepsilon). \tag{4.9.2}$$

The Lindstedt method calls for introducing a new time variable, which for convenience will be called s rather than t^+, by

$$s := \nu(\varepsilon)t, \tag{4.9.3}$$

and a new solution function $v(x, s, \varepsilon) := u(x, s/\nu(\varepsilon), \varepsilon)$; then v must satisfy

$$\begin{aligned}
&\nu(\varepsilon)^2 v_{ss} - v_{xx} + v = \varepsilon\nu(1 - v^2)v_s, \\
&v(0, s) = 0, \\
&v(\pi, s) = 0, \\
&v(x, s + 2\pi, \varepsilon) = v(x, s, \varepsilon).
\end{aligned} \tag{4.9.4}$$

We will, further, insist that 2π be the least period of v; this will be the case, for any periodic solution, if ν is chosen appropriately. Solutions of (4.9.4) will be sought in the form

$$\begin{aligned}
&v(s, x, \varepsilon) = v_0(x, s) + \varepsilon v_s(x, s) + \cdots, \\
&\nu(\varepsilon) = \nu_0 + \varepsilon\nu_1 + \cdots.
\end{aligned} \tag{4.9.5}$$

Upon substituting (4.9.5) into (4.9.4) one finds the following sequence of problems (see Exercise 4.9.1):

$$\begin{aligned}
&\nu_0^2 v_{0ss} - v_{0xx} + v_0 = 0, \\
&v_0(0, s) = v_0(\pi, s) = 0, \\
&v_0(x, s + 2\pi) = v_0(x, s) \quad \text{(least period)},
\end{aligned} \tag{4.9.6a}$$

$$v_0^2 v_{1ss} - v_{1xx} + v_1 = -2v_0 v_1 v_{0ss} + v_0(1 - v_0^2)v_{0s},$$

$$v_1(0, s) = v_1(\pi, s) = 0, \qquad\qquad\qquad (4.9.6b)$$

$$v_1(x, s + 2\pi) = v_1(x, s) \qquad \text{(not necessarily least period)}$$

$$\vdots$$

Problem (4.9.6a) can be solved by separation of variables. Writing

$$v_0(x, s) = X(x)S(s), \qquad\qquad (4.9.7)$$

one can write the partial differential equation as

$$v_0^2 \frac{S''}{S} = \frac{X''}{X} - 1 = -\lambda,$$

where λ is a separation constant. This leads to the two linear ordinary differential equation eigenvalue problems

$$X'' + (\lambda - 1)X = 0,$$
$$X(0) = X(\pi) = 0, \qquad\qquad (4.9.8)$$

and

$$S'' + \frac{\lambda}{v_0^2} S = 0,$$
$$S(s + 2\pi) = S(s). \qquad\qquad (4.9.9)$$

Problem (4.9.8) has nontrivial solutions only if $\lambda - 1 = n^2$ for $n = 1, 2, \ldots$; in this case the solution (up to a multiplicative constant) is $X(x) = \sin nx$. Problem (4.9.9) has solutions for which 2π is the least period only if $v_0^2 = \lambda$; in this case all solutions have this period, being linear combinations of $\cos s$ and $\sin s$. Together these facts imply that there is an infinite sequence of solutions for v_0 and v_0, namely

$$v_0^{(n)} = \sqrt{1 + n^2},$$
$$v_0^{(n)} = \left(A_0^{(n)} \cos s + B_0^{(n)} \sin s \right) \cos nx, \qquad n = 1, 2, \ldots, \qquad (4.9.10)$$

where the amplitudes $A_0^{(n)}$ and $B_0^{(n)}$ are not yet determined. It is not possible to sum over n in the second equation of (4.9.10) to form other solutions of (4.9.6a), as one usually does in the method of separation of variables; the reason is that $v_0^{(n)}$ only satisfies (4.9.6a) when $v_0 = v_0^{(n)}$, and v_0 can only take on one value at a time in (4.9.6a). Therefore the procedure from here is to take each $v_0^{(n)}$ as the starting point of a perturbation family

$v^{(n)}(x, s, \varepsilon)$ for (4.9.4) and attempt to determine $A_0^{(n)}$, $B_0(n)$, and ν_1 so that (4.9.6b) is solvable.

But at this point there is one more observation to make based on our experience with ordinary differential equations. We have seen that for an autonomous ordinary differential equation in the phase plane, any periodic solution must cross the y axis, so that by changing the origin of time (which is permissible for autonomous equations) the initial condition could be taken in the form $y(0) = \alpha(\varepsilon)$, $\dot{y}(0) = 0$; the first order perturbation solution then became $y_0 = \alpha_0 \cos t^+$ rather than $\alpha_0 \cos t^+ + \beta_0 \sin t^+$. This simplification was crucial, because it was only after reducing the number of undetermined constants in this way that the determining equation was able to have simple solutions. The same sort of thing happens with the present problem. It is too much to try to determine $A_0^{(n)}$, $B_0^{(n)}$, and ν_1 at the next stage of the computations; the determining equations would come out to be degenerate. Instead one observes that (4.9.4) is autonomous (in the sense that it does not explicitly depend upon s), so that any solution remains a solution if the origin of time is shifted. Then it is possible to assume that $B_0^{(n)} = 0$, since for any periodic solution, a change in the origin of time will make this true. (In other words, differentiating (4.9.10) with respect to s reveals that for any such solution, there is a time at which the velocity of the string is zero at every point. Take such a time as the origin.) Therefore, from here on we work with

$$\nu_0^{(n)} = \sqrt{1 + n^2},$$
$$v_0^{(n)} = A_0^{(n)} \cos s \cos nx. \tag{4.9.11}$$

Inserting (4.9.11) into the right hand side of (4.9.6b), the explicit equation for v_1 is found to be

$$(1 + n^2)v_{1ss} - v_{1xx} + v_1$$
$$= A\sqrt{1 + n^2}\{2\nu_1 \cos s \cos nx - \sin s \cos nx + A^2 \cos^2 s \sin s \cos^3 nx\}$$
$$= A\sqrt{1 + n^2}\{2\nu_1 \cos s \cos nx + \left(\tfrac{3}{16}A^2 - 1\right) \sin s \cos nx$$
$$+ \tfrac{1}{16}A^2 \sin s \cos 3nx + \tfrac{1}{16}A^2 \sin 3s \cos 3nx\}, \tag{4.9.12}$$

where $A = A_0^{(n)}$. The second version of the right hand side is obtained from Table E.1 as usual. Of course, in addition to (4.9.12), v_1 must satisfy the boundary and periodicity conditions in (4.9.6b). Solutions satisfying these conditions exist if and only if the first two terms in the right hand side of (4.9.12) vanish; these terms are in resonance with the free frequencies of the operator $L = (1 + n^2)\partial^2/\partial s^2 - \partial^2/\partial x^2 + 1$ and would lead to secular terms in v_1, which are incompatible with periodicity. Therefore

$$\nu_1 = 0,$$
$$A_0^{(n)} = \frac{\sqrt{3}}{4}. \qquad (4.9.13)$$

The theory involved here is called the *Fredholm alternative*: If L is self-adjoint with respect to an inner product on a space of functions satisfying certain side conditions, then an equation $Lv = f$ with those side conditions is solvable if and only if f is orthogonal to the solutions of $Lv = 0$ with the same side conditions; orthogonality here is with respect to the inner product. In the present case, the inner product is

$$\langle f, g \rangle = \frac{1}{2\pi^2} \int_0^{2\pi} \int_0^{\pi} f(x,s)g(x,s)dxds,$$

and the linearly independent solutions of $Lv = 0$ are (as already found by separation of variables) $\cos s \cos nx$ and $\sin s \cos nx$. Therefore the solvability conditions are

$$\langle f, \cos s \cos nx \rangle = 0,$$
$$\langle f, \sin s \cos nx \rangle = 0. \qquad (4.9.14)$$

With f equal to the right hand side of (4.9.12), these conditions just state that the first two terms vanish, leading to (4.9.13).

We will leave the problem at this point. The treatment has been purely formal, and we offer no details about the actual existence of periodic solutions. Frequently, nonlinear hyperbolic equations have solutions that cannot be continued smoothly beyond a certain time because of the formation of shock waves; a rigorous treatment must show that this does not happen within one period, and therefore does not happen at all.

Exercises 4.9

1. Explain why v_0 in (4.9.6) must have 2π as its least period and v_1 need not.

2. Find the general form of the determining equations when the right hand side of (4.9.1) is replaced by $\varepsilon h(u, u_t)$.

4.10. NOTES AND REFERENCES

The notion of secular terms is as old as its name, derived from celestial mechanics, implies. I believe there is some dispute over whether Lindstedt was the first to use the series named for him. I know of no source for the easy argument showing that Lindstedt series have the trade-off property, and so far as I know that property has not previously been named. However, it has often been taken for granted, and someone must have verified it. (On the other hand, it is also sometimes taken for granted in the setting of multiple scale expansions, and in this generality we will show later that it is false.)

The subject of nonlinear oscillations (Sections 4.3–4.7) is quite old, and one of the classical expositions (still valuable although dated) is

J. J. Stoker, *Nonlinear Vibrations in Mechanical and Electrical Systems*, Wiley-Interscience, New York, 1950.

A more modern (and more difficult) treatment, focusing on the existence and stability of periodic solutions, is given in Chapters 5–8 of

Jack K. Hale, *Ordinary Differential Equations*, Wiley-Interscience, New York, 1969.

(There is a slight error in the graph of the response curve for Duffing's equation, and the theorems concerned with almost periodic function have a missing hypothesis.) A reference containing many experimental as well as mathematical results (mostly without rigorous proof) is

Chihiro Hayashi, *Nonlinear Oscillations in Physical Systems*, McGraw-Hill, New York, 1964, reprinted by Princeton University Press, 1985.

The computational aspect of nonlinear oscillations is represented in all of Nayfeh's books, and in particular in

Ali Nayfeh and Dean Mook, *Nonlinear Oscillations*, Wiley, New York, 1979.

The Hopf bifurcation, briefly touched on in Section 4.4, is extensively treated in

J. E. Marsden and M. McCracken, *The Hopf Bifurcation and Its Applications*, Springer-Verlag, New York, 1976.

Finally, the most modern point of view on nonlinear oscillations, that of dynamical systems theory, is represented in

John Guckenheimer and Philip Holmes, *Nonlinear Oscillations, Dynamical Systems, and Bifurcations of Vector Fields*, second printing (revised and corrected), Springer-Verlag, New York, 1983.

The entrainment theorem (Theorem 4.5.1 and its generalizations) has a long and somewhat obscure history. Much of the literature takes it for granted that any forced oscillation must be entrained; occasionally it is suggested that to prove this is an unsolved problem. In fact, there are several proofs in existence. The one given here, covering the case (4.5.7) in which the time-dependent term is separate from the state-dependent terms, was found by one of my graduate students but is so simple that it has probably been discovered many times. I have developed a generalization of this proof to handle the general case of (4.5.4) and a wide variety of other forced evolution equations (including delay differential equations); the result is stated in the paragraph following equation (4.5.8) above, and the proof can be found in

James Murdock, Frequency entrainment for almost periodic evolution equations, *Proc. Amer. Math. Soc.* **96** (1986), 626–628.

At the time I submitted this paper I expressed concern that the result might be known, and indeed the editor was able to locate the following somewhat obscure reference, which contains this result among others, with a different proof (valid for periodic ordinary differential equations only, not general evolution equations):

J. Massera, Observaciones sobre las soluciones periodicas de ecuaciones diferenciales, *Bol. Fac. Ingen. Montevideo* **4** (1950–53), 37–45.

Perhaps the earliest fairly general entrainment theorem, valid only for asymptotically stable periodic solutions, is the one given in

E. Trefftz, Zu den Grundlagen der Schwingungstheorie, *Mathematische Annalen* **95** (1926), 307–312.

The example in Exercise 4.5.1 for which entrainment does not hold is given in

N. Forbat, *Analytische Mechanik der Schwingungen*, V.E.B. Deutscher Verlag der Wissenschaft, Berlin, 1966.

The distinction between critical and noncritical cases of forcing is quite clear for the second order equations (or two-dimensional systems) presented here: Either *all* unperturbed solutions have a period in common with the forcing, or *none* of them do. Things become more complicated in higher-dimensional systems, where it is possible that *some subspace* is filled with periodic solutions having a period in common with the forcing, while solutions outside that subspace do not. The reference by Hale (mentioned at the beginning of this section) discusses this situation.

The self-sustained oscillation in a partial differential equation, treated in Section 4.9, occurs on page 419 of

M. Millman, *Perturbation solutions of some nonlinear boundary value problems, in Bifurcation Theory and Nonlinear Eigenvalue Problems* (J. Keller and S. Antman, eds.), Benjamin, New York, 1969.

CHAPTER 5

MULTIPLE SCALES

5.1. OVERVIEW OF MULTIPLE SCALES AND AVERAGING

This chapter and the next concern initial value problems of oscillatory type on long intervals of time. Until Section 5.4, we will study autonomous oscillatory second order initial value problems of the form

$$\ddot{y} + k^2 y = \varepsilon f(y, \dot{y}, \varepsilon),$$
$$y(0) = \alpha, \tag{5.1.1}$$
$$\dot{y}(0) = \beta.$$

In Section 5.4 it will be shown that this and many other systems, including periodically forced oscillators and systems of several coupled oscillators, can be put into *periodic standard form*

$$\dot{\mathbf{u}} = \varepsilon \mathbf{f}(\mathbf{u}, t, \varepsilon), \tag{5.1.2}$$

where $\mathbf{u} = (u_1, \ldots, u_N)$ is an N-dimensional vector variable and f is periodic in t. Because of its generality and simplicity, periodic standard form will be used in the rest of this chapter and most of the next. It may appear that changing an equation into standard form is an unnecessary step requiring extra work, but in fact the variables which form the components of $\mathbf{u}$ are often exactly those of greatest interest from the standpoint of applications. In addition, theoretical discussions can be given once for equations in standard form and then applied to a wide variety of problems, each of which would require separate treatment if standard form were not used.

For these reasons we consider standard form to be an advantage rather than a disadvantage. On the other hand, beginning with (5.1.1) makes for a smooth transition from previous chapters.

Of course, equation (5.1.1) has been studied in Section 4.4 by the Lindstedt method. However, due to the inherent limitations of that method, only periodic solutions could be found at that time. For instance, the limit cycle of a Van der Pol equation could be located, but it was not possible to find approximations of the nonperiodic orbits that approach the limit cycle. It is the purpose of the method of multiple scales and the method of averaging to remove this limitation. These methods enable the construction of an approximate solution with arbitrary initial conditions which no longer need to satisfy a determining equation. One feature of the Lindstedt method carries over to these new methods: the results are usually valid on an "expanding interval" of time. In most cases this expanding interval has length $\mathcal{O}(1/\varepsilon)$. Error estimates on longer intervals, such as $\mathcal{O}(1/\varepsilon^2)$ or even for all time, can be obtained under various special circumstances, but there are no general results of this type. In particular, the "trade-off" property of Lindstedt expansions expressed in equation (4.2.15) is definitely false for multiple scale and averaging methods in general.

The solution of the reduced problem of (5.1.1) can be written in several ways, such as

$$y = A \cos kt + B \sin kt \qquad (5.1.3)$$

and

$$y = \rho \cos(\psi - kt), \qquad (5.1.4)$$

where A, B, ρ, and ψ are constants of integration. The first of these, (5.1.3), will be called *Cartesian form*, and (5.1.4) will be called *polar form*, because the relations $A = \rho \cos \psi$ and $B = \rho \sin \psi$, which follow from (5.1.3) and (5.1.4), are the same as the relations between Cartesian and polar coordinates in the plane. (This will become more explicit in Section 5.4 and explains our preference for (5.1.4) over forms such as $y = a \sin(\delta + kt)$.) The solution of the reduced problem can be regarded as the "zeroth" approximation to the solution of the perturbed problem (5.1.1). According to the results of Chapter 2, this zeroth approximation has an error $\mathcal{O}(\varepsilon)$ uniformly for t in a finite interval $0 \le t \le T$, and is (in most cases) useless on longer intervals. The first task of both the method of multiple scales and the method of averaging is to improve this zeroth approximation to produce a "first approximation" having error $\mathcal{O}(\varepsilon)$ uniformly for t in an expanding interval of the form $0 \le t \le L/\varepsilon$. This is an essentially different problem from that of improving the accuracy to $\mathcal{O}(\varepsilon^2)$ on a finite interval, a task which is accomplished quite satisfactorily by regular perturbation theory.

The terms "zeroth" and "first" approximation, used in the last paragraph, have a slightly different significance here than in previous chapters. The phrase *zeroth approximation* always refers to *the solution of the reduced problem*. In earlier chapters, the solution of the reduced problem was also the leading term in the asymptotic series being constructed. However, in the methods of multiple scales and averaging, the leading term of the asymptotic solution is not the solution of the reduced problem, but is already an improvement over that solution. The phrase *first approximation* will always be used to refer to *the first improvement of the zeroth approximation*. Therefore, in regular perturbation theory, the "first approximation" consists of two terms of the asymptotic solution (the zeroth and first order terms, as measured by powers of ε); in Chapters 5 and 6, the "first approximation" will consist only of the leading term of the asymptotic series.

The method of multiple scales and the method of averaging both give the same answer to the problem of finding a first approximation to the solution of (5.1.1) on an expanding interval, although they obtain this answer by quite different reasoning. The first approximation given by both methods takes the same form as the zeroth approximation (5.1.3) or (5.1.4), except that the quantities A, B, ρ, and ψ, which were constants of integration, become slowly varying functions of time. More precisely, the approximations take the form

$$y \cong A(\varepsilon t) \cos kt + B(\varepsilon t) \sin kt \qquad (5.1.5)$$

or

$$y \cong \rho(\varepsilon t) \cos(\psi(\varepsilon t) - kt), \qquad (5.1.6)$$

where the functions of εt are the solutions of certain differential equations which will be given in the next section. Any function of εt can be considered as a slowly varying function of time, since the derivative of such a function contains a factor of ε due to the chain rule. (Another way to look at this is that t must increase a great deal before εt will change very much, when ε is small.) It is customary to write either $\tau := \varepsilon t$, or else $T_0 = t$ and $T_1 = \varepsilon t$, so that the approximations take the form

$$y \cong A(\tau) \cos kt + B(\tau) \sin kt = A(T_1) \cos kT_0 + B(T_1) \sin kT_0 \qquad (5.1.7)$$

or

$$y \cong \rho(\tau) \cos(\psi(\tau) - kt) = \rho(T_1) \cos(\psi(T_1) - kT_0). \qquad (5.1.8)$$

Here τ is called *slow time*, and the variables t and τ (or T_0 and T_1) are referred to as two *time scales*. When form (5.1.8) is used, one speaks of the *method of slowly varying amplitude and phase*, since ρ and ψ are the amplitude and phase of the simple harmonic motion (5.1.4) in the reduced case. (The "method" of slowly varying amplitude and phase is not actually a method in itself, but only a notation which can be used with several methods such as averaging and multiple scales.)

In most applications of the methods of multiple scales or averaging, the first approximation is sufficient. Therefore this approximation will be discussed at length, in both this chapter (Section 5.2) and the next (Sections 6.1, 6.2, and 6.3). To go beyond the first approximation can be quite difficult. Beyond the first approximation, the methods of averaging and of multiple scales do not always give the same results; in fact, each of these methods (multiple scales and averaging) has several forms, and these different forms do not give the same results beyond the first approximation. (This is possible because these are generalized asymptotic expansions and do not satisfy a uniqueness theorem like Theorem 1.8.3. The different results do not conflict with each other; each is a legitimate asymptotic approximation of the exact solution.) The various multiple scale methods have the reputation of being computationally simpler than the method of averaging after the first approximation, and this is no doubt true in many specific problems. On the other hand, the theoretical foundations of the method of averaging are much better understood than those of multiple scales, and it is possible to draw many conclusions about the behavior of the solutions, such as stability and periodicity, from the averaging calculations.

Higher order multiple scale methods for oscillatory problems can be grouped into three types:

1. Two-scale methods using $T_0 := t$ and $T_1 = \tau := \varepsilon t$. The solutions are written in the form $y_0(t, \tau) + \varepsilon y_1(t, \tau) + \cdots$.
2. Two-scale methods using a strained time $t^+ := (\nu_0 + \varepsilon \nu_1 + \varepsilon^2 \nu_2 + \cdots + \varepsilon^\ell \nu_\ell)t$ and a slow time $\tau := \varepsilon t$, with a suitable choice of $\nu_1, \ldots, \nu_\ell$. The solutions appear as $y_0(t^+, \tau) + \varepsilon y_1(t^+, \tau) + \cdots$.
3. Multiple scale methods using M scales $T_m := \varepsilon^m t$ for $m = 1, \ldots, M$. The solutions are written $y_0(T_0, T_1, \ldots, T_M) + \varepsilon y_1(T_0, T_1, \ldots, T_M) + \cdots$. A variation of this method (the "short form") omits one time scale in each successive term; for instance, a three-scale three-term solution would look like $y_0(T_0, T_1, T_2) + \varepsilon y_1(T_0, T_1) + \varepsilon^2 y_2(T_0)$.

The theory of higher order approximations by the first of these methods is fairly well understood. This form is applicable to a wide variety of problems and gives approximations to any order of accuracy which are valid on expanding intervals of length $\mathcal{O}(1/\varepsilon)$ but not (except under special circumstances) on longer intervals. The second and third forms are less satisfactory; they are intended to give approximations on expanding intervals longer than $\mathcal{O}(1/\varepsilon)$, but they do not always work (even formally), and there is not yet an adequate theory to explain when they are successful and why. The second form is motivated by the Lindstedt method; the idea is that many nonperiodic solutions have the form of damped oscillations in which the "periodic part" can be described using t^+ and the damping takes place on the time scale τ. The third form is the most general, since it includes the first two, the first by taking $M = 1$, and the second since t^+ (taken to ℓ terms) can be written as a function of the scales $T_0, \ldots, T_\ell$. In Section 5.3 (which may be omitted without loss of continuity) the first and third of these methods will be illustrated.

To conclude this introductory section we will study a certain linear problem of the form (5.1.1) "in reverse." That is, we will write down the exact solution, and then expand this solution into three different forms of multiple scale approximations and examine the error. This is, of course, "cheating," and is intended only to give insight; we are not actually using the method of multiple scales. (It is rather like "cheating" by using the quadratic formula in Chapter 1.) The example to be studied is

$$\ddot{y} + 2\varepsilon\dot{y} + (1 + \varepsilon)y = 0,$$
$$y(0) = \alpha, \qquad\qquad\qquad (5.1.9)$$
$$\dot{y}(0) = 0,$$

with exact solution

$$y = \alpha e^{-\varepsilon t} \cos \sqrt{1 + \varepsilon - \varepsilon^2}\, t + \frac{\varepsilon\alpha}{\sqrt{1 + \varepsilon - \varepsilon^2}} e^{-\varepsilon t} \sin \sqrt{1 + \varepsilon - \varepsilon^2}\, t. \quad (5.1.10)$$

There are two effects apparent here: The damping term in (5.1.9) causes an exponential decay of the solution "on the time scale εt," while both perturbation terms interact to produce a frequency shift from the free frequency 1. The shifted frequency can be expanded as

$$\sqrt{1 + \varepsilon - \varepsilon^2} \sim 1 + \frac{1}{2}\varepsilon - \frac{5}{8}\varepsilon^2 + \frac{15}{48}\varepsilon^3 + \cdots. \qquad (5.1.11)$$

The first form of multiple scale solution which will be developed is a two-time-scale expansion using t and τ. The technique for converting (5.1.10) into a two-time expansion is the same as that used in Section 4.7 for changing a Lindstedt expansion into a two-time expansion. First, using (5.1.10) and (5.1.11), and the time scales t and τ, write

$$y = \alpha e^{-\tau} \cos\left(t + \frac{1}{2}\tau - \frac{5}{8}\varepsilon\tau + \cdots\right)$$
$$+ \frac{\varepsilon\alpha}{\sqrt{1 + \varepsilon - \varepsilon^2}} e^{-\tau} \sin\left(t + \frac{1}{2}\tau + \cdots\right). \qquad (5.1.12)$$

Next, expand each term in ε holding t and τ constant as though τ were an independent variable rather than a function of ε. The result, through order ε, is

$$y \cong \alpha e^{-\tau} \cos\left(t + \frac{1}{2}\tau\right) + \varepsilon\alpha e^{-\tau}\left(\frac{5}{8}\tau + 1\right)\sin\left(t + \frac{1}{2}\tau\right). \qquad (5.1.13)$$

The same argument as in Section 4.7 shows that the error committed in passing from (5.1.10) to (5.1.13) is of order ε^2 on expanding intervals of order $1/\varepsilon$.

In fact, it is not hard to show that for this example the error is actually of order ε^2 for all $t > 0$. Consider the first term of (5.1.12). When this term is expanded in ε and two terms are retained, the error committed is bounded by a constant times $\varepsilon^2 t e^{-\varepsilon t}$. This expression reaches a maximum of $\varepsilon^2 e^{-\varepsilon}$ at $t = 1$, and since this maximum is less than ε^2, the error is of order ε^2 as claimed. A similar analysis can be done for the second term of (5.1.12). The reason that a solution using only two time scales is able to remain valid for all time, rather than only on an expanding interval of order $1/\varepsilon$, is that the exponential damping acts to reduce not only the size of the solution but also of the error as $t \to \infty$. The mechanism is similar to that by which the regular perturbation solution of a damped oscillation was found to be valid for all time in Chapter 4, the difference being that in Chapter 4 the damping was of order $\mathcal{O}(1)$ and here it is $\mathcal{O}(\varepsilon)$; such damping is too weak to rescue the regular expansion from the ravages of its secular terms.

The second type of multiple scale approximation to be derived from (5.1.10) is a two-scale approximation using the scales

$$t^+ = \left(1 - \frac{5}{8}\varepsilon^2\right)t,$$
$$\tau = \varepsilon t. \qquad (5.1.14)$$

Such an approximation can be obtained from (5.1.12) by dropping the dotted terms inside the trig functions:

$$y \cong \alpha e^{-\tau} \cos\left(t^+ + \frac{\tau}{2}\right) + \varepsilon\alpha e^{-\tau}\sin\left(t^+ + \frac{\tau}{2}\right). \qquad (5.1.15)$$

The final type of multiple scale expansion to be derived from (5.1.10) is a three-scale solution using t, τ, and $\sigma := \varepsilon^2 t$ (or T_0, T_1, and T_2). We leave it to the reader (Exercise 5.1.1) to check that the complete two-term, three-scale expansion of (5.1.10) is

$$y \cong \alpha e^{-\tau} \cos\left(t + \frac{1}{2}\tau - \frac{5}{8}\sigma\right)$$
$$+ \varepsilon\alpha e^{-\tau}\left(1 - \frac{15}{48}\sigma\right)\sin\left(t + \frac{1}{2}\tau - \frac{5}{8}\sigma\right). \qquad (5.1.16)$$

The "short form" of this three-scale expansion is obtained by dropping σ from the term of order ε; this is equivalent to (5.1.15). We have said that adding a third time scale is usually an attempt to gain validity on a longer time interval. Here it is not needed for that purpose, since as noted above, (5.1.13) is already uniformly valid for all time because of the damping.

Exercises 5.1

1. Introduce t, τ, and σ as "independent" variables in (5.1.10) and expand in ε to obtain (5.1.16).

5.2. THE FIRST ORDER TWO-SCALE APPROXIMATION

The idea of the two-scale method is to seek an approximate solution of the initial value problem

$$\ddot{y} + k^2 y = \varepsilon f(y, \dot{y}),$$
$$y(0) = \alpha, \qquad\qquad (5.2.1)$$
$$\dot{y}(0) = \beta$$

in the form

$$y \sim y_0(t, \tau) + \varepsilon y_1(t, \tau) + \varepsilon^2 y_2(t, \tau) + \cdots, \qquad (5.2.2)$$

where $\tau = \varepsilon t$. In this section, only the first term of (5.2.2) will be determined. (Recall from the last section that y_0 is called the first approximation, rather than the zeroth, because it is not equal to the solution of the reduced problem.) As in the Lindstedt method, finding y_0 requires that the equation for y_1 be written down as well; the equation for y_2 will also be found, since it is needed in the next section.

It must be understood that (5.2.2) is a rather peculiar way to write the solution of an ordinary differential equation. There are actually only two independent variables, t and ε, in (5.2.2); τ is a function of these two, and so is not independent. Nevertheless, the principal steps in finding the coefficients y_n are carried out as though t, τ, and ε were independent variables. This is one reason why these steps cannot be justified rigorously in advance, but are merely heuristic. Secondly, it must be remarked that (5.2.2) is a generalized asymptotic expansion, since ε enters both through the gauges (which are just the powers of ε) and also through the coefficients y_n by way of τ. Although there is no general theorem allowing the differentiation of a generalized asymptotic expansion term by term (see Section 1.8), it is nevertheless reasonable to construct the coefficients of (5.2.2) on the assumption that such differentiation is possible, and then to justify the resulting series by direct error estimation afterwards. The procedure is to take the total derivative of (5.2.2) with respect to t twice, termwise, and substitute the result into (5.2.1). At the time of taking these derivatives, τ is regarded as equal to εt, but afterwards, τ is treated as an independent variable. Since the total dependence of each term $y_n(t, \tau)$ upon t is a combination of its explicit dependence on t and its indirect dependence on t by way of τ, the first and second total derivatives of each term are given by the following formulas based on the chain rule:

$$\frac{d}{dt}y_n(t, \tau) = \frac{d}{dt}y_n(t, \varepsilon t) = \frac{\partial}{\partial t}y_n(t, \varepsilon t) + \varepsilon\frac{\partial}{\partial \tau}y_n(t, \tau)$$
$$= y_{nt}(t, \tau) + \varepsilon y_{n\tau}(t, \tau), \tag{5.2.3}$$

$$\frac{d^2}{dt^2}y_n(t, \tau) = y_{ntt}(t, \tau) + 2\varepsilon y_{nt\tau}(t, \tau) + \varepsilon^2 y_{n\tau\tau}(t, \tau).$$

These equations are easiest to remember in operator form:

$$\frac{d}{dt} = \frac{\partial}{\partial t} + \varepsilon\frac{\partial}{\partial \tau},$$
$$\frac{d^2}{dt^2} = \frac{\partial^2}{\partial t^2} + 2\varepsilon\frac{\partial^2}{\partial \tau \partial t} + \varepsilon^2\frac{\partial^2}{\partial \tau^2}. \tag{5.2.4}$$

Substituting the derivatives of (5.2.2), computed in this way, into (5.2.1) gives the following expressions for the differential equation and initial conditions:

$$(y_{0tt} + 2\varepsilon y_{0t\tau} + \varepsilon^2 y_{0\tau\tau}) + \varepsilon(y_{1tt} + 2\varepsilon y_{1t\tau} + \varepsilon^2 y_{1\tau\tau})$$
$$+ \varepsilon^2(y_{2tt} + 2\varepsilon y_{2t\tau} + \varepsilon^2 y_{2\tau\tau}) + \cdots + k^2(y_0 + \varepsilon y_1 + \varepsilon^2 y_2 + \cdots),$$
$$= \varepsilon f(y_0 + \varepsilon y_1 + \cdots, y_{0t} + \varepsilon y_{0\tau} + \varepsilon y_{1t} + \cdots), \tag{5.2.5}$$
$$y_0(0, 0) + \varepsilon y_1(0, 0) + \varepsilon^2 y_2(0, 0) + \cdots = \alpha,$$
$$y_{0t}(0, 0) + \varepsilon\{y_{0\tau}(0, 0) + y_{1t}(0, 0)\} + \varepsilon^2\{y_{1\tau}(0, 0) + y_{2t}(0, 0)\} + \cdots = \beta.$$

Expanding the right hand side of the differential equation in powers of ε and identifying coefficients of equal powers leads to the following sequence of initial value problems for the coefficients:

$$y_{0tt} + k^2 y_0 = 0,$$
$$y_0(0, 0) = \alpha, \tag{5.2.6a}$$
$$y_{0t}(0, 0) = \beta;$$

$$y_{1tt} + k^2 y_1 = f(y_0, y_{0t}) - 2y_{0t\tau},$$
$$y_1(0, 0) = 0, \tag{5.2.6b}$$
$$y_{1t}(0, 0) = -y_{0\tau}(0, 0);$$

$$y_{2tt} + k^2 y_2 = f_y(y_0, y_{0t})y_1 + f_{\dot{y}}(y_0, y_{0t})(y_{0\tau} + y_{1t})$$
$$- y_{0\tau\tau} - 2y_{1t\tau},$$
$$y_2(0, 0) = 0, \tag{5.2.6c}$$
$$y_{2t}(0, 0) = -y_{1\tau}(0, 0).$$

It should be remembered that even the step of equating coefficients of equal powers of ε, used in passing from (5.2.5) to (5.2.6), is not justified by any theorem about generalized asymptotic expansions (since there is no uniqueness theorem for such expansions). It is instead a heuristic assumption used to arrive at a candidate for an approximate solution, whose validity is to be determined afterwards by error analysis.

Since t and τ are being treated (temporarily) as independent, the differential equation in (5.2.6a) is actually a *partial* differential equation for a function y_0 of two variables t and τ. However, since no derivatives with respect to τ appear in (5.2.6a), it may be regarded instead as an *ordinary* differential equation for a function of t, regarding τ as merely an auxiliary parameter. Therefore the general solution of (5.2.6a) may be obtained from the general solution of the corresponding ordinary differential equation just by letting the arbitrary *constants* become arbitrary *functions* of τ:

$$y_0(t, \tau) = A(\tau)\cos kt + B(\tau)\sin kt = \rho(\tau)\cos(\psi(\tau) - kt). \qquad (5.2.7)$$

The initial conditions of (5.2.6a) impose the following restrictions upon the otherwise arbitrary functions in (5.2.7):

$$\begin{aligned}
A(0) &= \alpha, \\
B(0) &= \beta/k, \\
\rho(0) &= \sqrt{\alpha^2 + (\beta/k)^2}, \\
\psi(0) &= \arctan(\beta/k\alpha).
\end{aligned} \qquad (5.2.8)$$

Up to this point, the calculations are routine, although they require taking careful thought for the meaning of each step, so that, for instance, the total and partial derivatives with respect to t are not confused. Now, however, we have reached a crucial point, because it becomes clear that the initial value problem (5.2.6a) for y_0 does not completely determine y_0. We have used all of the information contained in (5.2.6a), and the functions A and B, or (alternatively) ρ and ψ, are still undetermined except for their initial values (5.2.8). Some new idea is necessary in order to complete the determination of these functions, and hence of y_0.

The new idea needed is actually not so new—it is the concept of eliminating secular terms, used repeatedly in Chapter 4. However, the notion of secular term required here is slightly different from that of Chapter 4. It can be motivated as follows. Our aim is that (5.2.2) should be a uniformly valid asymptotic expansion, on expanding intervals of (at least) order $1/\varepsilon$, of the exact solution of (5.2.1). In an asymptotic expansion, the error after the first term (that is, the difference between y_0 and the exact solution) is expected to be of the order of the second term, εy_1. Therefore to try to make the error be of order ε on expanding intervals of order $1/\varepsilon$, we should arrange things so that εy_1 is of order ε on such intervals, or in other words, that y_1 is bounded on these intervals. It is not guaranteed in advance that this will produce a valid asymptotic solution, but the plan we are adopting is the most likely to lead to the desired result. (In the language of Section 1.8, we know that in order to be *uniformly valid* on expanding intervals of order $1/\varepsilon$, the series must be *uniformly ordered* on these intervals, and that in turn requires that the coefficients must be bounded. On the other hand, achieving this uniform ordering does not automatically guarantee that the series is uniformly valid.) Since τ is bounded on expanding intervals of order $1/\varepsilon$, it is permissible for y_1 to contain so-called τ-*secular terms* (or T_1-*secular terms*) which involve factors of τ, but not t-secular terms which involve factors of t. (Both kinds of terms are secular in the sense of being unbounded on the positive t axis, but only the latter are unbounded on expanding intervals of order $1/\varepsilon$.)

With this idea in mind, we shall examine the differential equation (5.2.6b) for y_1 in order to see whether y_1 will contain t-secular terms. In view of (5.2.7), this differential equation takes the form

$$y_{1tt} + k^2 y_1 = f\big(A(\tau)\cos kt + B(\tau)\sin kt, -kA(\tau)\sin kt + kB(\tau)\cos kt\big)$$
$$+ 2kA'(\tau)\sin kt - 2kB'(\tau)\cos kt \qquad (5.2.9)$$

in Cartesian variables. Since the right hand side of (5.2.9) is periodic in t with period $2\pi/k$, it can be expanded in a Fourier series in t (for fixed τ). Then y_1 will be free of t-secular terms if and only if this Fourier series has no terms in $\sin kt$ and $\cos kt$; these terms, if present, would be linearly resonant with the free frequency k. Setting the coefficients of these terms equal to zero imposes two conditions on $A(\tau)$ and $B(\tau)$, namely, the following differential equations:

$$\frac{dA}{d\tau} = P(A, B) :=$$
$$-\frac{1}{2\pi}\int_0^{2\pi/k} f(A\cos kt + B\sin kt, -kA\sin kt + kB\cos kt)\sin kt\, dt,$$
$$\frac{dB}{d\tau} = Q(A, B) := \qquad (5.2.10)$$
$$+\frac{1}{2\pi}\int_0^{2\pi/k} f(A\cos kt + B\sin kt, -kA\sin kt + kB\cos kt)\cos kt\, dt.$$

For any specific function f, the integrals on the right hand side can (at least in principle, and often in fact) be evaluated, so that $P(A, B)$ and $Q(A, B)$ become known functions and (5.2.10) becomes an explicit set of differential equations for A and B. These equations are to be solved with the initial conditions given by (5.2.8); this completes the determination of A and B, and hence of y_0 (see (5.2.7)).

Since our purpose in this section is only to find the first approximation, we will not solve (5.2.9) for y_1. It was necessary to make use of the equation for y_1, but only in order to arrive at equations (5.2.10) for A and B. Briefly, the way to continue is as follows. Now that A and B are fixed, the equation (5.2.9) can be solved for y_1; but this equation does not completely determine y_1, rather it introduces further unknown functions that must be determined by eliminating resonant terms from the equation for y_2. This idea will be developed further in the next section.

In general, the functions $P(A, B)$ and $Q(A, B)$ in (5.2.10) are nonlinear, and at first it is not apparent how to solve these differential equations for A and B. In terms of the polar variables ρ and ψ, the corresponding equations are easier to solve. Therefore we will repeat the discussion of how to compute y_0, using these polar variables. When the polar form of (5.2.7) is substituted into (5.2.6b), the following equation is obtained, corresponding to (5.2.9):

$$y_{1tt} + k^2 y_1 = f\big(\rho(\tau)\cos(\psi(\tau) - kt), k\rho(\tau)\sin(\psi(\tau) - kt)\big)$$
$$- 2k\rho'(\tau)\sin\big(\psi(\tau) - kt\big) - 2k\rho(\tau)\psi'(\tau)\cos\big(\psi(\tau) - kt\big).$$
$$(5.2.11)$$

Recall that this equation is treated as an ordinary differential equation for y_1 as a function of t, with τ appearing as a parameter (and therefore treated as constant); the condition that y_1 have no t-secular terms is that the right hand side have no terms in $\sin kt$ and $\cos kt$ when expanded in a Fourier series. It is not difficult to show that this condition is equivalent to

$$\rho' = \frac{1}{2\pi} \int_0^{2\pi/k} f\big(\rho\cos(\psi - kt), k\rho\sin(\varphi si - kt)\big) \sin(\psi - kt)dt,$$
$$(5.2.12)$$
$$\psi' = \frac{1}{2\pi\rho} \int_0^{2\pi/k} f\big(\rho\cos(\psi - kt), k\rho\sin(\psi - kt)\big) \cos(\psi - kt)dt.$$

One way to see this is shown in Exercise 5.2.1.

Another way, which makes the result obvious but depends on theory that some readers may not know, is to observe that (for fixed τ) the functions $\sin(\psi(\tau) - kt)$ and $\cos(\psi(\tau) - kt)$ span the same vector space of functions as $\sin kt$ and $\cos kt$. Then (5.2.12) merely says that the projection of the right hand side of (5.2.11) onto these new basis vectors is zero, using the standard inner product for the space of functions of period $2\pi/k$.)

Next, (5.2.12) can be simplified by the substitution $\theta = \varphi - kt$ to the form

$$\rho' = F(\rho) := \frac{1}{2\pi k} \int_0^{2\pi} f(\rho\cos\theta, k\rho\sin\theta)\sin\theta \, d\theta,$$
$$(5.2.13)$$
$$\psi' = G(\rho) := \frac{1}{2\pi k\rho} \int_0^{2\pi} f(\rho\cos\theta, k\rho\sin\theta)\cos\theta \, d\theta.$$

These are the promised equations, which are simpler than the corresponding equations (5.2.10) in Cartesian variables since F and G depend upon
only one variable ρ, whereas P and Q depend upon both A and B. Although both (5.2.10) and (5.2.13) are in general nonlinear, the equations
in (5.2.13) are "solvable by quadrature," that is, the problem of solving
them is reducible to a problem of integration. Namely, the first equation
can be written in separated form $d\rho/F(\rho) = d\tau$ and integrated, using the
initial condition from (5.2.8); after the solution $\rho(\tau)$ is found, the second equation becomes $d\psi = G(\rho(\psi))d\psi$, where the right hand side is a
known function of ψ to be integrated. Provided these two integrations can
be done, the solution is complete.

Of course, since (5.2.10) is equivalent to (5.2.13) by a simple change
of variables, it also is solvable by quadrature although that is not obvious
from its form. This observation is useful, since the Cartesian form has
a definite advantage when it comes to computing higher approximations
(Section 5.3). The procedure for solving (5.2.10) by quadrature will be
developed in the examples below.

This completes the heuristic derivation of the first order two-scale approximation. It has already been mentioned that the same approximation
will be derived again, from an entirely different point of view, in Chapter
6. The question remains: Does this approximation accomplish its purpose?
The answer is yes. The following theorem is proved at the end of Section
5.5:

Theorem 5.2.1. *For fixed α and β there exist constants ε_1, c, and T such
that the exact solution $y(t, \varepsilon)$ of (5.2.1) satisfies*

$$|y(t,\varepsilon) - y_0(t, \varepsilon t)| \le c\varepsilon \quad \text{for} \quad 0 \le t \le T/\varepsilon, \quad 0 \le \varepsilon \le \varepsilon_1. \quad (5.2.14)$$

A similar theorem giving more information about T is proved in Chapter 6 (Theorem 6.2.2).

Example 5.2.1. A Linear Problem

As a first illustration of the method of multiple scales, we will compute
the first approximation to the (exactly solvable) linear problem discussed
in Section 5.1:

$$\ddot{y} + 2\varepsilon\dot{y} + (1 + \varepsilon)y = 0,$$
$$y(0) = \alpha, \quad (5.2.15)$$
$$\dot{y}(0) = 0.$$

The desired approximation has already been derived from the exact solution; it is the first term of (5.1.13). Here it will be obtained by following
the procedures described above. As always, it is more instructive to work

out each step of the procedure for each application, rather than to apply formulas such as (5.2.10) or (5.2.13) mechanically. Substitution of (5.2.2) into (5.2.15), using (5.2.4), leads to the following sequence of linear initial value problems (compare (5.2.6)):

$$
\begin{aligned}
&y_{0tt} + y_0 = 0, \\
&y_0(0,0) = \alpha, \\
&y_{0t}(0,0) = 0;
\end{aligned}
\tag{5.2.16a}
$$

$$
\begin{aligned}
&y_{1tt} + y_1 = -(y_0 + 2y_{0t} + 2y_{0t\tau}), \\
&y_1(0,0) = 0, \\
&y_{1t}(0,0) = -y_{0\tau}(0,0);
\end{aligned}
\tag{5.2.16b}
$$

$$
\begin{aligned}
&y_{2tt} + y_2 = -(2y_{0\tau} + y_{0\tau\tau} + y_1 + 2y_{1t} + 2y_{1t\tau}), \\
&y_2(0,0) = 0, \\
&y_{2t}(0,0) = -y_{1\tau}(0,0).
\end{aligned}
\tag{5.2.16c}
$$

We will need only the first two of these in this section, but the third will be used later. The solution of (5.2.16a) in "amplitude/phase" form is

$$
\begin{aligned}
&y_0 = \rho(\tau) \cos(\psi(\tau) - t), \\
&\rho(0) = \alpha, \\
&\psi(0) = 0,
\end{aligned}
\tag{5.2.17}
$$

where $\rho(\tau)$ and $\psi(\tau)$ are as yet undetermined except for their initial conditions. The equation for y_1 now becomes

$$
y_{1tt} + y_1 = -\rho(1 + 2\psi') \cos(\psi - t) - 2(\rho' + \rho) \sin(\psi - t).
\tag{5.2.18}
$$

Since ψ is a function of τ only, it is treated as a constant while solving the equation for y_1. Rather than expand the right hand side in a Fourier series in t, it is more convenient to expand it in $\psi - t$; in fact, it is already expressed as a Fourier series in $\psi - t$, which contains only two terms, both of them resonant with the free frequency 1. Therefore the heuristic rule that t-secular terms should be eliminated compels us to choose $\rho(\tau)$ and $\psi(\tau)$ to satisfy

$$
\begin{aligned}
&\rho' = -\rho, \\
&\psi' = -1/2.
\end{aligned}
\tag{5.2.19}
$$

The solution of these equations, with the initial conditions given in (5.2.17), are

$$
\begin{aligned}
&\rho(\tau) = \alpha e^{-\tau}, \\
&\psi(\tau) = -\tau/2,
\end{aligned}
\tag{5.2.20}
$$

leading to the complete first approximation,

$$y_0(t, \tau) = \alpha e^{-\tau} \cos(-\tau/2 - t) = \alpha e^{-\tau} \cos(t + \tau/2). \qquad (5.2.21)$$

Now we will repeat these calculations in Cartesian variables. As remarked above, it is more difficult to compute the first approximation in Cartesian variables than in polar, but for higher approximations the reverse is true. Therefore it is worthwhile to understand how to compute the first approximation in Cartesian variables. Beginning with (5.2.16a), the solution can be written

$$y_0 = A(\tau) \cos t + B(\tau) \sin t,$$
$$A(0) = \alpha, \qquad\qquad\qquad (5.2.22)$$
$$B(0) = 0.$$

The equation for y_1 is

$$y_{1tt} + y_1 = -(A + 2B + 2B') \cos t + (-B + 2A + 2A') \sin t. \qquad (5.2.23)$$

The right hand side is, once again, already expanded in a Fourier series (this time in t), having two terms, both of which are resonant. The nonsecularity conditions are

$$A' = -A + B/2,$$
$$B' = -A/2 - B. \qquad\qquad (5.2.24)$$

This is a linear system of two differential equations in two unknown functions, and can be solved as such, for instance by matrix methods. But it is linear only because the original equation (5.2.15) is linear; so to solve (5.2.24) by linear methods will not help with more general problems. Instead, observe that by the chain rule and by (5.2.24),

$$\frac{d}{d\tau}(A^2 + B^2) = 2AA' + 2BB' = -2(A^2 + B^2). \qquad (5.2.25)$$

This equation can be solved for the quantity $A^2 + B^2$:

$$A^2 + B^2 = [A(0)^2 + B(0)^2]e^{-2\tau} = (\alpha e^{-\tau})^2. \qquad (5.2.26)$$

(Notice that since $A^2 + B^2 = \rho^2$, we have actually introduced polar coordinates "on the sly" here. This is the secret of how to solve (5.2.10) by quadrature.) It follows that there exists $\xi(\tau)$ such that

$$A(\tau) = \alpha e^{-\tau} \cos \xi(\tau),$$
$$B(\tau) = \alpha e^{-\tau} \sin \xi(\tau). \qquad\qquad (5.2.27)$$

To compute ξ, substitute the first equation of (5.2.27) into the first equation of (5.2.24) to obtain

$$2A' + 2A - B = -\alpha(2\xi' + 1)\sin\xi = 0$$

or

$$\xi' = -1/2, \qquad\qquad (5.2.28)$$

so that $\xi(\tau) = \xi(0) - \tau/2$. (Compare (5.2.28) with the second equation of (5.2.19): ξ is another "secret" polar coordinate.) The initial value $\xi(0) = 0$ follows from (5.2.22) and (5.2.27), from which

$$
\begin{aligned}
A(\tau) &= \alpha e^{-\tau}\cos\frac{\tau}{2}, \\
B(\tau) &= -\alpha e^{-\tau}\sin\frac{\tau}{2},
\end{aligned}
\qquad\qquad (5.2.29)
$$

giving

$$y_0(t, \tau) = \alpha e^{-\tau}\left\{\cos\frac{\tau}{2}\cos t - \sin\frac{\tau}{2}\sin t\right\}, \qquad\qquad (5.2.30)$$

in agreement with (5.2.21).

Example 5.2.2. The Van der Pol Equation

For the next example, consider the Van der Pol equation

$$
\begin{aligned}
\ddot{y} + \varepsilon(y^2 - 1)\dot{y} + y &= 0, \\
y(0) &= \alpha, \\
\dot{y}(0) &= 0,
\end{aligned}
\qquad\qquad (5.2.31)
$$

which has been studied in Section 4.4. Since this is a nonlinear equation, the exact solution is not available for comparison with the approximation found by multiple scale methods. However, it has been seen that for small ε there is a unique limit cycle and it is stable. These features can be recognized in the approximate solution by two-timing, which is (see Exercise 5.2.2):

$$y \cong \frac{2\alpha}{\sqrt{\alpha^2 + (4 - \alpha^2)e^{-\varepsilon t}}}\cos t. \qquad\qquad (5.2.32)$$

It is possible to recognize some of the properties of the exact solutions of Van der Pol's equation in this approximate solution. For instance, as $t \to \infty$ in (5.2.32), the transient effect (on the time scale τ) dies out, leaving the periodic solution

$$y \cong 2\cos t \qquad\qquad (5.2.33)$$

which coincides with the Lindstedt approximation to the limit cycle found in Exercise 4.4.1. Thus the approximate solution exhibits a stable limit cycle close to the exact limit cycle of the Van der Pol equation. One might wonder why the two-timing approximation is this good; after all, it is only intended to be uniformly valid on an expanding interval, not for all time. It turns out that a careful analysis of the error in (5.2.33) shows that while the phase ψ is only approximated well on an expanding interval, the amplitude ρ is approximated well for all time. As in the previous example, this is due to damping (or more precisely, to the attracting character of the limit cycle). In this problem, damping only affects the error in a direction normal to the limit cycle, and "in-track error" continues to accumulate in the tangential direction just as it does for the Lindstedt approximation to the limit cycle. (See Section 4.3 for the notion of "in-track error.") It is not always permissible to draw conclusions about the behavior of a system as time approaches infinity from approximate solutions, but when the behavior in question is sufficiently "strong" in some sense, it will survive in the approximations. The further exploration of this topic is beyond this book; see Notes and References.

Exercises 5.2

1. Derive (5.2.12) from the nonsecularity condition on (5.2.11) by another method than that indicated in the text. Hint: Multiply the right hand side of (5.2.11) by $\sin kt$ and by $\cos kt$, and integrate from 0 to $2\pi/k$. Set the results equal to zero, and use trig identities.

2. Find the first order two-scale approximation for the Van der Pol equation (5.2.30). Do not use the general formulas, but carry out the procedure of substituting (5.2.2) into (5.2.30) and obtaining equations for y_0 and y_1. Use polar form for the solution for y_0, and fix $\rho(\tau)$ and $\psi(\tau)$ by eliminating resonant terms from the equation for y_1. After you have obtained your approximate solution, compare it with the solution to Exercise 2.4.4b. Repeat the calculation using Cartesian variables.

3. Find the first order two-scale approximation for the autonomous Duffing equation $\ddot{y} + y + \varepsilon y^3 = 0$, $y(0) = \alpha$, $\dot{y}(0) = 0$. Do the calculation in both polar and Cartesian variables.

4. Find a leading order approximation (by the multiple scale method) for $\ddot{y} + \varepsilon y \dot{y} + y = 0$, $y(0) = 1$, $\dot{y}(0) = 0$. Compare the result with Exercise 4.3.6.

5.3* HIGHER ORDER APPROXIMATIONS

In this section (which may be omitted without loss of continuity) we will investigate three strategies for finding higher order multiple scale approximations. These three strategies have already been mentioned at the end of Section 5.1, where three types of second order approximations were given for the linear initial value problem (5.1.9). Briefly stated, the three approaches are: (1) to continue as in Section 5.2 with the two time scales t and τ, computing higher order terms in the series (5.2.2); (2) to use two time scales, but replace t by a "strained" time scale similar to that used in the Lindstedt method; and (3) to add more time scales besides t and τ.

The purpose of the first strategy is to improve the accuracy of the first approximation, without attempting to increase the length of time $\mathcal{O}(1/\varepsilon)$ for which the approximation is valid. For problems of the form (5.1.1) it can be shown that the two-time-scale expansion using t and τ which was begun in the last section can be carried out to any order, and that the solution (5.2.2) taken through the term y_{k-1} has error $\mathcal{O}(\varepsilon^k)$ on expanding intervals of length $\mathcal{O}(1/\varepsilon)$. Thus, at least in theory, the first strategy is always successful at achieving its goal. However, to carry out the solution in practice requires solving certain differential equations in order to eliminate secular terms; these differential equations are in general nonlinear, and therefore may not have "closed form" solutions (that is, explicit solutions in terms of elementary functions). So the fact that the solutions are possible "in theory" does not always guarantee that the calculations are possible in practice, although in many cases they are.

The second and third strategies are more ambitious. Their aim is not only to improve the asymptotic order of the error estimate, but also to extend the validity of the approximations to "longer" intervals of time, that is, expanding intervals of length $\mathcal{O}(1/\varepsilon^2)$ or longer. In other words, these methods are an attempt to recapture some of the "trade-off" property enjoyed by Lindstedt expansions in the periodic case. These methods were originally developed by heuristic reasoning only, and there does not yet exist a fully adequate rigorous theory explaining their range of validity. There are a number of specific problems of the form (5.1.1) for which these methods work, or seem to work. There are others for which they definitely do not work. Because the situation is not yet fully understood, we will not attempt to state any general results for these strategies, but will illustrate the third strategy briefly and comment on the difficulties.

It should be mentioned that there is another approach to finding higher order approximations for these problems, namely the method of "higher order averaging" which will be developed in the next chapter. This method does not require specifying a set of time scales in advance. Instead, the necessary time scales appear automatically in the course of solving the problem. The approximations constructed by averaging are always valid at least on expanding intervals of order $1/\varepsilon$, and sometimes on longer intervals. It would appear that this method is the most powerful.

To explain the higher order two-time-scale method using t and τ, we will continue the solution of the linear example (5.2.15) from where it was left off in Section 5.2. At that time we had found y_0, either in the form

(5.2.21) or (5.2.30). In doing so, we had eliminated resonant terms from the right hand side of (5.2.16b) and had in fact found that the entire right hand side was resonant. Therefore we begin with (5.2.16b) in the form

$$
\begin{aligned}
y_{1tt} + y_1 &= 0, \\
y_1(0,0) &= 0, \\
y_{1t}(0,0) &= \alpha.
\end{aligned}
\tag{5.3.1}
$$

Although the polar form was simplest for finding y_0, the computational advantage seems to lie with the Cartesian form this time, so we write the solution of (5.3.1) as

$$
\begin{aligned}
y_1(t,\tau) &= C(\tau)\cos t + D(\tau)\sin t, \\
C(0) &= 0, \\
D(0) &= \alpha.
\end{aligned}
\tag{5.3.2}
$$

As before, the solution is not uniquely determined by the information available, and it is necessary to adopt heuristic reasoning to fix the functions C and D completely. Since the error in the approximation $y \cong y_0 + \varepsilon y_1$ is expected to behave like $\varepsilon^2 y_2$, the procedure is to eliminate t-secular terms from y_2 so that it remains bounded on expanding intervals of order $1/\varepsilon$. To this end we examine equation (5.2.16c), which can be written out as

$$
\begin{aligned}
y_{2tt} + y_2 = &-\{A'' + 2A' + 2D' + C + 2D\}\cos t \\
&-\{B'' + 2B' - 2C' + D - 2C\}\sin t,
\end{aligned}
\tag{5.3.3}
$$

where A and B are as in (5.2.29), and C and D are as yet unknown. Both terms on the right hand side of (5.3.3) are resonant and must be set equal to zero. Making use of the expressions for A and B, the resulting nonsecularity conditions are

$$
\begin{aligned}
C' &= -C + \frac{1}{2}D + \frac{5}{8}\alpha e^{-\tau}\sin\frac{\tau}{2}, \\
D' &= -\frac{1}{2}C - D + \frac{5}{8}\alpha e^{-\tau}\cos\frac{\tau}{2}.
\end{aligned}
\tag{5.3.4}
$$

This inhomogeneous linear system of ordinary differential equations is solvable, with the initial conditions given in (5.3.2):

$$
\begin{aligned}
C(\tau) &= \alpha e^{-\tau}\left(\frac{5}{8}\tau + 1\right)\sin\frac{\tau}{2}, \\
D(\tau) &= \alpha e^{-\tau}\left(\frac{5}{8}\tau + 1\right)\cos\frac{\tau}{2}.
\end{aligned}
\tag{5.3.5}
$$

When this is substituted into (5.3.2), it is easy to see that $y_0 + \varepsilon y_1$ coincides with the approximation (5.1.13) developed from the exact solution.

If we had solved (5.3.1) in polar form $y = \rho_1 \cos(\psi_1 - t)$, the nonsecularity conditions and initial conditions to be satisfied by $\rho_1(\tau)$ and $\psi_1(\tau)$ would have been (after some calculation)

$$\rho' + \rho = \frac{5}{8} \alpha e^{-\tau} \sin\left(\psi + \frac{\tau}{2}\right),$$

$$\rho(2\psi' + 1) = \frac{5}{4} \alpha e^{-\tau} \cos\left(\psi + \frac{\tau}{2}\right),$$

$$\rho(0) = \alpha,$$

$$\psi(0) = \frac{\pi}{2}.$$

It is possible to solve these by clever inspection; the solution is

$$\rho = \alpha e^{-\tau}\left(\frac{5}{8}\tau + 1\right),$$

$$\psi = \frac{\pi}{2} - \frac{\tau}{2}.$$

However there does not seem to be any technique behind this, and if one attempts to solve the original problem (5.2.15) with initial velocity β instead of zero, the polar equation becomes intractable, whereas the analog of (5.3.4) is still an inhomogeneous linear system. This is why we said that the Cartesian form seems preferable for this problem. Nevertheless, it must be emphasized that (5.3.4) is linear only because the original problem (5.3.4) is linear, and there is no general procedure for solving the nonsecularity conditions that arise in the course of the two-timing method. As mentioned before, the problem is a practical one, not a theoretical one: The nonsecularity conditions in the two-timing method (using "unstrained" fast time t and slow time τ) are always initial value problems of a sort which, in theory, have solutions, although they may not be solvable in "closed form" using elementary functions.

The method to be investigated next has been used to solve a number of applied problems and is quite popular in the engineering literature. However, it rests on quite shaky foundations from a mathematical point of view. The purpose of the following example is not so much to teach the method, as to discuss it and point out some of the difficulties with it. Anyone wishing to learn the method for practical use will have to study many examples from the literature to see how these difficulties are dealt with on an ad hoc basis in different situations.

The problem we will study is once again the linear example (5.2.15). The solution will be sought in the form

$$y \cong y_0(t, \tau, \sigma) + \varepsilon y_1(t, \tau, \sigma) + \varepsilon^2 y_2(t, \tau, \sigma), \tag{5.3.6}$$

where $\tau = \varepsilon t$ and $\sigma = \varepsilon^2 t$. Working with the operator equation

$$\frac{d}{dt} = \frac{\partial}{\partial t} + \varepsilon \frac{\partial}{\partial \tau} + \varepsilon^2 \frac{\partial}{\partial \sigma},$$

one obtains the following sequence of differential equations and initial conditions:

$$y_{0tt} + y_0 = 0,$$
$$y_0(0, 0, 0) = \alpha, \tag{5.3.7a}$$
$$y_{0t}(0, 0, 0) = 0;$$

$$y_{1tt} + y_1 = -(y_0 + 2y_{0t} + 2y_{0t\tau}),$$
$$y_1(0, 0, 0) = 0, \tag{5.3.7b}$$
$$y_{1t}(0, 0, 0) = -y_{0\tau}(0, 0, 0);$$

$$y_{2tt} + y_2 = -(2y_{0\tau} + y_{0\tau\tau} + 2y_{0t\sigma} + y_1 + 2y_{1t} + 2y_{1t\tau}),$$
$$y_2(0, 0, 0) = 0, \tag{5.3.7c}$$
$$y_{2t}(0, 0, 0) = -y_{1\tau}(0, 0, 0) - y_{0\sigma}(0, 0, 0).$$

These equations are very similar to (5.2.16), the only differences being that the initial conditions occur at $(0, 0, 0)$ rather than $(0, 0)$ and that (5.3.7c) has additional terms involving derivatives with respect to σ. Therefore the solution of these equations proceeds very much like that of (5.2.16) until (5.3.7c) is reached. It is only necessary to remember that each function is allowed to depend on σ. Thus, the solution of (5.3.7a) is

$$y_0 = A(\tau, \sigma) \cos t + B(\tau, \sigma) \sin t, \tag{5.3.8}$$

with

$$A(0, 0) = \alpha,$$
$$B(0, 0) = 0;$$

compare (5.2.22). Then it is possible to follow through the steps from (5.2.23) to (5.2.26) almost exactly, except that A' and B' must be written A_τ and B_τ because they depend also on σ, and in (5.2.25) one has $\frac{\partial}{\partial \tau}$. In place of (5.2.27), one gets

$$A(\tau, \sigma) = K(\sigma) e^{-\tau} \cos \xi(\tau, \sigma),$$
$$B(\tau, \sigma) = K(\sigma) e^{-\tau} \sin \xi(\tau, \sigma),$$

where

$$K(0) = \alpha,$$
$$\xi(0, 0) = 0.$$

As in (5.2.28), $\xi_\tau = -1/2$; therefore

$$\xi(\tau, \sigma) = \eta(\sigma) - \frac{\tau}{2}$$

for some function $\eta(\sigma)$ satisfying $\eta(0) = 0$. Putting (5.3.8) together with the results just obtained, one finds that at this point y_0 is given by

$$y_0(t, \tau, \sigma) = K(\sigma)e^{-\tau} \cos\left(t + \frac{\tau}{2} - \eta(\sigma)\right), \qquad (5.3.9)$$

where K and η are as yet unspecified functions satisfying

$$K(0) = \alpha,$$
$$\eta(0) = 0.$$

Here we see the first serious difference between this method and the two-scale method: in the latter, y_0 is completely specified when the resonant terms are eliminated from the y_1 equation, while in the present method there is a dependence on σ which is not yet determined.

Having eliminated the resonant terms from the y_1 equation, one is left with

$$y_{1tt} + y_1 = 0,$$
$$y_1(0, 0, 0) = 0,$$
$$y_{1t}(0, 0, 0) = \alpha,$$

resembling (5.3.1). In solving (5.3.1) by (5.3.2), we used $\cos t$ and $\sin t$ as a fundamental set of solutions, but in the present case later calculations will be simplified if we use the cosine and sine of the quantity $t + \tau/2 - \eta(\sigma)$ appearing in (5.3.9); this is permissible since for fixed τ and σ these are linearly independent solutions of the differential equation. Therefore we write

$$y_1(t, \tau, \sigma) = C(\tau, \sigma)\cos(t + \frac{\tau}{2} - \eta(\sigma)) + D(\tau, \sigma)\sin(t + \frac{\tau}{2} - \eta(\sigma)). \quad (5.3.10)$$

It is easy to check (using the fact that $\eta(0) = 0$ and the initial conditions for y_1) that

$$C(0, 0) = 0,$$
$$D(0, 0) = \alpha.$$

Until now the calculations have been routine. We are now rapidly approaching the point at which the fundamental nature of the method (and its difficulties) will become clear. Substituting (5.3.9) and (5.3.10) into the right hand side of (5.3.7c), one obtains the following equation for y_2:

$$y_{2tt} + y_2 = \left[\left(\frac{5}{4} - 2\eta'\right) K e^{-\tau} - 2\left(D + D_\tau\right)\right] \cos\left(t + \frac{\tau}{2} - \eta\right)$$

$$+ \left[2K' e^{-\tau} + 2\left(C + C_\tau\right)\right] \sin\left(t + \frac{\tau}{2} - \eta\right). \tag{5.3.11}$$

Here, of course, the prime denotes a derivative with respect to σ, since it is applied to functions of σ alone. In order that y_2 will contain no t-secular terms (terms proportional to t), the resonant terms must be eliminated from the right hand side of (5.3.11), as usual. Since the entire right hand side is resonant, the coefficients of the cosine and sine terms must vanish, leading to the equations

$$\begin{aligned} C_\tau &= -C - K' e^{-\tau}, \\ D_\tau &= -D + \left(\tfrac{5}{8} - \eta'\right) K e^{-\tau}. \end{aligned} \tag{5.3.12}$$

This provides differential equations for C and D once K and η are known, but is of no help in determining the latter functions which are still unknown. It is clear that a new principle is required, besides the avoidance of t-secular terms, which will serve to determine K and η.

Recall the reason for avoiding t-secular terms: A term proportional to t will grow with time (assuming the other factors do not approach zero) and will become of order $1/\varepsilon$ at the end of an expanding interval of length $1/\varepsilon$. If a term of formal order ε^n is t-secular, its actual order on the expanding interval is ε^{n-1}, and the series fails to be uniformly ordered. Now the aim of a three-scale method is to achieve uniform validity on an expanding interval of length $1/\varepsilon^2$. On an interval of this length, a τ-secular term will produce the same type of disordering that a t-secular term produces on the shorter expanding interval. Now it is clear from (5.3.9) that τ-secular terms cannot arise in y_0. (And in any case, according to the definition of uniform ordering in Section 1.8, the leading term of a perturbation series does not matter.) So the first possibility for such a term occurs in y_1; and in fact y_1 will contain a τ-secular term if and only if C or D contains such a term. So we propose the following rule: $K(\sigma)$ and $\eta(\sigma)$ should be chosen in such a way that the solutions C and D of (5.3.12) contain no τ-secular terms.

We will now test this proposed rule by examining the solutions of (5.3.12). Remember that the coefficients K' and $(5/8 - \eta')$ are functions of σ alone and therefore may be treated as constants in solving (5.3.12) for C and D as functions of τ. Now it is apparent that the "forcing" terms proportional to $e^{-\tau}$ produce a response proportional to $\tau e^{-\tau}$; this can be seen by explicitly solving the equations, or by the fact that the associated homogeneous equations have a solution proportional to $e^{-\tau}$ and so the forced equations respond with an additional factor of τ. Now a term of the form $\tau e^{-\tau}$ is often considered as a τ-secular term. In fact, we shall show below that this is not the case; but for the moment let us regard this term as something to be avoided, because it contains a factor of τ. In order to avoid this term it is necessary to take

$$K' = 0,$$
$$\eta' = \tfrac{5}{8}. \tag{5.3.13}$$

With the initial conditions, this implies

$$K(\sigma) = \alpha,$$
$$\eta(\sigma) = \tfrac{5}{8}\sigma.$$

Finally, then, y_0 is completely determined:

$$y_0 = \alpha e^{-\tau} \cos\left(t + \frac{\tau}{2} - \frac{5\sigma}{8}\right), \tag{5.3.14}$$

in agreement with the first term of the expansion (5.1.16) derived from the exact solution.

Although it appears that we have solved this problem successfully, there is a difficulty: As hinted above, *the term $\tau e^{-\tau}$ is not actually a τ-secular term*. To understand why, recall the precise meaning of a secular term: a term that is unbounded (on the domain of interest) and therefore causes the perturbation series to be not uniformly ordered. Now $\tau e^{-\tau}$ attains a maximum and then decreases to zero as τ approaches infinity. Therefore it is not secular, either on an expanding interval or indeed on the entire real line. It follows that in fact there is no need to impose the conditions (5.3.13) which we used to determine K and η. It is *convenient* to do so, because these conditions simplify the equations (5.3.12). But we would have obtained just as good a solution (as far as the presence of secular terms is concerned) had we made any choice at all for K and η. The ultimate reason that K and η are not unique in this problem is that three scales are not actually needed to solve it: We have already pointed out in Section 5.1 that the two-scale solution for this problem is in fact valid for all time, because of the damping.

In the literature, one finds at least three justifications given for requiring (5.3.13) in this problem. One is that it simplifies the equations (5.3.12). Another is that it eliminates the term $\tau e^{-\tau}$, which is considered as a bad term merely because it *looks* secular. The third is that it makes the ratio y_1/y_0 bounded. If y_1 contains a term proportional to $\tau e^{-\tau}$, then indeed this ratio is unbounded; but this is not harmful. The demand that y_{n+1}/y_n be bounded comes from the idea that the $n + 1$st term of a perturbation series should be small compared to the nth term. But a careful study of the requirements for uniform asymptotic validity (see the discussion surrounding Theorem 1.8.1) shows that the correct concept of uniform ordering involves the boundedness of the coefficients and not of their ratios. So *in the present problem* the only valid justification of (5.3.12) is the first: that it simplifies the equations.

In the foregoing discussion we have emphasized repeatedly that *in this problem* the requirement to eliminate τ-secular terms does not determine K and η uniquely. *In general*, there are three a priori possibilities, all of which do in fact occur in specific problems: There may be a unique solution that is free of τ-secular terms; there may be many solutions that are free of these terms (as in the problem treated above); or there may be no solution that is free of these terms. Clearly nonuniqueness of the solution (as in our case) is not a serious drawback for a method. What is most disconcerting about the three-scale method is the existence of problems for which the equations corresponding to (5.3.12) have no solutions that are free of τ-secular terms. When this is encountered, the method simply fails. An example of such a problem is the following equation, which combines a "Duffing term" of order ε with a "Van der Pol term" of order ε^2:

$$\ddot{y} + \varepsilon^2(y^2 - 1)\dot{y} + y + \varepsilon y^3 = 0. \qquad (5.3.15)$$

The ambitious reader may wish to attempt a three-scale solution of this problem, although the calculations are formidable (and are best handled with a notation that uses complex variables). But the fact is that at the end one finds there is no way to eliminate secular terms. (A reference is given in Section 5.7.) At this point there is no general theory that would explain why the method fails for some problems and succeeds with others.

In cases for which the three-scale method works (formally), the resulting series is uniformly ordered on expanding intervals of length $1/\varepsilon^2$. This in itself does not imply that the series is uniformly valid on these intervals; as usual, uniform ordering of the terms in the series says nothing about the error. But it is at least plausible that such solutions are uniformly valid. We will not address the question of error estimation for the methods discussed in this section, but will prove the validity of the first order two-scale approximation in Section 5.5.

5.4. PERIODIC STANDARD FORM

All of the differential equations considered in Chapter 4 and in this chapter, and a great many others, can be put into the form

$$\dot{x} = \varepsilon f(x, t, \varepsilon) = \varepsilon f_1(x, t) + \varepsilon^2 f_2(x, t) + \cdots, \qquad (5.4.1)$$

where x is an N-dimensional vector and f is an N-vector-valued function that is periodic in t with period independent of ε. Usually the period is assumed to be 2π; this can always be achieved by changing the unit of time.

This section is entirely concerned with the procedure for achieving the form (5.4.1), which we call *periodic standard form*, and which is frequently referred to in the literature as *standard form for the method of averaging*. (There are actually several standard forms used in averaging and multiple scale methods, some of which will appear later, and we distinguish these by calling them *periodic standard form, quasiperiodic standard form, angular standard form*, and so forth.) The section is organized into examples; each example is a general class of equations, rather than a concrete instance. The first few examples are types of equations which have been considered before. After these, new types of problems are introduced, including some (for illustrative purposes) which cannot be put into standard form.

Example 5.4.1. The Autonomous Oscillator

There are several ways to put an autonomous oscillator

$$\ddot{y} + k^2 y = \varepsilon f(y, \dot{y}) \tag{5.4.2}$$

into periodic standard form. Begin by introducing phase plane variables

$$u = y, \qquad v = \dot{y}/k \tag{5.4.3}$$

to obtain

$$\begin{aligned}
\dot{u} &= kv, \\
\dot{v} &= -ku + \frac{\varepsilon}{k} f(u, kv).
\end{aligned} \tag{5.4.4}$$

When $\varepsilon = 0$, the solutions of this system rotate clockwise at angular velocity k on circles centered at the origin. If a new set of coordinate axes is introduced which rotates clockwise at this same rate, the unperturbed solutions will appear in the new coordinate system to be at rest. Letting a and b be the new rotating coordinates, the transformation from (u, v) to (a, b) is given by

$$\begin{aligned}
u &= a \cos kt + b \sin kt, \\
v &= -a \sin kt + b \cos kt.
\end{aligned} \tag{5.4.5}$$

When this transformation is applied to (5.4.4), the result will be in periodic standard form. To carry out the calculation, differentiate (5.4.5) with respect to time and substitute into (5.4.4), obtaining

$$\dot{a} \cos kt + \dot{b} \sin kt = 0,$$

$$-\dot{a} \sin kt + \dot{b} \cos kt = \frac{\varepsilon}{k} f(a \cos kt + b \sin kt, -ak \sin kt + bk \cos kt). \tag{5.4.6}$$

Now solve this system for $\dot{a}$ and $\dot{b}$ (by, for example, multiplying the first equation by $\cos kt$ and the second by $-\sin kt$ and adding) to obtain

$$\dot{a} = -\frac{\varepsilon}{k}f(a\cos kt + b\sin kt, -ak\sin kt + bk\cos kt)\sin kt,$$

$$\dot{b} = +\frac{\varepsilon}{k}f(a\cos kt + b\sin kt, -ak\sin kt + bk\cos kt)\cos kt. \qquad (5.4.7)$$

The right hand side of this system is periodic in t with period $2\pi/k$, and contains a factor of ε; it is therefore in periodic standard form with $\mathbf{x} = (a, b)$. Note that the function $\mathbf{f}$ is constructed from f as specified on the right hand side of (5.4.7); in particular, the lightface letter f does not denote the magnitude (norm or length) of the boldface vector $\mathbf{f}$, a convention common in physics. Instead we regard $\mathbf{f}$ as an entirely separate letter from f, and define it as needed in each example. (In this book the norm of a vector $\mathbf{v}$ is denoted $\|\mathbf{v}\|$, as usual in mathematics.)

Another way to put (5.4.2) into standard form is to introduce polar coordinates into the phase plane before rotating. Setting

$$u = r\cos\theta,$$

$$v = r\sin\theta \qquad (5.4.8)$$

in (5.4.4) leads to

$$\dot{r} = \frac{\varepsilon}{k}f(r\cos\theta, kr\sin\theta)\sin\theta,$$

$$\dot{\theta} = -k + \frac{\varepsilon}{kr}f(r\cos\theta, kr\sin\theta)\cos\theta. \qquad (5.4.9)$$

(A specific example of this computation is given below; see the steps from (5.4.17) to (5.4.18).) To rotate polar coordinates at angular velocity k in the clockwise direction it is only necessary to change the angular variable, leaving the radial variable as it is: setting

$$\theta = \varphi - kt \qquad (5.4.10)$$

yields

$$\dot{r} = \frac{\varepsilon}{k}f(r\cos(\varphi - kt), kr\sin(\varphi - kt))\sin(\varphi - kt),$$

$$\dot{\varphi} = \frac{\varepsilon}{kr}f(r\cos(\varphi - kt), kr\sin(\varphi - kt))\cos(\varphi - kt), \qquad (5.4.11)$$

which is in periodic standard form with $\mathbf{x} = (r, \varphi)$. (Do not forget, in checking that an equation is in periodic standard form, to check both that it is periodic and that it contains the factor ε.)

The crucial step in arriving at the periodic standard forms (5.4.7) and (5.4.11) is to make a change of variables which "removes" the unperturbed solution, that is, renders the unperturbed solution constant (in the new variables). In the present example, it is easy to see how to remove the unperturbed solution: Since that solution is a rotation, it is only necessary to rotate the coordinates at the same rate. When it is not obvious how to proceed, the following strategy is useful: First find the general solution of the reduced problem, which will contain arbitrary constants, and then reinterpret this solution as a change of variables (the new variables being the "constants"). In the present case, the solution of (5.4.4) with $\varepsilon = 0$ is (5.4.5) with a and b constant, or else

$$u = r\cos(\varphi - kt),$$
$$v = r\sin(\varphi - kt), \qquad (5.4.12)$$

with r and φ constant. Using (5.4.5) as a coordinate change (with a and b as the new variables) leads to (5.4.7), and using (5.4.12) leads to (5.4.11).

The idea of using the unperturbed solution to define a change of variables is closely related to the method of variation of constants (or variation of parameters). In Appendix C, it is shown that the usual method of variation of constants (for second order inhomogeneous linear differential equations) can be regarded as a change of variables. We will now show that the changes of variables used above can be regarded as a variation of constants, and that in this way it is possible to achieve both of the periodic standard forms (5.4.7) and (5.4.11) directly from the second order equation (5.4.2) without first introducing the phase plane variables u and v. We will not make much use of this method, since we regard the previous methods as more conceptually clear, but variation of parameters is commonly seen in the literature. One begins with the solution of (5.4.2) when $\varepsilon = 0$, which can be written in the form

$$y = a\cos kt + b\sin kt, \qquad (5.4.13)$$

where a and b are constants. The derivative of (5.4.13) is, of course,

$$\dot{y} = -ak\sin kt + bk\cos kt. \qquad (5.4.14)$$

The idea of variation of constants is to look for the solution of the perturbed problem (5.4.2) in the form (5.4.13) by allowing a and b to become functions of t rather than constants. Since it requires two conditions to determine the two unknown functions a and b, it is permissible to impose (5.4.14) as a condition on a and b in addition to the requirement that (5.4.13) solve (5.4.2); this is equivalent to

$$\dot{a}\cos kt + \dot{b}\sin kt = 0. \qquad (5.4.15)$$

Substituting (5.4.13) and (5.4.14) into (5.4.2) leads directly to (5.4.7), by algebraic steps similar to those carried out above. A similar analysis beginning with $y = r\cos(\varphi - kt)$ leads to (5.4.11).

In order to understand the procedures leading to periodic standard form, it is important to follow the steps in some specific examples. We will therefore put the Van der Pol equation

$$\ddot{y} + \varepsilon(y^2 - 1)\dot{y} + y = 0 \tag{5.4.16}$$

into periodic standard form. First write (5.4.16) as a system with $u = y$, $v = \dot{y}$:

$$\begin{aligned} \dot{u} &= v, \\ \dot{v} &= -u + \varepsilon(1 - u^2)v. \end{aligned} \tag{5.4.17}$$

Then differentiate (5.4.8) and substitute into (5.4.17), obtaining

$$\dot{r}\cos\theta - r\dot{\theta}\sin\theta = r\sin\theta,$$
$$\dot{r}\sin\theta + r\dot{\theta}\cos\theta = -r\cos\theta + \varepsilon(1 - r^2\cos^2\theta)r\sin\theta.$$

Multiplying these equations by $\cos\theta$ and $\sin\theta$, and adding or subtracting, gives

$$\begin{aligned} \dot{r} &= \varepsilon(1 - r^2\cos^2\theta)r\sin^2\theta, \\ \dot{\theta} &= -1 + \varepsilon(1 - r^2\cos^2\theta)\sin\theta\cos\theta. \end{aligned} \tag{5.4.18}$$

(Notice the division by r in obtaining the second equation.) Now (5.4.10) gives the periodic standard form (remember $k = 1$):

$$\begin{aligned} \dot{r} &= \varepsilon\left(1 - r^2\cos^2(\varphi - t)\right)r\sin^2(\varphi - t), \\ \dot{\varphi} &= \varepsilon\left(1 - r^2\cos^2(\varphi - t)\right)\sin(\varphi - t)\cos(\varphi - t). \end{aligned} \tag{5.4.19}$$

Additional practice with these ideas will be found in Exercises 5.4.1, 5.4.2, and 5.4.3.

Example 5.4.2. Harmonic Resonance

The periodically forced nearly linear oscillator is

$$\ddot{y} + k^2 y = \varepsilon f(y, \dot{y}, \omega(\varepsilon)t), \tag{5.4.20}$$

where f is periodic of period 2π in its third argument and therefore of period $2\pi/\omega(\varepsilon)$ in t. Consider the *harmonic resonance* case (defined in Section 4.5) for equation (5.4.20), that is, the case in which the forcing frequency reduces to the free frequency when $\varepsilon = 0$:

$$\omega(\varepsilon) \sim k + \varepsilon\omega_1 + \varepsilon^2\omega_2 + \cdots. \tag{5.4.21}$$

Since the unperturbed form of (5.4.20) is the same as that of (5.4.2), it is tempting to try the same coordinate transformations (5.4.5) and (5.4.12) which lead to periodic standard form in the previous example. However, the result of applying (5.4.5) to (5.4.20) is

$$\dot{a} = -\frac{\varepsilon}{k}f\,(a\cos kt + b\sin kt, -ak\sin kt + bk\cos kt, \omega(\varepsilon)t)\sin kt,$$

$$\dot{b} = +\frac{\varepsilon}{k}f\,(a\cos kt + b\sin kt, -ak\sin kt + bk\cos kt, \omega(\varepsilon)t)\cos kt,$$

which is not in periodic standard form because the right hand side is not periodic in t (unless $\omega(\varepsilon) \equiv k$). This periodicity can be recaptured by modifying the angular velocity at which the coordinates are rotated when $\varepsilon \neq 0$. Namely, if (5.4.5) is replaced by

$$u = a\cos\omega(\varepsilon)t + b\sin\omega(\varepsilon)t,$$
$$v = -a\sin\omega(\varepsilon)t + b\cos\omega(\varepsilon)t, \tag{5.4.22}$$

which reduces to (5.4.5) when $\varepsilon = 0$, then the result of transforming (5.4.20) is

$$\dot{a} = -b\,(\omega(\varepsilon) - k) - \frac{\varepsilon}{k}f\,(u, kv, \omega(\varepsilon)t)\sin\omega(\varepsilon)t,$$

$$\dot{b} = +a\,(\omega(\varepsilon) - k) + \frac{\varepsilon}{k}f\,(u, kv, \omega(\varepsilon)t)\cos\omega(\varepsilon)t, \tag{5.4.23}$$

where u and v are being used not as separate variables but as "stand-ins" for the expressions given by (5.4.22), in order to shorten the equations. This system is nearly in periodic standard form, except that the period of the right hand side is $2\pi/\omega(\varepsilon)$, which depends upon ε. For some purposes this is satisfactory, but in order to eliminate the ε dependence from the period it is only necessary to introduce the strained time

$$t^+ = \omega(\varepsilon)t. \tag{5.4.24}$$

Then (5.4.23) becomes

$$\frac{da}{dt^+} = -b\frac{\omega(\varepsilon) - k}{\omega(\varepsilon)}$$
$$\quad - \frac{\varepsilon}{k\omega(\varepsilon)}f(a\cos t^+ + b\sin t^+, -ak\sin t^+ + bk\cos t^+, t^+)\sin t^+,$$

$$\frac{db}{dt^+} = +a\frac{\omega(\varepsilon) - k}{\omega(\varepsilon)} \tag{5.4.25}$$
$$\quad + \frac{\varepsilon}{k\omega(\varepsilon)}f(a\cos t^+ + b\sin t^+, -ak\sin t^+ + bk\cos t^+, t^+)\cos t^+,$$

which can be expanded, using (5.4.21), into

$$\frac{da}{dt^+} =$$

$$-\frac{\varepsilon}{k} \left\{ b\omega_1 + \frac{1}{k} f(a\cos t^+ + b\sin t^+, -ak\sin t^+ + bk\cos t^+, t^+)\sin t^+ \right\}$$

$$+ \mathcal{O}(\varepsilon^2),$$

$$\frac{db}{dt^+} = \tag{5.4.26}$$

$$+\frac{\varepsilon}{k} \left\{ a\omega_1 + \frac{1}{k} f(a\cos t^+ + b\sin t^+, -ak\sin t^+ + bk\cos t^+, t^+)\cos t^+ \right\}$$

$$+ \mathcal{O}(\varepsilon^2).$$

This is in periodic standard form. It would not have been correct to expand in ε prior to introducing t^+, because this would have destroyed the periodicity (since the period in t depends upon ε). The essential point in achieving periodic standard form for this problem is to use (5.4.22). In Example 5.4.1, the coordinate change that leads to periodic standard form is the solution of the unperturbed equation. In the present example, (5.4.22) has built into it both the solution of the unperturbed equation (to which it reduces when $\varepsilon = 0$) and the perturbation frequency $\omega(\varepsilon)$. If variation of constants is used, it is necessary to modify the postulated form of the solution in the same way. (Exercise 5.4.4.)

Similarly, it is possible to obtain a periodic standard form in polar coordinates. Setting

$$u = r\cos(\varphi - \omega(\varepsilon)t),$$
$$v = r\sin(\varphi - \omega(\varepsilon)t) \tag{5.4.27}$$

results in

$$\dot{r} = \frac{\varepsilon}{k} f(u, kv, \omega(\varepsilon)t)\sin(\varphi - \omega(\varepsilon)t),$$

$$\dot{\varphi} = (\omega(\varepsilon) - k) + \frac{\varepsilon}{kr} f(u, kv, \omega(\varepsilon)t)\cos(\varphi - \omega(\varepsilon)t) + \mathcal{O}(\varepsilon^2), \tag{5.4.28}$$

where this time u and v stand for the expressions given in (5.4.27). (See Exercise 5.4.5.) Upon introducing t^+ and expanding in ε, this gives

$$\frac{dr}{dt^+} = \frac{\varepsilon}{k^2} f(r\cos(\varphi - t^+), kr\sin(\varphi - t^+), t)\sin(\varphi - t^+) + \mathcal{O}(\varepsilon^2),$$

$$\tag{5.4.29}$$

$$\frac{d\varphi}{dt^+} = \frac{\varepsilon}{k} \left\{ \omega_1 + \frac{1}{kr} f(r\cos(\varphi - t^+), kr\sin(\varphi - t^+), t)\cos(\varphi - t^+) \right\}$$

$$+ \mathcal{O}(\varepsilon^2),$$

which is in periodic standard form.

As an illustration of periodic standard form for a forced oscillator in harmonic resonance, consider the weakly forced and damped Duffing equation

$$\ddot{y} + \varepsilon \delta_1 \dot{y} + y + \varepsilon \lambda_1 y^3 = \varepsilon \gamma_1 \cos \omega(\varepsilon) t. \tag{5.4.30}$$

Here $k = 1$, so $u = y$ and $v = \dot{y}$, and by comparison with (5.4.20),

$$f(u, v, t^+) = \gamma_1 \cos t^+ - \lambda_1 u^3 - \delta_1 v.$$

Therefore the periodic standard form in polar coordinates (5.4.24) is in this case

$$\frac{dr}{dt^+} =$$
$$\varepsilon \left\{ \gamma_1 \cos t^+ \sin(\varphi - t^+) - \lambda_1 r^3 \cos^3(\varphi - t^+) \sin(\varphi - t^+) - \delta_1 r \sin^2(\varphi - t^+) \right\}$$
$$+ \mathcal{O}(\varepsilon^2),$$

$$\frac{d\varphi}{dt^+} = \tag{5.4.31}$$
$$\varepsilon \left\{ \omega_1 + \frac{\gamma_1}{r} \cos t^+ \cos(\varphi - t^+) - \lambda_1 r^2 \cos^4(\varphi - t^+) - \delta_1 \cos(\varphi - t^+) \sin(\varphi - t^+) \right\}$$
$$+ \mathcal{O}(\varepsilon^2).$$

The reader should obtain these equations by working through the steps leading to (5.4.29) beginning with (5.4.30). See Exercise 5.4.6. Another way to obtain a periodic standard form for (5.4.20) is given in Exercise 5.4.7.

Similar periodic standard forms are possible for subharmonic, superharmonic, and supersubharmonic resonances. In nonresonant cases, when $\omega(0)/k$ is not rational, it is not possible to achieve periodic standard form. In this case either *quasiperiodic standard form* or *angular standard form* may be used; these forms will be mentioned again in the next example, but serious consideration of them is deferred to Chapter 6.

Example 5.4.3. Coupled Autonomous Oscillators

Consider the following system of coupled autonomous oscillators:

$$\ddot{y}_i + k_i^2 y_i = \varepsilon f_i(y_1, \ldots, y_N, \dot{y}_1, \ldots, \dot{y}_N) \tag{5.4.32}$$

for $i = 1, \ldots, N$. When $\varepsilon = 0$, these oscillators are uncoupled, each having its own free frequency k_i. Following the ideas in Example 5.4.1, phase plane variables

$$u_i := y_i, \qquad v_i := \dot{y}_i / k_i \tag{5.4.33}$$

may be introduced for each oscillator, and polar coordinates r_i and θ_i defined by

$$u_i = r_i \cos \theta_i,$$
$$v_i = r_i \sin \theta_i. \tag{5.4.34}$$

In these coordinates the system looks like

$$\dot{r}_i = \frac{\varepsilon}{k_i} f_i(r_1 \cos \theta_1, \ldots, r_N \cos \theta_N, k_1 r_1 \sin \theta_1, \ldots, k_N r_N \sin \theta_N) \sin \theta_i,$$

$$\dot{\theta}_i = -k_i + \frac{\varepsilon}{k_i r_i} f_i(r_1 \cos \theta_1, \ldots, k_N r_N \sin \theta_N) \cos \theta_i. \tag{5.4.35}$$

This is a form which will appear again in Chapter 6, and will be called *angular standard form*. If we continue as in Example 5.4.1, rotating each phase plane at its own angular velocity k_i by introducing

$$\varphi_i := \theta_i + k_i t, \tag{5.4.36}$$

the result is

$$\dot{r}_i = \frac{\varepsilon}{k_i} f_i(r_1 \cos(\varphi_1 - k_1 t), \ldots, k_N r_N \sin(\varphi_N - k_N t)) \sin(\varphi_i - k_i t),$$

$$\dot{\varphi}_i = \frac{\varepsilon}{k_i r_i} f_i(r_1 \cos(\varphi_1 - k_1 t), \ldots, k_N r_N \sin(\varphi_N - k_N t)) \cos(\varphi_i - k_i t). \tag{5.4.37}$$

The right hand side is not periodic unless all of the free frequencies k_i are equal, or at least are all integer multiples of a single frequency K:

$$k_i = m_i K. \tag{5.4.38}$$

In this case, (5.4.37) is in periodic standard form with period $2\pi/K$. Otherwise, the right hand side of (5.4.37) is what is called a *quasiperiodic function*, and (5.4.37) is said to be in *quasiperiodic standard form*. A quasiperiodic function is one which results from the combination of finitely many periodic components with different periods. (This is a special case of an *almost periodic function*, which allows for infinitely many different periods.) We will not discuss multiple scale methods for quasiperiodic or angular standard forms, but averaging methods for these cases will be introduced in Chapter 6.

Example 5.4.4. The Duffing Equation with Strong Forcing

Recall from Section 4.7 that the general form of Duffing's equation in natural (but already nondimensionalized) parameters is

$$\ddot{y} + \delta\dot{y} + y + \lambda y^3 = \gamma\cos\omega t,$$
$$y(0) = \alpha, \qquad\qquad\qquad (5.4.39)$$
$$\dot{y}(0) = \beta.$$

Until now, the perturbation families that we have considered for (5.4.39) have always involved weak forcing, that is, γ has been made into a function of ε in such a way that $\gamma(0) = 0$. Usually we have taken $\gamma(\varepsilon) = \varepsilon\gamma_1$ for simplicity, although it is equally possible to take $\gamma(\varepsilon) = \varepsilon\gamma_1 + \varepsilon^2\gamma_2 + \cdots$, as in (4.7.2). Now we will consider what happens under the assumption of strong forcing, that is, $\gamma(\varepsilon) = \gamma_0 + \varepsilon\gamma_1 + \cdots$ with $\gamma_0 \neq 0$; for simplicity, we will assume $\gamma(\varepsilon) \equiv \gamma_0$. The parameters δ and λ will be taken to be small as usual, so the differential equation to be considered is

$$\ddot{y} + \varepsilon\delta_1\dot{y} + y + \varepsilon\lambda_1 y^3 = \gamma_0\cos\omega(\varepsilon)t. \qquad (5.4.40)$$

The reduced problem for this equation is

$$\ddot{y} + y = \gamma_0\cos\omega(0)t, \qquad\qquad (5.4.41)$$

which behaves quite differently according to whether $\omega(0) = 1$ (the harmonic resonance case) or $\omega(0) \neq 1$. In the former case, all solutions of (5.4.41) are unbounded (since they contain a term with a factor of t). It is not possible to attain periodic standard form in this case.

In the latter case, all solutions of (5.4.41) are bounded and are the sum of two periodic functions, one having frequency 1 and the other $\omega(0)$. Such solutions are periodic if these two periodic terms have a common period; this happens whenever $\omega(0) = p/q$ is a rational number other than 1. Under these circumstances, that is, in the nonharmonic resonance case, periodic standard form can be achieved. As an illustration, consider the *third subharmonic* problem

$$\omega(\varepsilon) = 3 + \varepsilon\omega_1. \qquad\qquad (5.4.42)$$

In this case the general solution of the reduced problem (5.4.41) can be written

$$y = a\cos t + b\sin t - \frac{\gamma_0}{8}\cos 3t, \qquad (5.4.43)$$

with a and b constant. Our previous experience suggests that something similar to (5.4.43) should be used as a variation-of-constants solution for the perturbed problem (5.4.40); more precisely, the correct form should reduce to (5.4.43) when $\varepsilon = 0$ but should have period depending on ε in such a way as to be entrained to the forcing frequency ω with entrainment index $\ell = 3$; in other words the period should equal $6\pi/\omega$. In order to find the correct form, we will use the strategy developed for the harmonic case in Exercise 5.4.7. Namely, write (5.4.40) in the form

$$\ddot{y} + \frac{\omega^2}{9}y = \gamma_0 \cos \omega t + \frac{\omega^2 - 9}{9}y - \varepsilon\delta_1\dot{y} - \varepsilon\lambda_1 y^3, \qquad (5.4.44)$$

where ω is given by (5.4.42). All of the terms on the right hand side, except for the first, are small of order ε, and the new "free frequency" on the left hand side matches the expected solution frequency. Now set $\varepsilon = 0$ to obtain the reduced equation in the form

$$\ddot{y} + \frac{\omega^2}{9}y = \gamma_0 \cos \omega t. \qquad (5.4.45)$$

This is exactly the same as (5.4.41), considering that $\omega = 3$ when $\varepsilon = 0$; however, by retaining ω (and not replacing it by 3) the general solution of (5.4.45) will still have the correct frequency when $\varepsilon \neq 0$. (If this all seems slightly mystical, remember that once the correct change of variables is found, it will justify itself.) The general solution of (5.4.45) is

$$y = a \cos \frac{\omega}{3}t + b \cos \frac{\omega}{3}t - \frac{9\gamma_0}{8\omega^2} \cos \omega t. \qquad (5.4.46)$$

Again, this is actually the same as (5.5.43) when $\varepsilon = 0$, but now it has the correct form for a variation-of-constants solution when $\varepsilon \neq 0$. (Surprisingly, perhaps, one cannot get the right form just by modifying the frequencies in (5.5.43); there is also a change in the coefficient of the third term.)

From here the sailing is smooth, although tedious. One allows a and b to become variables, while imposing the usual condition that the first derivative of (5.4.46) be the same as if a and b were constants; that is to say, one sets

$$\dot{a} \cos \frac{\omega}{3}t + \dot{b} \sin \frac{\omega}{3}t = 0$$

and proceeds to differentiate (5.4.46) twice and use (5.4.44) to obtain a second equation for $\dot{a}$ and $\dot{b}$. The final result is that

$$\dot{a} = -\varepsilon f(a, b, t, \varepsilon) \sin \frac{\omega(\varepsilon)}{3}t + \mathcal{O}(\varepsilon^2),$$
$$\dot{b} = +\varepsilon f(a, b, t, \varepsilon) \cos \frac{\omega(\varepsilon)}{3}t + \mathcal{O}(\varepsilon^2), \qquad (5.4.47)$$

where

$$f(a, b, t, \varepsilon) = \frac{2}{3}\omega_1 \left(a \cos \frac{\omega}{3}t + b \sin \frac{\omega}{3}t - \frac{\gamma_0}{8} \cos \omega t \right)$$
$$- \delta_1 \left(-a \sin \frac{\omega}{3}t + b \cos \frac{\omega}{3}t + \frac{\gamma_0}{8} \sin \omega t \right)$$
$$- \lambda_1 \left(a \cos \frac{\omega}{3}t + b \sin \frac{\omega}{3}t - \frac{\gamma_0}{8} \cos \omega t \right)^3, \qquad (5.4.48)$$

with (of course) ω again given by (5.4.42). In obtaining (5.4.48) we have replaced ω by 3 everywhere except inside the trig functions, since the error committed by doing so belongs to the $\mathcal{O}(\varepsilon^2)$ terms in (5.4.47). Now (5.4.47) is in periodic standard form, except that one may still wish to introduce a strained time t^+ to make the period independent of ε.

To understand the idea of a third subharmonic, consider what it means for the forcing to undergo three cycles while the solution goes through one. Imagine, for instance, pushing a child on a swing. One usually pushes the child with a harmonic resonance, pushing once for each period. But an eccentric swing-pusher might try a cycle of push–pull–push as the child goes forward, pull–push–pull as the child swings back. Since the push predominates during the forward phase, and the pull predominates during the return phase, the overall input of energy acts to sustain the swinging motion against air resistance or other damping forces. Therefore it seems reasonable that a third subharmonic problem such as (5.4.42) might have periodic solutions. It has been shown in Exercise 4.7.3 that there are no periodic solutions to the third subharmonic problem when the forcing is small, at least when damping is present. Using the periodic standard form presented above, together with the methods of Section 6.3 below, it can be shown that periodic solutions do exist when the forcing is strong.

Exercises 5.4

1. Put the Van der Pol equation (5.4.16) into the periodic standard form (5.4.7) in Cartesian coordinates. Do this by following through the steps leading to (5.4.7), as we did in (5.4.17)–(5.4.19) for the polar form.

2. Put the free Duffing equation $\ddot{y} + y + \varepsilon y^3 = 0$ into periodic standard form in both Cartesian and Polar coordinates.

3. Repeat the derivation of (5.4.19) from (5.4.16), this time using variation of constants, beginning with (5.4.13) and (5.4.14) with $k = 1$.

4. Derive (5.4.26) by variation of constants. Begin by looking for a solution of (5.4.20) in the form $y = a \cos \omega(\varepsilon)t + b \sin \omega(\varepsilon)t$, where a and b are functions of t. (When a and b are constants and $\varepsilon = 0$, this expression reduces to the solution of the unperturbed equation; it also has the

correct frequency when $\varepsilon \neq 0$, since $\omega(\varepsilon)$ has been used in place of k.) Remember that an additional condition must be imposed on a and b when they become functions of t.

5. Carry out the details of the derivations of (5.4.26) and (5.4.29), beginning from (5.4.20). If hints are needed beyond those in the text, follow the patterns of Example 5.4.1.

6. Put the forced Duffing equation (5.4.30) into periodic standard form in polar coordinates (5.4.31) by working through the steps in the derivation of (5.4.29). Also obtain the Cartesian form corresponding to (5.4.26).

7. An alternate way to obtain a periodic standard form for (5.4.20) is to add $\omega^2 y$ to both sides, obtaining

$$\ddot{y} + \omega(\varepsilon)^2 y = \left(\omega(\varepsilon)^2 - k^2\right) y + \varepsilon f\left(y, \dot{y}, \omega(\varepsilon)t\right).$$

Introduce phase plane coordinates $U = y$, $V = \dot{y}/\omega(\varepsilon)$, then introduce rotating Cartesian coordinates A, B by an equation similar to (5.4.22). Introduce t^+ as in (5.4.24), show from (5.4.21) that $\omega^2 - k^2 = 2\varepsilon\omega_1 + \mathcal{O}(\varepsilon^2)$, and obtain a periodic standard form slightly different from (5.4.26). Then do the same thing in polar coordinates. (This is related to the discussion of "detuning" in Section 4.5; see (4.5.14).)

5.5. THE FIRST ORDER TWO-SCALE APPROXIMATION IN PERIODIC STANDARD FORM AND ITS JUSTIFICATION

This section has three objectives. The first order two-scale approximation will be obtained for an arbitrary system in periodic standard form. Then this approximation will be applied to an autonomous oscillator using the periodic standard form obtained in Example 5.4.1. Finally it will be shown that for any system in periodic standard form this first approximation has error $\mathcal{O}(\varepsilon)$ on expanding intervals of order $1/\varepsilon$. The method of error estimation used is similar to that in Section 3.3, but takes advantage of special features of the periodic standard form and of the two-scale approximation. As a corollary of the general result, Theorem 5.2.1 will be proved at the end of this chapter; this theorem was stated without proof in Section 5.2, and justifies the direct application of the two-scale method to second order equations without finding the periodic standard form.

Consider the initial value problem

$$\dot{x} = \varepsilon f(x, t, \varepsilon),$$
$$x(0) = \alpha,$$

$$(5.5.1)$$

where $\mathbf{x}$ is an N-vector and $\mathbf{f}$ is periodic in t with period 2π. This is to be solved by a two-time expansion

$$\mathbf{x} \sim \mathbf{x}_0(t, \tau) + \varepsilon \mathbf{x}_1(t, \tau) + \cdots, \tag{5.5.2}$$

where $\tau = \varepsilon t$. The total time derivative of (5.5.2) is (or, since this is heuristic, is supposed to be)

$$\dot{\mathbf{x}} \sim \mathbf{x}_{0t} + \varepsilon\{\mathbf{x}_{1t} + \mathbf{x}_{0\tau}\} + \cdots. \tag{5.5.3}$$

Substituting into (5.5.1), the equations for the first two coefficients are

$$\mathbf{x}_{0t} = 0,$$
$$\mathbf{x}_0(0, 0) = \alpha; \tag{5.5.4a}$$

$$\mathbf{x}_{1t} = \mathbf{f}(\mathbf{x}_0, t, 0) - \mathbf{x}_{0\tau},$$
$$\mathbf{x}_1(0, 0) = \mathbf{0}. \tag{5.5.4b}$$

The differential equation in (5.5.4a) tells us that $\mathbf{x}_0$ is a function of τ alone, and so we write simply $\mathbf{x}_0 = \mathbf{x}_0(\tau)$ with $\mathbf{x}_0(0) = \alpha$. The function $\mathbf{x}_0(\tau)$ must be determined so that the solution $\mathbf{x}_1$ of (5.5.4b) contains no t-secular terms.

The differential equation in (5.5.4b) now reads

$$\mathbf{x}_{1t}(t, \tau) = \mathbf{f}(\mathbf{x}_0(\tau), t, 0) - \mathbf{x}_{0\tau}(\tau), \tag{5.5.5}$$

and is to be thought of as an ordinary differential equation containing τ as a parameter; in fact, it can be solved simply by integrating both sides with respect to t. The right hand side is periodic in t and so can be expanded in a Fourier series. This is a first order equation, unlike the second order equations considered in Section 5.2, and for first order equations, secular terms do not arise from resonance between a Fourier component of the right hand side and a free frequency on the left. Instead, it is the constant term in the Fourier series that matters, since the integral of a constant is a linear function of time. In fact, all that is necessary to solve (5.5.5) is to expand $\mathbf{f}(\mathbf{x}_0(\tau), t, 0)$ in a Fourier series in t having vector-valued coefficients which are functions of τ:

$$\mathbf{f}(\mathbf{x}_0(\tau), t, 0) = \sum_{m=-\infty}^{\infty} \mathbf{a}_m(\tau)e^{imt}. \tag{5.5.6}$$

Then (5.5.5) becomes

$$\mathbf{x}_{1t}(t, \tau) = \{\mathbf{a}_0(\tau) - \mathbf{x}_0'(\tau)\} + \sum_{m \neq 0} \mathbf{a}_m(\tau)e^{imt}. \tag{5.5.7}$$

Upon integrating with respect to t, each term in the summation remains periodic and hence bounded, while the term in braces becomes multiplied by t, producing a secular term unless $x'_0(\tau) = a_0(\tau)$. Since $a_0(\tau)$, being the constant term of a Fourier series, is merely the average of the periodic function, this nonsecularity condition can be written

$$\frac{dx_0}{d\tau} = \frac{1}{2\pi} \int_0^{2\pi} f(x_0(\tau), t, 0) dt$$

or

$$\frac{dx_0}{d\tau} = \bar{f}(x_0), \tag{5.5.8}$$

where

$$\bar{f}(x) := \frac{1}{2\pi} \int_0^{2\pi} f(x, t, 0) dt. \tag{5.5.9}$$

(If $f(x, t, \varepsilon)$ has a fixed period other than 2π, (5.5.9) is replaced by the average over the appropriate period.) Since x_0 is in fact a function of only one time scale, we can just as well change (5.5.8) back into terms of t and write

$$\frac{dx_0}{dt} = \varepsilon \bar{f}(x_0). \tag{5.5.10}$$

This has the following interpretation: To find the differential equation (5.5.10) for the first order two time approximation for (5.5.1), it is only necessary to average the right hand side of (5.5.1) over one period in t, with x held constant. This is just the recipe which will be given for the first approximation by the method of averaging, to be considered in the next chapter.

For the remainder of this section, we will work with this recipe for the first approximation. All of the heuristic reasoning leading from (5.5.1) to (5.5.10) may be set aside, including the notation x_0 and the existence of the series (5.5.2); we will merely consider the idea that a good approximation to the solution of (5.5.1) might happen to be given by the solution of

$$\begin{aligned} \dot{z} &= \varepsilon \bar{f}(z), \\ z(0) &= \alpha. \end{aligned} \tag{5.5.11}$$

The only equations appearing above that will be used in the remainder of this section are (5.5.1), (5.5.9), and (5.5.11). The reason for the letter z is so that solutions of (5.5.1) and (5.5.11) can be compared without confusion. When one is only interested in the approximate solutions, it is more convenient to write (5.5.11) as $\dot{x} \cong \varepsilon \bar{f}(x)$, without introducing another variable; but then there is no notational distinction between the exact and approximate solutions.

Before examining the error in the approximation given by (5.5.11), we stop to consider what these equations look like in the familiar case of an autonomous oscillator. The standard form for an autonomous oscillator is given by either (5.4.7) in Cartesian coordinates or by (5.4.11) in polar coordinates. We repeat these here as

$$
\dot{a} = -\frac{\varepsilon}{k} f(a \cos kt + b \sin kt, -ak \sin kt + bk \cos kt, \varepsilon) \sin kt,
$$
$$
\dot{b} = +\frac{\varepsilon}{k} f(a \cos kt + b \sin kt, -ak \sin kt + bk \cos kt, \varepsilon) \cos kt
$$

(5.5.12)

and

$$
\dot{r} = \frac{\varepsilon}{k} f(r \cos(\varphi - kt), kr \sin(\varphi - kt), \varepsilon) \sin(\varphi - kt),
$$
$$
\dot{\varphi} = \frac{\varepsilon}{kr} f(r \cos(\varphi - kt), kr \sin(\varphi - kt), \varepsilon) \cos(\varphi - kt).
$$

(5.5.13)

Averaging (5.5.12) over one period and changing the names of the variables from $\mathbf{x} = (a, b)$ to $\mathbf{z} = (A, B)$ gives

$$
\frac{dA}{dt} =
$$
$$
-\frac{\varepsilon}{2\pi} \int_0^{2\pi/k} f(A \cos kt + B \sin kt, -Ak \sin kt + Bk \cos kt) \sin kt \, dt,
$$
$$
\frac{dB}{dt} =
$$
$$
+\frac{\varepsilon}{2\pi} \int_0^{2\pi/k} f(A \cos kt + B \sin kt, -Ak \sin kt + Bk \cos kt) \cos kt \, dt.
$$

(5.5.14)

Recall that the original second order autonomous oscillator (5.4.2) was changed into periodic standard form by using (5.4.3) and (5.4.5). Therefore the relationship between the solution y of the second order equation and the solution (a, b) of the first order system (5.5.12) is

$$
y = a \cos kt + b \sin kt.
$$

(5.5.15)

According to our present ideas, the solution (A, B) of (5.5.14) should give the first approximation to (a, b), and therefore the first approximation to the solution of the autonomous oscillator in the original variables should be given by

$$
y \cong A \cos kt + B \sin kt.
$$

(5.5.16)

When we applied the method of multiple scales directly to the autonomous oscillator, in Section 5.2, the first approximation y_0 was found to have the form (5.5.16) with A and B being functions of τ satisfying (5.2.10). But (5.2.10) is exactly the same as (5.5.14) when the latter is expressed in terms of τ. Therefore our current procedure gives the same results as the method of Section 5.2.

A similar analysis can be made of the polar form. Namely, if (5.5.13) is averaged over t, the variable names are changed from $\mathbf{x} = (r, \varphi)$ to $\mathbf{z} = (\rho, \psi)$, and the result is expressed in terms of τ, then the equations obtained are exactly (5.2.12).

Next we turn to the question of error estimation for the first order two-scale approximation. The theorem stated below will be proved again in Chapter 6 as Theorem 6.2.2, by quite different methods.

The solution of (5.5.1) will be denoted $\mathbf{x}(t, \alpha, \varepsilon)$. A similar expression could be used for the solution of (5.5.11), but in Chapter 6 it is more convenient to use a slightly different notation, and we will introduce it here to avoid conflict. Notice that if $\tau := \varepsilon t$ is introduced in (5.5.11) the result is $d\mathbf{z}/d\tau = \bar{\mathbf{f}}(\mathbf{z})$; this is the same as (5.5.8) and does not contain the perturbation parameter ε. The solution of this equation will be written $\mathbf{z}(\tau, \alpha)$; of course, it does not depend upon ε. Then the solution of (5.5.11) is $\mathbf{z}(\varepsilon t, \alpha)$ (where $\mathbf{z}$ is the same function). Writing $\mathbf{z}(\varepsilon t, \alpha)$ for the solution of (5.5.11) reveals the manner of dependence upon ε more precisely than writing $\mathbf{z}(t, \alpha, \varepsilon)$ would do. Of course, $\mathbf{x}(t, \alpha, \varepsilon)$ cannot be written as $\mathbf{x}(\varepsilon t, \alpha)$ since it depends upon ε in a more complicated way than through the combination εt.

Theorem 5.5.1. *Let* $\mathbf{f}$ *be smooth. There exist positive constants* ε_0, T, *and* c *such that the solutions of (5.5.1) and (5.5.11) satisfy*

$$\|\mathbf{x}(t, \alpha, \varepsilon) - \mathbf{z}(\varepsilon t, \alpha)\| < c\varepsilon \qquad \text{for} \ \ 0 \leq t \leq \frac{T}{\varepsilon} \ \ \text{and} \ \ 0 \leq \varepsilon \leq \varepsilon_0. \ \ (5.5.17)$$

Proof. Choose a closed ball K of some positive radius δ centered at α, and let $\varepsilon_0 > 0$. Let M be an upper bound for both $\|\bar{\mathbf{f}}(\mathbf{z})\|$ and $\|\mathbf{f}(\mathbf{z}, t, \varepsilon)\|$ for $\mathbf{z}$ in K and $0 \leq \varepsilon \leq \varepsilon_0$; such a bound exists because $\mathbf{f}$ is periodic in t. Then both solutions $\mathbf{x}$ and $\mathbf{z}$ begin at the center of K and reach at most a distance $\varepsilon M t$ from the center by time t, as long as they remain in K (assuming $0 \leq \varepsilon \leq \varepsilon_0$). But on the other hand, they cannot leave K until they have travelled a distance δ. It follows that they must remain in K at least until time T/ε, where $T = \delta/M$. In the remainder of the proof, various bounds and Lipschitz constants will be defined using the set K; these are now automatically known to hold for the interval $0 \leq t \leq T/\varepsilon$.

Since α is fixed throughout the proof, it will be omitted from the expressions $\mathbf{x}(t, \alpha, \varepsilon)$ and $\mathbf{z}(\varepsilon t, \alpha)$ for convenience. Write $\mathbf{x}(t, \varepsilon) = \mathbf{z}(\varepsilon t) + \mathbf{R}(t, \varepsilon)$, so that $\mathbf{R}$ is the remainder (or error) to be estimated. Then (5.5.1) implies $\dot{\mathbf{z}} + \dot{\mathbf{R}} = \varepsilon \mathbf{f}(\mathbf{z} + \mathbf{R}, t, \varepsilon)$. Combining this with (5.5.11) gives the following equation satisfied by the remainder:

$$\dot{\mathbf{R}} = \varepsilon \left[\mathbf{f}(\mathbf{z}(\varepsilon t) + \mathbf{R}, t, \varepsilon) - \bar{\mathbf{f}}(\mathbf{z}(\varepsilon t)) \right],$$

$$\mathbf{R}(0, \varepsilon) = 0. \tag{5.5.18}$$

Integrating from 0 to t, using the initial condition, yields

$$\mathbf{R}(t, \varepsilon) = \varepsilon \int_0^t \left[\mathbf{f}(\mathbf{z}(\varepsilon s) + \mathbf{R}(s, \varepsilon), s, \varepsilon) - \bar{\mathbf{f}}(\mathbf{z}(\varepsilon s)) \right] ds. \qquad (5.5.19)$$

Notice that the dummy variable of integration s (not t) now appears in the argument of $\mathbf{z}$ and $\mathbf{R}$ inside the integrals; from here on these arguments will be omitted for brevity. Adding and subtracting $\mathbf{f}(\mathbf{z}, s, \varepsilon)$ and using the triangle inequality and the fact that the norm of an integral is less than or equal to the integral of the norm, we have

$$\|\mathbf{R}(t, \varepsilon)\| \leq \varepsilon \int_0^t \|\mathbf{f}(\mathbf{z} + \mathbf{R}, s, \varepsilon) - \mathbf{f}(\mathbf{z}, s, \varepsilon)\| ds$$

$$+ \varepsilon \left\| \int_0^t \left[\mathbf{f}(\mathbf{z}, s, \varepsilon) - \bar{\mathbf{f}}(z) \right] ds \right\|. \qquad (5.5.20)$$

Notice that only in the first term has the norm been taken inside the integral. Let L be a Lipschitz constant for $\mathbf{f}$ with respect to its first (vector) argument, valid in K. Then the first term on the right in (5.5.20) is bounded by the integral of $\varepsilon L \|\mathbf{R}\|$. The second term is more difficult to estimate, and is the heart of the matter; it is here that the average of $\mathbf{f}$ enters, therefore it is here that the form of (5.5.11) comes into play. The key result is that there is a constant c_0 such that the last term in (5.5.20) is bounded by $c_0 \varepsilon$ as long as $0 \leq t \leq T/\varepsilon$; this will be proved below as Lemma 5.5.2. Assuming this result for the moment, (5.5.20) implies

$$\|\mathbf{R}(t, \varepsilon)\| \leq \varepsilon L \int_0^t \|\mathbf{R}(s, \varepsilon)\| ds + c_0 \varepsilon. \qquad (5.5.21)$$

From here the argument follows the familiar Gronwall pattern. Let

$$S(t, \varepsilon) = \int_0^t \|\mathbf{R}(s, \varepsilon)\| ds; \qquad (5.5.22)$$

then (5.5.21) can be written

$$\|\mathbf{R}(t, \varepsilon)\| = \frac{dS}{dt} \leq \varepsilon L S + c_0 \varepsilon, \qquad (5.5.23)$$
$$S(0) = 0.$$

Solving this differential inequality in the usual way (see, for instance, the solution of (4.2.13)) gives

$$S(t, \varepsilon) \leq \frac{c_0}{L} \left(e^{\varepsilon L t} - 1 \right) \leq \frac{c_0}{L} \left(e^{L T} - 1 \right), \qquad (5.5.24)$$

where the last expression uses $t \leq T/\varepsilon$; recall that the entire argument only holds up to this time. Finally, putting (5.5.24) back into (5.5.23) gives

$$\|\mathbf{R}(t, \varepsilon)\| \leq \varepsilon c_0 e^{LT}. \tag{5.5.25}$$

This is the statement (5.5.17) of the theorem. ∎

The missing step in the above proof is the estimate for the last term in (5.5.20). The crucial fact about the integrand $\mathbf{f}(\mathbf{z}, s, \varepsilon) - \bar{\mathbf{f}}(\mathbf{z})$ is that for fixed $\mathbf{z}$, it has zero average (with respect to s). Therefore, if $\mathbf{z}$ were fixed during the integration, the integral would be bounded for all time. Of course, $\mathbf{z}$ is not fixed during the integration, but is a slowly varying function of s; specifically, it is $\mathbf{z}(\varepsilon s)$. The crucial point is that because this is only slowly varying, the integral is still bounded, not for all time, but on the expanding interval $0 \leq t \leq T/\varepsilon$. Namely:

Lemma 5.5.2. *Let $\boldsymbol{\varphi}(\mathbf{z}, t, \varepsilon)$ be a smooth function, periodic in t with period 2π for each fixed $\mathbf{z}$ and ε, and having zero mean with respect to t. Let $\mathbf{z}(\varepsilon t)$ be the solution of (5.5.11), which remains in K for $0 \leq t \leq T/\varepsilon$. Then there exists a constant c_0 such that*

$$\left\| \int_0^t \boldsymbol{\varphi}\,(\mathbf{z}(s, \varepsilon), s, \varepsilon)\,ds \right\| \leq c_0 \quad \text{for} \quad 0 \leq t \leq \frac{T}{\varepsilon} \quad \text{and} \quad 0 \leq \varepsilon \leq \varepsilon_0. \tag{5.5.26}$$

Proof. The proof given here requires that $\boldsymbol{\varphi}$ have at least two continuous derivatives; it uses Fourier series estimates and integration by parts. There is another proof which requires only continuity (in fact even weaker hypotheses defined in terms of measure theory will suffice), but the computations used in that proof are less natural. (See Notes and References.)

Since $\boldsymbol{\varphi}$ is periodic with zero mean and has two continuous derivatives, it can be written as a uniformly convergent Fourier series

$$\boldsymbol{\varphi}(\mathbf{z}, t, \varepsilon) = \sum_{n \neq 0} \mathbf{a}_n(\mathbf{z}, \varepsilon) e^{int} \tag{5.5.27}$$

without a constant term, where the $\mathbf{a}_n$ are vector-valued coefficients satisfying $\|\mathbf{a}_n(\mathbf{z}, t, \varepsilon)\| \leq k/n^2$ for some constant k and for all $\mathbf{z}$ in K, all t, and for $0 \leq \varepsilon \leq \varepsilon_0$. Now substitute $\mathbf{z}(\varepsilon t)$ into (5.5.27). To shorten the notation, we will omit ε as an argument from $\boldsymbol{\varphi}$ and $\mathbf{a}_n$ from here on. Since the convergence is uniform, the resulting series can be integrated termwise by parts to obtain

$$\int_0^t \boldsymbol{\varphi}\,(\mathbf{z}(\varepsilon s, s))\,ds = \sum_{n \neq 0} \frac{\mathbf{a}_n\,(\mathbf{z}(\varepsilon t))\,e^{int} - \mathbf{a}_n\,(\mathbf{z}(0))}{in}$$
$$- \int_0^t \frac{e^{ins}}{in} \mathbf{a}_n'\,(\mathbf{z}(\varepsilon s)) \frac{d}{ds}\mathbf{z}(\varepsilon s)\,ds. \tag{5.5.28}$$

Here $\mathbf{a}'_n$ is the matrix of partial derivatives of $\mathbf{a}_n$ with respect to $\mathbf{z}$; since these are the Fourier coefficients of $\partial\boldsymbol{\varphi}/\partial\mathbf{z}$, which has at least one continuous derivative, they satisfy $\|\mathbf{a}'_n\| \leq \ell/n$ for $\mathbf{z}$ in K, for some constant ℓ, using a matrix norm. Since $d\mathbf{z}(\varepsilon s)/ds = \varepsilon\bar{\mathbf{f}}(\mathbf{z})$, its norm is bounded by εM (where M is as in the first paragraph of the proof of Theorem 5.5.1). Putting all of this together gives the conclusion:

$$
\left\| \int_0^t \boldsymbol{\varphi}\left(\mathbf{z}(\varepsilon s, s)\right) ds \right\| \leq \sum_{n\neq 0} \frac{\|\mathbf{a}_n\left(\mathbf{z}(\varepsilon t)\right)\| + \|\mathbf{a}_n\left(\mathbf{z}(0)\right)\|}{n}
$$

$$
+ \sum_{n\neq 0} \int_0^t \frac{1}{n}\|\mathbf{a}'_n\left(\mathbf{z}(\varepsilon s)\right)\| \varepsilon M ds
$$

$$
\leq \sum_{n\neq 0} \frac{2k}{n^3} + \sum_{n\neq 0} \frac{\ell\varepsilon M^2}{n^2} t
$$

$$
\leq 2k\left(\sum_{n\neq 0}\frac{1}{n^3}\right) + \ell M^2 T\left(\sum_{n\neq 0}\frac{1}{n^2}\right) =: c_0.
$$

(5.5.29)

In the next to the last step we have used $t \leq T/\varepsilon$. ∎

Theorem 5.5.1 justifies the first order two-scale method for systems in periodic standard form. But it is easy to apply this theorem to obtain a justification also for the first order two-scale method applied directly to second order scalar equations without first changing them into periodic standard form. We will now illustrate how this is done by using Theorem 5.5.1 to prove Theorem 5.2.1, justifying the first order two-scale method for autonomous oscillators.

Proof of Theorem 5.2.1. It was shown above (in discussing equation (5.5.16)) that the approximate solution y_0 for an autonomous oscillator constructed in Section 5.2 by applying the two-scale method directly is the same as the one constructed in this section by first putting the oscillator into periodic standard form. For the periodic standard form, Theorem 5.5.1 implies that $a = A + \mathcal{O}(\varepsilon)$ and $b = B + \mathcal{O}(\varepsilon)$ for $0 \leq t \leq T/\varepsilon$. It is only necessary to carry this estimate over to the original variable y. We have $y = a\cos kt + b\sin kt = (A + \mathcal{O}(\varepsilon))\cos kt + (B + \mathcal{O}(\varepsilon))\sin kt = A\cos kt + B\sin kt + \mathcal{O}(\varepsilon) = y_0 + \mathcal{O}(\varepsilon)$ for $0 \leq t \leq T/\varepsilon$; the next-to-the-last equality in this sequence holds because $\cos kt$ and $\sin kt$ are bounded for all time. This proves Theorem 5.2.1. A similar estimate can be proved for the velocity; see Exercise 5.5.1. ∎

Exercises 5.5

1. Let y be the exact solution of the autonomous oscillator and let y_0 be the two-scale solution constructed in Section 5.2. Prove from Theorem

5.5.1 that $\dot{y} = y_{0t} + \mathcal{O}(\varepsilon)$ on an expanding interval. Hint: $\dot{y} = v$; see (5.4.7). Remark: Taking the total derivative of $y \cong y_0$ with respect to t suggests (heuristically) that $\dot{y} \cong y_{0t} + \varepsilon y_{0\tau}$. The second term is formally of order $\mathcal{O}(\varepsilon)$, the same as the expected order of the error, so it is reasonable to drop it. This exercise shows that this heuristic conclusion is in fact correct.

2. Find the first approximation for the free Duffing equation $\ddot{y} + y + \varepsilon y^3 = 0$, $y(0) = \alpha$, $\dot{y}(0) = 0$ using periodic standard form. Work the problem in both polar and Cartesian form, and express the solutions in the original variable y. See Exercise 5.4.2, and compare your result with Exercise 5.2.3.

5.6* PARTIAL DIFFERENTIAL EQUATIONS

The partial differential equation

$$u_{tt} - u_{xx} + u = 0 \qquad (5.6.1)$$

is called a *linear dispersive wave equation* and has as one of its particular solutions (this can be checked easily) the function

$$u = \sin(kx + \omega t), \qquad (5.6.2)$$

where k is arbitrary and

$$\omega := \sqrt{1 + k^2}. \qquad (5.6.3)$$

Throughout this section ω will be used as in (5.6.3); that is, it is not an independent constant but merely an abbreviation. The solution (5.6.2) satisfies the following initial conditions at $t = 0$:

$$\begin{aligned} u(x, 0) &= \sin kx, \\ u_t(x, 0) &= \omega \cos kx. \end{aligned} \qquad (5.6.4)$$

As an example of the multiple scale method in partial differential equations, we will determine the first order two-scale approximation to the solution of the following perturbation of (5.6.1):

$$u_{tt} - u_{xx} + u + \varepsilon u^3 = 0. \qquad (5.6.5)$$

The solution to be found will be the one having the same initial conditions (5.6.4) as the solution (5.6.2) of (5.6.1).

At the very beginning, there is a pesky notational difficulty that does not arise in ordinary differential equations; it is a minor nuisance, but is not serious. Based on our experience with ordinary differential equations in Section 5.2, we write

$$\tau = \varepsilon t \qquad (5.6.6)$$

and represent the solution in the two-scale form

$$u(x, t, \varepsilon) = u_0(x, t, \tau) + \varepsilon u_1(x, t, \tau) + \cdots . \qquad (5.6.7)$$

At this point we want to apply the usual differentiation rules

$$\frac{d}{dt} = \frac{\partial}{\partial t} + \varepsilon \frac{\partial}{\partial \tau},$$
$$\frac{d^2}{dt^2} = \frac{\partial^2}{\partial t^2} + 2\varepsilon \frac{\partial^2}{\partial \tau \partial t} + \varepsilon^2 \frac{\partial^2}{\partial \tau^2}, \qquad (5.6.8)$$

which were stated before in (5.2.4). The difficulty is that in order to find u_{tt}, the left hand side of (5.6.7) must be *partially* differentiated twice with respect to t, whereas the left hand sides of (5.6.8) involve *ordinary* derivatives. If we change the ordinary derivatives in (5.6.8) to partial derivatives, the result is nonsense, because the partial derivatives on the right hand side have a different meaning from those on the left. To resolve the difficulty it is only necessary to think of what (5.6.8) are really intended to mean. In the original application, there was no x in (5.6.7), and what was needed was to express a derivative with respect to the single time variable t in terms of derivatives with respect to two time variables t and τ. This is done by equations (5.6.8), with the ordinary derivatives on the left being derivatives with respect to t alone (when there is no τ in the equation), and the partial derivatives with respect to t on the right being derivatives with respect to t when τ is present and held constant. It is only necessary to preserve this interpretation even when the variable x is present. That is, equations (5.6.8) are still correct if we temporarily regard d/dt as a *partial derivative with respect to t with only x held constant,* and $\partial/\partial t$ as a *partial derivative with respect to t with both x and τ held constant.* With no further ado, then, we apply (5.6.8) to (5.6.7) and obtain the following sequence of problems, where the subscript t (when attached to one of the u_n rather than to u itself) means the derivative holding x and τ constant:

$$u_{0tt} - u_{0xx} + u_0 = 0,$$
$$u_0(x, 0, 0) = \sin kx, \qquad (5.6.9a)$$
$$u_{0t}(x, 0, 0) = \omega \cos kx;$$

$$u_{1tt} - u_{1xx} + u_1 = -u_0^3 - 2u_{0t\tau},$$
$$u_1(x, 0, 0) = 0, \qquad\qquad (5.6.9b)$$
$$u_{1t}(x, 0, 0) = -u_{0\tau}(x, 0, 0),$$

$$\vdots$$

Now the first of these problems, (5.6.9a), is the same as (5.6.1) and (5.6.4) except for the possibility of dependence on τ, so its solution is the same as (5.6.2) except for amplitude and phase factors that may depend on τ:

$$u_0(x, t, \tau) = a(\tau)\sin(kx + \omega t + \phi(\tau)). \qquad (5.6.10)$$

The initial conditions imply that

$$a(0) = 1,$$
$$\phi(0) = 0. \qquad\qquad (5.6.11)$$

Substituting this (partial) solution for u_0 into (5.6.9b) and replacing $\sin^3$ by its expansion from Table E.1 gives

$$\begin{aligned}
u_{1tt} - u_{1xx} + u_1 = {} & \left(-\frac{3}{4}a^3 + 2a\omega\phi'\right)\sin(kx + \omega t + \phi) \\
& - 2a'\omega\cos(kx + \omega t + \phi) \\
& + \frac{1}{4}a^3\sin 3(kx + \omega t + \phi). \qquad (5.6.12)
\end{aligned}$$

The first harmonics on the right hand side of (5.6.12) have the same form as the solutions of the associated homogeneous equation. Therefore they are in resonance with the free solutions and lead to t-secular terms. (Partial differential equations of this type behave similarly to ordinary differential equations in this respect.) Therefore the method of multiple scales calls for completing the determination of $a(\tau)$ and $\phi(\tau)$ by setting the coefficients of these first harmonics equal to zero, leading t:

$$a' = 0,$$
$$\phi' = +\frac{3a^2}{8\omega}. \qquad\qquad (5.6.13)$$

Solving (5.6.13) with initial conditions (5.6.11) gives

$$a(\tau) = 1,$$
$$\phi(\tau) = +\frac{3a^2}{8\omega}\tau,$$

so that the complete first order two-scale approximation for the desired solution $u(x, t, \varepsilon)$ is

$$u_0(x, t, \varepsilon t) = \sin\left(kx + \omega t + \frac{3a^2}{8\omega}\varepsilon t\right).$$

5.7. NOTES AND REFERENCES

The method of multiple scales is emphasized in the books by Nayfeh and the one by Kevorkian and Cole which are listed in the Preface. Kevorkian and Cole prefer the version using slow time τ and strained fast time t^+, whereas Nayfeh gives more attention to the version using $T_0, T_1, \ldots, T_M$. As explained in Sections 5.1 and 5.3, these two versions of multiple scales are only needed when attempting to gain validity for intervals longer than $1/\varepsilon$. The simpler version, using t and τ, is given in Chapters 3 and 4 of the book by Smith, along with error estimation (which is absent in the other books).

For the relationship between multiple scales and averaging, the essential references are

> Lawrence M. Perko, Higher order averaging and related methods for perturbed periodic and quasiperiodic systems, *SIAM J. Appl. Math.* **17** (1968), 698–724

and

> Willy Sarlet, On a common derivation of the averaging method and the two-time-scale method, *Celestial Mech.* **17** (1978), 299–312.

The proof of validity of the first order two-scale method for systems in periodic standard form, presented in Section 5.5, is due to J. G. Besjes, who presented it as a proof of the validity of first order averaging. (Actually his proof is the one mentioned in the first line of our proof and does not assume two continuous derivatives.) It is to be found in

> J. G. Besjes, On the asymptotic methods for non-linear differential equations, *J. Méchanique* **8** (1969), 357–372.

For additional comments and references concerning this proof, see Section 6.7.

The example (5.3.14), along with some others for which the multiple scale method fails on time scales longer than $1/\varepsilon$, is given in

> L. A. Rubenfeld, On a derivative-expansion technique and some comments on multiple scaling in the asymptotic approximation of solutions of certain differential equations, *SIAM Rev.* **20** (1978), 79–105.

A different situation causing such a failure is discussed in Section 6.5 below.

A lengthy discussion of the nonlinear wave problem in Section 5.6, along with many variations, is given in Section 4.4.1 of the book by Kevorkian and Cole (listed in the Preface).

CHAPTER 6

AVERAGING

6.1. FIRST ORDER AVERAGING

The method of averaging is a way of finding approximate solutions to systems of differential equations which have (or can be put into) one of several *standard forms*. Much of this chapter will deal with systems in *periodic standard form* introduced in Section 5.4, although later sections will also concern *angular standard form*. The present chapter depends on Section 5.1, which places the method of averaging in context, and Section 5.4, which explains how to put systems into periodic standard form. Otherwise it can be read independently of Chapter 5, although some of the results overlap.

The simplest form of the method of averaging, called *first order averaging*, consists of replacing the system of differential equations

$$\dot{\mathbf{x}} = \varepsilon \mathbf{f}(\mathbf{x}, t, \varepsilon), \tag{6.1.1}$$

in which $\mathbf{f}$ is 2π-periodic in t, with the simpler system

$$\dot{\mathbf{z}} = \varepsilon \bar{\mathbf{f}}(\mathbf{z}), \tag{6.1.2}$$

where

$$\bar{\mathbf{f}}(z) = \frac{1}{2\pi} \int_0^{2\pi} \mathbf{f}(\mathbf{z}, t, 0) dt. \tag{6.1.3}$$

The averaged system (6.1.2) is simpler than (6.1.1) because it is autonomous, but in general it is still nonlinear, and is not necessarily solvable.

275

On the other hand, in many common applications (especially in two dimensions) it can be solved explicitly using elementary functions, and it is then reasonable to use solutions of (6.1.2) as approximations to solutions of (6.1.1). It will be shown (Theorems 6.2.2 and 6.2.3) that for a long time, a solution of (6.1.2) remains close to the solution of (6.1.1) with the same initial condition, the error being $\mathcal{O}(\varepsilon)$ on expanding time intervals of length $\mathcal{O}(1/\varepsilon)$. But even when the averaged system cannot be solved explicitly, it is still possible to draw conclusions about the exact solutions from the averaged equations. For instance, the method of averaging gives a way of finding determining equations for periodic solutions, and therefore gives an alternate justification for some of the results in Chapter 4. At the same time, it allows determination of stability or instability of the periodic solutions, a topic which could only be handled in Chapter 4 for the case of self-sustained oscillations in the plane. These aspect of averaging will be covered in Section 6.3.

There are at least three ways to motivate the first order method of averaging, and at least three ways to justify the method. The three motivations, briefly stated, are

1. The simple guess that a physical system's response to external influences is determined more by the average influence than by fluctuations about the average.
2. The multiple scale method for systems in periodic standard form leads to the averaged equations for the first approximation. (This was seen in Section 5.5.)
3. The attempt to make (6.1.1) autonomous by a change of coordinates leads in a natural way to (6.1.2). This will be seen in Section 6.2.

The three ways to justify averaging are

A. Comparison of the solutions of (6.1.2) with experimental results or with numerical solutions of (6.1.1).
B. Direct estimation of the error $\|x - z\|$. (This was done in Section 5.5 for the first approximation.)
C. A less direct estimation of the same error by means of the coordinate transformation mentioned in item 3 above. This will be done in Section 6.2.

Any of the "motivations" (1, 2, or 3) may be paired with the "justifications" A or B, but justification C makes sense only in connection with motivation 3. Clearly, motivation 1 and justification A fit together nicely (and belong with a strong applications orientation). Motivation 2 and justification B have been presented together in Section 5.5. One may wonder, then, why the third pair (motivation 3 and justification C, given in Section 6.2) is needed at all. The answer is that much more follows from this third approach than just another way to prove an error estimate for first order averaging. In the first place, the coordinate transformation given by this approach is the key to finding periodic solutions (and their determining equations) by the method of averaging (Section 6.3). In the second place, this approach will suggest a natural way to extend the method of averaging to higher order approximations (Section 6.5).

Before turning to the theory of averaging, we will present an example of how averaging is used informally both to solve a system approximately and to draw conclusions about the behavior of the system.

Example 6.1.1. Mathieu's Equation

Mathieu's equation is

$$\ddot{y} + (k^2 + \varepsilon \cos 2t)y = 0. \tag{6.1.4}$$

This equation is linear, but is not solvable in terms of elementary functions because the coefficient of y is not constant. The forcing term in (6.1.4) is commonly called a *parametric excitation*, because it is thought of as a periodic variation of the restoring force parameter k^2.

> One way of arriving at this equation is to consider a pendulum whose length varies periodically in time, for instance by having the string pass through a hole in a horizontal board which is raised and lowered periodically. The equation of motion of such a pendulum is $M\ddot{y} + \sqrt{G/L(t)} \sin y = 0$, where M is the (dimensional) mass, G the gravitational constant, $L(t)$ the time-dependent length, and y the angle from the vertical (in radians, hence already nondimensional). If $\sin y$ is replaced by y, a linearized equation suitable for small oscillations is obtained; in nondimensional form, it is $\ddot{y} + f(t)y = 0$, which is called "Hill's equation." Then (6.1.4) is the special case $f(t) = k^2 + \varepsilon \cos 2t$. The factor 2 is merely a matter of convention.

From the discussion of forced oscillations in Section 4.5, one would suspect that in order to obtain the full range of phenomena associated with (6.1.4) one should replace the forcing frequency 2 by an ε-dependent frequency $\omega(\varepsilon)$. But for this particular equation it has become customary to regard k^2 as a function of ε instead. This has the same effect, and one form can be changed into the other by a rescaling of time. Therefore we assume

$$k^2(\varepsilon) = 1 + \varepsilon\lambda_1 + \mathcal{O}(\varepsilon^2); \tag{6.1.5}$$

the choice of $k^2(0) = 1$ singles out a particular resonance to be studied. With these choices, Mathieu's equation can be written

$$\ddot{y} + y = -\varepsilon(\lambda_1 + \cos 2t)y. \tag{6.1.6}$$

In order to study (6.1.6) by the method of averaging, it is first necessary to write it as a system of first order differential equations in periodic standard form. We will do this by variation of constants. (See Section 5.4 for a complete discussion.) The general solution of (6.1.6) for $\varepsilon = 0$ can be written

$$y = a \cos t + b \sin t, \tag{6.1.7a}$$

with a and b being constant; the derivative of this solution is

$$\dot{y} = -a \sin t + b \cos t. \tag{6.1.7b}$$

We wish to write the solution of (6.1.6), and its derivative, in the same form even when $\varepsilon \neq 0$, by allowing a and b to become functions of t. Calculation reveals that this will be the case, provided a and b are taken to satisfy

$$\begin{aligned} \dot{a} &= \varepsilon(\lambda_1 + \cos 2t)(a \cos t + b \sin t) \sin t, \\ \dot{b} &= \varepsilon(\lambda_1 + \cos 2t)(a \cos t + b \sin t) \cos t. \end{aligned} \tag{6.1.8}$$

This system is in periodic standard form with period 2π.

A convenient way to average the system (6.1.8) is to write $\cos 2t = \cos^2 t - \sin^2 t$, multiply out the right hand sides, and use Table E.1 in Appendix E; replacing each product of sines and cosines by the constant term in its Fourier series gives the average. The result, using new variables $\mathbf{z} = (A, B)$ in place of $\mathbf{x} = (a, b)$, is

$$\begin{aligned} \dot{A} &= \varepsilon \frac{2\lambda_1 - 1}{4} B, \\ \dot{B} &= -\varepsilon \frac{2\lambda_1 + 1}{4} A. \end{aligned} \tag{6.1.9}$$

If the initial conditions for (6.1.4) are

$$\begin{aligned} y(0) &= \alpha, \\ \dot{y}(0) &= \beta, \end{aligned} \tag{6.1.10}$$

then (6.1.7) shows that the initial conditions for a and b are

$$\begin{aligned} a(0) &= \alpha, \\ b(0) &= \beta. \end{aligned} \tag{6.1.11}$$

The same initial conditions will be used for the averaged system (6.1.9). It is not difficult to compute the solution of (6.1.9) with initial conditions (6.1.11); see Exercise 6.1.1. The results are as follows. Define the constants $\mu := \sqrt{1 - 4\lambda_1^2/4}$, $\sigma := \sqrt{(1 - 2\lambda_1)/(1 + 2\lambda_1)}$, $\nu := \sqrt{4\lambda_1^2 - 1/4}$, and

$\tau := \sqrt{(2\lambda_1 - 1)/(2\lambda_1 + 1)}$. Then, if $4\lambda_1^2 < 1$, the solutions are exponential and may be written using hyperbolic trigonometric functions as

$$A(t) = \alpha \cosh \mu\varepsilon t + \sigma\beta \sinh \mu\varepsilon t,$$
$$B(t) = \frac{\alpha}{\sigma} \sinh \mu\varepsilon t + \beta \cosh \mu\varepsilon t. \tag{6.1.12}$$

If $4\lambda_1^2 > 1$, then the solutions are oscillatory and have the form

$$A(t) = \alpha \cos \nu\varepsilon t + \tau\beta \sin \nu\varepsilon t,$$
$$B(t) = -\frac{\alpha}{\tau} \sin \nu\varepsilon t + \beta \cos \nu\varepsilon t. \tag{6.1.13}$$

If $4\lambda_1^2 = 1$, then there are two cases. If $\lambda_1 = 1/2$, then

$$A(t) = \alpha,$$
$$B(t) = \beta - \frac{\alpha}{2}\varepsilon t. \tag{6.1.14}$$

If $\lambda_1 = -1/2$ then

$$A(t) = \alpha - \frac{\beta}{2}\varepsilon t,$$
$$B(t) = \beta. \tag{6.1.15}$$

The results given in (6.1.12)–(6.1.15) are the exact solutions of the averaged equations (6.1.9). They are also the approximate solutions given by the method of averaging for the equations (6.1.8) in periodic standard form. But we are actually most interested in an approximate solution of the second order equation (6.1.4) in the original variable y. In order to obtain this, A and B must be substituted into (6.1.7a) in place of a and b. This gives

$$y(t, \varepsilon) \cong A(t) \cos t + B(t) \sin t. \tag{6.1.16}$$

Of course, equations (6.1.12), (6.1.13), (6.1.14), or (6.1.15) should be used in (6.1.16) according to the value of λ_1. This completes the computational portion of the first order method of averaging for this problem.

Next we will consider the qualitative properties of these approximate solutions. When $4\lambda_1^2 < 1$ it follows from (6.1.12) that A and B are unbounded. Therefore the approximate solution (6.1.16) is also unbounded. When $4\lambda_1^2 > 1$ it follows from (6.1.13) that A and B are periodic with period $2\pi/\nu$, but this does not imply that (6.1.16) is periodic, since that expression involves both of the periods $2\pi/\nu$ and 2π. However it does follow that (6.1.16) is bounded and quasiperiodic, and it will be periodic if $2\pi/\nu$ and 2π have a common integer multiple, that is, if ν is a rational number. But the most interesting case arises when $4\lambda_1^2 = 1$, that is, when $\lambda_1 = \pm 1/2$. If $\lambda_1 = 1/2$ then according to (6.1.14), A and B are constant when $\alpha = 0$, and the approximate solution is $y = \beta \sin t$. If $\lambda_1 = -1/2$ then A and B are constant when $\beta = 0$, and the approximate solution is $y = \alpha \cos t$. Both of these solutions are periodic with period 2π.

It is natural to ask whether these qualitative features of the approximate solutions carry over to the exact solutions. For instance, do there exist solutions of period 2π when $\lambda_1 = \pm 1/2$? Unfortunately this is not exactly the case. One can show (with the aid of a method called Floquet theory) that there exist two functions $k^2(\varepsilon)$ for which (6.1.4) has periodic solutions of period 2π, and that when expanded in the form (6.1.5) these functions satisfy $\lambda_1 = \pm 1/2$. These curves can be graphed in the (k^2, ε) plane (see Fig. 6.1.1). The method of averaging gives the correct tangent lines to these curves at their crossing point, but whereas the averaged equations have periodic solutions for *any* curve tangent to these lines, the exact equations have such solutions only for *one particular* such curve (for each line). Incidentally, solutions corresponding to points between the two curves in Fig. 6.1.1 are unbounded, those outside this region are bounded; once

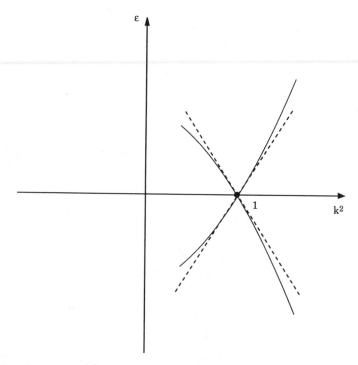

FIG. 6.1.1. Values of (k^2, ε) near $k^2 = 1$ for which Mathieu's equation has periodic solutions. The dotted straight lines are the tangents to these curves. The averaged equations determine these tangents correctly, but falsely suggest that periodic solutions exist for any curve tangent to these lines.

again, averaging gives a roughly correct picture, but only as far as finding the tangent lines of the curves separating bounded and unbounded solutions.

Return for a moment to the averaged equations (6.1.9), and notice that when $\lambda_1 = \pm 1/2$, the right-hand side has nonsimple zeroes. (Specifically, when $\lambda_1 = 1/2$ the line $A = 0$ is a line of zeroes, and when $\lambda_1 = -1/2$, $B = 0$ is such a line.) In Section 6.3 it will be shown that *simple* zeroes of the averaged equation correspond to periodic solutions of the exact equation. Part of the reason that the averaging method is slightly incorrect in predicting the periodic solutions of (6.1.4) is that the zeroes are not simple in this problem.

It should be stressed that the averaged equations (6.1.9) are linear only because the original equation (6.1.4) is linear. One cannot always expect the averaged equations to be solvable. This is in sharp contrast to the methods of Chapters 2 and 4, which always yield linear equations for the approximate solutions to any order even when the original equations are (weakly) nonlinear.

Exercises 6.1

1. Solve the linear system (6.1.9) with initial conditions (6.1.11). Either use matrix methods, or else derive the equation $d^2 A/dt^2 + \varepsilon^2 (4\lambda_1^2 - 1)y/16 = 0$ and a similar equation for B.

6.2. JUSTIFICATION OF FIRST ORDER AVERAGING

This section is devoted to providing a foundation for the first order method of averaging. In addition to providing error estimates, the work in this section is crucial to the applications and extensions of averaging dealt with in later sections of this chapter. For convenience the equations defining the first order method of averaging will now be repeated from the last section. The method consists of replacing a system in periodic standard form

$$\dot{\mathbf{x}} = \varepsilon \mathbf{f}(\mathbf{x}, t, \varepsilon) \qquad (6.2.1)$$

by the autonomous system

$$\dot{\mathbf{z}} = \varepsilon \bar{\mathbf{f}}(\mathbf{z}), \qquad (6.2.2)$$

where

$$\bar{\mathbf{f}}(z) = \frac{1}{2\pi} \int_0^{2\pi} \mathbf{f}(\mathbf{z}, t, 0) dt. \qquad (6.2.3)$$

Probably the oldest "justification" of averaging is not a mathematical argument at all, but might be said to belong to philosophy of science. One of the aims of science is to find models of reality which are both simple enough and accurate enough to be useful for making predictions. Any real system can only be described in terms of differential equations by ignoring many small influences. Whenever an equation is too hard to solve, it is legitimate to simplify the equation (by leaving out some terms, for example) until it is solvable, and to compare the solution with experimental data. Of course, there should be some reason for thinking that the factors omitted are less important than the ones retained. Examples of this procedure are the ignoring of air resistance in simple treatments of projectile motion, and treating molecules as hard spheres in the kinetic theory of gasses. From this point of view, simplifying (6.2.1) to obtain (6.2.2) is no different than other simplifications which must of necessity already have been made in obtaining (6.2.1) itself. The ultimate test of (6.2.2) is not whether its solutions are close to those of (6.2.1), since even (6.2.1) is not reality but only another model of reality. The real test is comparison with experiment.

From a mathematical standpoint, justification of averaging refers to the comparison of (6.2.1) and (6.2.2). Why should the solutions of these two equations be close?

It is a natural idea that a physical (or other) system should respond "on the average" to the average of the influences acting on it. In fact this idea is so natural that one might think it would hold for systems in any form. However, it is absolutely essential for this that the original system be in standard form (that is, periodic standard form or one of several similar standard forms that will be introduced later). To see this, consider the following initial value problem, which is not quite in standard form:

$$\dot{x}_1 = 1 \qquad\qquad x_1(0) = 0$$
$$\dot{x}_2 = \varepsilon \cos(x_1 - t) \qquad x_2(0) = 0. \qquad (6.2.4)$$

This has exact solution $x_1(t) = t$, $x_2(t) = \varepsilon t$. If the right hand side of (6.2.4) is averaged over t, the result is

$$\dot{z}_1 = 1 \qquad z_1(0) = 0$$
$$\dot{z}_2 = 0 \qquad z_2(0) = 0, \qquad (6.2.5)$$

with exact solution $z_1(t) = t$, $z_2(t) = 0$. The error in the second component is εt, which equals 1 when $t = 1/\varepsilon$; if averaging were valid for this system (in the way that it is for systems in standard form), the error at this time would be of order ε. What went wrong? Because of the first equation in (6.2.4), x_1 varies with time in such a way that $\cos(x_1 - t)$ remains at its maximum value 1 and is not well approximated by its average value 0; this creates a difference of ε between the right hand sides of the second equations of (6.2.4) and (6.2.5). Although this discrepancy is small, it leads to an error in z_2 which grows to 1 in time $1/\varepsilon$. If $\dot{x}_1$ were ε instead of 1, (6.2.4) would be in standard form. In this case, x_1 would vary slowly, and during one period $0 \leq t \leq 2\pi$ the term $\cos(x_1 - t)$ would be almost the same as if x_1 were constant: It would perform a nearly sinusoidal oscillation and would be much better approximated by its average value 0 than in the case of the actual (6.2.4). Thus the importance of periodic standard form (6.2.1) is that for small ε, x is nearly constant, so that it makes sense to average the right hand side over one period with x held constant, as done in (6.2.3). It is not valid to average a term over time in an equation such as

$$\ddot{y} + k^2 y = \varepsilon f(y, \dot{y}, t). \tag{6.2.6}$$

This equation can be changed into a system in periodic standard form and then averaged (assuming f has period $2\pi/k$ in t), but averaging the right hand side of (6.2.6) directly leads to incorrect results.

As indicated in Section 6.1, the method of averaging will be developed "from scratch" in this section by a method completely unrelated to that of Chapter 5. A change of variables from x to y will be performed in equation (6.2.1) in an attempt to eliminate the dependence upon t. It will be discovered that this is not quite possible, but it is possible to eliminate t from the terms of first order in ε. The result of doing this looks like (6.2.2) plus higher order terms, and one obtains the averaged equation (6.2.2) by dropping these terms. Thus one passes from (6.2.1) to (6.2.2) in two steps instead of one: first a coordinate change, then a truncation. Since each of these steps is better understood mathematically than the single step of "averaging," it becomes possible to understand in great detail the relation between solutions of (6.2.1) and (6.2.2).

The first step is to consider a certain type of change of variables called a *near-identity transformation*. In general, a near-identity transformation is any change of variables which reduces to the identity when $\varepsilon = 0$. When working with periodic standard form, a near-identity transformation is also required to be periodic in t (with the same period as the system of differential equations). The general form for such a transformation is

$$\mathbf{x} = \mathbf{y} + \varepsilon \mathbf{u}(\mathbf{y}, t, \varepsilon), \tag{6.2.7}$$

where $\mathbf{u}$ is 2π-periodic in t. Since $\mathbf{u}$ is assumed to be smooth, (6.2.7) can be expanded in an asymptotic Taylor series in ε:

$$\mathbf{x} \sim \mathbf{y} + \varepsilon \mathbf{u}_1(\mathbf{y}, t) + \varepsilon^2 \mathbf{u}_2(\mathbf{y}, t) + \cdots. \tag{6.2.8}$$

(Warning: The notation here is chosen so that the exponents match the subscripts in (6.2.8). Comparing (6.2.8) with (6.2.7), it follows that $\mathbf{u} = \mathbf{u}_1 + \varepsilon \mathbf{u}_2 + \cdots$; here the exponents do not match the subscripts.) In the present section we only need transformations of the restricted class

$$\mathbf{x} = \mathbf{y} + \varepsilon \mathbf{u}_1(\mathbf{y}, t); \tag{6.2.9}$$

that is, $\mathbf{u}_1 = \mathbf{u}$ and $\mathbf{u}_n = \mathbf{0}$ for $n \geq 2$, so $\mathbf{u}$ is independent of ε. A near-identity transformation is written as a mapping from the (new) y-space to the (old) x-space. In order to be a valid coordinate change, this mapping must be invertible. It is trivially invertible when $\varepsilon = 0$, since in that case it is the identity. It can be shown, using the implicit function theorem and some point set topology, that for any compact set in x- or y-space there is an ε_0 such that (6.2.7) is invertible on that compact set for $0 \leq \varepsilon < \varepsilon_0$; the argument is like the proof of Theorem B.3 given in Appendix B.

Having defined what is meant by a near-identity transformation, we now compute the effect of such a transformation when used as a change of variables for equation (6.2.1). For this purpose it is best to expand (6.2.1) in powers of ε, and since we are concerned only with the first order this expansion can be written

$$\dot{\mathbf{x}} = \varepsilon \mathbf{f}_1(\mathbf{x}, t) + \varepsilon^2 \hat{\mathbf{f}}(\mathbf{x}, t, \varepsilon), \tag{6.2.10}$$

where $\hat{\mathbf{f}}$ denotes a remainder. Since the transformation to the new variable y is smooth, the transformed equation is smooth and can also be expanded in ε:

$$\dot{\mathbf{y}} = \varepsilon \mathbf{g}_1(\mathbf{y}, t) + \varepsilon^2 \hat{\mathbf{g}}(\mathbf{y}, t, \varepsilon). \tag{6.2.11}$$

Notice that (6.2.11) has no zero order term in ε, being in this respect exactly like (6.2.10). In order to see that this is correct, recall that when $\varepsilon = 0$, then $\dot{\mathbf{x}} = 0$ and $\mathbf{x} = \mathbf{y}$. It follows that $\dot{\mathbf{y}} = 0$, so the right hand side of (6.2.11) must vanish when $\varepsilon = 0$; therefore there is no term of order zero. In order to compute the effect of (6.2.9) on (6.2.10) to first order, it is only necessary to find an expression for the function $\mathbf{g}_1$ in (6.2.11) in terms of $\mathbf{f}_1$ and $\mathbf{u}_1$. To do this, differentiate (6.2.9) with respect to t using the fact that x satisfies (6.2.10) and y satisfies (6.2.11). Thus

$$\dot{\mathbf{x}} = \dot{\mathbf{y}} + \varepsilon \left\{ \frac{\partial \mathbf{u}_1}{\partial y} \dot{\mathbf{y}} + \frac{\partial \mathbf{u}_1}{\partial t} \right\} = \varepsilon \left\{ \mathbf{g}_1(\mathbf{y}, t) + \frac{\partial \mathbf{u}_1}{\partial t}(\mathbf{y}, t) \right\} + \mathcal{O}(\varepsilon^2). \tag{6.2.12}$$

It is important to understand that (6.2.12) is a somewhat peculiar formula which mixes variables in a way that is improper for most purposes: It is a formula for $\dot{\mathbf{x}}$ expressed in terms of y. The meaning of (6.2.12) is that to find $\dot{\mathbf{x}}$ at a given point x, the right hand side of (6.2.12) should be evaluated at the point y corresponding to x under (6.2.9). Therefore the right hand side of (6.2.12) is not to be equated directly to the right hand side of (6.2.10); first, (6.2.10) must also be expressed in the y variables. Substituting (6.2.9) into (6.2.10) and expanding in ε leads to

$$\dot{\mathbf{x}} = \varepsilon \mathbf{f}_1(\mathbf{y} + \varepsilon \mathbf{u}_1, t) + \mathcal{O}(\varepsilon^2) = \varepsilon \mathbf{f}_1(\mathbf{y}, t) + \mathcal{O}(\varepsilon^2). \tag{6.2.13}$$

Now it is permissible to equate the right hand sides of (6.2.12) and (6.2.13) order by order in ε, obtaining to first order

$$\mathbf{f}_1(\mathbf{y}, t) = \mathbf{g}_1(\mathbf{y}, t) + \frac{\partial \mathbf{u}_1}{\partial t}(\mathbf{y}, t). \qquad (6.2.14)$$

Thus if a specific near-identity transformation (6.2.9) is given, the effect of this transformation is to change the differential equation (6.2.10) into (6.2.11), where $\mathbf{g}_1$ is given by (6.2.14).

In the previous discussion it was assumed that $\mathbf{u}_1$ was given. We now ask the question: Is it possible to choose $\mathbf{u}_1$ in such a way that $\mathbf{g}_1$ is independent of time? In other words, can a near-identity transformation be used to simplify an equation in periodic standard form by making it "autonomous to first order?" It turns out that this is possible, but only by taking $\mathbf{g}_1(\mathbf{y})$ to be the time average of $\mathbf{f}_1(\mathbf{y}, t)$. Once again, as in Chapter 5, an independent line of reasoning apparently having nothing to do with the idea of averaging has brought us around to recognizing the importance of the average of this expression (which is the same as $\mathbf{f}(\mathbf{y}, t, 0)$).

In order to formulate the question in the last paragraph carefully, we repeat the equations (6.2.9), (6.2.10), (6.2.11), and (6.2.14) in a slightly different form to emphasize the new point of view: Given the system

$$\dot{\mathbf{x}} = \varepsilon \mathbf{f}_1(\mathbf{x}, t) + \varepsilon^2 \hat{\mathbf{f}}(\mathbf{x}, t, \varepsilon), \qquad (6.2.15)$$

we seek a transformation of the form

$$\mathbf{x} = \mathbf{y} + \varepsilon \mathbf{u}_1(\mathbf{y}, t) \qquad (6.2.16)$$

in which $\mathbf{u}_1$ is 2π-periodic in t and which carries (6.2.15) into

$$\dot{\mathbf{y}} = \varepsilon \mathbf{g}_1(\mathbf{y}) + \varepsilon^2 \hat{\mathbf{g}}(\mathbf{y}, t, \varepsilon), \qquad (6.2.17)$$

where $\mathbf{g}_1(\mathbf{y})$ is independent of t. This will be accomplished provided $\mathbf{u}_1$ is chosen so as to satisfy

$$\frac{\partial \mathbf{u}_1}{\partial t}(\mathbf{y}, t) = \mathbf{f}_1(\mathbf{y}, t) - \mathbf{g}_1(\mathbf{y}). \qquad (6.2.18)$$

Equation (6.2.18) can be viewed as an ordinary differential equation for $\mathbf{u}_1$ as a function of t, which contains $\mathbf{y}$ as a parameter; it is solvable by integrating with respect to t while holding $\mathbf{y}$ constant. The general solution is

$$\mathbf{u}_1(\mathbf{y}, t) = \int_0^t \{\mathbf{f}_1(\mathbf{y}, s) - \mathbf{g}_1(\mathbf{y})\} \, ds + \mathbf{c}_1(\mathbf{y}), \qquad (6.2.19)$$

where $c_1(y)$ is the "constant" of integration. This solution is 2π-periodic if and only if the integral in (6.2.19) vanishes when $t = 2\pi$ (Exercise 6.2.1), which is equivalent to

$$\mathbf{g}_1(\mathbf{y}) = \frac{1}{2\pi} \int_0^{2\pi} \mathbf{f}_1(\mathbf{y}, s) ds. \tag{6.2.20}$$

Thus (as claimed above) it is possible to make $\mathbf{g}_1$ independent of t, but only by taking it to be the average of $\mathbf{f}_1$. The transformed equation is then

$$\dot{\mathbf{y}} = \varepsilon \bar{\mathbf{f}}(\mathbf{y}) + \varepsilon^2 \hat{\mathbf{g}}(\mathbf{y}, t, \varepsilon). \tag{6.2.21}$$

Here $\hat{\mathbf{g}}$ is computable from $\mathbf{f}_1$, $\hat{\mathbf{f}}$, and $\mathbf{u}_1$, although for most purposes it is unnecessary to compute this; it is sufficient to regard (6.2.21) as

$$\dot{\mathbf{y}} = \varepsilon \bar{\mathbf{f}}(\mathbf{y}) + \mathcal{O}(\varepsilon^2) \tag{6.2.22}$$

uniformly in $\mathbf{y}$ and t, remembering that the $\mathcal{O}(\varepsilon^2)$ term is not autonomous. The solution for $\mathbf{u}_1$ (and hence also the expression for $\hat{\mathbf{g}}$) is not unique, since (6.2.19) contains the arbitrary "constants" $c_1(\mathbf{y})$.

In order to understand the solution (6.2.19) for $\mathbf{u}_1$ in more detail, it is helpful to expand the periodic function $\mathbf{f}_1(\mathbf{y}, t)$ as a Fourier series in t with coefficients depending upon $\mathbf{y}$. Using the complex version of Fourier series, this expansion takes the form

$$\mathbf{f}_1(\mathbf{y}, t) = \sum_{n=-\infty}^{\infty} \mathbf{a}_n(\mathbf{y}) e^{int}. \tag{6.2.23}$$

Since the average value of a Fourier series is simply its constant term, (6.2.20) can be written

$$\mathbf{g}_1(\mathbf{y}) = \mathbf{a}_0(\mathbf{y}), \tag{6.2.24}$$

and (6.2.18) becomes

$$\frac{\partial \mathbf{u}_1}{\partial t}(\mathbf{y}, t) = \sum_{n \neq 0} \mathbf{a}_n(\mathbf{y}) e^{int}. \tag{6.2.25}$$

If this equation is integrated from zero to t (as in (6.2.19)), the result is

$$\mathbf{u}_1(\mathbf{y}, t) = \sum_{n \neq 0} \frac{\mathbf{a}_n(\mathbf{y})}{in} \left(e^{int} - 1 \right) + c_1(\mathbf{y}), \tag{6.2.26}$$

which is not in the form of a Fourier series. On the other hand, a simpler solution can be found by taking the termwise indefinite integral of (6.2.25), obtaining

$$\mathbf{u}_1(\mathbf{y}, t) = \sum_{n \neq 0} \frac{\mathbf{a}_n(\mathbf{y})}{in} e^{int} + \mathbf{d}_1(\mathbf{y}). \tag{6.2.27}$$

(The operation of termwise indefinite integration of a series is not generally permissible, even when the series is uniformly convergent, but it can be justified in this case because the coefficients of (6.2.27) are smaller in norm than those of (6.2.23), implying by the comparison test that (6.2.27) is also uniformly convergent.) The relationship between the two integration constants is clearly

$$\mathbf{c}_1(\mathbf{y}) = \mathbf{d}_1(\mathbf{y}) + \sum_{n \neq 0} \frac{\mathbf{a}_n(\mathbf{y})}{in}. \tag{6.2.28}$$

Now there are two common choices for the integration constants. One is to choose $\mathbf{d}_1(\mathbf{y}) = \mathbf{0}$, which means that the Fourier series (6.2.27) for $\mathbf{u}_1$ has no constant term and therefore $\mathbf{u}_1$ has zero average. The second is to choose $\mathbf{c}_1(\mathbf{y}) = \mathbf{0}$, which makes the integral expression (6.2.19) simple but makes the Fourier series more complicated. The strong advantage in favor of the second choice is that when $\mathbf{c}_1 = \mathbf{0}$, it follows from (6.2.19) that

$$\mathbf{u}_1(\mathbf{y}, 2\pi n) = \mathbf{0} \tag{6.2.29}$$

for each integer n. This means that the near-identity transformation (6.2.16) reduces to the identity at time $t = 0$ and at all subsequent *stroboscopic times* $t = 2\pi n$. Thus, if the two systems were observed by a stroboscopic light flashing at these times, the solutions of (6.2.15) and (6.2.17) with the same initial conditions would appear to coincide.

In other words, the *period map* (see (D.15)) of the systems (6.2.15) and (6.2.17) are the same when (6.2.29) holds. The method of averaging using $\mathbf{c}_1 = \mathbf{0}$ is called *stroboscopic averaging*. There are, occasionally, reasons to make a different choice of the arbitrary constants in (6.2.19) or (6.2.27) besides the two that we have considered.

The results obtained so far can be summarized as a theorem.

Theorem 6.2.1. *The system (6.2.15), smooth and 2π-periodic, can be transformed into (6.2.22) by a near-identity transformation of the form (6.2.16), which can be taken to reduce to the identity at $t = 0$ and all other stroboscopic times.*

For easy reference in the following discussion, the fundamental equations will be repeated along with the names that will be used to describe them.

exact system	$\dot{\mathbf{x}} = \varepsilon \mathbf{f}_1(\mathbf{x}, t) + \varepsilon^2 \hat{\mathbf{f}}(\mathbf{x}, t, \varepsilon),$	(6.2.30)
transformation	$\mathbf{x} = \mathbf{y} + \varepsilon \mathbf{u}_1(\mathbf{y}, t),$	(6.2.31)
transformed system	$\dot{\mathbf{y}} = \varepsilon \bar{\mathbf{f}}(\mathbf{y}) + \varepsilon^2 \hat{\mathbf{g}}(\mathbf{y}, t, \varepsilon),$	(6.2.32)
averaged system	$\dot{\mathbf{z}} = \varepsilon \bar{\mathbf{f}}(\mathbf{z}),$	(6.2.33)
guiding system	$\dfrac{d\mathbf{z}}{d\tau} = \bar{\mathbf{f}}(\mathbf{z}) \qquad (\tau = \varepsilon t).$	(6.2.34)

The *transformed system* is obtained from the *exact system* by the near-identity change of variables (the *transformation*); the *averaged system* is obtained from the transformed system by deleting the $\mathcal{O}(\varepsilon^2)$ terms and introducing the new variable $\mathbf{z}$. This variable $\mathbf{z}$ is not related either to $\mathbf{x}$ or to $\mathbf{y}$ by any transformation, but is used merely to distinguish the solutions of (6.2.32) and (6.2.33). The *guiding system* is obtained from the averaged system by changing the *independent* variable from t to $\tau = \varepsilon t$. Notice that ε has completely disappeared from (6.2.34). Therefore the solution of the guiding system with initial condition $\mathbf{z} = \alpha$ at $\tau = 0$ can be denoted $\mathbf{z}(\tau, \alpha)$ and is not a function of ε; this solution is called the *guiding center*. The solution of (6.2.33) with the initial condition α can then be written $\mathbf{z}(\varepsilon t, \alpha)$, where $\mathbf{z}$ denotes the same mathematical function. In contrast to this, the solutions of (6.2.30) and (6.2.32) with the same initial condition must be written $\mathbf{x}(t, \alpha, \varepsilon)$ and $\mathbf{y}(t, \alpha, \varepsilon)$ because the dependence of these solutions upon t and ε does not occur solely through the combination εt. The most important of these solution functions will now be summarized and named.

exact solution	$\mathbf{x}(t, \alpha, \varepsilon),$	(6.2.35)
guiding center	$\mathbf{z}(\tau, \alpha),$	(6.2.36)
first approximation	$\mathbf{z}(\varepsilon t, \alpha).$	(6.2.37)

There is a technical point which should be mentioned here; it is not of great importance in this section, but will become important for higher order averaging (Section 6.5). It is clear from (6.2.31) that the exact solution $\mathbf{x}(t, \alpha, \varepsilon)$ is related to the solution $\mathbf{y}(t, \alpha, \varepsilon)$ of (6.2.32) by

$$\mathbf{x}(t, \alpha, \varepsilon) = \mathbf{y}(t, \alpha, \varepsilon) + \varepsilon \mathbf{u}_1(\mathbf{y}(t, \alpha, \varepsilon), t). \qquad (6.2.38)$$

Since (6.2.32) differs only slightly from (6.2.33), it is reasonable to think that $\mathbf{z}(\varepsilon t, \alpha)$ should be a good approximation to $\mathbf{y}(t, \alpha, \varepsilon)$. Therefore it would appear that to obtain the best possible approximation to $\mathbf{x}$ using $\mathbf{z}$, one should substitute $\mathbf{z}$ into (6.2.38) in place of $\mathbf{y}$ and obtain

$$\mathbf{x}(t, \alpha, \varepsilon) \cong \mathbf{z}(\varepsilon t, \alpha) + \varepsilon \mathbf{u}_1(\mathbf{z}(\varepsilon t, \alpha), t). \qquad (6.2.39)$$

This is known as the *improved first approximation*; in fact, it is a slightly better approximation than the *first approximation* $x(t, \alpha, \varepsilon) \cong z(t, \alpha, \varepsilon)$. However, the difference is not significant asymptotically. The additional term εu_1 in the improved first approximation is of asymptotic order ε, which is the same as the order of the error in the first approximation (on a suitable expanding interval); therefore there is no harm, asymptotically speaking, in omitting the correction term. This is a great advantage in computation, because it is unnecessary to compute the function u_1 in using first order averaging. That is, for first order averaging the entire apparatus of the near-identity transformation is needed only for the theory, not for the applications: In practice, one just averages (6.2.1) to obtain (6.2.2). In higher order averaging things are not this simple, because the near-identity transformation (or at least part of it) must be computed and used in each application.

The improved first approximation helps to clarify why the first approximation is sometimes called the "guiding center." The second term εu_1 in (6.2.39) represents a rapid small oscillation superimposed upon the slow drift given by the first term; "rapid" here means "on the time scale t," and "slow" means "on the time scale εt." Therefore the first term may be thought of as a center about which small oscillations take place, and which guides the oscillating motion as it drifts. Of course we are still speaking only of the improved first approximation here, not the exact solution, but the exact solution exhibits the same feature, with the additional fact that the oscillations may drift away from the guiding center after a time of order $1/\varepsilon$ has passed (since this is the length of the expanding interval for which the averaging approximations remain valid). The overall motion, then, has the appearance shown in Fig. 6.2.1. In this book the term "guiding center" is used for the first approximation *when it is expressed in terms of the slow time* $\tau := \varepsilon t$; when expressed in terms of t it is merely called the first approximation.

Our aim now is to establish the fact that the exact solution and the first approximation remain within order $\mathcal{O}(\varepsilon)$ of each other over a time interval of length $\mathcal{O}(1/\varepsilon)$. In order for the theorems to be useful, the hypotheses should involve only known or computable functions; the function $\mathbf{f}$

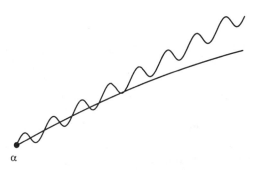

FIG. 6.2.1. Schematic illustration of a guiding center. The first order averaged solution moves slowly along the smooth arc. The exact solution makes rapid small oscillations about this arc and also gradually drifts away.

is known, and it is reasonable to assume that the guiding center z is computable, since otherwise the first approximation is of no practical use. The theorems to follow impose hypotheses only on these two functions (and not, for instance, on the exact solution, which is not assumed to be computable).

Theorem 6.2.2. *Assume that (6.2.30) is smooth and periodic in t. Suppose that the guiding system (6.2.34) has a solution $z(\tau, \alpha)$ which exists on the interval $0 \leq \tau \leq T$. Then there exist constants $\varepsilon_1 > 0$ and $c > 0$ such that the solution of the exact system (6.2.30) with initial condition α exists on the expanding interval $0 \leq t \leq T/\varepsilon$, and the following error estimate holds:*

$$\|x(t, \alpha, \varepsilon) - z(\varepsilon t, \alpha)\| < c\varepsilon \quad \text{for } 0 \leq t \leq T/\varepsilon, \quad 0 \leq \varepsilon \leq \varepsilon_1.$$

$$(6.2.40)$$

Theorem 6.2.2 will be proved as a corollary of the following more general theorem, which includes a careful statement of the type of uniformity which the error estimate possesses with respect to the initial condition α.

Theorem 6.2.3. *Assume that (6.2.30) is smooth and periodic in t. Let K be a compact subset of $\mathbf{R}^N$ and let V be a larger compact subset containing K in its interior. (For instance, K could be a large closed ball around the origin and V a larger such ball.) Let ε_0 be such that the near-identity transformation (6.2.31) is valid (invertible) for y in V and $0 \leq \varepsilon < \varepsilon_0$. Let $T_0 > 0$ be given arbitrarily, and for each α in K let $T(\alpha)$ be the largest real number less than or equal to T_0 such that $z(\tau, \alpha)$ belongs to K for $0 \leq \tau \leq T(\alpha)$. Then there exist constants $c > 0$ and ε_1 in the interval $0 < \varepsilon_1 \leq \varepsilon_0$ such that the following estimate holds for all α in K:*

$$\|x(t, \alpha, \varepsilon) - z(\varepsilon t, \alpha)\| < c\varepsilon \quad \text{for } 0 \leq t \leq T(\alpha)/\varepsilon, \quad 0 \leq \varepsilon \leq \varepsilon_1.$$

$$(6.2.41)$$

Included in this statement is the existence of $x(t, \alpha, \varepsilon)$ on the indicated expanding interval.

Proof of Theorem 6.2.3. Since V is compact, all functions are smooth, and all functions of t occurring on the right hand sides of (6.2.30)–(6.2.34) are periodic, it follows that all of these functions are bounded and have Lipschitz constants on V. These constants will be freely introduced as needed in the following arguments. First, consider the solution (6.2.36) of the guiding system for any fixed α in K. Either this solution remains in K for all τ-time or leaves K at some time $\tau = T_1(\alpha)$. In the former case, let $T(\alpha) = T_0$ (where T_0 is the given constant in the theorem statement), and in the latter case, let $T(\alpha)$ be the minimum of T_0 and $T_1(\alpha)$. Then for $0 \leq \tau \leq T(\alpha)$, $z(\tau, \alpha)$ belongs to K, and so for $0 \leq t \leq T(\alpha)/\varepsilon$, $z(\varepsilon t, \alpha)$ belongs to K. In order to understand the construction of this time interval, notice that if α is near to the boundary of K then $T(\alpha)$ is likely to be less than T_0 because the solution of the guiding system may run out of K before τ reaches T_0, whereas if α is deeply inside K, then $T(\alpha)$ is likely to equal T_0, since this absolute limit on τ-time will probably be reached before z reaches the boundary. If all solutions of the guiding system pass through K in τ-time less than T_0, then all values of $T(\alpha)$ will be less than T_0. If on the other hand the slow vector field points into K at all boundary points, then all solutions of the slow equation remain in K for all time. In this case, $T(\alpha)$ will equal T_0 for all α. Even in this case it is not (in general) possible to extend the error estimate (6.2.41) to all time, and although T_0 is arbitrary, it is necessary to have such a limit in order to obtain suitable constants ε_1 and c.

The next step in the proof is to estimate the difference between y and z. Let L be a Lipschitz constant for $\bar{f}$ on V, and let B be an upper bound for $\|\hat{g}(y, t, \varepsilon)\|$ for y in V, $0 \leq t \leq 2\pi$, and $0 \leq \varepsilon \leq \varepsilon_0$. Then from (6.2.32), (6.2.33), the triangle inequality, and the "running-away" inequality (Lemma 3.3.1) it follows that *for as long as* y *and* z *remain in* V, the following inequalities hold:

$$\frac{d}{dt}\|y - z\| \leq \|\dot{y} - \dot{z}\| \leq \varepsilon\|\bar{f}(y) - \bar{f}(z)\| + \varepsilon^2\|\hat{g}(y, t, \varepsilon)\|$$
$$\leq \varepsilon L\|y - z\| + \varepsilon^2 B.$$

It is an easy Gronwall argument to conclude from this that

$$\|y(t, \alpha, \varepsilon) - z(\varepsilon t, \alpha)\| \leq \varepsilon\frac{B}{L}(e^{\varepsilon L t} - 1) \qquad (6.2.42)$$

for as long as y and z remain in V. Since z remains in K (and hence in V) for $0 \leq t \leq T(\alpha)/\varepsilon$, and since the right hand side of (6.2.42) is increasing in t, it follows that

$$\|y(t, \alpha, \varepsilon) - z(\varepsilon t, \alpha)\| \leq \varepsilon\frac{B}{L}(e^{LT(\alpha)} - 1) \leq \varepsilon\frac{B}{L}(e^{LT_0} - 1) = c_1\varepsilon, \quad (6.2.43)$$

as long as both y remains in V and $0 \leq t \leq T(\alpha)/\varepsilon$, with $c_1 = \frac{B}{L}(e^{LT_0} - 1)$. Now choose ε_1 so that $0 \leq \varepsilon_1 \leq \varepsilon_0$ and $\varepsilon_1 \leq D/c_1$, where D is the minimum distance between the boundaries of K and V. For ε in the interval $0 \leq \varepsilon \leq \varepsilon_1$, consider the solutions for y and z beginning at α in K. The estimate (6.2.43) holds from $t = 0$ until $t = T(\alpha)/\varepsilon$ or until y leaves V, whichever occurs first. But since z belongs to K until $t = T(\alpha)/\varepsilon$, and since $c_1\varepsilon < D$, (6.2.43) itself prevents y from deviating sufficiently from z to leave V before time $T(\alpha)/\varepsilon$. Therefore (6.2.43) holds for $0 \leq t \leq T(\alpha)/\varepsilon$ and $0 \leq \varepsilon \leq \varepsilon_1$. (If this argument seems circular, see Exercise 6.2.4 for a slightly different formulation using proof by contradiction. The argument is not actually circular.) From the general existence theory of differential equations, the solution for y can only cease to exist after leaving the compact set V. (See Appendix D.) Therefore this solution exists at least until time $T(\alpha)/\varepsilon$.

The final step in the proof of Theorem 6.2.3 is to compare $\mathbf{x}$ with $\mathbf{y}$. But this is simple, since they are related by the coordinate transformation (6.2.31). Let c_2 denote the maximum value of $\|\mathbf{u}_1(\mathbf{y}, t)\|$ for $\mathbf{y}$ in V and $0 \leq t \leq 2\pi$. Then it is clear that

$$\|\mathbf{x}(t, \alpha, \varepsilon) - \mathbf{y}(t, \alpha, \varepsilon)\| \leq c_2 \varepsilon \qquad (6.2.44)$$

as long as $\mathbf{y}$ belongs to V, for $0 \leq \varepsilon \leq \varepsilon_0$. But we already know that $\mathbf{y}$ belongs to V when $0 \leq t \leq T(\alpha)/\varepsilon$ and $0 \leq \varepsilon \leq \varepsilon_1$. Letting $c = c_1 + c_2$ and putting (6.2.43) and (6.2.44) together gives (6.2.41). ∎

Proof of Theorem 6.2.2. Recall that in this theorem, α and T are given and fixed.

Let K be the set of points traced out by $\mathbf{z}(\tau, \varepsilon)$ for $0 \leq \tau \leq T$, and let V be a compact set containing K in its interior. Then T is precisely $T(\alpha)$ as defined in Theorem 6.2.3, for the specific point α with which we are concerned; that it, T is the amount of τ-time for which the solution starting at α remains in K. Therefore (6.2.40) is merely a special case of (6.2.41). ∎

Exercises 6.2

1. Show that (6.2.19) is 2π-periodic in t if and only if (6.2.20) holds.

2. Average the free Duffing equation (in periodic standard form) by the method of this section. Find the near-identity transformation, the first approximation, and the improved first approximation, and express these approximations in terms of the original variable y of Duffing's equation. (See Exercises 5.2.3, 5.4.2, and 5.5.2). Hint: For computational purposes it is best to work with all Fourier series in the real (sine and cosine) form rather than in the complex form used in the text. In working this problem you should work through the steps in real form (rather than attempt to translate the complex series from the text into real form). All of the Fourier series that arise in this problem are finite. Observe the difference between the stroboscopic and zero-mean forms of the near-identity transformation, and how this choice affects the improved first approximation.

3. Repeat Exercise 2 for the free Van der Pol equation.

4. It has been proved above that (6.2.43) holds for $0 \leq t \leq T(\alpha)/\varepsilon$, provided that $\mathbf{y}$ remains in V for this time. Assume that $\mathbf{y}$ leaves V during this time for some solution $\mathbf{y}(t, \alpha, \varepsilon)$ with $0 \leq \varepsilon \leq \varepsilon_1$ (where ε_1 is as defined above), and deduce a contradiction. This shows that for $0 \leq \varepsilon \leq \varepsilon_1$, the requirement that $\mathbf{y}$ remains in V is automatically satisfied.

6.3. EXISTENCE AND STABILITY OF PERIODIC SOLUTIONS

It has been shown in the last section that for a system in periodic standard form, the exact solution tends to make small oscillations around a slowly moving guiding center, possibly drifting away from this center after a long time has passed. When the averaged system has a rest point, that is, a point where $\bar{\mathbf{f}}(\mathbf{z}) = \mathbf{0}$, then the guiding center does not move, and the exact solution tends to oscillate about this motionless point until it drifts away. The theme of this section is that *if the zero of $\bar{\mathbf{f}}$ is simple, then the exact solution does not drift away, but is actually periodic.* The periodic solutions found in this way are *small* oscillations, very close to the rest point of the averaged system; more precisely, their amplitude tends to zero as $\varepsilon \to 0$. But in most applications, a *small* oscillation in the "standard form variables" corresponds to a *nearly sinusoidal* large oscillation in the original variables. (For instance the solution of an oscillator problem may take the form $y = a \cos t + b \sin t$ where a and b are standard form variables. If a and b are constant, the solution is a sinusoidal oscillation; if a and b undergo small oscillations around a fixed position, with period 2π, then y is still periodic and is close to a sinusoidal oscillation.) It will be seen in this section that the equation $\bar{\mathbf{f}}(\mathbf{z}) = \mathbf{0}$ functions as a *determining equation* for periodic solutions (in the sense of Chapter 4), and provides an easy way of establishing some results stated without proof in that chapter.

In addition to establishing the existence of periodic solutions, the method of averaging often gives an easy way of determining whether they are stable. A periodic solution is called *asymptotically stable* if all solutions close to the periodic solution remain close to it and tend toward the periodic solution as $t \to \infty$. (The precise definition of this and several closely related kinds of stability is rather technical.) In practice, this means that if a physical system is executing a stable periodic motion, any slight disturbance of the system will not disrupt the motion drastically since the disturbance will tend to die out. In this sense it is often claimed that only stable periodic motions can actually be observed in physical systems, since unstable motions will not survive the inevitable disruptions that occur in real systems; therefore to show the "physical existence" of a periodic motion one must show both its mathematical existence and its stability. (This statement is not entirely correct. For instance, it is to some extent still an open question whether the various periodic and quasiperiodic motions of the planets are stable, yet these motions have been observed since prehistoric times. In this case it is a matter of time scales: It can be shown that any existing instabilities would require an extremely long time to become apparent.)

The results of this section have two important advantages over those of Section 6.2. First, the results are valid for all time. That is, periodicity is by definition a property that lasts forever, and so is stability; so these results are not limited to expanding intervals. Secondly, *it is not necessary to be able to solve the averaged system in order to use the results of this section.* It is only necessary to be able to find zeroes of the averaged vector field and to examine the matrix of partial derivatives of the vector field at these points. We have already pointed out that the averaged equations cannot necessarily be solved, so it is of great practical importance that one can obtain useful information from the averaged equations even in this case.

A precise statement of these results is given in the following theorem. Only part of the theorem is proved here, since the rest of the proof depends on advanced topics in differential equations. The remainder of this section concerns the application of this theorem to situations introduced in Chapter 4. Other applications will occur in Sections 6.4 and 6.6.

Theorem 6.3.1. *Suppose that the system*

$$\dot{\mathbf{x}} = \varepsilon \mathbf{f}(\mathbf{x}, t, \varepsilon) \tag{6.3.1}$$

is smooth and 2π-periodic. Let the first order averaged system

$$\dot{\mathbf{z}} = \varepsilon \bar{\mathbf{f}}(\mathbf{z}) \tag{6.3.2}$$

have a rest point at $\mathbf{z} = \mathbf{z}_0$, that is,

$$\bar{\mathbf{f}}(\mathbf{z}_0) = \mathbf{0}, \tag{6.3.3}$$

and let the matrix of partial derivatives of $\bar{\mathbf{f}}$ at this rest point be denoted by

$$A := \bar{\mathbf{f}}_{\mathbf{z}}(\mathbf{z}_0). \tag{6.3.4}$$

If A is nonsingular (that is, the rest point is simple) then there exists a unique ε-dependent initial condition $\alpha(\varepsilon)$, defined for ε in some interval $|\varepsilon| < \varepsilon_1$, such that $\alpha(0) = \mathbf{z}_0$ and such that the solution $\mathbf{x}(t, \alpha(\varepsilon), \varepsilon)$ of (6.3.1) with this initial condition is periodic with period 2π.

The following additional results are true only for the positive interval $0 < \varepsilon < \varepsilon_1$. If the eigenvalues of A lie in the left half of the complex plane (have real parts less than zero) then the periodic solution $\mathbf{x}(t, \alpha(\varepsilon), \varepsilon)$ is asymptotically stable (all nearby solutions remain close to, and approach, the periodic solution as $t \to \infty$). If at least one eigenvalue lies in the right half-plane the periodic solution is unstable (some solutions starting arbitrarily near the periodic solution diverge from it as t increases). The only remaining possibility is that some eigenvalues are on the imaginary axis and the rest are in the left half-plane; in this borderline case, no conclusion about the stability of the periodic solution is possible from the first order averaged equation alone. In the first case (eigenvalues in the left half-plane), the basic error estimate $\|\mathbf{x} - \mathbf{z}\| < c\varepsilon$ given in (6.2.40) holds for all future time ($t \geq 0$, and not merely $0 \leq t \leq T/\varepsilon$) for any solution that is attracted to the periodic solution as $t \to \infty$.

Partial Proof. Recall the basic equations of the method of averaging given in (6.2.30)–(6.2.34). The pattern of the proof is first to establish the properties of the averaged system (the z-system), then those of the transformed system (the y-system), and finally to carry these over to the exact system (the x-system) by means of the transformation. Throughout the proof, it will be assumed that the stroboscopic version of averaging is used. In fact, neither the hypotheses nor the conclusions of the theorem make any reference to the y-system (which is the only place where the "version" of averaging used has any effect), so we are at liberty to make any convenient choice.

I. Behavior of the z-system: The behavior of the averaged system (6.3.2) is easy to describe under the hypotheses of the theorem. First, there is a rest point (for all ε) at z_0. Next, in a neighborhood of this rest point, the averaged system can be written as $\dot{w} = Aw + \mathcal{O}(\|w\|^2)$, where $w = z - z_0$. Systems of this form are discussed in Appendix D; if A has no eigenvalues on the imaginary axis, then the linear part determines the behavior of solutions near $w = 0$. It follows that all solutions of the averaged system beginning near z_0 tend toward z_0 as $t \to \infty$, provided that all eigenvalues of A are in the left half-plane; if at least one eigenvalue is in the right half-plane, then some solutions diverge from z_0; and otherwise no conclusion is possible, since in this case the behavior depends on the nonlinear terms.

II. Behavior of the y-system: Next, we will establish that the y-system has a periodic solution (not necessarily a rest point!) close to z_0 if A is nonsingular. Let $y(t, \alpha, \varepsilon)$ denote, as usual, the solution of (6.2.32) with arbitrary initial condition α. Since the differential equation is 2π-periodic, a solution will be 2π-periodic if and only if it returns to its initial value at $t = 2\pi$; in this case, both the solution and the vector field (defined by the differential equation) have returned to their original state simultaneously and must repeat themselves thereafter (see Appendix D). Therefore the condition for periodicity can be written

$$0 = y(2\pi, \alpha, \varepsilon) - \alpha = \int_0^{2\pi} \dot{y}(t, \alpha, \varepsilon)dt$$

$$= \int_0^{2\pi} \left\{ \varepsilon \bar{f}(y(t,\alpha,\varepsilon)) + \varepsilon^2 \hat{g}(y(t,\alpha,\varepsilon), t, \varepsilon) \right\} dt =: \Phi(\alpha, \varepsilon).$$

$$(6.3.5)$$

The equation $\Phi(\alpha, \varepsilon) = 0$ immediately suggests using the implicit function theorem (in vector form) to find a solution $\alpha(\varepsilon)$. But the implicit function theorem does not apply, because when $\varepsilon = 0$, (6.3.5) is identically zero and so has no simple solutions. To correct this, it is only necessary to remove a factor of ε from the second integral occurring in (6.3.5). That is, define

$$\Psi(\alpha, \varepsilon) := \int_0^{2\pi} \left\{ \bar{\mathbf{f}}(\mathbf{y}(t, \alpha, \varepsilon)) + \varepsilon \hat{\mathbf{g}}(\mathbf{y}(t, \alpha, \varepsilon), t, \varepsilon) \right\} dt. \tag{6.3.6}$$

When $\varepsilon = 0$ this reduces to

$$\Psi(\alpha, 0) = \int_0^{2\pi} \bar{\mathbf{f}}(\mathbf{y}(t, \alpha, 0)) \, dt = 2\pi \bar{\mathbf{f}}(\alpha). \tag{6.3.7}$$

Since $\alpha = \mathbf{z}_0$ is by hypothesis a simple zero of this equation, the implicit function theorem implies the existence of a unique function $\alpha(\varepsilon)$ defined for sufficiently small ε, such that $\alpha(0) = \mathbf{z}_0$ and $\Psi(\alpha, \varepsilon) = \mathbf{0}$. Therefore also $\Phi(\alpha, \varepsilon) = \varepsilon \Psi(\alpha, \varepsilon) = \mathbf{0}$, implying (by (6.3.5)) that $\mathbf{y}(t, \alpha(\varepsilon), \varepsilon)$ is periodic.

The stability properties of this family of periodic solutions will not be proved rigorously, since this depends upon the notion of the variational equations of a periodic solution and upon Floquet theory. We will only say that the matrix A continues to have the dominant influence, as long as it has no eigenvalues on the imaginary axis, so that the stability properties of the periodic solution of the y-system are the same as for the rest point of the z system. (In the indeterminate case, when A does have eigenvalues on the imaginary axis, the system is sensitive to small neglected terms, and so the term of order ε^2 that is present in the y-system but not in the z-system can cause the two to have different stability properties.)

III. Behavior of the x-system: Next we turn to the properties of the exact system. Since the transformation (6.2.31) is taken to be stroboscopic, the solution of the exact system corresponding to the periodic solution of the transformed system will have the same initial condition $\alpha(\varepsilon)$; that is,

$$\mathbf{x}(t, \alpha(\varepsilon), \varepsilon) = \mathbf{y}(t, \alpha(\varepsilon), \varepsilon) + \varepsilon \mathbf{u}_1 (\mathbf{y}(t, \alpha(\varepsilon), \varepsilon), t). \tag{6.3.8}$$

Since all functions on the right hand side of (6.3.8) are 2π-periodic, so is the left hand side. This proves the existence of a family of periodic solutions for the exact system. In the same way, the stability properties of the periodic y-solution are carried over to x by the transformation (6.2.31). That is, if all solutions starting near the periodic solution in the y-variables approach the periodic solution, the same will be true for the images of these solutions under the transformation to the x-variables.

The remaining statement in the theorem, concerning the error estimates for a solution attracted to an asymptotically stable periodic solution, will not be proved, except to say that it is a case of the general principle that damping improves error estimates. (An easier illustration of this idea has occurred already in Chapter 3. In the present case, it is the distance from the given solution to the periodic solution that is "damped.") See Section 6.7 for references to the literature. ∎

In Chapter 4, periodic solutions were studied for the following three types of second order equations:

$$\ddot{y} + k^2 y = \varepsilon f(y), \tag{6.3.9}$$

$$\ddot{y} + k^2 y = \varepsilon f(y, \dot{y}), \tag{6.3.10}$$

and

$$\ddot{y} + k^2 y = \varepsilon f(y, \dot{y}, \omega(\varepsilon)t). \tag{6.3.11}$$

The first type is conservative, and for any initial conditions the solution is periodic provided ε is small enough. The second type is the general autonomous oscillator and frequently has limit cycles that can be located by a determining equation. It is also possible to decide the stability of these limit cycles in an elementary way. These topics were treated in detail in Chapter 4. For the third type, the general forced oscillator, the determining equation for periodic solutions was found in Chapter 4 (for the case of harmonic resonance, $\omega(0) = k$) but was not proved, and no attempt was made to discuss stability. From the standpoint of the method of averaging, the difficulty of these problems is exactly reversed: The third type is the easiest to handle and will be treated first. This will complete the work begun in Section 4.6. After that, the first two cases will be examined by the method of averaging, not to obtain new results but to explain why these problems are more difficult by the present method. The ideas encountered in this discussion will appear again in Section 6.6, where they will be used in the study of coupled systems of oscillators.

The first step in handling (6.3.11) by the method of averaging is to put it into periodic standard form. This has been accomplished in (5.4.26).

$$\frac{da}{dt^+} =$$
$$- \frac{\varepsilon}{k} \left\{ b\omega_1 + \frac{1}{k} f(a \cos t^+ + b \sin t^+, -ak \sin t^+ + bk \cos t^+, t^+) \sin t^+ \right\}$$
$$+ \mathcal{O}(\varepsilon^2),$$

$$\frac{db}{dt^+} = \tag{6.3.12}$$
$$+ \frac{\varepsilon}{k} \left\{ a\omega_1 + \frac{1}{k} f(a \cos t^+ + b \sin t^+, -ak \sin t^+ + bk \cos t^+, t^+) \cos t^+ \right\}$$
$$+ \mathcal{O}(\varepsilon^2).$$

To locate the points from which periodic solution families originate, it is only necessary to set the average of the order ε terms in (6.3.12) equal to zero:

$$b\omega_1 + \frac{1}{2\pi k}$$
$$\times \int_0^{2\pi} f(a\cos t^+ + b\sin t^+, -ak\sin t^+ + bk\cos t^+, t^+)\sin t^+\, dt^+ = 0,$$
$$\tag{6.3.13}$$
$$a\omega_1 + \frac{1}{2\pi k}$$
$$\times \int_0^{2\pi} f(a\cos t^+ + b\sin t^+, -ak\sin t^+ + bk\cos t^+, t^+)\cos t^+\, dt^+ = 0.$$

Theorem 6.3.1 guarantees that every simple zero (a_0, b_0) of these equations gives rise to a periodic solution family $(a(t^+, \varepsilon), b(t^+, \varepsilon))$ of (6.3.12) which reduces to the rest point (a_0, b_0) when $\varepsilon = 0$. In order to interpret this result it is necessary to trace it back to the original variables occurring in (6.3.11). First, according to (5.4.22) the periodic solution becomes $(a(\omega(\varepsilon)t, \varepsilon), b(\omega(\varepsilon)t, \varepsilon))$ in the original time variable. Then according to (5.4.18),

$$u = a\,(\omega(\varepsilon)t, \varepsilon)\cos\omega(\varepsilon)t + b\,(\omega(\varepsilon)t, \varepsilon)\sin\omega(\varepsilon)t,$$
$$v = -a\,(\omega(\varepsilon)t, \varepsilon)\sin\omega(\varepsilon)t + b\,(\omega(\varepsilon)t, \varepsilon)\cos\omega(\varepsilon)t;$$
$$\tag{6.3.14}$$

that is, recalling that $u = y$ and $v = \dot{y}/k$, the periodic solution of (6.3.11) and its derivative are given by

$$y = a\,(\omega(\varepsilon)t, \varepsilon)\cos\omega(\varepsilon)t + b\,(\omega(\varepsilon)t, \varepsilon)\sin\omega(\varepsilon)t,$$
$$\dot{y} = -ka\,(\omega(\varepsilon)t, \varepsilon)\sin\omega(\varepsilon)t + kb\,(\omega(\varepsilon)t, \varepsilon)\cos\omega(\varepsilon)t.$$
$$\tag{6.3.15}$$

It is important to check that (6.3.14) and (6.3.15) are in fact periodic; this is the case because both the transformation to periodic standard form, and the solution in standard form, have the same period. In terms of the variables a and b, the periodic solution is merely a small oscillation around the point (a_0, b_0) from which the solution family originates. In terms of the phase plane variables u and v, it is a large nearly circular motion, reducing exactly to a circle when $\varepsilon = 0$. In the original scalar variable y it is a nearly sinusoidal oscillation with modulated amplitudes, reducing to the sinusoidal oscillation $a_0\cos kt + b_0\sin kt$ when $\varepsilon = 0$. Finally, to compare the present result with Section 4.6, observe that the initial values of y and $\dot{y}$ in (6.3.15) when $\varepsilon = 0$ are $\alpha_0 = a_0$ and $\beta_0 = kb_0$; the notation here is that of (4.6.4). It is then easy to see that the equations (6.3.13) for the points of origination of periodic families are the same as the determining equations (4.6.14). This completes the proof of the results stated in Section 4.6.

It is possible, and often more convenient, to carry out the same analysis beginning from the periodic standard form in polar coordinates, (5.4.24), in place of (5.4.23). This leads immediately to the following alternative set of determining equations:

$$\frac{1}{2\pi} \int_0^{2\pi} f\left(r\cos(\varphi - t^+), kr\sin(\varphi - t^+), t^+\right) \sin(\varphi - t^+)dt^+ = 0,$$

$$\omega_1 + \frac{1}{2\pi kr} \int_0^{2\pi} f\left(r\cos(\varphi - t^+), kr\sin(\varphi - t^+), t^+\right) \cos(\varphi - t^+)dt^+ = 0.$$

$$(6.3.16)$$

For each simple solution (r_0, φ_0) of these equations, there exists a periodic solution family of (6.3.11) for small ε which is close to the sinusoidal oscillation $y = r_0 \cos(\varphi_0 - \omega(\varepsilon)t)$ (see (5.4.20)).

At first sight it would seem that the autonomous oscillator (6.3.10) would be as easy to handle by the method of averaging as the forced oscillator (6.3.11), since the periodic standard forms for the two cases are almost identical. The reason for the difficulty is that when the original system is autonomous, it is not possible for the averaged equations to have a simple zero, and therefore Theorem 6.3.1 can never be applied. To see this, consider the periodic standard form of (6.3.10) in polar coordinates, as given in (5.4.11):

$$\dot{r} = \frac{\varepsilon}{k} f\left(r\cos(\varphi - kt), kr\sin(\varphi - kt)\right) \sin(\varphi - kt),$$

$$\dot{\varphi} = \frac{\varepsilon}{kr} f\left(r\cos(\varphi - kt), kr\sin(\varphi - kt)\right) \cos(\varphi - kt).$$

$$(6.3.17)$$

Since averaging the right hand side over t is the same as averaging over $\theta := \varphi - kt$, the averaged equations can be written (using new variables ρ and ψ corresponding to $\mathbf{z}$) as

$$\dot{\rho} = \varepsilon F(\rho) := \frac{\varepsilon}{2\pi k} \int_0^{2\pi} f(\rho\cos\theta, k\rho\sin\theta)\sin\theta \, d\theta,$$

$$\dot{\psi} = \varepsilon G(\rho) := \frac{\varepsilon}{2\pi k\rho} \int_0^{2\pi} f(\rho\cos\theta, k\rho\sin\theta)\cos\theta \, d\theta.$$

$$(6.3.18)$$

(See also (5.2.12) and (5.2.13).) The essential feature of these equations is that the right hand sides depend only upon ρ, not upon ψ. That is what makes these equations solvable by quadratures, as pointed out in Section 5.2, so that it is possible to find the exact solution and apply the error estimates from Section 6.2; this is not usually possible in the forced case. But by the same token, if there is a value ρ_0 such that $F(\rho_0) = G(\rho_0) = 0$, then all points (ρ_0, ψ) with arbitrary ψ are rest points. Since these rest points form a circle, none of them are isolated, and therefore certainly none of them are simple. Analytically speaking, the Jacobian $\partial(F,G)/\partial(\rho,\psi)$ is always zero because F and G do not depend upon ψ. Therefore the hypotheses of Theorem 6.3.1 can never be satisfied. However, we know from Section 4.4 that there is a usable determining equation for this problem, and there should be a way to find it by averaging. In fact there is, but only after obtaining a different periodic standard form for these equations.

As a clue to understanding this situation, recall the way autonomous equations were treated in Section 4.4. Because *solutions* of autonomous equations lie on *orbits*, and different solutions lying on the same orbit differ only by a shift in time, it was sufficient to consider only initial conditions lying on the positive u-axis in the phase plane. The determining equations were then found by considering a point on this axis and following its orbit once around the origin until it again intersected the positive u-axis. If the point returned to itself, its orbit was periodic and had period 2π in terms of the polar angle variable θ, although its period in terms of t was not predictable without further computation and in general depended upon ε. This points out two reasons why Theorem 6.3.1, applied to (6.3.17), cannot be expected to give the correct determining equations: The dimension is wrong (there should only be one determining equation, not two), and the period is wrong (Theorem 6.3.1 gives only solutions with time period 2π). Both of these problems are correctable in one step. Recall the discussion of how to put (6.3.10) into standard form using polar coordinates. The first step (equations (5.4.8) and (5.4.9)) was to introduce polar coordinates r and θ into the phase plane, obtaining

$$\dot{r} = \frac{\varepsilon}{k} f(r\cos\theta, kr\sin\theta)\sin\theta,$$

$$\dot{\theta} = -k + \frac{\varepsilon}{kr} f(r\cos\theta, kr\sin\theta)\cos\theta. \tag{6.3.19}$$

At this point, instead of rotating coordinates, let us divide the first equation of (6.3.19) by the second. This is permissible because time does not appear explicitly in these equations (they are autonomous). The result is

$$\frac{dr}{d\theta} = -\frac{\varepsilon}{k^2} f(r\cos\theta, kr\sin\theta)\sin\theta + \mathcal{O}(\varepsilon^2). \tag{6.3.20}$$

This equation has already been derived and used in Section 4.4; see (4.4.5). The solutions of this equation are functions $r(\theta)$ giving the orbits of (6.2.19), that is, the geometrical loci of the solutions without regard to time. In Section 4.4 this equation was used to calculate the Poincaré map from the positive u-axis to itself and thus to find the determining equation for periodic orbits. Here we will use (6.3.20) in a different way to obtain the same determining equation. Namely, (6.3.20) is an instance of a system in periodic standard form! Unlike the previous periodic standard form (6.3.17) for the same equation (6.3.10), this "system" is one-dimensional (it is a single scalar equation); also, its independent variable is θ, not t. Nonetheless, Theorem 6.3.1 may be applied to it, with the result that if the average of the right hand side (after dropping the second order terms) has a simple zero at ρ_0, then there exists a periodic solution family of (6.3.20) originating at ρ_0, having period 2π in θ. This solution corresponds to a periodic orbit in the phase plane, carrying an infinite number of periodic solutions of (6.3.19) differing by a time translation. These periodic solutions have an unspecified period in time, and the result coincides exactly with that obtained in Section 4.4. The determining equation is

$$-\frac{1}{2\pi} \int_0^{2\pi} f(\rho_0 \cos \theta, k\rho_0 \sin \theta) \sin \theta \, d\theta = 0, \tag{6.3.21}$$

which is the same as (4.4.8) except for the minus sign, which we have retained (from the right hand side of (6.3.20)) in order that the stability considerations in the next paragraph will come out correctly.

The stability criterion in Theorem 6.3.1 calls for examining the eigenvalues of the matrix of partial derivatives of (6.3.21) at a solution. Since in the present case this matrix is one-by-one, being simply the derivative of (6.3.21) with respect to ρ, the condition for asymptotic stability is just that this derivative be negative. This is the same result obtained previously by a heuristic argument in Theorem 4.4.1. Asymptotic stability in this case must be interpreted with some care. Since Theorem 6.3.1 is applied to (6.3.20), it is the periodic solution *of that equation* that is asymptotically stable. This means that as $\theta \to \infty$, solutions $r(\theta)$ near the periodic solution approach the periodic solution. Since these solutions are *orbits* of (6.3.19), this says that *orbits* near the limit cycle approach the limit cycle as they wind around it, or in other words, the periodic orbit of (6.3.19) attracts all nearby orbits. On the other hand, a periodic *solution* of (6.3.19) does not attract all nearby *solutions* as $t \to \infty$, since (for instance) even two nearby periodic solutions lying on the limit cycle do not approach each other, but remain separated by a constant time lag. In other words, the attracting power of the limit cycle is in the radial direction, not in the angular direction. The technical expression is that periodic solutions of (6.3.19) are not asymptotically stable, but are *asymptotically orbitally stable with asymptotic phase*; we will leave the precise definition to a course in differential equations.

Finally, it is necessary to say a word about the conservative case (6.3.9). In Section 4.3 it was seen that all initial conditions for this equation give periodic solutions if ε is small enough, and there is no need for a determining equation at all. Since (6.2.9) is a special case of (6.2.10), it is necessary to ask why the results just obtained for autonomous equations do not apply in the conservative case. The answer is, once again, that the determining equation (even in the form appropriate for autonomous equations) cannot have simple solutions. In fact, the averaged equation for $\dot{\rho}$ vanishes identically in the conservative case, and no conclusions are possible. Therefore we will not attempt to reproduce the results found in Section 4.3 here.

The ideas of this section will be explained further by way of examples in the following Section 6.4, and the exercises will be found at the end of that section.

6.4. FORCED DUFFING AND VAN DER POL EQUATIONS

In this section the methods of Section 6.3 will be used to study the periodic solutions of three perturbation problems: the harmonically forced Duffing equation; the harmonically forced Van der Pol equation; and the third subharmonic for the Duffing equation.

Example 6.4.1. Harmonically Forced Duffing Equation

This problem has already been extensively treated in Section 4.7, where the periodic solutions were found by the Lindstedt method, and in Example 5.4.2, where it was put into periodic standard form.

Here we will begin with the periodic standard form, apply Theorem 6.3.1, and obtain in that way the same determining equations found in Section 4.7. This time, however, we will be able to discuss the stability of the periodic solutions as well as their existence.

The harmonic resonance problem for Duffing's equation has been specified in equation (4.3.7) as

$$\ddot{y} + \varepsilon\delta_1\dot{y} + y + \varepsilon\lambda_1 y^3 = \varepsilon\gamma_1\cos(1 + \varepsilon\omega_1)t. \qquad (6.4.1)$$

The periodic standard form is given in (5.4.31) as

$$\frac{dr}{dt^+} =$$
$$\varepsilon\left\{\gamma_1\cos t^+\sin(\varphi - t^+) - \lambda_1 r^3\cos^3(\varphi - t^+)\sin(\varphi - t^+) - \delta_1 r\sin^2(\varphi - t^+)\right\}$$
$$+ \mathcal{O}(\varepsilon^2),$$

$$\frac{d\varphi}{dt^+} = \qquad\qquad\qquad\qquad\qquad\qquad\qquad (6.4.2)$$
$$\varepsilon\left\{\omega_1 + \frac{\gamma_1}{r}\cos t^+\cos(\varphi - t^+) - \lambda_1 r^2\cos^4(\varphi - t^+) - \delta_1\cos(\varphi - t^+)\sin(\varphi - t^+)\right\}$$
$$+ \mathcal{O}(\varepsilon^2),$$

where

$$t^+ = \omega(\varepsilon)t = (1 + \varepsilon\omega_1)t. \qquad (6.4.3)$$

Recall from (5.4.27) that the variables in (6.4.1) and (6.4.2) are related by

$$
\begin{aligned}
y &= u = r\cos(\varphi - t^+), \\
\dot{y} &= v = r\sin(\varphi - t^+).
\end{aligned}
\tag{6.4.4}
$$

The first step is to average the right hand sides of (6.4.2) with respect to t^+. Most of the terms contain t^+ only in the combination $\varphi - t^+$, and in averaging these terms it is convenient to observe the fact that for any 2π-periodic function f,

$$
\frac{1}{2\pi}\int_0^{2\pi} f(\varphi - t^+)dt^+ = \frac{1}{2\pi}\int_0^{2\pi} f(x)dx.
$$

This saves expanding a good many trig functions by the sum and difference rules; for instance, $\cos^4(\varphi - t^+)$ can be written $\cos^4 x$ and averaged by Table E.1 to give 3/8. But the first term in the first equation, and the second term in the second equation, must be expanded and averaged over t^+. The result of all of this is that the averaged equations are

$$
\begin{aligned}
\frac{dr}{dt^+} &\cong \varepsilon\left\{\frac{1}{2}\gamma_1\sin\varphi - \frac{1}{2}\delta_1 r\right\}, \\
\frac{d\varphi}{dt^+} &\cong \varepsilon\left\{\omega_1 + \frac{\gamma_1}{2r}\cos\varphi - \frac{3\lambda_1}{8}r^2\right\}.
\end{aligned}
\tag{6.4.5}
$$

As is customary in applications, we have not introduced different names for the variables after averaging.

According to Section 6.2, simple rest points of (6.4.5) correspond to periodic solutions of (6.4.2) and to nearly sinusoidal periodic solutions of (6.4.1). Therefore the determining equations for these periodic solutions are

$$
\begin{aligned}
F(r,\varphi) &= \frac{1}{2}\gamma_1\sin\varphi - \frac{1}{2}\delta_1 r = 0, \\
G(r,\varphi) &= \omega_1 + \frac{\gamma_1}{2r}\cos\varphi - \frac{3\lambda_1}{8}r^2.
\end{aligned}
\tag{6.4.6}
$$

These are clearly equivalent to the determining equations (4.7.4) found by Lindstedt's method.

There are two slight differences between these equations and (4.7.4). The first is that we have not simplified (6.4.6) by, for instance, multiplying the top equation by 2. The reason for this is that we must retain the exact averaged form in order to compute the matrix of partial derivatives correctly in the stability analysis below. The second difference is the occurrence of θ instead of φ in (4.7.4). This is slightly more than a notational difference. In (4.7.4), θ is the polar angle in the phase plane. In deriving the periodic standard form, we rotated the phase plane at angular velocity $\omega(\varepsilon) = 1 + \varepsilon\omega_1$, defining φ by $\theta = \varphi - t^+$. (See for instance (5.4.10).) But recall that in the Lindstedt method, the determining equations were intended to locate the *initial conditions* for the periodic solution (in the limit as $\varepsilon \to 0$). At the initial time $t^+ = 0$, the angles φ and θ are equal, and the two sets of determining equations are equivalent. Under the present interpretation, the determining equations pick out any points in the *rotating* plane which are constant solutions for the averaged equations. As the plane spins, these fixed points trace out circles in the actual phase plane, whose projections onto the y- (or u-) axis are sinusoidal approximate solutions (which are limits as $\varepsilon \to 0$ of the exact solutions). Therefore from this point of view, the determining equations pertain to the entire orbit and not only to the initial conditions.

The graphical interpretation of these determining equations has been discussed at length in Section 4.7; see equations (4.7.7) and Fig. 4.7.2, and Exercise 4.7.1. We will not repeat this discussion here, except to record the following equations for later use: From the first equation of (6.4.6), one finds

$$\sin\varphi = \frac{\delta_1 r}{\gamma_1} \quad \text{and} \quad \cos^2\varphi = 1 - \frac{\delta_1^2 r^2}{\gamma_1^2}, \tag{6.4.7a}$$

while the second equation of (6.4.6) implies

$$\cos\varphi = \frac{3\lambda_1 r^3}{4\gamma_1} - \frac{2\omega_1 r}{\gamma_1}. \tag{6.4.7b}$$

Together they imply

$$\left(\omega_1 - \frac{3}{8}\lambda_1 r^2\right)^2 = \frac{\gamma_1^2}{4r^2} - \frac{\delta_1^2}{4}, \tag{6.4.8}$$

which is the same as (4.7.7a). It is this equation which is graphed in Fig. 4.7.2; this graph is repeated here as Fig. 6.4.1, with the additional feature that *the doubled-back section of the curve between the vertical tangents is shown as a lighter line than the rest, indicating that these points correspond to unstable solutions whereas the others are stable.* It will be our task in the remainder of Example 6.4.1 to prove this fact.

In order to carry out the stability analysis according to Theorem 6.3.1, it is first necessary to compute the matrix of partial derivatives of F and G in (6.4.6) and to evaluate these at the rest points. Then we should compute the eigenvalues of this matrix at each rest point and determine the sign of the real part. Since the matrix in question is two-by-two, the following lemma greatly simplifies the argument.

Lemma 6.4.1. *If A is a two-by-two matrix, then both eigenvalues have negative real parts if and only if the trace of A is negative and the determinant is positive.*

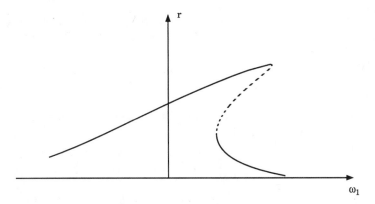

FIG. 6.4.1. A response curve for the harmonically forced damped Duffing's equation showing the unstable periodic solutions by a dotted line between the vertical tangent points. (Compare Fig. 4.7.2.)

Proof. Recall that if

$$A = \begin{bmatrix} a & b \\ c & d \end{bmatrix},$$

then the trace of A is $T = a + d$ and the determinant is $D = ad - bc$. The characteristic equation for the eigenvalues is then $\lambda^2 - T\lambda + D = 0$, with roots

$$\lambda = \frac{T \pm \sqrt{T^2 - 4D}}{2}.$$

Now if $D < 0$ then the square root is real and is larger than T, so the two values of λ have opposite sign. If $D = 0$, one of the eigenvalues is zero. But if $D > 0$ then the square root is either real and less than T, or is complex. In either case the sign of T determines the sign of the real part of both eigenvalues. So the only way for both eigenvalues to have a negative real part is to have $D > 0$ and $T < 0$. ∎

Now the matrix of partial derivatives of F and G at an arbitrary point is

$$A = \begin{bmatrix} -\frac{1}{2}\delta_1 & \frac{1}{2}\gamma_1 \cos \varphi \\ -\frac{\gamma_1}{2r^2} \cos \varphi - \frac{3}{4}\lambda_1 r & -\frac{\gamma_1}{2r} \sin \varphi \end{bmatrix}.$$

The trace of this matrix is $-\delta_1/2 - (\gamma_1/2r) \sin \varphi$. At any of the rest points, we may replace $\sin \varphi$ by $\delta_1 r/\gamma_1$ according to (6.4.7a), with the result that the trace becomes $-\delta_1$, which is negative (assuming, as we do, that the damping is positive). Therefore according to Lemma 6.4.1, the rest point will be stable provided that the determinant is positive. But

$$\det A = \frac{\delta_1 \gamma_1}{4r} \sin \varphi + \frac{\gamma_1^2}{4r^2} \cos^2 \varphi + \frac{3}{8} \lambda_1 \gamma_1 r \cos \varphi.$$

Substituting for $\sin \varphi$ and $\cos^2 \varphi$ from (6.4.7a), and for $\cos \varphi$ from (6.4.7b), one finds that at any rest point,

$$\det A = \frac{1}{4r^2} \left\{ \gamma_1^2 - 3\lambda_1 r^4 \left(\omega_1 - \frac{3}{8}\lambda_1 r^2 \right) \right\}. \tag{6.4.9}$$

So the periodic solutions are stable when this quantity is positive.

In order to interpret this result, differentiate (6.4.8) implicitly (regarding r as a function of ω_1) to obtain

$$\frac{dr}{d\omega_1} = \frac{4(\omega_1 - \frac{3}{8}\lambda_1 r^2)r^3}{3\lambda_1 r^4(\omega_1 - \frac{3}{8}\lambda_1 r^2) - \gamma_1^2}. \tag{6.4.10}$$

From this one sees that the denominator of (6.4.10) is equal to $\det A$ as given by (6.4.9), up to a nonzero factor. Since vertical tangents occur where the denominator of (6.4.10) changes sign, it follows that $\det A$ changes sign at the vertical tangents in Fig. 6.4.1, so that these vertical tangents are the places where the stability changes. (A general version of this argument is outlined in Exercise 6.4.3.) If Fig. 6.4.1 is interpreted as a bifurcation diagram, showing the changes in the number of rest points (of the averaged equation, or the number of periodic solutions of the original equation) as ω_1 is varied, then the vertical tangent points are pair bifurcations. In the terminology of differential equations these are called *saddle-node bifurcations*, because the unstable rest points emerging from the bifurcation are of saddle type and the stable points are nodes.

Example 6.4.2. Harmonically Forced Van der Pol Equation

The harmonically forced Van der Pol equation, with small forcing and small "Van der Pol term," can be written

$$\ddot{y} + \varepsilon(y^2 - 1)\dot{y} + y = \varepsilon \gamma_1 \cos(1 + \varepsilon\omega_1)t. \tag{6.4.11}$$

The calculations for this problem are similar to those for the Duffing equation above, although the final results are rather different. After putting the system in polar periodic standard form (5.4.29) and averaging, one obtains (Exercise 6.4.1)

$$\frac{dr}{dt^+} = \varepsilon \left[\frac{1}{2}\gamma_1 \sin \varphi + \frac{1}{2}r - \frac{1}{8}r^3 \right],$$

$$\frac{d\varphi}{dt^+} = \varepsilon \left[\omega_1 + \frac{\gamma_1}{2r} \cos \varphi \right]. \tag{6.4.12}$$

The equations for the rest points are therefore

$$F(r, \varphi) = \frac{1}{2}\gamma_1 \sin \varphi + \frac{1}{2}r - \frac{1}{8}r^3 = 0,$$

$$G(r, \varphi) = \omega_1 + \frac{\gamma_1}{2r} \cos \varphi = 0.$$

(6.4.13)

These equations together with $\sin^2 \varphi + \cos^2 \varphi = 1$ imply

$$\left(\frac{1}{4}r^3 - r\right)^2 + 4r^2 \omega_1^2 = \gamma_1^2,$$

(6.4.14)

which is the response curve giving the approximate amplitude r of the periodic solutions that exist for given forcing coefficients ω_1 and γ_1. It will prove convenient to draw these curves in slightly different coordinates, which will be defined after carrying out the stability analysis.

The matrix of partial derivatives of (6.4.13) is

$$\begin{bmatrix} F_r & F_\varphi \\ G_r & G_\varphi \end{bmatrix} = \begin{bmatrix} \frac{1}{2} - \frac{3}{8}r^2 & \frac{\gamma_1}{2} \cos \varphi \\ -\frac{\gamma_1}{2r^2} \cos \varphi & -\frac{\gamma_1}{2r} \sin \varphi \end{bmatrix} = \begin{bmatrix} \frac{1}{2} - \frac{3}{8}r^2 & -r\omega_1 \\ \frac{\omega_1}{r} & -\frac{1}{8}r^2 + \frac{1}{2} \end{bmatrix}.$$

(6.4.15)

The second form in (6.4.15) is obtained from the first by using (6.4.13) to eliminate the trig functions; therefore the second form is valid only at the rest points. Now according to Lemma 6.4.1, the condition for stability is that the trace of (6.4.15) be negative and the determinant be positive. The trace is $1 - r^2/2$, which is negative if $r > \sqrt{2}$. The determinant is

$$\left(\frac{1}{2} - \frac{3}{8}r^2\right)\left(\frac{1}{2} - \frac{1}{8}r^2\right) + \omega_1^2,$$

which is a quadratic in r^2 and is most easily analyzed by introducing a new variable equal or proportional to r^2. For some reason it has become customary to take this variable in the form $R := r^2/4$. In terms of this variable, the trace and determinant are

trace $= 1 - 2R$,   determinant $= \frac{1}{4}(1 - 3R)(1 - R) + \omega_1^2$. (6.4.16)

Now in the (ω_1, R) plane one can draw the curves trace $= 0$ and determinant $= 0$; the former is the line $R = 1/2$, and the latter is the ellipse

$$9\left(R - \frac{2}{3}\right)^2 + 12\omega_1^2 = 1.$$

(6.4.17)

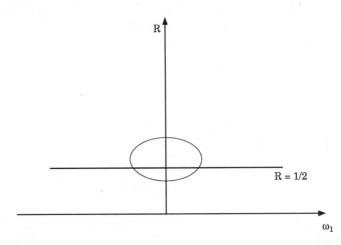

FIG. 6.4.2. The curves trace $= 0$ and determinant $= 0$ for the forced Van der Pol equation.

The details of the figures in the following discussion are given in Exercise 6.4.2. The relative positions of the ellipse and the line are shown in Fig. 6.4.2. It is also possible to express the response curves (6.4.14) in terms of the variable R; these curves (for various γ_1) are shown in Fig. 6.4.3. Notice that depending on γ_1 the curve is either a single arc with no vertical tangents (Fig. 6.4.3a) or two arcs, one of which is a closed curve with two vertical tangents (Fig. 6.4.3b). (There is one exceptional curve which intersects itself and divides the other two types. It is shown in Fig. 6.4.3c.)

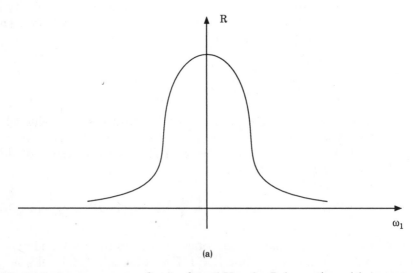

(a)

FIG. 6.4.3. Response curves for the forced Van der Pol equation. (a) A response curve with one component. (b) A response curve with two components. (c) The unique response curve forming the transition between those of the form (a) and (b).

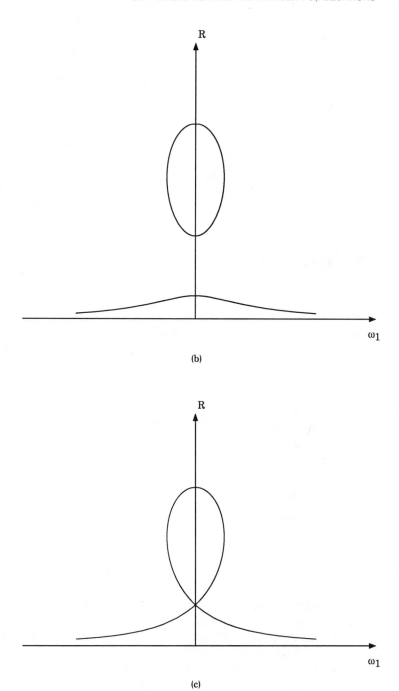

(b)

(c)

FIG. 6.4.3 (*Continued*)

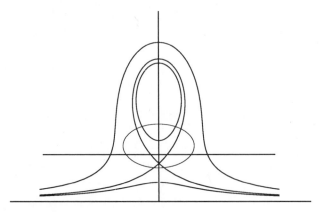

FIG. 6.4.4. Superposition of the curves in Figs. 6.2.2 and 6.2.3. The ellipse is the locus of vertical tangents of the response curves. Solutions above the straight line and outside the ellipse are stable.

From Exercise 6.4.3 it follows that (as in the case of Duffing's equation) the vertical tangents occur at the intersections of the response curves with the curve determinant = 0 (which is the ellipse). Fig. 6.4.4 shows the response curves overlaid on the line and the ellipse. The conditions for the stability of a point on any response curve is that it lie above the line and outside the ellipse. Therefore in this case there are two kinds of stability changes: those which occur at vertical tangents (and represent saddle-node bifurcations), and those which occur across the line $R = 1/2$, which represent a change of stability without a bifurcation.

Example 6.4.3. Third Subharmonic of Duffing's Equation

The third subharmonic problem for Duffing's equation has been defined in (5.4.39) together with (5.4.41); it is

$$\ddot{y} + \varepsilon\delta_1\dot{y} + y + \varepsilon\lambda_1 y^3 = \gamma_0 \cos(3 + \varepsilon\omega_1)t. \qquad (6.4.18)$$

The periodic standard form is given in (5.4.46) and (5.4.47) in Cartesian form. Averaging this is a lengthy process but is purely routine. The period is $6\pi/\omega$. Instead of averaging with respect to t over this period, the same result is obtained by setting $\theta = \omega t/3$ and averaging with respect to θ over period 2π. This is most easily achieved in the following steps: First, fully expand (5.4.46) so that each equation has 16 terms (this involves cubing the last term in f). Next, discard the terms that are clearly odd, since these average to zero. Of the remaining terms, a few are recognizable from Table E.1, and the average can be read off from the constant term in the Fourier series. The remaining terms, such as $\cos\theta\sin^2\theta\cos 3\theta$, can be averaged by using the complex exponential expressions for the sine and cosine functions, multiplying out, and (again) retaining only the constant term. The averaged equations are

$$\dot{a} = \varepsilon P(\omega_1, a, b),$$
$$\dot{b} = \varepsilon Q(\omega_1, a, b),$$

$$(6.4.19)$$

where

$$
\begin{aligned}
P(\omega_1, a, b) = & -\frac{1}{3}\omega_1 b - \frac{1}{2}\delta_1 a + \frac{3}{8}\lambda_1 a^2 b + \frac{3}{8}\lambda_1 b^3 \\
& + \frac{3}{32}\lambda_1 \gamma_0 ab + \frac{3}{256}\lambda_1 \gamma_0^2 b, \\
Q(\omega_1, a, b) = & \frac{1}{3}\omega_1 a - \frac{1}{2}\delta_1 b - \frac{3}{8}\lambda_1 a^3 - \frac{3}{8}\lambda_1 ab^2 \\
& + \frac{3}{64}\lambda_1 \gamma_0 a^2 - \frac{3}{64}\lambda_1 \gamma_0 b^2 - \frac{3}{256}\lambda_1 \gamma_0^2 a.
\end{aligned}
$$

$$(6.4.20)$$

The determining equations for periodic solutions are then $P = 0$, $Q = 0$. These equations simplify considerably in polar coordinates. Let $a = r \cos \varphi$, $b = r \sin \varphi$, and then simplify by computing $P \cos \varphi + Q \sin \varphi$ and $Q \cos \varphi - P \sin \varphi$; the result is

$$F(r, \varphi) = \frac{1}{2}\delta_1 - \frac{3}{64}\lambda_1 \gamma_0 r \sin 3\varphi = 0,$$

$$G(\omega_1, r, \varphi) = \frac{1}{3}\omega_1 - \frac{3}{8}\lambda_1 r^2 + \frac{3}{64}\lambda_1 \gamma_0 r \cos 3\varphi - \frac{3}{256}\lambda_1 \gamma_0^2 = 0.$$

$$(6.4.21)$$

It is actually easier to arrive at this result by setting up the periodic standard form in polar coordinates in the first place, but we leave this to the reader (Exercise 6.4.4). The first two examples of this section were worked in polar coordinates, and we did this one in Cartesian just to show how it looks and why polar is usually better.

These equations have a threefold symmetry: If (r, φ) satisfies (6.4.21), so do $(r, \varphi + 2\pi/3)$ and $(r, \varphi + 4\pi/3)$. In the undamped case ($\delta_1 = 0$), the first equation reduces to $\sin 3\varphi = 0$, so the "basic" solutions can be taken to be those with $\varphi = 0$, and the rest can be obtained from these by the symmetry. Putting $\varphi = 0$ into the second equation, the response curve becomes

$$\omega_1 = \frac{9}{8}\lambda_1 \left(r^2 - \frac{\gamma_0}{8}r + \frac{\gamma_0^2}{32} \right).$$

$$(6.4.22)$$

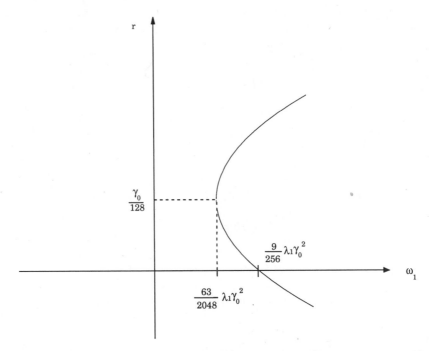

FIG. 6.4.5. Response curve for the third subharmonic of Duffing's equation without damping. Each point on the graph corresponds to three periodic solutions, so that for each ω_1 there are either six periodic solutions or none.

This curve is a parabola, shown in Fig. 6.4.5. For $\omega_1 < 63\lambda_1\gamma_0^2/2048$ there are no solutions, while for larger ω_1 there are two; with the symmetrical solutions this makes six in all. Figure 6.4.6a shows the arrangement of these solutions in the polar (r, φ) plane when both values of r are positive, and Fig. 6.4.6b shows the placement when one r is negative (that is, for ω_1 large enough that one side of the parabola in Fig. 6.4.5 has crossed into the lower half-plane where $r < 0$). It is perhaps surprising that there are no solutions for $\omega_1 = 0$, since this corresponds to exact subharmonic resonance. In other words, a solution with period three times that of the forcing cannot be produced when the forcing period is exactly one-third of the free period, but only when the forcing period is slightly shorter. To put it more dramatically, one cannot subharmonically excite an oscillation having the free frequency!

When damping is present, it is no longer easy to solve for φ, but one can eliminate φ from (6.4.21) by solving the first equation for $\sin 3\varphi$ and then using $\cos 3\varphi = \pm\sqrt{1 - \sin^2 3\varphi}$. The result is

$$\omega_1 = \frac{9}{8}\lambda_1\left(r^2 \mp \frac{1}{24}\sqrt{9\gamma_0^2 r^2 - 1024\delta_1^2/\lambda_1^2} + \frac{\gamma_0^2}{32}\right). \qquad (6.4.23)$$

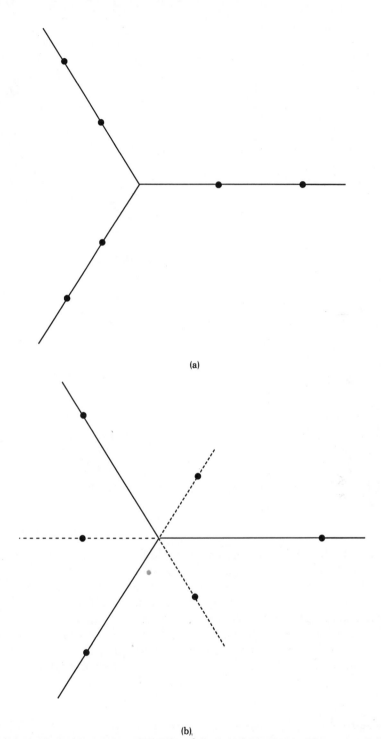

(a)

(b)

FIG. 6.4.6. The arrangement of the six periodic solutions (see Figure 6.4.5) in the polar (r, ϕ) plane: (a) when both r values are positive, (b) when one r value is negative.

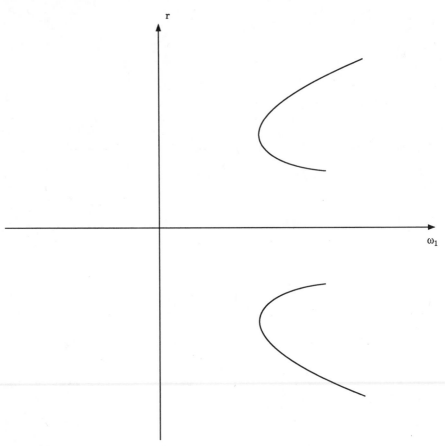

FIG. 6.4.7. A response curve for the third subharmonic of Duffing's equation with damping.

It is evident that only one choice of the ambiguous sign can be correct here, since the second equation of (6.4.21) can hold for only one value of the cosine (for each choice of the other variables). Since (6.4.23) should reduce to (6.4.22) when $\delta_1 = 0$, it is evident that the $\mp$ in (6.4.23) should be chosen so that the term has the same sign as $-\gamma_0 r$. An interesting feature of (6.4.23) is that the radical becomes imaginary when r is too small (for specified values of the other parameters, and assuming $\delta_1 > 0$). This means that the graph of (6.4.23) ceases to exist in an interval around the ω_1 axis. A typical response curve is shown in Fig. 6.4.7.

Exercises 6.4.4

1. Carry out the details leading from (6.4.11) to (6.4.12); that is, put the forced Van der Pol equation in polar periodic standard form and average.

2. Check that (6.4.16) implies (6.4.17); then show that Fig. 6.4.2 shows the correct positioning of the ellipse and the line. In particular, check that the ellipse is entirely in the upper half-plane and is intersected by the line below its center. Then write equation (6.4.14) in terms of the variable R and analyze the level curves for various γ to show that Figs. 6.4.3 and 6.4.4 are correct. Hint: After introducing R, find the critical points of the left hand side and classify them (saddle, maximum, or minimum). Show that there is a saddle at the lowest point on the ellipse.

3. Suppose that the determining equations have the form $F(r, \varphi) = 0$, $G(\omega_1, r, \varphi) = 0$. Show that vertical tangent points on the response curve occur where the Jacobian determinant $\partial(F, G)/\partial(r, \varphi)$ vanishes. This generalizes the result in the last paragraph of Example 6.4.1. Hint: Regard the determining equations as defining ω_1 and φ as implicit functions of r, and calculate the derivative $d\omega_1/dr$ (see Appendix B).

4. Beginning with the third subharmonic Duffing equation in the form (6.4.18), put it into polar periodic standard form and average it to obtain (6.2.21).

5. Carry out a stability analysis for the third subharmonic of Duffing's equation by examining the trace and determinant of the matrix of partial derivatives of (6.2.21) at the zeroes. Draw a response curve showing stable solutions by a heavy line and unstable ones by a light line.

6.5* HIGHER ORDER AVERAGING

In this section the method of averaging for systems in periodic standard form will be extended beyond the first approximation. The primary purpose of this extension is to gain higher orders of accuracy on expanding intervals of length $\mathcal{O}(1/\varepsilon)$; in this the method is always successful in theory, and in practice is limited only by the difficulty of the computations. A secondary purpose is to extend the asymptotic length of the expanding intervals of validity; this purpose can only be achieved under certain circumstances, and both positive and negative results will be stated, including some that are rather new. Besides these *quantitative* uses of higher order averaging (improving the results of Section 6.2), there are important *qualitative* uses (improving the results of Section 6.3). For instance, it is possible to use higher order averaging to decide the stability of some periodic solutions which "fall through the cracks" of Theorem 6.3.1 (by having eigenvalues on the imaginary axis). These qualitative results will not be developed here (but see Notes and References, Section 6.7).

It was shown in Section 6.2 that first order averaging can be understood as a two-step process: First perform a near-identity transformation which renders the system autonomous to first order, then truncate the new system at the first order. Averaging of order k follows the same pattern, except that the near-identity transformation must now make the system autonomous through order k, and the truncation occurs at order k. In order to achieve this, the near-identity transformation must contain terms through order k. The process can be summarized as follows:

$$\text{exact system} \qquad \dot{\mathbf{x}} = \varepsilon \mathbf{f}_1(\mathbf{x}, t) + \cdots + \varepsilon^k \mathbf{f}_k(\mathbf{x}, t) + \varepsilon^{k+1} \hat{\mathbf{f}}(\mathbf{x}, t, \varepsilon),$$

(6.5.1)

$$\text{transformation} \qquad \mathbf{x} = \mathbf{y} + \varepsilon \mathbf{u}_1(\mathbf{y}, t) + \cdots + \varepsilon^k \mathbf{u}_k(\mathbf{y}, t), \qquad (6.5.2)$$

$$\text{transformed system} \qquad \dot{\mathbf{y}} = \varepsilon \mathbf{g}_1(\mathbf{y}) + \cdots + \varepsilon^k \mathbf{g}_k(\mathbf{y}) + \varepsilon^{k+1} \hat{\mathbf{g}}(\mathbf{y}, t, \varepsilon),$$

(6.5.3)

$$\text{averaged system} \qquad \dot{\mathbf{w}} = \varepsilon \mathbf{g}_1(\mathbf{w}) + \cdots + \varepsilon^k \mathbf{g}_k(\mathbf{w}), \qquad (6.5.4)$$

$$\text{guiding system} \qquad \frac{d\mathbf{z}}{d\tau} = \mathbf{g}_1(\mathbf{z}) = \bar{\mathbf{f}}(\mathbf{z}) \qquad (\tau = \varepsilon t). \qquad (6.5.5)$$

Of course, the functions $\mathbf{u}_n$ and $\mathbf{g}_n$ appearing in these equations will have to be constructed from the given functions $\mathbf{f}_n$. The function $\mathbf{g}_1$ is the average of $\mathbf{f}_1$, which is the same as the average $\bar{\mathbf{f}}$ which appears in the first order method of averaging; see (6.1.3). But the higher order terms $\mathbf{g}_n$ are not the averages of the corresponding $\mathbf{f}_n$. Rather, they are averages of complicated combinations of $\mathbf{f}_n$ together with derivatives of $\mathbf{f}_j$ for $j < n$ and of $\mathbf{u}_j$ for $j \leq n$. The *guiding system* is defined to be the same as the guiding system for first order averaging. Since the letter $\mathbf{z}$ has already been used for the guiding center, a new letter $\mathbf{w}$ will be used for the averaged system (6.5.4). (In the first order method of averaging, the guiding system was obtained by substituting the slow time τ into the averaged equation. It is not useful to introduce τ into (6.5.4), since the resulting system does not turn out to be independent of ε when $k > 1$. But the guiding system from first order averaging is still useful in doing error estimates for higher order averaging.)

As in Section 6.2, the first order of business is to establish the existence of the near-identity transformation (6.5.2).

Theorem 6.5.1. *Let* $\mathbf{f}$ *be sufficiently smooth, and periodic in* t *with period* 2π. *Then there exists a transformation of the form (6.5.2),* 2π-*periodic in* t, *carrying (6.5.1) into (6.5.3). It is possible to choose the* $\mathbf{u}_n$ *to have mean value zero, or to choose them stroboscopically, that is, in such a way that (6.5.2) reduces to the identity at times* $t = 0$, $t = 2\pi$, *etc.*

Proof. We will sketch the proof for second order averaging ($k = 2$). For the third order case see Exercise 6.5.1; this should make the general case clear.

First one argues (as in Section 6.2) that for any choice of the u_n ($n = 1, \ldots, k$) the near-identity transformation (6.5.2) is invertible on compact sets for small ε and is therefore a valid coordinate transformation, and also that it carries (6.5.1) into a system having the form of (6.5.3) except that the $\mathbf{g}_n$ will in general depend upon t. Knowing this, it is permissible to differentiate (6.5.2) with respect to t and use (6.5.3) to obtain an expansion for $\dot{x}$ in terms of $\mathbf{y}$; another expansion for the same function may be found by substituting (6.5.2) into (6.5.1) and expanding. By equating coefficients of like powers of ε, one obtains the following formulas for the functions $\mathbf{g}_1(\mathbf{y}, t)$ and $\mathbf{g}_2(\mathbf{y}, t)$ arising when an arbitrary near-identity transformation (6.5.2) is applied to (6.5.1):

$$\begin{aligned}
\mathbf{g}_1 &= \mathbf{f}_1 - \mathbf{u}_{1t}, \\
\mathbf{g}_2 &= \mathbf{f}_2 + \mathbf{f}_{1x}\mathbf{u}_1 - \mathbf{u}_{1y}\mathbf{g}_1 - \mathbf{u}_{2t},
\end{aligned} \tag{6.5.6}$$

where all functions are evaluated at $(\mathbf{y}, t)$.

Letter subscripts in (6.5.6) denote partial derivatives; $\mathbf{u}_{1t}$ is a column vector of partial derivatives with respect to t, whereas $\mathbf{f}_{1x}$ is an $N \times N$ matrix of partial derivatives $\partial f_{1i}/\partial x_j$. Derivatives of $\mathbf{f}_n$ with respect to its vector argument are denoted with a subscript $\mathbf{x}$ even when the function is evaluated at $\mathbf{y}$, because $\mathbf{x}$ is the letter most often associated with the first variable in $\mathbf{f}_n$. In other words, $\mathbf{f}_{nx}(\mathbf{y}, t)$ means "first differentiate $\mathbf{f}_n(\mathbf{x}, t)$ with respect to $\mathbf{x}$ and then evaluate the resulting function at $\mathbf{y}$."

Reversing the point of view, (6.5.6) may be regarded as a set of differential equations for the $\mathbf{u}_n$ if the $\mathbf{g}_n$ are given. The goal, then, is to choose $\mathbf{g}_n(\mathbf{y})$ to be independent of t and such that the equations

$$\begin{aligned}
\mathbf{u}_{1t}(\mathbf{y}, t) &= \mathbf{f}_1(\mathbf{y}, t) - \mathbf{g}_1(\mathbf{y}), \\
\mathbf{u}_{2t}(\mathbf{y}, t) &= (\mathbf{f}_2 + \mathbf{f}_{1x}\mathbf{u}_1 - \mathbf{u}_{1y}\mathbf{g}_1) - \mathbf{g}_2(\mathbf{y})
\end{aligned} \tag{6.5.7}$$

have solutions $\mathbf{u}_n(\mathbf{y}, t)$ which are periodic in t; the functions in the parenthesis are evaluated at $(\mathbf{y}, t)$. Exactly as in Section 6.2, the first equation is solvable for $\mathbf{u}_1$ provided $\mathbf{g}_1$ is taken to be the average of $\mathbf{f}_1$, and $\mathbf{u}_1$ may be chosen either with mean zero or "stroboscopically." Once $\mathbf{u}_1$ is chosen, the parenthesis in the second equation is well-defined, and the second equation is solvable provided that $\mathbf{g}_2$ is taken to be the average of the parenthesis. Once again, $\mathbf{u}_2$ may be taken to have mean zero or stroboscopically. ∎

Since the near-identity transformation exists, the entire scheme of equations (6.5.1)–(6.5.5) exists. As before, the solutions of these equations with initial condition α will be denoted $\mathbf{x}(t, \alpha, \varepsilon)$, $\mathbf{y}(t, \alpha, \varepsilon)$, $\mathbf{w}(t, \alpha, \varepsilon)$, and $\mathbf{z}(\tau, \alpha)$. Here one expects that $\mathbf{w}$ is a good approximation to $\mathbf{y}$, and since x is related to y by (6.5.2), the best approximation to x that can be obtained from w should be

$$\tilde{x}_{imp}(t, \alpha, \varepsilon) := w + \varepsilon u_1(w, t) + \cdots + \varepsilon^k u_k(w, t), \qquad (6.5.8)$$

where w stands for $w(t, \alpha, \varepsilon)$. This is called the *improved kth approximation* to the exact solution x. Since the error in this approximation turns out to be of order ε^k on intervals of length $\mathcal{O}(1/\varepsilon)$, it is permissible to leave out the last term in (6.5.8) and define what is called the *kth approximation*,

$$\tilde{x}(t, \alpha, \varepsilon) := w + \varepsilon u_1(w, t) + \cdots + \varepsilon^{k-1} u_{k-1}(w, t). \qquad (6.5.9)$$

Here lies a major difference between first and higher order averaging: In first order averaging, the "last term" in the transformation is the only term, and when it is omitted, there is no need to use the transformation at all; therefore the transformation is only of importance theoretically. In higher order averaging, the transformation must be used at least through the next-to-the-last term in order to achieve the gain in accuracy that is desired. However, it is not necessary to calculate the last term, and that is some saving of labor.

Now we are ready to state the main theorem on the asymptotic validity of higher order averaging. The proof is left to Exercise 6.5.2.

Theorem 6.5.2. *The error in the kth approximation (6.5.9) is $\mathcal{O}(\varepsilon^k)$ uniformly for t in an expanding interval of length $\mathcal{O}(1/\varepsilon)$ and uniformly for α in compact sets. More precisely, let compact sets K and V be as in Theorem 6.2.3, and let $T(\alpha)$ be defined as in that theorem, using the same first order guiding center (6.5.5). Then there exist constants c and ε_0 such that*

$$\|x(t, \alpha, \varepsilon) - \tilde{x}(t, \alpha, \varepsilon)\| \leq c\varepsilon^k$$

for α in K, $0 \leq \varepsilon \leq \varepsilon_0$, and $0 \leq t \leq T(\alpha)/\varepsilon$.

Next we turn to the question of whether the method of averaging can be extended beyond times of order $1/\varepsilon$. There are, first of all, two extreme cases. It has already been mentioned in Section 6.3 (see Theorem 6.3.1) that even first order averaging is valid for all (future) time when a solution is approaching a suitable type of attracting orbit. On the other hand, it will be seen later in this section that for certain problems (containing what are called *saddle points*), averaging to any order at all will never be valid past a time of order $1/\varepsilon$. Between these strong positive and negative results there is considerable room for manipulation. We will state two types of results. The first gives conditions under which the improved accuracy of higher order averaging can be "traded off" for longer intervals of validity as in the Lindstedt method (see Section 4.2). The second type of result concerns the previously mentioned case of saddle points. It turns out that even though the approximate solutions of initial value problems lose their validity as they pass close to a saddle point, there is another way to interpret these solutions in which they retain their validity; it is necessary to give up the idea of solving initial value problems and to replace it by a concept of "shadowing."

For a system in periodic standard form (6.5.1), it is generally expected that the point $\mathbf{x}$ will cover a given finite distance in a time of order $1/\varepsilon$, since the velocity $\dot{\mathbf{x}}$ appears to be of order ε. However, it can happen that the actual time taken to cover a finite distance is longer than this. To see how this can come about, suppose that $\mathbf{f}_1(\mathbf{x}, t)$ is a function with mean value zero for every $\mathbf{x}$; that is, suppose that $\mathbf{g}_1(\mathbf{z}) = \bar{\mathbf{f}}(\mathbf{z}) \equiv \mathbf{0}$. Then, according to (6.5.3), the point $\mathbf{y}$ moves at most at a rate proportional to ε^2, and it follows from (6.5.2) that the same is true of the overall, large-scale motion of $\mathbf{x}$ (considering that the functions $\mathbf{u}_n$ represent small rapid vibrations around the motion of $\mathbf{y}$ and do not contribute to the covering of any substantial distance). Under these circumstances it takes a time of order $1/\varepsilon^2$ for $\mathbf{x}$ to cover a substantial distance (that is, a finite distance which is large compared to the amplitude of the rapid small oscillations). In this situation, error estimates are possible for a time interval of length $\mathcal{O}(1/\varepsilon^2)$. Similarly, if both $\mathbf{g}_1$ and $\mathbf{g}_2$ vanish identically, estimates are possible for a time of length $1/\varepsilon^3$. In general, *error estimates for initial value problems are possible for as long as it takes the solution to cover a substantial finite distance, provided that the reason for the decreased speed of travel is the vanishing of low order averages.* (Decreased speed due to passing a saddle point, discussed below, does not qualify.) The accuracy of the error estimate in such a case is determined by trade-off: If kth order averaging is used, so that the error on the usual interval of length $\mathcal{O}(1/\varepsilon)$ is $\mathcal{O}(\varepsilon^k)$, and if the first r averages vanish (with, of course, $r < k$), then it is possible to trade off up to r orders of accuracy for increased time. That is, the error on an interval of length $\mathcal{O}(1/\varepsilon^{1+j})$ will be $\mathcal{O}(\varepsilon^{k-j})$, for any $j = 0, 1, \ldots, r$.

Precise results are given in Theorem 6.5.3 below and in Exercise 6.5.4. (The proof of Theorem 6.5.3 is Exercise 6.5.3). The "modified guiding system," defined precisely in the statement of the theorem, is obtained by truncating the averaged equations at the first nonzero order (which is order ε^{r+1}) and then introducing a slow time variable which eliminates ε (this will be $\tau := \varepsilon^{r+1} t$, called the *natural time scale* for the problem).

Theorem 6.5.3. *Suppose that the quantities $\mathbf{g}_1, \ldots, \mathbf{g}_r$ in the averaged system (6.5.4) are identically zero. Let K and V be as in Theorem 6.2.3. Let $T_0 > 0$ be fixed arbitrarily, and let $T(\alpha)$ be the largest τ-time less than or equal to T_0 for which the solution of the "modified guiding system" $d\mathbf{z}/d\tau = \mathbf{g}_{r+1}(\mathbf{z})$ beginning at α remains in K. Let j be an integer, $0 \leq j \leq r$. Then there exist constants c and ε_1 such that*

$$\|\mathbf{x}(t, \alpha, \varepsilon) - \tilde{\mathbf{x}}(t, \alpha, \varepsilon)\| < c\varepsilon^{k-j} \quad \text{for} \quad 0 \leq t \leq T(\alpha)/\varepsilon^{1+j}, \quad 0 \leq \varepsilon \leq \varepsilon_1,$$
$$(6.5.10)$$

for all α in K.

The remainder of this section is an informal discussion, without proofs, of why averaging (regardless of the order of averaging) breaks down after time $1/\varepsilon$ in the presence of saddle points, and what may be done about this. (For additional information see Notes and References.) In Section 6.3, the matrix

$$A := \bar{\mathbf{f}}_\mathbf{z}(\mathbf{z}_0) \tag{6.5.11}$$

was introduced, where $\mathbf{z}_0$ is a rest point of the guiding system. If A has no eigenvalues on the imaginary axis, then A completely determines the stability properties of both the rest point $\mathbf{z}_0$ of the guiding system and the corresponding periodic solution of the exact system; the matrix A, the rest point, and the periodic solution are all said to be *hyperbolic* in this case. If all the eigenvalues are in the left half-plane, the rest point and periodic solution are called *sinks* or *attractors*; if they are all in the right half-plane, *sources* or *repellors*; and if there is at least one eigenvalue in each half-plane, *saddles*. According to Theorem 6.3.1, a sink is asymptotically stable. (This applies both to the averaged system, where the sink is a rest point, and to the exact system, where it is a periodic solution). Saddles and sources are unstable; a saddle is distinguished by having some solutions which approach it and others which diverge from it as $t \to \infty$, whereas all solutions diverge from the neighborhood of a source. As an easy example of a saddle rest point (the example does not involve averaging), consider the linear two-dimensional system

$$\begin{aligned} \dot{x} &= x, \\ \dot{y} &= -y. \end{aligned} \tag{6.5.12}$$

These equations are in fact uncoupled, and the solution through the point (x_0, y_0) is $(x_0 e^t, y_0 e^{-t})$. Any initial point on the y-axis approaches the rest point at the origin as $t \to \infty$, whereas any initial point on the x-axis diverges from it; the y-axis is called the *stable manifold* of the rest point, and the x-axis the *unstable manifold*. Now consider an *approximate* solution of (6.5.12) beginning on the y-axis. However close the approximate solution remains to the exact solution, it cannot be expected to remain exactly on the y-axis. Therefore (see Fig. 6.5.1) it will eventually come to a point near the origin but on one side or the other of the stable manifold; it will in fact come close to the unstable manifold, and eventually be accelerated exponentially away from the rest point and hence away from the exact solution. Therefore *any approximation method for initial value problems applied to a problem with saddle points has a limited time of validity unless the method is capable of computing stable manifolds exactly.* Since no approximation method is able to do this, no approximate solutions in the presence of saddle points can last forever.

In the case of a problem in periodic standard form, it is possible to be more specific. We have seen that the time taken to cover a finite distance for a problem in standard form is of order $1/\varepsilon$. Therefore this is

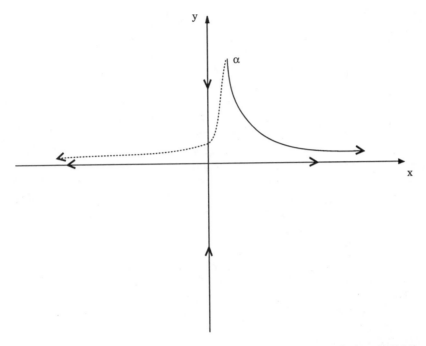

FIG. 6.5.1. The presence of a saddle point can cause an exact solution (solid line) and an approximate solution (dotted line) to diverge exponentially.

the length of time required for an orbit approaching a saddle (that is, an orbit lying on the stable manifold of a saddle) to reach a small neighborhood of the saddle. (Here one must think of a saddle periodic orbit, rather than a rest point, and the pictures are more complicated, but the principle is the same.) This is also the amount of time for which the basic error estimates are valid either for the method of averaging or the method of multiple scales. After this time the exact and approximate solutions begin to separate, due to the effect of the saddle, and by a time of order $1/\varepsilon^2$ it is possible for these solutions to be exponentially far apart (that is, their distance can be greater than a constant times $e^{1/\varepsilon}$). This great a separation is unusual; more often, the exact and approximate solutions will approach different sinks after they separate, and their eventual distance apart will be finite. But in either case it is clear that the distance between the exact and approximate solutions is not small with ε at time $1/\varepsilon^2$, so there is no useful error estimate valid at that time. This conclusion does not depend upon the details of the method; it is of no help to go to higher orders of averaging or to add more time scales. We have reached here a fundamental limitation of asymptotic methods. (The careful reader will notice that we have said nothing about "uniformly with respect to what" in speaking of these asymptotic orders, and it is of course here that the important details lie. We are not trying to do more than give a general impression. The positive result in the next paragraph will be stated more precisely and will give an idea of exactly what is at stake here.)

As hinted above, the picture is not quite this bleak: There is a way to recover a useful result, but it requires giving up the attempt to solve initial value problems. The idea will be explained using first order averaging. Consider, then, the exact system (6.2.30) and the guiding system (6.2.34) for first order averaging, and suppose that the guiding system has a saddle. For simplicity we will assume the system is two-dimensional; see Fig. 6.5.2. Consider a short arc running up to the stable manifold at a finite distance from the saddle. The solutions of the guiding system beginning on this arc come arbitrarily close to the saddle and fill up a region (shaded in Fig. 6.5.2) called an *elbow neighborhood*; we will cut this neighborhood off by a similar arc running up to the unstable manifold. The theorem we wish to state is an error estimate that is uniformly valid for the solutions running through this neighborhood.

Theorem 6.5.4. *Given an elbow neighborhood U for the guiding system, there exist constants c and ε_0 with the following property: For any solution $\mathbf{z}(\tau)$ of the guiding system running through U, there is a family of exact solutions $\mathbf{x}(t, \varepsilon)$ such that $\|\mathbf{x}(t, \varepsilon) - \mathbf{z}(\varepsilon t)\| \leq c\varepsilon$ for all t such that $\mathbf{z}(\varepsilon t)$ lies in U.*

This theorem is misleadingly simple, and it is necessary to explain the features of it that are unusual. First of all, the exact solutions $\mathbf{x}$ do not

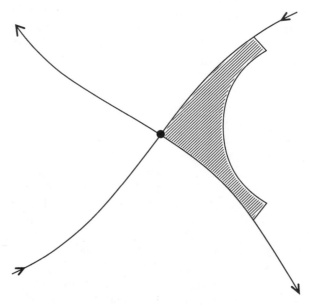

FIG. 6.5.2 An elbow neighborhood of a saddle.

necessarily have the same initial values as the approximate solutions z with which they are compared. Because of this, we do not begin with a specific exact solution (identified by its initial conditions) and then proclaim that a certain solution of the averaged equations approximates it. Instead, the order is reversed: A solution of the averaged equations is selected, and it is stated that there exists some exact solution that remains close to this solution (or *shadows* it) for as long as it remains in U. In other words, *the solutions of the averaged system are significant from an applied standpoint*, in that they accurately reflect possible behaviors of the exact system; it is just that the error is "spread out" over the entire time, instead of being zero at the beginning and gradually growing. The second unusual feature of this theorem is the nature of the times involved; they are unbounded, but are not infinite and are not "expanding" (in the usual sense). The fact is that the τ-time taken by a guiding solution to pass through U (and for which the theorem is valid) depends upon how close the solution passes to the saddle, and approaches infinity as one considers solutions closer to the saddle. If we were only concerned with a single solution in U, there would be no need to invoke a new theorem: The τ-time taken *by any single solution* in passing through U is finite, therefore the t-time is of order $1/\varepsilon$, and the standard averaging estimate suffices. (Even the correct initial conditions can be used.) It is the desire to have a single c and ε_0 valid *at once for all solutions in U* that calls for the shadowing approach. Everything becomes touchier as the solutions pass closer to the saddle; not only do the solutions take longer to pass, but the presence of error is more dangerous (it takes less error to find yourself on the wrong side of the stable manifold and be spun off in the wrong direction). Needless to say, the techniques needed to prove this theorem are quite different from anything considered in this book; they come from the modern "theory of dynamical systems" and have something in common with the now popular study of "chaotic motions," in which the ideas of hyperbolicity, exponential separation of nearby orbits, and shadowing all appear.

Exercises 6.5

1. Carry out the details of the proof of Theorem 6.5.1 in the case $k = 3$. In particular, find the differential equation for u_3 and the expression which must be averaged to find g_3.

2. Prove Theorem 6.5.2. Follow the pattern of Theorem 6.2.3.

3. Prove Theorem 6.5.3. Follow the pattern of Theorem 6.2.3.

4. Under the hypotheses of Theorem 6.5.3, prove the following result. Let T be any positive real number, and let $1 \leq j < r$. (Note the strict inequality.) Then there exist constants c and ε_2 such that $\|\tilde{x}(t, \alpha, \varepsilon) - x(t, \alpha, \varepsilon)\| \leq c\varepsilon^{k-j}$ for $0 \leq t \leq T/\varepsilon^{1+j}$, for $0 \leq \varepsilon \leq \varepsilon_2$, and for all α in K. The point of this result is that when trading off less than the maximum

possible amount of accuracy an arbitrary fixed T may be used in place of the function $T(\alpha)$. Hint: Because the point x moves at a speed of order ε^r, it cannot cross the gap between K and V in the time interval with which we are concerned, if ε is small enough.

6.6* ANGULAR STANDARD FORM, COUPLED OSCILLATORS, AND SMALL DIVISORS

Many systems involving periodic phenomena which cannot be put into periodic standard form can be put into one of the forms

$$\dot{\mathbf{r}} = \varepsilon \mathbf{f}(\mathbf{r}, \boldsymbol{\theta}, \varepsilon),$$
$$\dot{\boldsymbol{\theta}} = \boldsymbol{\omega} + \varepsilon \mathbf{g}(\mathbf{r}, \boldsymbol{\theta}, \varepsilon) \tag{6.6.1}$$

or

$$\dot{\mathbf{r}} = \varepsilon \mathbf{f}(\mathbf{r}, \boldsymbol{\theta}, \varepsilon),$$
$$\dot{\boldsymbol{\theta}} = \boldsymbol{\Theta}(\mathbf{r}) + \varepsilon \mathbf{g}(\mathbf{r}, \boldsymbol{\theta}, \varepsilon). \tag{6.6.2}$$

In both of these forms, called *angular standard forms*, r is an N-dimensional vector called the *vector of amplitudes* or simply *amplitude vector*, and $\boldsymbol{\theta}$ is an M-dimensional vector called the *vector of angles* or *angle vector*. As the name indicates, each component θ_j of $\boldsymbol{\theta}$ is an angle, and therefore the functions **f** and **g** must be 2π-periodic in each component (with all other variables held constant). The vector $\boldsymbol{\omega}$ in (6.6.1) or $\boldsymbol{\Theta}(\mathbf{r})$ in (6.6.2) is called the *vector of frequencies* or *frequency vector*, and the only difference between the first and second form is that in the first, the frequencies are constant while in the second they depend on the amplitudes.

Angular standard form arises naturally when considering systems of coupled oscillators; this has already been pointed out in Section 5.4, for instance in equations (5.4.34), which are the basis for the exercises at the end of this section. Another source of many equations in angular standard form is the "Hamilton–Jacobi theory" in classical mechanics. This theory produces what are called "integrable Hamiltonian systems in action/angle variables," which are equations having the form (6.6.2) with $N = M$ and $\varepsilon = 0$. Then any small perturbation of these "integrable" equations will have the form of (6.6.2). Finally, it is not difficult to change periodic standard form into angular standard form: It suffices to introduce a single angle θ and write (6.1.1) as

$$\dot{\mathbf{x}} = \varepsilon \mathbf{f}(\mathbf{x}, \theta, \varepsilon),$$
$$\dot{\theta} = 1. \tag{6.6.3}$$

This justifies considering angular standard form as a generalization of periodic standard form.

The first step in studying (6.6.1) is to understand the motion when $\varepsilon = 0$. In this case the amplitudes r_i, $i = 1, \ldots, N$ are constant, and the angles θ_j, $j = 1, \ldots, M$ rotate at constant angular velocities (or frequencies) ω_j. This is the zeroth approximation to the motion for small ε. Next, consider the functions $\mathbf{f}$ and $\mathbf{g}$. Since these are periodic in the angles, and since these angles are rotating rapidly (in comparison to any possible change in the amplitudes), it seems reasonable to average $\mathbf{f}$ and $\mathbf{g}$ over the angles in order to achieve simplified equations to give a first approximation (an improvement of the zeroth approximation) to the motion. This would entirely remove the dependence of the right hand side of (6.6.1) upon the angles. However, it was found in Section 6.2 that a better way to remove the dependence of a function on a fast variable (upon which it depends periodically) is to perform a near-identity transformation (or change of variables) which "pushes the dependence on the fast variables to a higher order." To this end, therefore, we attempt to find a change of variables

$$\mathbf{r} = \boldsymbol{\rho} + \varepsilon\mathbf{u}_1(\boldsymbol{\rho}, \boldsymbol{\psi}),$$
$$\theta = \boldsymbol{\psi} + \varepsilon\mathbf{v}_1(\boldsymbol{\rho}, \boldsymbol{\psi}) \tag{6.6.4}$$

which will carry the following expanded form of (6.6.1)

$$\dot{\mathbf{r}} = \varepsilon\mathbf{f}_1(\mathbf{r}, \boldsymbol{\theta}) + \mathcal{O}(\varepsilon^2),$$
$$\dot{\boldsymbol{\theta}} = \boldsymbol{\omega} + \varepsilon\mathbf{g}_1(\mathbf{r}, \boldsymbol{\theta}) + \mathcal{O}(\varepsilon^2) \tag{6.6.5}$$

into the form

$$\dot{\boldsymbol{\rho}} = \varepsilon\mathbf{F}_1(\boldsymbol{\rho}) + \mathcal{O}_F(\varepsilon^2),$$
$$\dot{\boldsymbol{\psi}} = \boldsymbol{\omega} + \varepsilon\mathbf{G}_1(\boldsymbol{\rho}) + \mathcal{O}_F(\varepsilon^2), \tag{6.6.6}$$

where the formal big-oh symbols represent functions depending on both $\boldsymbol{\rho}$ and $\boldsymbol{\psi}$. (In those cases when the transformation (6.6.4) turns out actually to exist, there is no difficulty in verifying that the formal order symbols are rigorously of the indicated order, uniformly for compact sets of amplitudes and for all values of the angles.)

Now a simple calculation (see Exercise 6.6.1) shows that in order to carry (6.6.5) into (6.6.6), (6.6.4) must satisfy

$$\mathbf{u}_{1\boldsymbol{\psi}}(\boldsymbol{\rho}, \boldsymbol{\psi})\boldsymbol{\omega} = \mathbf{f}_1(\boldsymbol{\rho}, \boldsymbol{\psi}) - \mathbf{F}_1(\boldsymbol{\rho}),$$
$$\mathbf{v}_{1\boldsymbol{\psi}}(\boldsymbol{\rho}, \boldsymbol{\psi})\boldsymbol{\omega} = \mathbf{g}_1(\boldsymbol{\rho}, \boldsymbol{\psi}) - \mathbf{G}_1(\boldsymbol{\rho}). \tag{6.6.7a}$$

In order to be certain that these equations are understood, we will write them out again in scalar notation, recalling that in (6.6.7a) a vector function with a vector subscript is a matrix of partial derivatives, and that $\boldsymbol{\omega}$ is a column vector. The scalar form, then, is

$$
\sum_{k=1}^{M} \omega_k \frac{\partial u_{1i}}{\partial \psi_k} = f_{1i} - F_{1i} \qquad \text{for} \quad i = 1, \ldots, N,
$$

$$
\sum_{k=1}^{M} \omega_k \frac{\partial v_{1j}}{\partial \psi_k} = g_{1j} - G_{1j} \qquad \text{for} \quad j = 1, \ldots, M.
$$

(6.6.7b)

These, then, are partial differential equations to be solved for $\mathbf{u}_1$ and $\mathbf{v}_1$ as functions of $\boldsymbol{\psi}$; the $\boldsymbol{\rho}$ may be considered as parameters.

In order to solve these partial differential equations, we will write the given functions $\mathbf{f}_1$ and $\mathbf{g}_1$ and the unknown functions $\mathbf{u}_1$ and $\mathbf{g}_1$ as multiple Fourier series in the angles. This is a simple step, but in order to make certain that the notation is understood, we will develop it in some detail. In full scalar notation, we have for the ith component of $\mathbf{f}_1$ the Fourier series

$$
f_{1i}(\rho_1, \ldots, \rho_N, \psi_1, \ldots, \psi_M)
$$

$$
= \sum_{\nu_1=-\infty}^{+\infty} \cdots \sum_{\nu_M=-\infty}^{+\infty} a_{1i\nu_1\ldots\nu_M}(\rho_1, \ldots, \rho_N) e^{\sqrt{-1}(\nu_1\psi_1\ldots\nu_M\psi_M)}.
$$

(6.6.8)

(We have used $\sqrt{-1}$ rather than i in order to avoid confusing this with the many indices occurring here. The coefficients a have the subscript 1 and i as in f_{1i}, the 1 indicating the first order term in (6.6.4), and the i indicating the component; the a also have the indices ν_1 through ν_M indicating the harmonics in the various angles.) This complicated notation can be greatly shortened. First, let $\boldsymbol{\nu} = (\nu_1, \ldots, \nu_M)$ be an M-dimensional integer vector (that is, a vector containing only integer entries), and let it be regarded as a row vector (while all other vectors are column vectors as usual). Then according to the rules of matrix multiplication

$$
\boldsymbol{\nu}\boldsymbol{\psi} = \nu_1\psi_1 + \cdots + \nu_M\psi_M.
$$

Summation over all integer vectors $\boldsymbol{\nu}$ can be indicated by a single summation sign. Then (6.6.8) takes the form

$$
f_{1i}(\boldsymbol{\rho}, \boldsymbol{\psi}) = \sum_{\nu} a_{1i\nu}(\boldsymbol{\rho}) e^{\sqrt{-1}\nu\psi}.
$$

Finally, the subscript i may be eliminated by introducing the vector $\mathbf{a} = (a_1, \ldots, a_N)$. This finally brings the notation down to a convenient form, and the Fourier series of the functions we require may be written as

$$\mathbf{f}_1(\boldsymbol{\rho}, \boldsymbol{\psi}) = \sum_{\nu} \mathbf{a}_{1\nu}(\boldsymbol{\rho}) e^{\sqrt{-1}\nu\boldsymbol{\psi}},$$

$$\mathbf{g}_1(\boldsymbol{\rho}, \boldsymbol{\psi}) = \sum_{\nu} \mathbf{b}_{1\nu} e^{\sqrt{-1}\nu\boldsymbol{\psi}},$$

$$\mathbf{u}_1(\boldsymbol{\rho}, \boldsymbol{\psi}) = \sum_{\nu} \mathbf{c}_{1\nu} e^{\sqrt{-1}\nu\boldsymbol{\psi}},$$

$$\mathbf{v}_1(\boldsymbol{\rho}, \boldsymbol{\psi}) = \sum_{\nu} \mathbf{d}_{1\nu} e^{\sqrt{-1}\nu\boldsymbol{\psi}}.$$

(6.6.9)

As a check on the understanding of the notation in (6.6.9), the reader should check (see Exercise 6.6.2) that the matrix of partial derivatives of $\mathbf{u}_1$ is given by

$$\mathbf{u}_{1\boldsymbol{\psi}} = \sum_{\nu} \sqrt{-1}\mathbf{c}_{1\nu}(\boldsymbol{\rho})\nu e^{\sqrt{-1}\nu\boldsymbol{\psi}}.$$

(6.6.10)

The product of the N-dimensional column vector $\mathbf{c}_{1\nu}$ and the M-dimensional row vector ν, taken in the indicated order, gives an $N \times M$ matrix, as expected for this matrix of partial derivatives; this result shows that derivatives of expressions in the form (6.6.9) may be taken according to the customary rules, provided that in "bringing down" the row vector coefficient ν from the exponential, it is placed *after* the column vector coefficient.

With this notational background, we are ready to return to the solution of the partial differential equations (6.6.7). At this point we will proceed heuristically; the steps which will be carried out next are formal and contain some "errors" which the reader might try to detect before they are pointed out (and corrected) later. Notice first that the constant term of (6.6.10), that is, the term with $\nu = 0$, vanishes. Therefore the first equation of (6.6.7) can hold only if $\mathbf{F}_1(\boldsymbol{\rho})$ equals (and hence cancels) the constant term of $\mathbf{f}_1(\boldsymbol{\rho}, \boldsymbol{\psi})$. Treating the second equation of (6.6.7) similarly, and using (6.6.9), we have

$$\mathbf{F}_1(\boldsymbol{\rho}) = \mathbf{a}_{10}(\boldsymbol{\rho}),$$

$$\mathbf{G}_1(\boldsymbol{\rho}) = \mathbf{b}_{10}.$$

(6.6.11)

According to the formula for Fourier coefficients, this says that $\mathbf{F}_1$ and $\mathbf{G}_1$ must be the averages of $\mathbf{f}_1$ and $\mathbf{g}_1$ over all of the angles simultaneously, just as expected. Now equations (6.6.7) may be written as

$$\sum_{\nu \neq 0} \sqrt{-1}\mathbf{c}_{1\nu}(\boldsymbol{\rho})\nu\boldsymbol{\omega} e^{\sqrt{-1}\nu\boldsymbol{\psi}} = \sum_{\nu \neq 0} \mathbf{a}_{1\nu}(\boldsymbol{\rho}) e^{\sqrt{-1}\nu\boldsymbol{\psi}},$$

$$\sum_{\nu \neq 0} \sqrt{-1}\mathbf{d}_{1\nu}(\boldsymbol{\rho})\nu\boldsymbol{\omega} e^{\sqrt{-1}\nu\boldsymbol{\psi}} = \sum_{\nu \neq 0} \mathbf{c}_{1\nu}(\boldsymbol{\rho}) e^{\sqrt{-1}\nu\boldsymbol{\psi}},$$

which immediately imply

$$\sqrt{-1}\mathbf{c}_{1\nu}(\boldsymbol{\rho})\nu\boldsymbol{\omega} = \mathbf{a}_{1\nu}(\boldsymbol{\rho}),$$
$$\sqrt{-1}\mathbf{d}_{1\nu}(\boldsymbol{\rho})\nu\boldsymbol{\omega} = \mathbf{b}_{1\nu}(\boldsymbol{\rho}).$$

On the left hand side of these equations the product of ν and $\boldsymbol{\omega}$, which is a scalar, may be taken before the product with $\mathbf{c}$ or $\mathbf{d}$. Therefore these equations may be readily solved by dividing by this scalar to obtain (for $\nu \neq \mathbf{0}$)

$$\mathbf{c}_{1\nu}(\boldsymbol{\rho}) = \frac{\mathbf{a}_{1\nu}(\boldsymbol{\rho})}{\sqrt{-1}\nu\boldsymbol{\omega}},$$
$$\mathbf{d}_{1\nu}(\boldsymbol{\rho}) = \frac{\mathbf{b}_{1\nu}(\boldsymbol{\rho})}{\sqrt{-1}\nu\boldsymbol{\omega}}.$$

$$(6.6.12)$$

Substituting these into (6.6.9) gives the following formulas for the desired transformations:

$$\mathbf{u}_1(\boldsymbol{\rho}, \boldsymbol{\psi}) = \sum_{\nu \neq 0} \frac{\mathbf{a}_{1\nu}(\boldsymbol{\rho})}{\sqrt{-1}\nu\boldsymbol{\omega}} e^{\sqrt{-1}\nu\boldsymbol{\psi}},$$
$$\mathbf{v}_1(\boldsymbol{\rho}, \boldsymbol{\psi}) = \sum_{\nu \neq 0} \frac{\mathbf{b}_{1\nu}(\boldsymbol{\rho})}{\sqrt{-1}\nu\boldsymbol{\omega}} e^{\sqrt{-1}\nu\boldsymbol{\psi}}.$$

$$(6.6.13)$$

In writing these equations we have adopted the convention that the average of $\mathbf{u}_1$ and $\mathbf{v}_1$ over the angles should be zero; actually any value may be used for the constant term (the term with $\nu = \mathbf{0}$), just as in Section 6.2. However there is nothing corresponding to "stroboscopic" averaging in the present context, so the only convention that suggests itself is the simplest, namely to omit the constant terms altogether.

As already suggested, two "errors" have been committed in the previous paragraph. One of them is quite blatant and should have been noticed by every reader; the second is more subtle. They are: (1) Some of the quantities $\nu\boldsymbol{\omega}$ in the denominators of (6.6.12) and (6.6.13) might be zero. In this case the series in question may not even be well-defined. (2) We have claimed that (6.6.13) defines the transformation functions, but have not checked that the series in these equations converge. The fact is that even when none of the denominators in (6.6.13) are zero, some of them may be sufficiently "small" that they create large coefficients in (6.6.13) and destroy the convergence of the series. The difficult question of "small denominators" will be discussed at some length below. But first let us consider the question of zero denominators.

Suppose first that $M = 2$ and that the constant vector of frequencies is $\boldsymbol{\omega} = (1, \sqrt{2})$. Then $\nu\boldsymbol{\omega} = \nu_1 + \nu_2\sqrt{2}$, which can equal zero only if $\sqrt{2} = -\nu_1/\nu_2$. Since $\sqrt{2}$ is irrational, this cannot happen, and none of the denominators in (6.6.13) are zero. Therefore this formal Fourier series at least *exists*; whether or not it *converges* is another question. But now suppose $\boldsymbol{\omega} = (1, 2/3)$. Then $\boldsymbol{\nu}\boldsymbol{\omega} = 0$ whenever $2\nu_1 = -3\nu_2$, which happens for infinitely many pairs of integers $\boldsymbol{\nu}$. The same thing happens any time that the ratio of the two frequencies ν_1/ν_2 is rational; this situation is called *internal resonance*, because it is a resonance occurring between two frequencies *within the system* rather than between a frequency of the system (a "free frequency") and an external frequency (a "forcing frquency"). Now there are two possibilities: On the one hand, *if for every ν such that $\boldsymbol{\nu}\boldsymbol{\omega} = 0$ it is the case that both* $\mathbf{a}_{1\nu} = \mathbf{0}$ *and* $\mathbf{b}_{1\nu} = \mathbf{0}$, then there is no difficulty. It is only necessary to run back over the argument leading to (6.6.13) and observe that if any of the $\mathbf{a}$ or $\mathbf{b}$ are zero then the corresponding $\mathbf{b}$ or $\mathbf{c}$ may be taken to be zero even if $\boldsymbol{\nu}\boldsymbol{\omega} = 0$. (Exercise 6.6.3.) Then (6.6.13) may be written down. Of course it still remains to investigate the convergence. On the other hand, *if there are one or more ν such that $\boldsymbol{\nu}\boldsymbol{\omega} = 0$ and also either* $\mathbf{a}_{1\nu} \neq \mathbf{0}$ *or* $\mathbf{b}_{1\nu} \neq \mathbf{0}$, then either the first or the second series in (6.6.13) cannot be written down, and in this case it is clearly impossible to carry out the near-identity transformation to averaged form.

These arguments clearly generalize to $M > 2$. The only difference is that it is no longer so easy to distinguish between the cases in which $\boldsymbol{\nu}\boldsymbol{\omega}$ can equal zero and those in which it cannot; it is no longer merely a question of whether the ratio of frequencies is rational or irrational. For instance, if $\boldsymbol{\omega} = (1, \sqrt{2}, \sqrt{3})$ then $\boldsymbol{\nu}\boldsymbol{\omega}$ can never equal zero, whereas if $\boldsymbol{\omega} = (\sqrt{2}, \sqrt{3}, \sqrt{2} - 2\sqrt{3})$, then there exist infinitely many ν for which $\boldsymbol{\nu}\boldsymbol{\omega} = 0$. The question is one of algebraic number theory. (For those with some exposure to the algebraic theory of fields: $\sqrt{2}$ and $\sqrt{3}$ are algebraically independent over the rationals, so the field formed by adjoining these to the rationals is a three-dimensional vector space over the rationals. In the first case the entries of $\boldsymbol{\omega}$ are linearly independent in this vector space, and in the second case they are not.) But aside from the difficulty of deciding between the cases, the same three cases exist:

1. *No internal resonances.* There are no integer row vectors ν such that $\boldsymbol{\nu}\boldsymbol{\omega} = 0$. In this case the series (6.6.13) can be written down, and if these series converge to smooth functions then the transformation to averaged form is possible. However, the convergence question has difficulties which will be discussed below.

2. *No activated internal resonances.* There are integer row vectors ν such that $\boldsymbol{\nu}\boldsymbol{\omega} = 0$, but for each such ν it is the case that both $\mathbf{a}_{1\nu} = \mathbf{0}$ and $\mathbf{b}_{1\nu} = \mathbf{0}$. The summation in (6.6.13) should be restricted to those ν for which $\boldsymbol{\nu}\boldsymbol{\omega} \neq 0$, and the convergence question again must be addressed.

3. *Activated internal resonances.* There are integer row vectors ν such that some coefficients in (6.6.13) actually have zero denominators

with nonzero numerators. *In this case it is not possible to carry out the transformation to the totally averaged form (6.6.6).* There are other "partially averaged" forms that can be achieved in this situation; they are not as simple as (6.6.6) but are still useful. We will investigate one such case, the *completely resonant case*, below.

In order to complete the discussion of cases 1 and 2, it remains to consider convergence of (6.6.13). In this book we have not usually been concerned with convergence, because our series are most often merely asymptotic. The present case is different; these series do not involve the small parameter ε, and are being used to define functions which must be proved to exist and be smooth. Therefore it is important to know whether the series converge. Since the functions $\mathbf{f}$ and $\mathbf{g}$ are smooth, their Fourier series (the first two series in (6.6.9)) converge uniformly because their coefficients decay sufficiently rapidly. (See Appendix E.) Now the coefficients of (6.6.13) are the same as these, except for the denominators $\sqrt{-1}\nu\boldsymbol{\omega}$. If these denominators are "large" (that is, if they are bounded away from zero) then the coefficients of (6.6.13) still decay rapidly enough and the series converge. But if there are "small" denominators (often called *small divisors*), dividing by them will magnify the coefficients, and if there are enough denominators far out in the series that are small enough, the convergence will be destroyed. In order to bring out the nature of this difficulty, consider the example, mentioned above, in which $M = 2$ and $\boldsymbol{\omega} = (1, \sqrt{2})$. We have seen that the denominator $\nu_1 + \nu_2\sqrt{2}$ vanishes only if $\sqrt{2} = -\nu_1/\nu_2$, which is impossible. But on the other hand, $\sqrt{2}$ can easily be approximated arbitrarily accurately by rational numbers such as $14/10$, $141/100$, $1414/1000$, and so forth. Therefore there exist infinitely many choices of ν_1 and ν_2 for which $\nu_1 + \nu_2\sqrt{2}$ is arbitrarily small, and the smaller values of this denominator tend to occur "further out" in the series where they can cause the most harm to the convergence.

The simplest solution to this small divisor difficulty, which is adequate for some but not all purposes, is to restrict consideration to functions $\mathbf{f}$ and $\mathbf{g}$ whose Fourier series are finite, that is, have finitely many nonzero coefficients. Since we have pointed out that whenever $\mathbf{a}_{1\nu}$ or $\mathbf{b}_\nu$ are zero it is possible to take $\mathbf{c}_{1\nu}$ or $\mathbf{d}_{1\nu}$ to be zero also, it then follows that the series (6.6.13) are finite, and there is no need to worry about convergence at all. Under this assumption, then, cases 1 and 2 are finished: The transformation to averaged form exists. Once this is known, the error estimation follows the same lines as in Section 6.2. We will not give details, but merely summarize with the following statement:

Theorem 6.6.1. *Suppose that system (6.6.1) is smooth, that the Fourier series of* $\mathbf{f}$ *and* $\mathbf{g}$ *are finite, and that there are either no internal resonances, or* *no activated internal resonances. Then the solutions of the averaged system*

$$\dot{\mathbf{R}} = \varepsilon \mathbf{F}_1(\mathbf{R}),$$
$$\dot{\boldsymbol{\theta}} = \boldsymbol{\omega} + \varepsilon \mathbf{G}_1(\mathbf{R}) \qquad (6.6.14)$$

approximate the solutions of (6.6.1) with the same initial conditions with accuracy $\mathcal{O}(\varepsilon)$ on expanding intervals of length $\mathcal{O}(1/\varepsilon)$.

The manner in which one handles the small divisor difficulty is closely tied up with one's philosophy of applied mathematics. Suppose that one is actually trying to model a physical system that can be expressed in the angular standard form (6.6.1). In achieving this form, it is necessary to make two crucial decisions that will affect the subsequent theory: A value must be chosen for $\boldsymbol{\omega}$, and functions must be chosen for $\mathbf{f}$ and $\mathbf{g}$. If the components of $\boldsymbol{\omega}$ are arrived at by measurement, then they will always be rational numbers, because measurement can only produce rational approximations to real numbers. If the components of $\boldsymbol{\omega}$ are rational, the system is completely resonant, and there will always be integer vectors $\boldsymbol{\nu}$ such that $\boldsymbol{\nu\omega} = 0$; the only question is whether the resonances are activated. Furthermore, if $\mathbf{f}$ and $\mathbf{g}$ are arrived at by measurement, this is often done by measuring the Fourier coefficients $\mathbf{a}_{1\nu}$ and $\mathbf{b}_{1\nu}$ in some indirect way. Since these coefficients decay, it is difficult to measure coefficients with large subscripts (that is, coefficients for which $\boldsymbol{\nu}$ contains large integer entries). Therefore in practice one is often forced to truncate the Fourier series and arrive at a model in which all Fourier series are finite. When working with such a model, no small divisor problems can arise. The question of whether any resonances are activated now becomes a question of how large the entries are in the vectors $\boldsymbol{\nu}$ for which $\boldsymbol{\nu\omega}$ are zero. If the only resonances holding among the $\boldsymbol{\omega}$ involve $\boldsymbol{\nu}$ so large that the corresponding Fourier coefficients cannot be measured and so are dropped, then according to the model, no resonances are activated and the averaging is possible. Therefore "low order" resonances—those for which there are small $\boldsymbol{\nu}$—are more important that "high order" resonances, which could only produce an effect by interacting with a high order harmonic in $\mathbf{f}$ or $\mathbf{g}$ that cannot be measured. Rather than a strict mathematical distinction between resonant and nonresonant cases, one has a practical question of "more-or-less resonant" situations. It may happen that a model fails to work because some high harmonic in $\mathbf{f}$ or $\mathbf{g}$ that is too small to measure becomes magnified by being divided by a very small divisor. (This divisor may actually be zero in the model but not in reality; remember that resonances are introduced when $\boldsymbol{\omega}$ is approximated by rational numbers.) An example of this is some early models of planetary motion which failed to take into account a resonance with Jupiter that has a substantial effect on the motion of the earth, an effect known as the "great inequality." It is possible to take the attitude that the small divisor question is entirely a matter of juggling models and trying to measure the quantities $\boldsymbol{\omega}$, $\mathbf{f}$, and $\mathbf{g}$ accurately enough to discover all the important resonances that affect the motion.

However, another attitude is possible toward exactly the same situation. One can reason as follows: Since high harmonics are presumably present even when they cannot be measured, and since the actual frequencies, like "most" real numbers, are presumably irrational, one should assume the worst, that is, that the Fourier series are infinite and there are small divisors causing possible divergences. According to this philosophy, it is important to develop theories covering the small denominator situation. Such theories exist, and they take different forms according to whether the system in question is "damped" (dissipative) or "conservative" (Hamiltonian). Under suitable damping conditions the theory is not terribly difficult, but in the conservative case, the theory is very technical and deserves to be considered one of the high points of twentieth-century mathematics. It is called the Kolmogorov–Arnol'd–Moser theory (frequently abbreviated to KAM) and is far too difficult to discuss in any detail here. Some references are given in Section 6.7.

Next we turn to the resonant case, or more precisely, the case in which there are activated internal resonances (case 3). Recall that in this case it is impossible to transform (6.6.1) into (6.6.6) because the series for the transformation has zero denominators. Therefore the solutions of (6.6.14) cannot be expected to give a good approximation to the solutions of (6.6.1). It turns out that although it is impossible to average over *all* the angles, it is still possible to average over *some* angles, which simplifies the system to a certain degree although not as much as complete averaging. The idea is to introduce a new modified set of angular variables, in terms of which some angles rotate rapidly and others slowly; then it is permissible to average over the rapidly rotating angles (called simply the *fast angles*). We will treat only one case in detail, the so-called completely resonant case. In the completely resonant case the components of $\boldsymbol{\omega}$ are integer multiples of a single real number λ, that is,

$$\omega_j = n_j\lambda. \tag{6.6.15}$$

This happens, for instance, if all the entries of $\boldsymbol{\omega}$ are rational; then λ can be taken as one over the common denominator of these entries. For simplicity we will assume that all of the frequencies ω_j are equal, so that each n_j in (6.6.15) equals 1; the general case of (6.6.15) goes very similarly. The case of equal frequencies is especially important because it arises when a collection of identical oscillators are coupled to one another; that is, when all the k_i are equal in (5.4.34). See Exercise 6.6.6 for an application of the following discussion to coupled oscillators.

When all of the internal free frequencies are equal, (6.6.1) written in components takes the form

$$\begin{aligned}
\dot{r}_i &= \varepsilon f_i(\mathbf{r}, \boldsymbol{\theta}, \varepsilon) \quad \text{for} \quad i = 1, \dots, N, \\
\dot{\theta}_j &= \lambda + \varepsilon g_j(\mathbf{r}, \boldsymbol{\theta}, \varepsilon) \quad \text{for} \quad i = 1, \dots, M.
\end{aligned} \tag{6.6.16}$$

Introduce $M - 1$ new angles ξ_j, defined by

$$\xi_j := \theta_j - \theta_M, \qquad j = 1, \dots, M - 1. \tag{6.6.17}$$

Substituting these angles in place of $\theta_1, \ldots, \theta_{M-1}$ in (6.6.16), keeping θ_M as it was, yields

$$\dot{r}_i = \varepsilon f_i(\mathbf{r}, \xi_1 + \theta_M, \ldots, \xi_{M-1} - \theta_{M-1}, \theta_M, \varepsilon) \quad \text{for } i = 1, \ldots, N,$$

$$\dot{\xi}_j = \varepsilon \left[g_j(\mathbf{r}, \xi_1 + \theta_M, \ldots, \xi_{M-1} - \theta_{M-1}, \theta_M, \varepsilon) \right.$$

$$\left. - g_M(\mathbf{r}, \xi_1 + \theta_M, \ldots, \xi_{M-1} - \theta_{M-1}, \theta_M, \varepsilon) \right] \quad \text{for } j = 1, \ldots, M-1, \tag{6.6.18}$$

$$\dot{\theta}_M = \lambda + g_M(\mathbf{r}, \xi_1 + \theta_M, \ldots, \xi_{M-1} - \theta_{M-1}, \theta_M, \varepsilon),$$

which can be rewritten using new functions as

$$\dot{r}_i = \varepsilon \tilde{f}_i(\mathbf{r}, \xi_1, \ldots, \xi_{M-1}, \theta_M),$$

$$\dot{\xi}_j = \varepsilon \tilde{g}_j(\mathbf{r}, \xi_1, \ldots, \xi_{M-1}, \theta_M), \tag{6.6.19}$$

$$\dot{\theta}_M = \lambda + \varepsilon \tilde{g}_M(\mathbf{r}, \xi_1, \ldots, \xi_{M-1}, \theta_M).$$

The distinguishing feature of (6.6.19) is that the equations for the ξ_j have factors of ε on the right hand side; that is, the ξ_j, like the r_i, are slowly varying quantities. The only rapidly varying angle remaining is θ_M. In fact (6.6.19) can be viewed as a new system in angular standard form, with the r_i and ξ_j together forming the amplitude vector of the new system, which is of dimension $N+M-1$, while θ_M is the only angle (or one-dimensional angle vector). Viewed in this way, system (6.6.19) is free of resonances and according to Theorem 6.6.1 may be averaged over the angle θ_M. That is, the solutions of (6.6.19) will be close, for an expanding interval of time, to those of the simplified system obtained by averaging the right hand side of (6.6.19) over θ_M.

But it is actually possible to say much more than this. If every equation in (6.6.19) except the last equation is divided by the last equation, the following system results:

$$\frac{dr_i}{d\theta_M} = \varepsilon \frac{\tilde{f}_i(\mathbf{r}, \xi_1, \ldots, \xi_{M-1}, \theta_M)}{\lambda + \varepsilon \tilde{g}_M(\mathbf{r}, \xi_1, \ldots, \xi_{M-1}, \theta_M)} \quad \text{for} \quad i = 1, \ldots, N,$$

$$\frac{\xi_j}{d\theta_M} = \varepsilon \frac{\tilde{g}_j(\mathbf{r}, \xi_1, \ldots, \xi_{M-1}, \theta_M)}{\lambda + \varepsilon \tilde{g}_M(\mathbf{r}, \xi_1, \ldots, \xi_{M-1}, \theta_M)} \quad \text{for} \quad j = 1, \ldots, M-1. \tag{6.6.20}$$

This is a system of $N+M-1$ dimensions in *periodic* standard form, with θ_M playing the role usually occupied by the time variable t. Therefore in the completely resonant case it is not only possible to obtain error estimates for the averaged equations (provided one averages only over the single fast angle), but it is also possible to recover all of the results obtained in Section 6.3 about existence and stability of periodic solutions. We have:

Theorem 6.6.2. *If the frequencies ω_j, $j = 1, \ldots, M$, in (6.6.1) are all equal to a single value λ, then the system may be put into the periodic standard form (6.6.20). For each simple zero of the right hand side of (6.6.20) averaged over θ_M, (6.6.20) has a family of periodic solutions for small ε. These solutions are of period 2π in θ_M, and when expressed in the original variables of (6.6.1) they are of period $2\pi/\lambda$ in time. Furthermore if the eigenvalues of the appropriate matrix lie in the left half-plane (see Exercise 6.6.4) then the solutions of (6.6.20) are asymptotically stable, and in the original variables they are asymptotically orbitally stable with asymptotic phase.*

(The concept of asymptotic orbital stability with asymptotic phase was introduced in Section 6.3 in connection with autonomous oscillators. The present discussion is a generalization of that one: Again the dimension drops by one, from $N + M$ to $N + M - 1$, and *solutions* of the lower-dimensional system correspond to *orbits* of the higher-dimensional one; solutions which attract in all directions correspond to orbits which only attract in the directions normal to the orbit.)

There is a geometrical interpretation of averaging in angular standard form which gives some additional insight to the preceding discussion. In the space of the variables $\mathbf{r}$ and $\boldsymbol{\theta}$, the set obtained by fixing $\mathbf{r}$ is an M-dimensional torus. Averaging the right hand side of (6.6.1) over this torus—that is, over all of the angles for fixed $\mathbf{r}$—is only reasonable if each unperturbed solution in some sense "almost" covers the entire torus. (The precise technical requirement is that the unperturbed solutions be "ergodic" in the torus.) This is in fact the case if there are no resonances among the free frequencies, that is, if there are no $\boldsymbol{\nu}$ for which $\boldsymbol{\nu}\boldsymbol{\omega} = 0$. In the extreme opposite case of complete resonance, the solutions of the unperturbed equations are periodic and hence cover only a closed curve embedded in the torus.

Thus it only makes sense to average over that closed curve and not over the entire torus. Since a point on a closed curve (a topological circle) is characterized by a single angle, that is exactly what is accomplished by the separation of fast and slow angles; averaging over the fast angle is averaging over the closed curve formed by the unperturbed solutions. In intermediate cases of partial resonance, the unperturbed orbits are ergodic in a torus of some intermediate dimension between 1 and M. In this case the angles can be separated by a change of variables into just the right number of fast and slow angles so that averaging over the fast angles corresponds to averaging over the closure of an unperturbed orbit.

So far, we have only considered the system (6.6.1), which has constant frequencies. What can be said about (6.6.2)? The fundamental difficulty here is that as $\mathbf{r}$ varies, the frequencies change, so that the system passes in and out of various states of resonance or nonresonance. Since the resonance or nonresonance of a system determines what type of averaging is appropriate for that system (whether averaging over all angles or only over one or more properly chosen fast angles), there is generally no single form of averaging that is suitable over the entire space. A great deal of work has been done on this problem, and it is much too extensive (and difficult) to report on in detail here. We will mention two approaches which are adequate for some problems.

The first approach is to restrict oneself to studying the motion near a particular point $\mathbf{r}^*$, using the form of averaging dictated by the frequencies at that point. Choose a fixed value $\mathbf{r}^*$ and introduce "dilated amplitudes" $\mathbf{R}$ by the equation

$$\mathbf{r} = \mathbf{r}^* + \sqrt{\varepsilon}\mathbf{R}. \qquad (6.6.21)$$

Substituting this into (6.2.1), setting $\mu = \sqrt{\varepsilon}$, and expanding in powers of μ, one finds

$$\dot{\mathbf{R}} = \mu \mathbf{f}(\mathbf{r}^*, \boldsymbol{\theta}, 0) + \mathcal{O}(\mu^2),$$
$$\dot{\boldsymbol{\theta}} = \boldsymbol{\Theta}(\mathbf{r}^*) + \mu \boldsymbol{\Theta}_\mathbf{r}(\mathbf{r}^*)\mathbf{R} + \mathcal{O}(\mu^2). \qquad (6.6.22)$$

This has the same form as (6.6.1), with $\boldsymbol{\omega} = \boldsymbol{\Theta}(\mathbf{r}^*)$, so it can be treated by the methods of this chapter (if the necessary hypotheses are satisfied). Any results obtained for a compact set $\|\mathbf{R}\| \leq c$ and an interval $0 \leq \mu \leq \mu_0$ hold for a corresponding neighborhood of $\mathbf{r}^*$ in accordance with (6.6.21).

The second approach is to average (6.6.2) over all of the angles making up the vector $\boldsymbol{\theta}$ to obtain an approximate system, regardless of the fact that resonances may occur, and then to attempt error estimates for the resulting system. In order to gain some insight into this, suppose that system (6.6.2) is first expanded in ε, as in (6.6.5) except that $\boldsymbol{\omega}$ is replaced by $\boldsymbol{\Theta}(\mathbf{r})$, and suppose that $\mathbf{f}_1$ and $\mathbf{g}_1$ are further expanded in Fourier series as in (6.6.9). Assume that these Fourier series are finite, so that there are only finitely many "active" ν for which $\mathbf{a}_{1\nu}$ or $\mathbf{b}_{1\nu}$ (or both) are nonzero. For each such active ν, the equation $\nu\boldsymbol{\Theta}(\mathbf{r}) = 0$ defines a resonance situation that is likely to disrupt the validity of the averaging process; therefore there are finitely many resonance possibilities that must be considered. Under "typical" circumstances (that is, unless the function $\boldsymbol{\Theta}$ satisfies restrictive special conditions), the equations $\nu\boldsymbol{\Theta}(\mathbf{r}) = 0$ define hypersurfaces in the space of variables $\mathbf{r}$, which are called *resonance surfaces*. It can be shown that the averaged equations (averaged over all θ) are valid (with the usual accuracy of $\mathcal{O}(\varepsilon)$ on expanding intervals of order $\mathcal{O}(1/\varepsilon)$) as long as the solutions do not cross these resonance surfaces. It can further be shown that if the solutions cross the resonance surfaces rapidly enough, they remain "in resonance" for a short enough time that the approximations are still of value, although the accuracy is diminished. If this accuracy is not sufficient it is sometimes possible to obtain more accurate solutions by using the "fully averaged" equations away from the resonance surfaces and the "local averages" obtained from (6.22) when crossing a resonance surface at a point $\mathbf{r}^*$. Finally, if there are infinitely many ν that are active, the resonance surfaces will crisscross the space densely and there will be no open regions in which the fully averaged equations are valid in the usual sense. But even in this case, they can still be of some value. It is possible to single out the most important resonances for treatment in the above manner and ignore the rest (with a further loss of accuracy). This is as much as we will say about these matters. (See Notes and References for an indication of some survey texts.)

Exercises 6.6

1. Derive equations (6.6.7). Hint: Compute $\dot{r}$ and $\dot{\theta}$ as functions of ρ and ψ in two ways, first by differentiating (6.6.4) and using (6.6.6), second by substituting (6.6.4) into the right hand side of (6.6.5) and expanding. If more help is needed, compare the derivation in Section 6.2.

2. Derive equations (6.6.10). Hint: Write out the Fourier series for $\mathbf{u}_1$ in scalar form and compute the partial derivatives of an arbitrary component of $\mathbf{u}_1$ with respect to an arbitrary component of ψ.

3. Repeat in detail the derivation of (6.6.13) under the assumption that some of the coefficients $\mathbf{a}_{1\nu}$ or $\mathbf{b}_{1\nu}$ are zero; show that in this case the corresponding coefficients $\mathbf{c}_{1\nu}$ or $\mathbf{d}_{1\nu}$ in (6.6.13) may be taken to be zero, instead of the indeterminate form $0/0$ which they appear to have.

4. Write out the components of the matrix which plays the crucial role in Theorem 6.6.2. (It must be nonsingular in order for the theorem to say that periodic solutions exist, and its eigenvalues must lie in the left half-plane for the theorem to say that these solutions will be stable.)

5. Consider the system of coupled Van der Pol equations $\ddot{y}_1 + \varepsilon(y_1^2 - 1)\dot{y}_1 + y_1 = \varepsilon y_2$, $\ddot{y}_2 + \varepsilon(y_2^2 - 1)\dot{y}_2 + 2y_2 = \varepsilon y_1$. Put this system in angular standard form and observe that it is nonresonant and that its Fourier series are finite, so that total averaging is permissible. Find the averaged system. Is this system explicitly solvable? Are there any solutions of the averaged system for which the amplitudes are constant? Are there any periodic or quasiperiodic solutions of the averaged system? (Solutions containing periodic functions of different periods are called quasiperiodic. Quasiperiodic solutions of the averaged system do not necessarily imply quasiperiodic solutions of the exact system; this question is much more difficult than that of periodic solutions. In general if the averaged equations have a quasiperiodic solution the most that can be said about the exact equations is that they have a solution that seems to be quasiperiodic for a long time.)

6. Suppose the second Van der Pol equation in the previous exercise is modified so that its free frequency is 1 (that is, the term $2y_2$ is replaced by y_2). Put the system in angular standard form and observe that it is completely resonant. Express the system in periodic standard form (with

one less dimension). Do there exist any periodic solutions of the exact system according to Theorem 6.6.2? What can you say about stability?

6.7. NOTES AND REFERENCES

The subject of averaging is vast, and it is possible to read four or five books entirely devoted to averaging and find very little overlap in the material which they cover. The fundamental approach taken here, based on the use of a near-identity transformation, was originated by Bogoliubov and Mitropolski (whose name appears with various spellings) and is presented in the following classic text which is still of value:

> N. N. Bogoliubov and Y. A. Mitropolskii, *Asymptotic Methods in the Theory of Nonlinear Oscillations*, Gordon and Breach, New York, 1961.

There exist many extensions of the theory of averaging to situations other than periodic standard form. For instance, there exists a first order (but not a higher order) averaging for equations in the form $x = \varepsilon f(x, t)$ where f is not necessarily periodic but has an average in the sense that the limit

$$\lim_{T \to \infty} \frac{1}{T} \int_0^T f(x, t) dt$$

exists. Stronger results hold if f is almost periodic. In order to handle a variety of cases, a number of different proofs of the validity of averaging have been developed. In this book we have presented two: that of Besjes in Section 5.5, and the proof by near-identity transformation in this chapter. Others appear in the references below. A particularly interesting recent proof for higher order averaging in periodic standard form, which combines features of the Besjes proof with the method of near-identity transformations, is given by

> J. A. Ellison, A. W. Saenz, and H. S. Dumas, Improved Nth order averaging theory for periodic systems, *J. Diff. Eq.* **84** (1990), 383–403.

This proof has the advantage (over that of Section 6.5) that it requires fewer hypotheses concerning smoothness, does not use the inverse of the near-identity transformation, and does not use the last term of the near-identity transformation even for the theoretical parts of the investigation. (Recall that our traditional approach uses the last term in the theory but not in the calculations.) These advantages make it possible that the validity of averaging might become provable for certain applications to partial differential equations, which until now have resisted theoretical treatment. (Averaging has been applied to some partial differential equations, but in the absence of a good theory there has been no advantage to this over a treatment by multiple scales.)

Three comprehensive surveys exist which, between them, cover most of the more recent work on averaging. Although all three of these references pursue topics from Section 6.6 far beyond what is said here, there is essentially no overlap in the actual material presented. They are:

J. A. Sanders and F. Verhulst, *Averaging Methods in Nonlinear Dynamical Systems*, Springer-Verlag, New York, 1985.

P. Lochak and C. Meunier, *Multiphase Averaging for Classical Systems*, Springer-Verlag, New York, 1988.

J. Murdock, Qualitative Theory of Nonlinear Resonance by Averaging and Dynamical Systems Methods, in *Dynamics Reported*, Vol. 1 (U. Kirchgraber and H. O. Walther, eds.), Wiley, New York, 1988.

The book by Sanders and Verhulst presents first order averaging for systems in periodic standard form by two methods, the one used here and another, due to Wiktor Eckhaus, that does not involve a near-identity transformation. (It is related to the Besjes proof but involves a new idea.) Then the second method is extended to certain systems that are not periodic. In Chapter 4 it is shown that first order averaging is valid for all time in suitably damped systems, a result which we stated without proof in Section 6.3. Chapter 5 deals with some issues addressed here in Section 6.6; in particular, it indicates a way of handling orbits that pass through resonance surfaces by combining the fully averaged solution away from the resonance surface with a local averaged solution at the resonance. (See the last paragraph of Section 6.6 above.)

The book by Lochak and Meunier deals primarily with recent Russian work pertaining to the material in Section 6.6. For instance it shows how to estimate the error in using the fully averaged equations in the presence of infinitely many resonance surfaces.

My own article (listed above) is chiefly concerned with extending the qualitative results in Section 6.3. It contains a complete proof of Theorem 6.3.1 and explains how to use higher order averaging to decide the stability of periodic solutions for which Theorem 6.3.1 fails. (This question was briefly mentioned in the first paragraph of Section 6.5 above.) It also treats systems of the form (6.6.2) in the way indicated by (6.6.21) and shows how to deduce the full orbit structure of the exact equations from the averaged equations under certain circumstances. The principal application is to problems of passage through resonance for which first order averaging is not sufficient to determine the topology of the orbits but second order averaging is. This article is a survey of my research on averaging up to the date of its publication. I have written two papers on averaging since that date. The first of these,

J. Murdock, On the length of validity of averaging and the uniform approximation of elbow orbits, with an application to delayed passage through resonance, *J. Appl. Math. Phys. (ZAMP)* **39** (1988), 586–596,

contains a proof of Theorem 6.5.4 and the details of the example for which validity cannot be extended beyond time $1/\varepsilon$. The other,

J. Murdock, A shadowing approach to passage through resonance, *Proceedings of the Royal Society of Edinburgh* **116A** (1990), 1–22,

contains a second (easier) proof of Theorem 6.5.4 and an application of these ideas to the passage through resonance problem; it also contains an exposition of the ideas of Sanders on passage through resonance mentioned above in connection with Chapter 5 of his book with Verhulst.

A number of special algebraic and analytic techniques have been developed to facilitate the computation of averaged systems under various circumstances. As seems to be customary in perturbation theory, each such computational device is often presented as if it were an entirely new method, rather than just a new way of carrying out averaging. Thus one hears of the method of Lie transforms, the Von Zeipel method, Struble's method, and so forth, all of which are essentially equivalent to averaging but differ in the "bookkeeping" details. (This is not at all meant to minimize the contributions of these methods; it is after all just these bookkeeping details that make averaging difficult in practice, and often a problem becomes solvable only because one of these methods makes the calculations feasible.) Most of these methods are expounded (from a formal point of view) in Chapter 5 of Nayfeh's *Perturbation Methods*, which also contains many references to specific applications.

Finally, some references should be given for the subject of small divisors. A good informal discussion is found in Appendix 8 of

V. I. Arnol'd, *Mathematical Methods of Classical Mechanics*, Springer-Verlag, New York, 1978.

A more detailed discussion with proofs (definitely not for the beginner) is given in Chapters 1, 2, and 5 of

J. Moser, Stable and Random Motions in Dynamical Systems, *Annals of Mathematics Studies No. 77*, Princeton University Press, Princeton, N.J., 1973.

Additional discussion is given in the book by Lochak and Meunier listed above.

PART III

TRANSITION LAYERS

CHAPTER 7

INITIAL LAYERS

7.1. INTRODUCTION TO INITIAL LAYERS

The remaining three chapters (Part III) of this book concern differential equations of the form

$$\varepsilon \ddot{y} + b(t)\dot{y} + c(t)y = 0$$

and other equations or systems of equations for which the order drops when $\varepsilon = 0$. This is a class of problems quite different from the oscillatory problems studied in Part II (Chapters 4, 5, and 6). In those chapters, the primary challenge was to find solutions valid for long intervals of time; regular perturbation theory gave solutions valid for finite intervals, but more subtle methods were able to retain validity for expanding intervals, or in some cases for all time. In the problems to be studied now, it is difficult even to find solutions that are valid for finite intervals, because the solutions undergo rapid transitions as they pass from one part of their domain to another. It is often fairly easy to find approximate solutions valid in one part of the domain or another; the problem is then to fit these solutions together to create an approximation valid over the entire domain. We will group these problems into three types:

1. *Initial layer problems*: initial value problems in which the solution makes a sudden, almost discontinuous jump immediately after being started at the initial point, and after this jump, continues as a smooth solution having a quite different character.
2. *Boundary layer problems*: boundary value problems in which the solution makes a sudden jump near one or more points at the boundary of its domain.

3. *Internal layer problems*: initial or boundary value problems in which the solution makes a sudden jump in the interior of its domain.

Together, these problems will be called *transition layer problems*.

There is one way in which all of these types of problems are very different from those studied until now: The exact solution usually does not exist when $\varepsilon = 0$. Until now, the *reduced problem* has always been a simpler, exactly solvable special case of the *perturbed problem*. In transition layer theory, the problem obtained by merely setting $\varepsilon = 0$ is generally self-contradictory, and therefore is not suitable as a starting point for developing a perturbation series. This difficulty is resolved by dropping some of the side conditions (that is, initial conditions or boundary conditions) when $\varepsilon = 0$ in order to define a suitable *reduced problem* from which to begin.

The concept of an initial layer will be introduced in this section by focusing on a very simple "model problem" that is, in fact, exactly solvable. In order to show how physical considerations are helpful in setting up singular perturbation problems correctly, this model problem will be discussed from the ground up, beginning with the physical derivation of the equations of motion. Then the problem will be treated by an elementary *patching method*, which is a good preparation for the more sophisticated *matching methods* and *correction methods* to be introduced in later sections. Finally the exact solution of the model problem will be found, and the patching approximation will be checked for accuracy against the exact solution. The reason for using a solvable example at first is the same as the reason for studying quadratic equations in Chapter 1: to extract every possible insight into a new topic from an example in which these insights can be tested without complicated error estimation procedures. Direct error estimation will not be carried out for the patching method, since it is intended only as a step towards the matching and correction methods of later sections. These methods will be introduced using the same model problem as in this section, but will then be extended to cover problems for which exact solutions in closed form do not exist.

Consider, then, a mass of M kg suspended from the ceiling by a spring. Until time $T = 0$ the mass is at rest; at $T = 0$ it is struck by a hammer, imparting to the mass an initial momentum (or impulse) of C kg m/sec. (Here capital letters stand for dimensional quantities, and T is measured in seconds.) Since momentum equals mass times velocity, the initial velocity is C/M. The reason for specifying the initial *momentum* rather than the initial *velocity* is that we wish to think of a specific mechanical device which strikes the mass, delivering a given impulse regardless of the amount of mass present. We are interested in the case in which M is very small; the smaller M is, the larger will be the initial velocity C/M. For the class of problems we are considering, C is regarded as a constant, whereas M is treated as a variable. M is almost the perturbation parameter in the problem, since it is small, but it will be seen that there is a better, nondimensional small parameter lurking behind the scenes.

The spring will be assumed to be linear, with a Hooke's law constant of K kg/sec^2. When a mass is hung from such a spring, gravity causes the spring to extend until the restoring force in the spring exactly balances the force exerted by gravity on the mass. This is the equilibrium position of the spring–mass system. If the spring is now stretched or compressed further by a distance Y from this equilibrium, then the total force acting on the mass will equal $-KY$. (This expression represents the net result of the gravitational and restoring forces.) Here Y is taken to be positive when the spring is extended and negative when it is compressed, that is, the Y axis points downwards; the force $-KY$ has the correct sign, since it is positive (i.e., directed downward) when Y is negative (i.e., when the spring is compressed). The law of motion for the spring–mass system as described so far is, by Newton's second law,

$$M\frac{d^2Y}{dT^2} = -KY. \tag{7.1.1}$$

One point to be noticed in this discussion is that if M is changed, the position of the equilibrium changes. Therefore Y cannot be regarded as measured from a fixed point in space, but rather from a point which depends upon M.

One more term must be added to (7.1) to obtain the equation which we will actually study. This is a friction or dissipation term. Actual frictional forces are not as simple to describe as gravitational and Hooke's law forces. Energy dissipation is due to a variety of effects such as conversion of kinetic energy to heat in the spring itself, and air resistance. Our assumption of a frictional force proportional to the velocity is dictated by mathematical simplicity more than physical reality; but then, the same can be said of Hooke's law itself. (Any actual spring will have deviations from linearity, as discussed in connection with Example 2.2.1.) In any case, the final complete description of the system we will study is given by the initial value problem

$$M\frac{d^2Y}{dT^2} + H\frac{dY}{dT} + KY = 0,$$
$$Y(0) = 0, \tag{7.1.2}$$
$$\frac{dY}{dT}(0) = \frac{C}{M}.$$

Our goal is to study (7.1.2) for small values of M, with C, H, and K fixed. If M were taken as a perturbation parameter, then a solution could be sought in the form of an asymptotic expansion such as

$$Y \sim Y_0 + Y_1 M + Y_2 M^2 + \cdots, \tag{7.1.3}$$

where the coefficients Y_n are functions of T, H, K, and C if an ordinary asymptotic expansion is sufficient, and also of M if a generalized expansion is necessary. (The latter turns out to be the case.) There is, of course, a disadvantage to this approach: If one were to change units in (7.1.3), each coefficient would have to be changed according to a different dimensional formula. Since Y has units of length (m), Y_0 has units of length, whereas Y_1 has units of length/mass (m/kg), and Y_2 is in units of length per mass squared (m/kg^2). It is awkward to work with such a series. If the perturbation parameter is taken to be nondimensional, then each coefficient in the asymptotic series will have the same dimensions, which is much more natural.

But there are much stronger reasons for nondimensionalizing the variables in (7.1.2). As pointed out in Chapter 2, introduction of nondimensional variables often forces one to recognize significant combinations of variables that would not otherwise stand out as important. In the present example, the effect of nondimensionalizing is almost miraculous, because it leads to the recognition that there are two distinct and quite different perturbation problems associated with (7.1.2), each of which turns out to be suitable over a different portion of the domain (that is, a different interval of time). These two problems are related to each other by a simple rescaling which can be discovered entirely on mathematical grounds, but it is remarkable that one is led directly to both of the significant formulations of the problem merely by dimensional analysis.

A dimensional analysis of our system begins with a table of the quantities appearing in (7.1.2) together with their dimensions (which we write in the metric system):

variable	units
Y	m
T	sec
M	kg
H	kg/sec
K	kg/sec^2
C	kg m/sec

The first two entries in this table are coordinates, and the last four are natural parameters. The next step is to form combinations of the natural parameters having the same dimensions as the coordinates. There is one simple combination of the parameters which has the dimensions of length, and two which have the dimensions of time:

variable	units
C/H	m
H/K	sec
M/H	sec

Thus the quantity C/H can be taken as a "characteristic length" for the system, and there are two "characteristic times," H/K and M/H. (These are not unique, but are the simplest choices, and all others can be expressed in terms of these. For example, H^3/K^2M has the dimensions of time and could be taken as a characteristic time, but it is more complicated than H/K and M/H and can be written as $(H/K)^2/(M/H)$.) Nondimensional lengths and times are defined as the ratios of the coordinates Y and T to the characteristic quantities with the same dimensions. Therefore we define

$$y := HY/C,$$
$$t := KT/H, \qquad (7.1.4)$$
$$\tau := HT/M.$$

It has already been pointed out that the quantity M, which at first seems to be the perturbation parameter for our problem, should be nondimensionalized, and therefore we seek a combination of the remaining three natural parameters having the dimensions of mass. There is one such quantity, namely H^2/K. Therefore we introduce the nondimensional perturbation parameter

$$\varepsilon := KM/H^2. \qquad (7.1.5)$$

It is not possible to form any further nondimensional combinations of the six quantities appearing in (7.1.2), other than functions of the four defined in (7.1.4) and (7.1.5), so these four should be sufficient to express (7.1.2) in nondimensional form. In fact there are two ways to do this, depending upon whether t or τ is used as the time variable. From the chain rule one can calculate from (7.1.4) that $dY/dT = (CK/H^2)(dy/dt)$ and $d^2Y/dT^2 = (CK^2/H^3)(d^2y/dt^2)$. Substituting these into (7.1.2) and using (7.1.5) leads to

$$\varepsilon \frac{d^2y}{dt^2} + \frac{dy}{dt} + y = 0,$$
$$y(0) = 0, \qquad (7.1.6)$$
$$\frac{dy}{dt} = \frac{1}{\varepsilon}.$$

Similar steps using τ in place of t yield

$$\frac{d^2y}{d\tau^2} + \frac{dy}{d\tau} + \varepsilon y = 0,$$
$$y(0) = 0, \qquad (7.1.7)$$
$$\frac{dy}{d\tau}(0) = 1.$$

As a check on the accuracy of these calculations, the reader should verify that (7.1.4) and (7.1.5) imply

$$\tau = t/\varepsilon, \qquad (7.1.8)$$

and that the change of independent variables given by (7.1.8) carries (7.1.6) into (7.1.7). Both of the systems (7.1.6) and (7.1.7) will play an important role in what follows, and it is important to realize that if (7.1.6) alone is given (for instance, by an incomplete dimensional analysis), the alternate form (7.1.7) can be obtained directly by the rescaling of time shown in (7.1.8). This rescaling can be found mathematically by setting $\tau = \varepsilon^{\nu} t$ and choosing ν so that the small parameter no longer multiplies the highest derivative. (The techniques of rescaling will be developed further in later sections.) Notice that (7.1.8) is equivalent to $t = \varepsilon\tau$, the reverse of the equation $\tau = \varepsilon t$ used in Chapter 5. These notations appear to be firmly entrenched in their respective applications, and it is hopeless to try to adopt a uniform usage.

It is worth pausing to consider all that has been accomplished by the process of nondimensionalizing in this problem. First, the number of parameters has been reduced from four (M, H, K, and C) to one (ε); there are not even any control parameters in the equations (7.1.6) and (7.1.7), only the perturbation parameter. (In most applications, some nondimensional control parameters will remain, but they will generally be fewer in number than the original parameters.) The significant small quantity has been found to be not M as originally postulated, but $\varepsilon = KM/H^2$. In other words, it does not matter whether M is small, K is small, or H is large; any of these situations lead to ε being small, and therefore the same perturbation problem applies to all of them. Finally, nondimensionalizing caused us to realize that there are two natural time scales in the problem, and led in a natural way to the two formulations using t-time and τ-time. As to the meaning of these two time scales, notice from (7.1.4) that if M is small (while K and H are not) then τ is large compared to "real time" T, whereas t is of the same order of magnitude as T. (More precisely: If M is allowed to approach zero while H and K are held constant, then a fixed interval $0 \leq \tau \leq \tau_0$ corresponds to a shrinking interval in terms of T, while a fixed interval of t-time corresponds to a fixed interval of T.) Thus when M is small, t is a "normal" time scale, and τ is a "fast" scale. This will be our primary understanding of these variables. (However, it has been pointed out that the perturbation problem is the same if K is small instead of M. In that case τ is a "normal" time and t is "slow." Finally, if H is large, t is slower than normal and τ is faster than normal. In all interpretations, τ is faster than t.)

Equation (7.1.6) is rather strange: Setting $\varepsilon = 0$ causes the second order differential equation to drop to first order, so that its solutions cannot satisfy two initial conditions; at the same time, the second initial condition becomes physically meaningless (although graphically an infinite derivative could mean a vertical tangent). We have already pointed out that the reduced problem in such cases must be defined carefully and is not obtained merely by setting $\varepsilon = 0$. But we need not face this issue immediately, since equation (7.1.7) is of the form studied in Chapter 2, and we can begin by finding out what regular perturbation theory applied to (7.1.7) is able to reveal.

The two-term perturbation series solution of (7.1.7) will be denoted

$$y^i(\tau, \varepsilon) = y_0^i(\tau) + \varepsilon y_1^i(\tau). \tag{7.1.9}$$

The superscript i in (7.1.9) is not a numerical index, but stands for the word "inner"; for reasons that will appear shortly, (7.1.7) is called the *inner equation*, and its perturbation series solution is called the *inner solution*. We will always use y^i to denote *the approximate solution of the inner equation taken to some fixed number of terms*, and not an infinite series. In the present instance we are taking two terms, and it follows from regular perturbation theory that for any fixed τ_0 the exact solution y satisfies

$$y(\tau, \varepsilon) = y^i(\tau, \varepsilon) + \mathcal{O}(\varepsilon^2) \quad \text{uniformly for} \quad 0 \leq \tau \leq \tau_0. \tag{7.1.10}$$

To compute y^i, substitute (7.1.9) into (7.1.7), ignoring ε^2, and use the methods of Chapter 2 to obtain the sequence of problems

$$\frac{d^2 y_0^i}{d\tau^2} + \frac{dy_0^i}{d\tau} = 0,$$
$$y_0^i(0) = 0, \tag{7.1.11a}$$
$$\frac{dy_0^i}{d\tau}(0) = 1$$

and

$$\frac{d^2 y_1^i}{d\tau^2} + \frac{dy_1^i}{d\tau} = -y_0^i,$$
$$y_1^i(0) = 0, \tag{7.1.11b}$$
$$\frac{dy_1^i}{d\tau}(0) = 0.$$

The solution of (7.1.11a) is $y_0^i = 1 - e^{-\tau}$; putting this into (7.1.11b) gives an inhomogeneous problem with solution $y_1^i = 2 - \tau - (2 + \tau)e^{-\tau}$. Thus the final inner solution is

$$y^i(\tau, \varepsilon) = (1 - e^{-\tau}) + \varepsilon\{(2 - \tau) - (2 + \tau)e^{-\tau}\}. \tag{7.1.12}$$

It may appear at this point that when y^i has been found, we are finished solving the problem. But the shortcoming of the regular perturbation method lies in the interval of validity of (7.1.10). When the interval $0 \leq \tau \leq \tau_0$ is expressed in the time variable t (which, we recall, is a measure of "real" time under the primary interpretation of our problem, namely, when M is small) the result is $0 \le t \le \varepsilon\tau_0$, which is a very short interval (for small ε) lying "close *in* toward the initial condition." This explains why (7.1.10) is called an *inner solution*. In order to find a solution valid for a finite interval of t-time such as $0 \le t \le 1$, or even perhaps on an infinite interval $0 \le t < \infty$, it is necessary somehow to find an "outer solution," that is, one that is valid "*out* away from the initial condition," and then to combine the inner and outer solutions in some way. All of the methods which have been proposed for doing this are based on the expectation (which is merely heuristic at this point) that the outer solution should come from version (7.1.6) of the differential equation, since that version is expressed in the t-variable.

The difficulty of working with (7.1.6) has already been noted: The problem drops order when $\varepsilon = 0$, so it cannot satisfy two initial conditions, and furthermore one of the initial conditions becomes (almost) meaningless. The answer to these problems is now at least partly clear. Since (7.1.6) is going to be used to find an outer solution, valid away from the initial time, the solution should not be expected to satisfy any particular initial conditions at all at $t = 0$ (not even the first one, which remains well-defined at $\varepsilon = 0$). What is needed, instead, is to find the *general* solution of the *outer equation*

$$\varepsilon\frac{d^2y}{dt^2} + \frac{dy}{dt} + y = 0, \tag{7.1.13}$$

and then, by examining this general solution, somehow select the *particular* outer solution which fits best with the inner solution, and finally combine the two into a solution that works both near and away from the initial time. It is not at all obvious at this point how to do this; there are several approaches, which will be introduced in this and the next two sections, and each approach is based on a different idea. But they all begin by naively substituting $y \sim y_0 + \varepsilon y_1 + \cdots$ into the outer equation (7.1.13) without initial conditions and computing the general solution with arbitrary constants. As with y^i, we will adopt the convention that y^o (o for outer) stands for a *fixed number of terms* in the perturbation series for the outer equation; in the present case, we again use two terms:

$$y^o(t, \varepsilon) = y_0^o(t) + \varepsilon y_1^o(t). \tag{7.1.14}$$

Substituting a perturbation series into (7.1.13) should be done carefully, because the result (perhaps surprising at first) is a sequence of *first order* differential equations for the coefficients. One finds

$$\varepsilon\left\{\frac{d^2y_0^o}{dt^2} + \varepsilon\frac{d^2y_1^o}{dt^2} + \cdots\right\} + \left\{\frac{dy_0^o}{dt} + \varepsilon\frac{dy_1^o}{dt} + \cdots\right\} + \{y_0^o + \varepsilon y_1^o + \cdots\} = 0, \tag{7.1.15}$$

and after separating terms of order ε^0 and ε^1,

$$\frac{dy_0^o}{dt} + y_0^o = 0 \qquad (7.1.16a)$$

and

$$\frac{dy_1^o}{dt} + y_1^o = -\frac{d^2y_0^o}{dt^2}. \qquad (7.1.16b)$$

The general solution of (7.1.16a) is $y_0^o = Ae^{-t}$, and after substituting this into (7.1.16b), its general solution is $y_1^o = (-At + B)e^{-t}$. Thus the general outer solution is

$$y^o(t, \varepsilon) = Ae^{-t} + \varepsilon(-At + B)e^{-t}. \qquad (7.1.17)$$

From this point there are several ways to proceed. In this section, we will develop a *patching method*, which is seldom used but is perhaps the easiest method to understand. In Section 7.2, the method of *matching by Van Dyke's matching rules* will be presented. This method is the most popular and is computationally the simplest, but is known to give incorrect answers to some problems (which are more difficult than any problems treated in this book). In Section 7.3, two alternate methods will be sketched; they are more powerful than Van Dyke's rules and give correct results in at least some of the cases where Van Dyke's rules do not. Finally, in Section 7.4, a multiple scale method will be given, which is entirely equivalent to Van Dyke's rules although the computations proceed differently. In Section 7.5, this method will be generalized to certain nonlinear problems.

The idea of the patching method is to create a heuristic approximate solution by using the inner solution (7.1.12) for t in some interval $0 \leq t \leq \eta(\varepsilon)$ and then simply switching to an outer solution (7.1.17) for $t \geq \eta(\varepsilon)$; the values of A and B are chosen so that the inner and outer solutions coincide as nearly as possible at the chosen switching point $\eta(\varepsilon)$. (That is, we want *constant* values for A and B which make the difference between y^i and y^o as small as possible in the asymptotic sense as $\varepsilon \to 0$.) The fundamental problem is to determine a suitable place to make the switch from one solution to the other. The interval $0 \leq t \leq \eta(\varepsilon)$ is called the *inner layer*, and $t \geq \eta(\varepsilon)$ is the *outer layer*.

In order to work with both the inner and outer solutions at the same time, it is necessary to write both in terms of the same variable. Setting $\tau = t/\varepsilon$ in (7.1.12) gives the *inner solution in the outer variable*,

$$y^i = (1 - e^{-t/\varepsilon}) + \varepsilon\left\{2 - \frac{t}{\varepsilon} - \left(2 + \frac{t}{\varepsilon}\right)e^{-t/\varepsilon}\right\}$$

$$= \left\{1 - t - (1 + t)e^{-t/\varepsilon}\right\} + \varepsilon\{2 - 2e^{-t/\varepsilon}\}. \qquad (7.1.18)$$

The second form here is obtained by moving two terms into the zero order that belonged in the first order when the τ variable was used. Setting the inner and outer solutions (7.1.18) and (7.1.17) equal at an (as yet unspecified) switching point $t = \eta(\varepsilon)$ yields

$$\{1 - \eta - (1+\eta)e^{-\eta/\varepsilon}\} + \varepsilon\{2 - 2e^{-\eta/\varepsilon}\} = Ae^{-\eta} + \varepsilon(-A\eta + B)e^{-\eta}. \quad (7.1.19)$$

It is not possible to satisfy this equation exactly, but we will choose A and B so that it is satisfied to as high an order in ε as possible. If η were known, one would expand both sides of (7.1.19) as asymptotic expansions in ε and chose A and B to make the lowest order terms equal. Since η itself is one of the unknowns, it will be necessary to discover the conditions that should be imposed on η as we go along. The following discussion is based on the idea of *limit process expansions* developed in Section 1.8; see (1.8.17). First, the switching point η is intended to be a point where *both* the inner and outer solutions give a good approximation to the solution. It is known that the inner solution is good out to $\mathcal{O}(\varepsilon)$, and it is reasonable to expect that it should remain valid (with some loss of accuracy) a little farther out, since it should not lose all of its accuracy suddenly. So it is plausible to require that $\eta \gg \varepsilon$; that is, to take the inner layer to be asymptotically wider than ε. This guarantees that the terms involving $e^{-\eta/\varepsilon}$ are transcendentally small, and also that terms in η will dominate terms in ε. It follows that an asymptotic expansion of both sides of (7.1.19) must use the gauge functions $\delta_0(\varepsilon) = 1$ and $\delta_1(\varepsilon) = \eta(\varepsilon)$. From letting $\varepsilon \to 0$ in (7.1.19), it follows that $A = 1$. Now write $e^{-\eta} = 1 - \eta + \eta^2/2 + \cdots$, subtract 1 from both sides of (7.1.19), divide by the next gauge $\delta_1 = \eta$, and again let $\varepsilon \to 0$. The result is $-1 = -1$, which is automatically true; this stage in the asymptotic expansion does not suffice to determine B. After removing the terms that are already found from both sides of (7.1.19), we are left with $2\varepsilon + \cdots = \eta^2/2 + B\varepsilon + \cdots$, where the dots denote terms that are clearly of higher order than either ε or η^2. It is necessary now to decide the relative sizes of η^2 and ε. If η^2 is dominant we get $0 = 1/2$, meaning that the equation cannot be satisfied; if ε is dominant, we get $B = 2$, which is satisfactory. Therefore we impose the condition $\eta^2 \ll \varepsilon$, that is, $\eta \ll \varepsilon^{1/2}$, and conclude $B = 2$. (More precisely, under this hypothesis the next gauge must be $\delta_2 = \varepsilon$; dividing by ε and letting $\varepsilon \to 0$ implies $B = 2$.) The result of our analysis is

$$\varepsilon \ll \eta(\varepsilon) \ll \varepsilon^{1/2},$$
$$A = 1, \quad\quad\quad\quad\quad (7.1.20)$$
$$B = 2,$$

and the patched solution is

$$y^p(t,\varepsilon) = \begin{cases} 1 - t - (1+t)e^{-t/\varepsilon} + 2\varepsilon(1 - e^{-t/\varepsilon}) & \text{for} \quad 0 \leq t \leq \eta(\varepsilon) \\ e^{-t} + \varepsilon(-t+2)e^{-t} & \text{for} \quad t > \eta(\varepsilon). \end{cases}$$
$$(7.1.21)$$

A graph of this solution is shown (for a fixed ε) in Fig. 7.1.1. The solution is discontinuous, having a slight jump at η because it was impossible to satisfy (7.1.19) exactly but only up to order ε (the third of its natural gauges). Also, there is still some ambiguity in (7.1.21) because η has not been exactly determined but only narrowed down to being between the asymptotic orders of ε and $\varepsilon^{1/2}$. The simplest choice of η satisfying these restrictions is $\eta = \varepsilon^{2/3}$, since $2/3$ is the fraction having the smallest numerator and denominator among all fractions lying between $1/2$ and 1. (To make sure that you understand where the patching point can be placed, see Exercise 7.1.1. Actually, the fact that η is not fixed exactly but can be taken anywhere in an interval is quite important, because it is related to what is called the *overlap condition*. This means that the region in which the inner solution is a good approximation overlaps the region in which the outer solution is good, so that the switching point between the two solutions can be taken anywhere in this overlap region; but this will be developed more precisely later, in Section 7.3.)

Having obtained the patched solution (7.1.21), the next question is: How good is it? To answer this question (for this particular example) we will set aside perturbation methods and admit to ourselves that the initial value problem (7.1.6) is in fact exactly solvable by any undergraduate

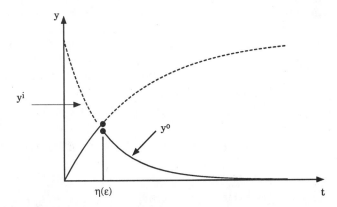

FIG. 7.1.1. An inner solution patched at $\eta(\varepsilon)$ to an outer solution. The dotted lines show the continuation of the inner and outer solutions beyond the domains in which they are useful. The point η moves toward zero as ε approaches zero, causing the initial layer (where the inner solution is valid) to shrink. The inner and outer solutions do not join together smoothly in the patching method, as they do in the matching methods of later sections.

student of differential equations, since it is a second order linear differential equation with constant coefficients. The exact solution is (see Exercise 7.1.2)

$$y = \frac{1}{\sqrt{1 - 4\varepsilon}} \left\{ e^{\lambda_1(\varepsilon)t} - e^{\lambda_2(\varepsilon)t} \right\},$$

$$\lambda_1 = \frac{-1 + \sqrt{1 - 4\varepsilon}}{2\varepsilon}, \tag{7.1.22}$$

$$\lambda_2 = \frac{-1 - \sqrt{1 - 4\varepsilon}}{2\varepsilon}.$$

Although this solution is exact, it is not simple; one cannot glance at this equation and "see" that it is going to exhibit an "initial layer effect." One might almost *prefer* the approximate solution (7.1.21), which is made up of recognizable terms showing the effect of two different time scales. Therefore we will begin from the exact solution and show how to derive from it the approximate solution (7.1.21), together with error estimates. The procedure is to expand the exact solution carefully and discard terms, estimating the magnitude of each discarded term. Since the terms have different magnitudes on different subdomains, we will be forced to divide the interval at a suitable point $\eta(\varepsilon)$ and arrive at two distinct approximations. We will carry this out only to leading order, and so will not obtain all of (7.1.21) but only its leading order terms; however, one could find the next order in the same way. (This is, of course, not a practical method for other problems, since it depends on the availability of an exact solution. However, it does cast a good deal of light on the meaning of an initial layer.)

To begin an analysis of the exact solution (7.1.22), we will assume that $0 < \varepsilon < 1/4$. Then λ_1 and λ_2 are negative real numbers, the coefficient $1/\sqrt{1 - 4\varepsilon}$ in y is equal to $1 + \mathcal{O}(\varepsilon)$, and the quantity in braces is bounded on $0 \le t < \infty$. The product of a bounded function of t with a term of order $\mathcal{O}(\varepsilon)$ is of order $\mathcal{O}(\varepsilon)$ uniformly in t. Therefore (7.1.22) implies

$$y = e^{\lambda_1(\varepsilon)t} - e^{\lambda_2(\varepsilon)t} + \mathcal{O}(\varepsilon) \quad \text{uniformly for} \quad 0 \le t < \infty. \tag{7.1.23}$$

Next, expand $\lambda_1 = -1 + \mathcal{O}(\varepsilon)$ and $\lambda_2 = -1/\varepsilon + 1 + \mathcal{O}(\varepsilon)$ and then write (7.1.23) as

$$y = e^{-t} \cdot e^{\mathcal{O}(\varepsilon) \cdot t} - e^{\left(-\frac{1}{\varepsilon} + 1\right)t} \cdot e^{\mathcal{O}(\varepsilon) \cdot t} + \mathcal{O}(\varepsilon) \quad \text{uniformly for} \quad 0 \le t < \infty.$$

Now if we restrict ourselves to a finite interval $0 \le t \le T$, then the expressions $e^{\mathcal{O}(\varepsilon) \cdot t}$ are uniformly $1 + \mathcal{O}(\varepsilon)$, so that

$$y = e^{-t} - e^{(1 - 1/\varepsilon)t} + \mathcal{O}(\varepsilon) \quad \text{uniformly for} \quad 0 \le t \le T. \tag{7.1.24}$$

The second term of this expression equals 1 at $t = 0$ and drops off very rapidly to zero after that, since $1 - 1/\varepsilon$ is a large negative number. Therefore this term represents a rapid transition occurring during a short period after the initial condition: an initial layer effect. It is possible to simplify this term without increasing the order of the error; it may be written as $e^t \cdot e^{-t/\varepsilon}$, and during the time that this term is important, t is sufficiently close to zero that e^t is close to 1 and may be dropped. For the necessary estimate, see Exercise 7.1.3. The result is

$$y = e^{-t} - e^{-t/\varepsilon} + \mathcal{O}(\varepsilon) \quad \text{uniformly for} \quad 0 \le t \le T. \tag{7.1.25}$$

The approximate solution

$$y^c = e^{-t} - e^{-t/\varepsilon} \tag{7.1.26}$$

is called the leading order *composite solution*; it will appear again in Section 7.2, where it will be obtained by another method. We have found it here by simplifying the exact solution, and we will continue simplifying and discarding terms from (7.1.26) until we have recovered the leading order inner, outer, and patched solutions as well.

Applying Taylor's theorem to the first term in (7.1.25), one sees that there exists a constant c such that $|e^{-t} - 1| < ct$ for $|t| \le T$. It follows that on $0 \le t \le \varepsilon$, $|e^{-t} - 1| < c\varepsilon$, while on $0 \le t \le \varepsilon^{1/2}$ this quantity is $< c\varepsilon^{1/2}$. This reasoning is perhaps more suggestive if written in the following form: Since $e^{-t} = 1 + \mathcal{O}(t)$ we have $e^{-t} = 1 + \mathcal{O}(\varepsilon)$ if t ranges from 0 up to ε, but only $1 + \mathcal{O}(\varepsilon^{1/2})$ if ε is allowed to run up to $\varepsilon^{1/2}$. (The use of the big-oh symbol here is somewhat careless, since sometimes t and sometimes ε is the small quantity; it is easy to make errors in uniformity this way. That is why the precise version with absolute values and constants was given first.) Thus (7.1.25) implies the following two estimates:

$$y = 1 - e^{-t/\varepsilon} + \mathcal{O}(\varepsilon) \quad \text{uniformly for} \quad 0 \le t \le \varepsilon, \tag{7.1.27}$$

and

$$y = 1 - e^{-t/\varepsilon} + \mathcal{O}(\varepsilon^{1/2}) \quad \text{uniformly for} \quad 0 \le t \le \varepsilon^{1/2}. \tag{7.1.28}$$

Notice that the approximation $1 + e^{-t/\varepsilon}$ involved in these estimates is nothing other than the inner solution y^i given by (7.1.12), truncated at the leading order and expressed in the variable t. The first estimate (7.1.27) is the same as the estimate given by regular perturbation theory for the leading order inner solution. (Recall that the inner solution was found by regular perturbation theory from the problem (7.1.7), and therefore its error estimate on finite intervals of τ-time is already known and was stated in (7.1.10) for the two-term inner solution; here we are dealing only with the one-term inner solution. The domain occurring in (7.1.27) is the same as a finite interval of τ-time.) The second estimate (7.1.28) is one that does not follow from regular perturbation theory. It is an estimate for the inner solution extended somewhat beyond its "natural" domain. It is important to remember that $0 \le t \le \varepsilon^{1/2}$ is a *longer* interval than $0 \le t \le \varepsilon$; these will be called the *long inner domain* and *short inner domain*, respectively.

So far we have given estimates on four different domains (the infinite interval $0 \le t < \infty$, the finite interval $0 \le t \le T$ for finite T, and the short and long inner domains). There is one additional domain to be considered. Namely, we will now return to (7.1.25) and simplify it on an interval of the form $\eta(\varepsilon) \le t \le T$. In the inner domains it was the first term of (7.1.25) that could be simplified; in the present "outer domain," it is the second. Since the function $e^{-t/\varepsilon}$ is decreasing in t for fixed ε, its maximum value on the interval in question occurs at the left-hand endpoint $\eta(\varepsilon)$ and is $e^{-\eta(\varepsilon)/\varepsilon}$. This quantity is transcendentally small (hence certainly $\mathcal{O}(\varepsilon)$) provided $\eta(\varepsilon) \gg \varepsilon$. Therefore

$$y = e^{-t} + \mathcal{O}(\varepsilon) \quad \text{uniformly for} \quad \eta(\varepsilon) \le t \le T. \tag{7.1.29}$$

Of course, the approximation e^{-t} is the one-term outer solution y^o. (Compare (7.1.17), the two-term outer solution, remembering that A is found to be 1 in (7.1.20).)

From (7.1.28) and (7.1.29) it is easy to create a leading-order patched solution and find its error estimate. Choose $\eta(\varepsilon)$ to lie asymptotically between ε and $\varepsilon^{1/2}$, that is, $\varepsilon \ll \eta(\varepsilon) \ll \varepsilon^{1/2}$; as we remarked before, $\eta = \varepsilon^{2/3}$ will do. Define

$$y^p = \begin{cases} 1 - e^{-t/\varepsilon} & \text{for} \quad 0 \le t \le \eta(\varepsilon) \\ e^{-t} & \text{for} \quad t > \eta(\varepsilon). \end{cases} \tag{7.1.30}$$

Then

$$y = y^p + o(1) \quad \text{uniformly for} \quad 0 \le t \le T. \tag{7.1.31}$$

With a little more work it could be shown that (7.1.31) actually holds uniformly on the infinite interval $0 \le t \le \infty$; this is true for some, but not all, initial layer problems. (We could have put the stronger estimate $\mathcal{O}(\varepsilon^{1/2})$ in place of $o(1)$, but at this point the important fact is merely that we have an asymptotic approximation. We cannot put $\mathcal{O}(\varepsilon)$, because (7.1.28) is not that strong. The error is $\mathcal{O}(\varepsilon)$ except in $\varepsilon < t < \eta(\varepsilon)$, because of (7.1.27) and (7.1.29).)

It is apparent from this discussion that (7.1.26) is better than (7.1.30), since the approximate solution is given by a single formula valid throughout its domain, has no discontinuities, and has a better error estimate (see (7.1.25)). In the next section we will see how to arrive at this composite solution from the inner and outer solutions by matching (in place of patching).

We will conclude this section with the answer to a question that may have puzzled the alert reader: Since (7.1.28) is the leading order of (7.1.12), and (7.1.29) is the leading order of (7.1.17), why isn't (7.1.30) the leading order of (7.1.21)? The leading order of (7.1.21) in the inner region contains the additional terms $-t - te^{-t/\varepsilon}$. These terms are of order $\varepsilon^{1/2}$ in the long inner domain, and so are of the same order as the error in (7.1.28) and can be included or omitted without harm; however, they must be included in (7.1.21) since they are larger than the $\mathcal{O}(\varepsilon)$ term which is retained there. It can be seen in equation (7.1.18) how these terms move back and forth between the leading order and the next order when the variable is changed between t and τ. If we had computed the inner and outer solutions to leading order only, these terms would never have appeared, and the leading order patched solution would be exactly (7.1.30). In the same way, if three-term inner and outer solutions were to be computed (that is, through order ε^2), additional terms would arise which move into the first two orders when expressed in terms of t. This phenomenon occurs in both matching and patching methods. In regular perturbation theory, the already calculated orders are not changed if higher orders are computed; in singular perturbation theory it is not always so.

Exercises 7.1

1. Is it possible to take the patching point to be $\eta(\varepsilon) = 2\varepsilon$? Explain the difference between (7.1.20) and $\varepsilon \leq \eta(\varepsilon) \leq \varepsilon^{1/2}$. (If you need help, reread Section 1.8.)

2. By solving the second order linear equation with constant coefficients and using the initial conditions to evaluate the constants, show that the exact solution of (7.1.6) is (7.1.22).

3. The difference between (7.1.24) and (7.1.25) is $f(t) = (e^t - 1)e^{-t/\varepsilon}$. Show that the maximum value of this function occurs at $t = \ln 1/(1 - \varepsilon)$ and is of order $\mathcal{O}(\varepsilon)$.

4. Show that the leading order inner and outer solutions to the model problem of this section, namely $y^i = 1 - e^{-t/\varepsilon}$ and $y^o = e^{-t}$, differ by $o(1)$ on an "overlap domain" $\eta_1(\varepsilon) \leq t \leq \eta_2(\varepsilon)$, where η_1 and η_2 are any points satisfying $\varepsilon \ll \eta_1(\varepsilon) \ll \eta_2(\varepsilon) \ll \varepsilon^{1/2}$. Show that the two-term inner and outer solutions occurring in (7.1.21) differ by $o(\varepsilon)$ on such an overlap domain. These "overlap conditions" help to explain why the patching point is not uniquely determined and also play a role in the foundations of the matching method (see Section 7.3).

7.2. MATCHING BY THE VAN DYKE RULES

In this section a matching method will be presented for a general class of second order ordinary differential equations with a small parameter multiplying the second derivative. In order to develop the method, we will at first continue studying the model problem introduced in the last section, representing a spring–mass system with small mass and an initial impulse. This system has the nondimensional equation of motion (with dot denoting d/dt)

$$\varepsilon \ddot{y} + \dot{y} + y = 0,$$
$$y(0) = 0, \tag{7.2.1}$$
$$\dot{y}(0) = 1/\varepsilon,$$

which in the fast time variable $\tau = t/\varepsilon$ takes the form (with $' = d/d\tau$)

$$y'' + y' + \varepsilon y = 0,$$
$$y(0) = 0, \tag{7.2.2}$$
$$y'(0) = 1.$$

It was found in the last section that the two-term perturbation solution of (7.2.2), called the *two-term inner solution*, is

$$y^i(\tau, \varepsilon) = \left(1 - e^{-\tau}\right) + \varepsilon\{(2 - \tau) - (2 + \tau)e^{-\tau}\}, \tag{7.2.3}$$

while the two-term perturbation solution of (7.2.1) without the initial conditions is

$$y^o(t, \varepsilon) = Ae^{-t} + \varepsilon(-At + B)e^{-t}. \tag{7.2.4}$$

This is the starting point from which we introduced the patching method in the last section; we return now to this starting point, and begin from here to explain the method of *matching by Van Dyke's matching rules*. As in the patching method, two objectives must be accomplished. First, it is necessary to determine the values of A and B in the outer solution. For this, the matching method will give the same result as the patching method although the reasoning leading to the result will be different. Second, it is necessary to combine the inner and outer solutions into a single approximation. Here matching gives a different result than patching. The new approximate solution y^c is called the *composite solution* and is a smooth blending of the inner and outer solutions rather than a sudden jump from one to the other. (A preview of the composite solution was given in equation (7.1.26) of the last section, where the leading order composite solution was derived from the exact solution.) Under the Van Dyke method of matching, both objectives—the evaluation of the constants in y^o and the

construction of y^c—are accomplished by using certain "expansion operators" which will now be introduced.

Let $y = \phi(t, \varepsilon)$ be any smooth function of t and ε. (Later we will take
this to be the exact solution of (7.2.1), but for the moment it is entirely
arbitrary.) Under the change of variable $t = \varepsilon\tau$, there is an associated
function $y = \psi(\tau, \varepsilon) := \phi(\varepsilon\tau, \varepsilon)$. Similarly, given any function $y = \psi(\tau, \varepsilon)$
there is an associated function $y = \phi(t, \varepsilon) := \psi(t/\varepsilon, \varepsilon)$ defined for $\varepsilon \neq 0$.
As long as $\varepsilon \neq 0$, then, it is convenient to regard any function y of either
(t, ε) or (τ, ε) as simultaneously a function of the other set of variables.
Nevertheless, the two functions $y = \phi(t, \varepsilon)$ and $y = \psi(\tau, \varepsilon)$ are given by
different formulas and have quite different power series when expanded
in ε. (You will see this very explicitly in Exercise 7.2.1). Expansion operators provide a way of naming these two distinct expansions of the "same"
function y, without having to make constant use of the function letters ϕ
and ψ. The *k-term inner expansion operator* is denoted I_k; when placed in
front of any function y of (t, ε) or (τ, ε) it means "express the function in
terms of the inner variable τ and then expand in powers of ε, keeping the
first k terms." The *k-term outer expansion operator* O_k means "express the
function in terms of the outer variable t and then expand in powers of ε,
keeping the first k terms." For example, $I_2 y$ stands for the first two terms
in the power series of $\psi(\tau, \varepsilon)$, whereas $O_2 y$ stands for the first two terms in
the power series of $\phi(t, \varepsilon)$. These power series must be thought of as limit
process expansions (Section 1.8) rather than as Taylor series, since it often
happens that when expressed in the outer variable, y is undefined at $\varepsilon = 0$.
The inner and outer expansion operations are often called "expanding in
ε with τ held constant" and "expanding in ε with t held constant," respectively. The subscript 2 stands for "two-term expansion"; that is, terms of
order ε^0 and ε^1 are retained. Thus the first term *omitted* by I_k or O_k is of
order ε^k.

Each of these expansions is *pointwise* asymptotically valid (when the appropriate variable is held constant); this is automatically the case for limit
process expansions because of the way they are constructed. More precisely,
$y \sim I_k y$ pointwise for $0 \leq \tau < \infty$ and $y \sim O_k y$ pointwise for $0 < t \leq T$.
Neither expansion is *uniformly* valid on $0 \leq t \leq T$. They are uniformly
valid in suitable ε-dependent inner and outer domains, which may or may
not overlap (depending on the nature of the function y), but the Van Dyke
heuristic does not concern itself with this question; we will return to it in
Section 7.3. The symbols I_p and O_q are also used in more general contexts
for p-term inner and q-term outer expansions with respect to any specified
sets of gauge functions. In some problems it is necessary to use different
gauges in the inner and outer regions; for instance the inner expansion may
proceed in powers of $\varepsilon^{1/2}$ and the outer in powers of ε. Then to achieve a
given accuracy may require a different number of terms in each expansion,
so one may wish to work with a value of p that is unequal to q. (In our examples $p = q$ and we call them both k.) When using gauges that are not integer
powers, p and q still denote the number of "terms" (meaning the number

of different gauges) retained, but this no longer equals the order of the first omitted term.

The Van Dyke matching method is based on the fact that for a large class of functions y, the inner and outer expansion operators commute. That is, it is frequently (although not always) the case that

$$O_k I_k y = I_k O_k y. \tag{VDH-1}$$

Expressed in words, this says that if you express a function y in terms of τ, expand in ε, truncate at k terms, express the result in terms of t, expand again in ε, and truncate at k terms, you will (often) get the same result as if you perform these operations in the opposite order, that is, first expanded with t held constant (keeping k terms) and then with τ held constant (keeping k terms). There is a theorem (called Fraenkel's theorem) giving the conditions on y under which this holds, but the theorem is of little practical use because when y is unknown (as, for instance, when y is the solution of a differential equation that cannot be solved exactly) there is no way to check these conditions. Therefore the Van Dyke method proceeds by making the *tentative heuristic hypothesis* that the exact solution y of the problem in question is one for which VDH-1 holds. The designation VDH-1 indicates that this is the first of three such *Van Dyke hypotheses* which form the basis for the Van Dyke matching method. (Specific examples of functions that do and do not satisfy VDH-1 will be pointed out later.)

The first Van Dyke hypothesis concerns only the exact solution y. The remaining two hypotheses relate the exact solution y with the inner and outer perturbation solutions. In order to state these hypotheses in a general form, let y^i denote the k-term inner solution (of any specific problem for which this makes sense) and let y^o denote the k-term outer solution. (Thus, for the only example we have introduced so far, $k = 2$, and the solutions in question are given by (7.2.3) and (7.2.4).) Then the Van Dyke hypotheses are

$$I_k y = y^i, \tag{VDH-2}$$

$$O_k y = y^o. \tag{VDH-3}$$

In words, these say that the *inner and outer expansions of the exact solution* are equal respectively to the *inner and outer perturbation solutions*, which are computed by formal perturbation methods (substituting series into the equations, equating terms with the same power of ε, and so forth) without knowledge of (or regard to) the exact solution. The exact interpretation of these hypotheses will depend upon the specific problem to which they are applied. For our model problem of the spring–mass system with an initial impulse, the interpretation is as follows. The inner solution y^i given by (7.2.3) is fully determined; also the expansion $I_2 y$ of the exact solution y is uniquely determined. Therefore VDH-2 simply says that these two functions are equal. But the outer solution y^o contains two undetermined coefficients A and B. Therefore VDH-3 must be understood as asserting the existence of unique choices of A and B such that y^o with these choices of A and B becomes equal to $O_2 y$. Throughout this chapter, the interpretation of VDH-2 and VDH-3 will be similar, but in Chapter 8 it will be seen that for boundary value problems there are usually undetermined constants in both y^i and y^o. In this case both statements must be taken as asserting the existence and uniqueness of appropriate values of these constants.

Although in any "real" application of the matching method the Van Dyke hypotheses are merely conjectural, in our model problem the exact solution is known and the hypotheses can be verified. This is done in Exercises 7.2.1 and 7.2.2. Seeing such an example worked out should give one the confidence that the Van Dyke hypotheses are plausible in other situations as well. Exercise 7.2.8 gives further illustrations of the first Van Dyke hypothesis. Example 7.6.1 in Section 7.6 involves a function for which the first Van Dyke hypothesis fails, and shows how this can cause Van Dyke matching to give an incorrect result. In that example the failure of the Van Dyke hypothesis can be traced to the existence of a "middle layer" between the inner and outer regions; when three layers are included (a "triple deck"), the Van Dyke hypotheses hold and matching is possible. A different (and more subtle) type of example in given in Exercise 7.2.9. Once again VDH-1 fails and matching gives an incorrect result, but this time there is no middle layer. Instead, the difficulty is a technical one which arises when logarithmic gauges are used to expand the inner and outer solutions.

For initial value problems of the type discussed in this chapter, VDH-2 is not actually a hypothesis, but can be shown to hold in all cases. In fact this is a consequence of Theorem 2.4.2. For the inner solution is simply the regular perturbation series solution of the inner differential equation, and in regular perturbation theory the computed approximate solution always agrees with the Taylor series of the exact solution. Since the outer solution is not the solution of any specific initial value problem (and since the outer equation does not fit any pattern discussed in Chapter 2) we have no such theorem for it, and VDH-3 is truly a hypothesis. In spite of these facts we have presented the three Van Dyke hypotheses on an equal footing, since in Chapter 8 (dealing with boundary value problems) it will be seen that neither VDH-2 nor VDH-3 follows from a general theorem, but both must be assumed as hypotheses.

It should be pointed out that the most general form of the Van Dyke hypotheses is somewhat more complicated than the version we have stated. The expansions represented by the inner and outer expansion operators need not be asymptotic power series but may be asymptotic series with respect to any specified sequence of gauge functions, and the gauge functions used, as well as the number of terms retained, for the inner solution need not be the same as for the outer. See Exercises 7.2.8 and 7.2.9. Also the definitions of the inner and outer expansions can be modified to meet a variety of situations arising in ordinary and partial differential equations where the inner and outer variables may be somewhat different than the t and τ in the present model problem.

Since all three of the Van Dyke hypotheses refer to the exact solution y, which (for any real problem in perturbation theory) is unknown, one might wonder how these hypotheses could be of any use. The answer is found by substituting VDH-2 and VDH-3 into VDH-1 to obtain

$$O_k y^i = I_k y^o, \qquad (7.2.5a)$$

an equation which (we hasten to say) is most often seen in the simpler looking but less precise form

$$y^{io} = y^{oi}. \qquad (7.2.5b)$$

The symbol y^{io} denotes the "outer expansion of the inner solution," called the *inner-outer solution*; similarly, y^{oi} is the inner expansion of the outer solution, called the *outer-inner solution*. In words, either form of (7.2.5) states that the k-term outer expansion of the k-term inner solution equals the k-term inner expansion of the k-term outer solution *when the undetermined constants in these solutions are chosen correctly*. The Van Dyke method uses equation (7.2.5) as a means of fixing these constants, and thereby achieving the "first objective" of matching. The "second objective," the creation of a composite solution, is then achieved by the formula

$$y^c := y^i + y^o - y^{io}. \qquad (7.2.6)$$

For the moment, we will take (7.2.6) as merely a definition of y^c. Later in this section it will be explained why y^c is a reasonable guess for an approximate solution valid over $0 \leq t \leq T$. Equations (7.2.5) and (7.2.6) express the essence of the Van Dyke method, and often (7.2.5) is presented as a hypothesis without mentioning what we have called VDH-1, 2, and 3. Although this is sufficient for problem solving, it leaves the origin of (7.2.5) a mystery.

To see how (7.2.5) and (7.2.6) are used, we will now work out the details for the model problem. First, in order to compute $y^{io} = O_2 y^i$, the inner solution (7.2.3) must be written in the outer variable, giving

$$\left(1 - e^{-t/\varepsilon}\right) + \varepsilon\left(2 - 2e^{-t/\varepsilon} - \frac{t}{\varepsilon} - \frac{t}{\varepsilon}e^{-t/\varepsilon}\right). \qquad (7.2.7)$$

Next this must be expanded in powers of ε to two terms. Since $e^{-t/\varepsilon}$ is not defined at $\varepsilon = 0$, a limit process expansion must be used. That is, let $\varepsilon \to 0$ to find the constant term $1 - t$; subtract this, divide by ε, and again let $\varepsilon \to 0$ to find that the next coefficient is 2. (In practice one may simply "discard" the terms containing $e^{-t/\varepsilon}$ as being transcendentally small and handle the rest as a Taylor series.) Therefore

$$y^{io} = (1 - t) + \varepsilon(2). \qquad (7.2.8)$$

Now, to compute $y^{oi} = I_2 y^o$, the outer solution (7.2.4) with its undetermined constants is to be expressed in the inner variable:

$$Ae^{-\varepsilon\tau} + \varepsilon(-A\varepsilon\tau + B)e^{-\varepsilon\tau},$$

and expanded to two terms:

$$y^{oi} = A + \varepsilon(-A\tau + B). \qquad (7.2.9)$$

According to (7.2.5), we should now set $y^{io} = y^{oi}$ and use this to determine A and B. First it is necessary to express both of these in the same variable. Choosing the outer variable t and setting $\tau = t/\varepsilon$, we find (see also Exercise 7.2.3)

$$y^{oi} = (A - At) + \varepsilon(B). \qquad (7.2.10)$$

Notice that as usual when changing from inner to outer variables, there is some shifting of terms between orders; in this case the term $-A\tau$ at order ε becomes $-At$ at order 1. It is natural to become anxious at this point and to worry about the term of order ε in (7.2.10): Should it "really" contain additional items that "drop down" from the order ε^2 term that has been omitted from (7.2.9)? The answer is a firm "no"; *there is no term of order ε^2 that has been omitted from (7.2.9)*. We have emphasized from the start that y^o denotes the outer perturbation solution *taken to a specific number of terms*. (One may think of first forming the infinite outer perturbation series and then omitting terms *from this series* to obtain y^o, but these terms are not omitted *from y^o* because they never belonged to it.) Another way to see this is to look back at the Van Dyke hypotheses and recall that O_k and I_k involve both expansion and truncation. The meaning of VDH-3 is precisely that these *expansion-and-truncation* operators commute. So it is right to set (7.2.8) equal to (7.2.10) and at once to read off

$$\begin{aligned} A &= 1, \\ B &= 2, \end{aligned} \qquad (7.2.11)$$

in agreement with (7.1.20). Notice that the quantity A is overdetermined; that is, in comparing (7.2.8) and (7.2.10) we first obtain $1 - t = A - At$ and $2 = B$ since terms of the same order in ε must be equal, and then $1 = A$ and $-1 = -A$ because two polynomials in t can only be equal if their coefficients are equal. Thus A is determined twice, and it seems to be "pure luck" that the two determinations are the same. Of course it is not luck; it comes about precisely because (for this model problem) the exact solution satisfies the Van Dyke hypotheses. That is, the Van Dyke hypotheses imply the existence of a solution to the overdetermined system of equations for the constants in y^o.

At this point y^o is fixed, and it is easy to write down the composite solution y^c according to the prescription (7.2.6), using (7.2.7), (7.2.4) with (7.2.11), and (7.2.8); for consistency, we have chosen to use the original or outer variable ("real" time) for each of the components. Notice that all of the terms in y^{io} cancel with terms in y^i. The final result is

$$y^c = \left(e^{-t} - e^{-t/\varepsilon} - te^{-t/\varepsilon}\right) + \varepsilon\left(-te^{-t} + 2e^{-t} - 2e^{-t/\varepsilon}\right). \qquad (7.2.12)$$

It remains to give a heuristic justification for (7.2.6). (The argument assumes the Van Dyke hypotheses, but uses additional intuitive reasoning and therefore is not rigorously valid even when the Van Dyke hypotheses are known to hold.) The inner expansion operator may be thought of as discarding from a function the terms that are negligible (to a specified order) when τ is held fixed and ε is small. Intuitively, this means that the inner expansion operator discards terms that are negligible "in the inner region." Similarly, the outer expansion operator may be thought of as discarding the terms that are negligible in the outer region. (In particular, the outer operator completely discards terms like $e^{-t/\varepsilon}$ that are exponentially small for any fixed $t \neq 0$ but are significant in a thin inner layer which becomes thinner as ε becomes smaller. It is not necessary to define the exact thickness of the inner layer in order to have the correct intuition here, but from Section 7.1 we know that for the model problem its thickness should be between orders ε and $\varepsilon^{1/2}$.) Therefore $y^i - y^{io}$, being the part discarded from y^i by the outer operator, should be negligible in the outer layer, and $y^o - y^{oi}$ should be negligible in the inner layer. Now using (7.2.6) together with $y^{io} = y^{oi}$, the composite solution can be written in two different ways:

$$y^c = y^i + (y^o - y^{oi}) = y^o + (y^i - y^{io}).$$

The first representation says that y^c is essentially the same as y^i in the inner layer; but y^i is supposed to be a good approximation to y in the inner layer, so y^c should be equally good. The second representation shows similarly that y^c is as good as y^o in the outer layer.

To what extent can these arguments be made precise? Specific estimates for $y^o - y^{oi}$ and $y^i - y^{io}$ can be given in any particular problem (see Exercise 7.2.4), since these are computable without knowing the exact solution. If error estimates are available for y^i and y^o in specific inner and outer regions that overlap, then the reasoning in the last paragraph shows that these four estimates taken together imply error estimates for y^c. But we will use this merely as heuristics and do the error estimates in a different way in Section 7.5.

Another common way to view (7.2.6) is to think of y^{io} as the "common part" of y^i and y^o. One can write the inner and outer solutions as $y^i = y^{io} + (y^i - y^{io})$ and $y^o = y^{io} + (y^o - y^{io})$; that is, each is equal to the common part plus a correction that is small outside of its own layer. Then adding $y^i + y^o$ includes both corrections for the inner and outer layers but counts the common part twice; therefore it must be subtracted off once in (7.2.6) to produce an approximation that should be good in both regions.

For reference the complete procedure illustrated above by the model problem will now be stated as a sequence of steps called the Van Dyke matching rules (VDMR). They will be stated in sufficient generality to apply to a variety of problems for which the notion of inner and outer variables makes sense. The actual matching procedure consists of steps 1 to 4; we have lumped some preliminaries (which may be quite difficult) into step 0.

VDMR-0: Using whatever insight is given by dimensional analysis, physical reasoning, or experience with similar problems, select inner and outer variables and suitable gauge functions for the inner and outer solutions. Then compute these solutions to the desired order. The number of terms p retained in the inner solution, and the number of terms q retained in the outer solution, should relate to the choice of gauges in such a way that the expected accuracy of the inner and outer solutions is the same. (In most examples the inner and outer gauges are the same and $p = q$.) The inner and outer solutions (with the specified number of terms) are denoted y^i and y^o, respectively, and one or both of these will contain undetermined constants.

VDMR-1: Compute $O_q y^i$. That is, express y^i in the outer variable and take its q-term limit process expansion using the outer gauges. The result is the inner-outer solution, denoted y^{io}.

VDMR-2: Compute $I_p y^o$. That is, express y^o in the inner variable and take its p-term limit process expansion using the inner gauges. The result is the outer-inner solution, denoted y^{oi}.

VDMR-3: Express y^{io} and y^{oi} in the same variable (either inner or outer), set them equal to one another, and see if it is possible to choose the undetermined constants so as to make these functions exactly equal. If this is possible, then y^i and y^o are said to *match*; if not, they *do not match* (in the Van Dyke sense). If the matching is not possible, this is a sign that the Van Dyke hypotheses are false for this problem (with this choice of inner and outer variables and gauges). If the matching is possible, this does not prove that the Van Dyke hypotheses are true, but one may proceed to the next step (with the understanding that occasionally, in particularly difficult problems, the results will be incorrect).

VDMR-4: Replace the arbitrary constants in y^i, y^o, and y^{io} by their values as found in step 3, and construct the composite solution by the formula $y^c = y^i + y^o - y^{io}$.

VDMR-5: Announce a conjecture to the effect that y^c (as computed in step 4) is a uniformly valid asymptotic expansion on the desired domain. If at all possible, prove this conjecture by direct error estimation, either for a specific problem or for a general class of problems.

Note: In difficult problems it may be impossible to guess the correct gauges in advance. In this case it is common to leave the gauges undetermined at first and fix them in step 3 so as to make the matching possible. In this case it is most convenient to proceed through the sequence of steps 1 to 4 several times. First look for $y^i = y_0^i \delta_0(\varepsilon)$ and $y^o = y_0^o \Delta_0(\varepsilon)$, with $p = q = 1$. After fixing these leading gauges and the leading order solutions by matching, return to step 1 and add the next undetermined gauges. If one solution (the inner or outer) falls behind the other in accuracy, it may be necessary to take two or more gauges at once in that solution in order to catch up. A wide variety of ad hoc techniques are used, especially in partial differential equations applications, to arrive at suitable variables and gauges. See Section 8.2 for an example.

In step 3, when we say that "matching is not possible" we mean that the matching condition implies contradictory values for the constants. In the opposite circumstance, when the matching condition appears to leave some of the constants unspecified (but without producing contradictions), it may be possible to complete the matching by applying the matching rules also to the derivatives of the solutions, or to strengthen the rules by other strategies. Although we will not attempt to consider all the possibilities, a few of them will appear in Section 7.6.

Up to this point, all of Chapter 7 has dealt with a single model problem, and that problem is exactly solvable without perturbation theory. But a general method has been developed, and it is time to begin applying the method to problems that are not exactly solvable. In previous chapters, most of the natural problems were "nearly linear," that is, nonlinear problems which become linear when $\varepsilon = 0$. (One exception was Mathieu's equation, Example 6.1.1, which was linear but with nonconstant coefficients.) Nonlinear problems with transition layers are often difficult and will be postponed to Section 7.7; here we confine our attention to second order linear initial value problems with initial impulses, but allow the coefficients to depend upon time so that the problems are not solvable exactly using known functions. Therefore we consider

$$\varepsilon \ddot{y} + b(t)\dot{y} + c(t)y = 0,$$
$$y(0) = \alpha, \tag{7.2.13}$$
$$\dot{y}(0) = \beta/\varepsilon + \gamma.$$

(If the first term is $\varepsilon a(t)\ddot{y}$, with $a(t) \neq 0$, we divide by $a(t)$ to achieve (7.2.13).) For such problems we will show that the Van Dyke matching method provides a formal solution, provided that $b(t) > 0$ on the interval on which the solution is sought (either a compact interval $0 \le t \le T$ or the infinite interval $0 \le t \le \infty$). Recall that a "formal solution" merely means that the computational procedures dictated by the method can be carried out (without encountering obstacles such as, for instance, dividing by zero). The hypothesis that $b(t) > 0$ is actually composed of two parts: At one point in the construction it is required that $b(t) \ne 0$ for all t in the interval, and at another point it is required that $b(0) > 0$. Together these are equivalent to $b(t) > 0$.

The most important part of the proof is to show that the inner and outer solutions match, and here the condition $b(0) > 0$ plays an essential role. (Exercise 7.2.5 gives an example in which this hypothesis is violated and matching is not possible.) The formal theorem given here says nothing about the validity of the approximate solution in regard to its accuracy. The error analysis will be carried out in Section 7.5, and it will be seen that the composite solution y^c is uniformly valid on a compact interval $0 \le t \le T$, provided that $b(t) > 0$ throughout that interval.

On an infinite interval the hypotheses of Theorem 7.2.1 suffice for the formal existence of the approximate solution, but not for its validity. It is important to recognize that the mere possibility of the matching does not guarantee uniform validity of the results. It is possible to prove validity of the solution for an infinite interval if *both* $b(t)$ *and* $c(t)$ are not only positive but bounded away from zero for $0 \le t < \infty$.

Theorem 7.2.1. *A formal approximate solution of (7.2.13) on an interval $0 \le t \le T$, $0 \le t < T$, or $0 \le t < \infty$ can be calculated using the Van Dyke matching rules, provided that $b(t) > 0$ throughout the interval in question.*

Proof. In the inner variable $\tau := t/\varepsilon$, (7.2.13) becomes a regular perturbation problem:

$$y'' + b(\varepsilon\tau)y' + \varepsilon c(\varepsilon\tau)y = 0,$$
$$y(0) = \alpha, \qquad\qquad (7.2.14)$$
$$y'(0) = \beta + \varepsilon\gamma.$$

Expand

$$b(\varepsilon\tau) = b_0 + \varepsilon b_1 \tau + \cdots,$$
$$c(\varepsilon\tau) = c_0 + \varepsilon c_1 \tau + \cdots,$$

where $b_0 = b(0)$, $b_1 = b'(0)$, etc., and the dots denote terms that are $\mathcal{O}(\varepsilon^2)$ on finite τ-intervals. Write the regular perturbation solution of (7.2.14) as

$y^i = y_0^i + \varepsilon y_1^i + \cdots$ (taken to some fixed finite order); then the leading term is found to satisfy the following initial value problem:

$$y_0^{i''} + b_0 y_0^{i'} = 0,$$
$$y_0^i(0) = \alpha, \qquad (7.2.15)$$
$$y_0^{i'}(0) = \beta.$$

This is a solvable problem with constant coefficients, and the solution is

$$y_0^i = -\frac{\beta}{b_0}e^{-b_0\tau} + \alpha + \frac{\beta}{b_0}. \qquad (7.2.16)$$

The higher order terms are known to be computable (Chapter 2). The outer problem is defined to be (7.2.13) with both initial conditions deleted. Write its perturbation solution in the form $y^o = y_0^o + \varepsilon y_1^o + \cdots$; then the leading order term is found to satisfy

$$b(t)\dot{y}_0^o + c(t)y_0^o = 0, \qquad (7.2.17)$$

which is solvable despite the nonconstant coefficients because it is a first order linear equation. (It is at this point that the hypothesis $b(t) \neq 0$ is needed.) The general solution is

$$y_0^o = A \exp\left[-\int_0^t \frac{c(s)}{b(s)}ds\right], \qquad (7.2.18)$$

where A is an arbitrary constant. We leave it to the reader to show that the higher order terms in y^o are computable and that each order introduces one new undetermined constant. (See Exercise 7.2.6.) In order to match the leading order terms (7.2.16) and (7.2.18), introduce the outer variable into (7.2.16) by setting $\tau = t/\varepsilon$; for constant $t \neq 0$ the first term is transcendentally small provided that $b_0 > 0$, and the leading order inner-outer solution is $\alpha + \beta/b_0$. If $b_0 < 0$, the inner solution expressed in the outer variable has no limit as $\varepsilon \to 0$ for constant t, and therefore a limit process expansion for this function does not exist; that is, $O_1 y^i$ does not exist, and it is not possible to continue the matching process. Assume, then, that $b_0 > 0$. Putting $t = \varepsilon\tau$ into (7.2.18) and taking the leading order yields A for the outer-inner solution. Equating these gives

$$A = \alpha + \frac{\beta}{b_0}, \qquad (7.2.19)$$

which fixes the undetermined constant in the outer solution. Finally, the composite solution to leading order is

$$y_0^c = -\frac{\beta}{b_0}e^{-b_0 t/\varepsilon} + \left(\alpha + \frac{\beta}{b_0}\right)\exp\left[-\int_0^t \frac{c(s)}{b(s)}ds\right]. \qquad (7.2.20)$$

In Exercise 7.2.6 the reader is invited to carry the calculation to the next order or beyond and see that no additional conditions beyond $b_0 > 0$ are necessary for the formal matching to be possible. ∎

Exercise 7.2.7 is an opportunity to practice the Van Dyke matching process on some specific problems. As always with such exercises, they should be reasoned out "from scratch" and not solved from a formula such as (7.2.20) or its higher order generalizations.

Exercises 7.2

1. Find the exact solutions $y = \phi(t, \varepsilon)$ of (7.2.1) and the exact solution $y = \psi(\tau, \varepsilon)$ of (7.2.2). Compute the first two terms of the power series expansion in ε of these functions. From the results, find $O_1 y$, $O_2 y$, $I_1 y$, and $I_2 y$. Compare these with the inner and outer solutions found in Section 7.2.1 and verify the first two Van Dyke hypotheses VDH-1 and VDH-2. (Hint: Some of the work for this problem was already done in the last section. It is easiest to work from known expansions of parts of the solutions, such as the exponential function and the binomial series for square roots, rather than apply Taylor's theorem or a limit process to the complete solutions. Notice that Taylor's theorem is not applicable to functions that are not defined when $\varepsilon = 0$, and that therefore some of your power series must be regarded as limit process expansions rather than Taylor series.)

2. Continuing from the results of the last exercise, verify the third Van Dyke hypothesis VDH-3 by expressing the inner and outer solutions in the "opposite" variables and re-expanding.

3. Carry out the matching to obtain (7.2.11) by expressing y^{io} and y^{oi} in terms of the inner variable, instead of the outer, before equating them.

4. For the model problem, using the specific expressions for y^i, y^o, and y^{io}, find big-oh estimates for $y^i - y^{io}$ in an outer layer $\eta(\varepsilon) \leq t \leq T$ and for $y^o - y^{io}$ in an inner layer $0 \leq t \leq \eta(\varepsilon)$, where $\varepsilon \ll \eta(\varepsilon) \ll \varepsilon^{1/2}$. Hint: Use the same kind of reasoning as used for the estimates at the end of Section 7.1.

5. Find the exact solution of the initial value problem $\varepsilon \ddot{y} - y = 0$, $y(0) = 0$, $\dot{y}(0) = 1/\varepsilon$. Graph the solution for small ε. Does it exhibit a boundary layer? Compute inner and outer solutions and show that they do not match. (Because the solution is exponentially large as $\varepsilon \to 0$ for any fixed $t \neq 0$, its behavior cannot be captured by any asymptotic series whose gauges are confined to polynomial growth. The proof of Theorem 7.2.1 explains why a matched solution cannot be constructed for this problem.)

6. Complete the proof of Theorem 7.2.1 at least as far as the term of order ε. Namely: Compute the terms of order ε in the inner and outer

solutions and show that they can be matched using the Van Dyke rules. Use the notations of (7.2.14). (If you wish, try to show that the nth order is constructible by setting up a suitable induction hypothesis about the structure of the solution up to the $(n-1)$st step.)

7. Find the two-term composite expansion for each of these problems by following the Van Dyke matching rules.

$$\text{(a)} \qquad \varepsilon \ddot{y} + (1+t)\dot{y} = 0,$$
$$y(0) = 0,$$
$$\dot{y}(0) = 1/\varepsilon;$$

$$\text{(b)} \qquad \varepsilon \ddot{y} + \dot{y} - ty = 0,$$
$$y(0) = 1,$$
$$\dot{y}(0) = 2 + 1/\varepsilon;$$

$$\text{(c)} \qquad \varepsilon \ddot{y} + (2-t)\dot{y} + ty = 0,$$
$$y(0) = \alpha,$$
$$\dot{y} = \gamma.$$

Remark: Since problem (c) has no impulse (that is, the second initial condition does not approach infinity as $\varepsilon \to 0$), the solution does not exhibit an initial layer in the leading order. However, there is an initial layer in the solution at order ε, and an initial layer in $\dot{y}$ at the leading order. Show that these features are present in your solution.

8. Let $y = t^2 + \varepsilon t + \varepsilon^{3/2} = \varepsilon^{3/2} + \varepsilon^2(\tau^2 + \tau)$, with $\tau = t/\varepsilon$. (The terms here are arranged in increasing order for fixed t and τ, respectively.) This problem and the next use the following generalized version of the first Van Dyke hypothesis. Let $\delta_i(\varepsilon)$, $i = 0, \ldots$ be a sequence of gauges used for inner expansions and let Δ_j be a possibly different sequence of gauges used for outer expansions. Then $I_p y = a_0(\tau)\delta_0(\varepsilon) + \cdots + a_{p-1}(\tau)\delta_{p-1}(\varepsilon)$ and $O_q y = b_0(t)\Delta_0(\varepsilon) + \cdots + b_{q-1}(t)\Delta_{q-1}(\varepsilon)$ denote the p-term and q-term limit process expansions with respect to these gauges (when such limit process expansions exist), and the Van Dyke hypothesis is the hypothesis that $I_p O_q y = O_q I_p y$.

(a) For inner expansions in this part of the problem, use the gauges $\varepsilon^3/2$ and ε^2; for outer expansions, use the gauges 1, ε, and $\varepsilon^3/2$. Check that with this choice of gauges, $I_p O_q y = O_q I_p y$ for $p = 1, 2$ and $q = 1, 2, 3$.

(b) Repeat the problem using gauges 1, $\varepsilon^{1/2}, \varepsilon, \varepsilon^{3/2}, \varepsilon^2$ for both inner and outer expansions, and taking $p, q = 1, \ldots, 5$. Notice that the Van Dyke hypotheses hold in spite of the fact that terms become interchanged in their order of importance when the variable is changed.

9. Let $y = \ln t / \ln \varepsilon$.

(a) Check that in the inner variable $\tau = t/\varepsilon$ this function becomes $y = 1 + \ln \tau / \ln \varepsilon$.

b) Take as gauges 1 and $-1/\ln \varepsilon$, for both inner and outer expansions. (Notice that $-1/\ln \varepsilon = 1/\ln(1/\varepsilon)$ is positive and monotone, that is, it is a gauge function, and $-1/\ln \varepsilon \ll 1$ although it is "closer" to 1 than any positive power of ε. See Exercise 7.2.8 for the definition of inner and outer expansion operators using gauge functions.) Show that $I_1 O_1 y = 0$ and $O_1 I_1 y = 1$; that is, the Van Dyke hypothesis fails for this function (with $p = q = 1$). On the other hand, $I_2 O_2 y = O_2 I_2 y$. (In fact, both are equal to y itself.)

(c) Suppose that some differential equation has as its exact solution the function y that we have been discussing. Suppose also that inner and outer solutions have been computed (using the same gauges as in part b) and that the results are $y^i = 1 + \ln \tau / \ln \varepsilon$ and $y^o = A + B \ln t / \ln \varepsilon$, where A and B are undetermined constants. Show that Van Dyke matching to leading order gives the incorrect result $A = 1$, and gives no warning that anything is amiss, but that Van Dyke matching to second order (where the Van Dyke hypothesis holds) gives the correct result $A = 0$, $B = 1$. ("Correct" and "incorrect" here mean that the result is, or is not, an asymptotic approximation of the exact function y to the indicated order with respect to the chosen gauges.) The type of error illustrated in this problem can be avoided by following the maxim "do not cut between logarithms." That is, in deciding the integers p and q to be used in I_p and O_q, do not chose integers which truncate the perturbation series between gauges that are logarithmically close to one another, such as 1 and $-1/\ln \varepsilon$.

7.3* MATCHING BY INTERMEDIATE VARIABLES AND OVERLAP DOMAINS

This section (which may be omitted without loss of continuity) presents an alternative approach to matching, which is more difficult to use in applications than Van Dyke matching but is perhaps somewhat more natural since it does not rest on a hypothesis about the commutativity of two expansion operators. Instead, the fundamental idea is that the undetermined constants should be chosen so that the inner and outer approximations agree asymptotically in some region where the inner and outer domains "overlap." This is very similar to the patching method of Section 7.1, in which the inner and outer approximations were made to agree asymptotically at a single transition point, but the present method allows the creation of a smooth transition between the inner and outer solutions in the form of a composite solution y^c as in the Van Dyke method. Another similarity between this and the Van Dyke method is that both use expansion operators, although in different ways. Most theoretical studies of the validity of matching use the method described here, because the overlap hypotheses seem to be more fundamental than the Van Dyke hypotheses, and because it is sometimes possible to prove the overlap conditions without knowing the exact solution (which does not seem to be the case for the Van Dyke hypotheses).

Return once again to the model problem

$$\varepsilon \ddot{y} + \dot{y} + y = 0,$$
$$y(0) = 0, \qquad\qquad\qquad\qquad (7.3.1)$$
$$\dot{y}(0) = 1/\varepsilon$$

of the spring–mass system with an initial impulse, with its inner and outer solutions

$$y^i(\tau, \varepsilon) = \left(1 - e^{-\tau}\right) + \varepsilon\{(2 - \tau) - (2 + \tau)e^{-\tau}\}, \qquad (7.3.2)$$

and

$$y^o(t, \varepsilon) = Ae^{-t} + \varepsilon(-At + B)e^{-t}. \qquad (7.3.3)$$

For the third time, we will determine the values of A and B and construct a combined solution.

Recall that the outer variable for (7.3.1) is t and the inner variable is $\tau := t/\varepsilon$. Introduce an *intermediate variable* ξ defined by

$$\xi := t/\varepsilon^\nu \qquad \text{with} \quad 0 < \nu < 1. \qquad (7.3.4)$$

Such a variable is "intermediate" between t and τ in the sense that it is stretched (or dilated) about the point 0, but not by as large a factor as τ is stretched. Every function of (t, ε) or of (τ, ε) can be regarded simultaneously as a function of (ξ, ε), and when expressed in the latter form will have an "expansion in ε with ξ held constant," which will (in general) differ from its inner and outer expansions (with τ or t, respectively, held constant). Thus, in addition to the expansion operators I_k and O_k of the last section, there is an "intermediate expansion operator" N_k. (N is the second letter of "intermediate," since the first letter already means "inner.") As before, the subscript k denotes the number of terms in the expansion, which is a limit process expansion with respect to some set of gauges chosen so that the required expansions exist. (As the example below shows, even if integer powers are the gauges for both y^i and y^o, it is in general necessary to use other gauges for the intermediate expansion. This is one of the reasons why matching with an intermediate variable is more difficult than Van Dyke matching.) In the same spirit as the Van Dyke matching rules (VDMR) of Section 7.2, but more concisely, we state the following intermediate variable matching rules (IVMR). Later these will be given an alternate explanation in terms of overlap domains.

IVMR: Let y^i be the p-term inner solution and y^o the q-term outer solution; take p and q equal if the same gauges are used for the inner and outer expansions, otherwise take them so that the expected error is of the same order in each case. Express y^i and y^o in the intermediate variable, and take their k-term expansions, where again k is chosen so that the expected error is of the same order. If possible, choose the undetermined constants (which appear in y^i, y^o, or both) in such a way as to make these intermediate expansions equal; that is, so that

$$N_k y^i = N_k y^o, \tag{7.3.5a}$$

or more briefly (using superscript n for intermediate expansion)

$$y^{in} = y^{on}. \tag{7.3.5b}$$

If this can be accomplished, define the composite solution by

$$y^c = y^i + y^o - y^{in}. \tag{7.3.6}$$

These rules will now be applied to the model problem. By beginning with the two-term inner and outer solutions (7.3.2) and (7.3.3) we have already selected $p = q = 2$. If we took $k = 2$ also, it turns out that the matching is not completely determined (A can be fixed but not B). Therefore we will take $k = 3$. First the inner solution is written in the intermediate variable by substituting $\tau = \xi/\varepsilon^{1-\nu}$ into (7.3.2). Since $1 - \nu > 0$ the exponentials are transcendentally small for fixed ξ as $\varepsilon \to 0$; when these are deleted, three terms remain, which form the 3-term intermediate expansion

$$N_3 y^i = 1 - \varepsilon^\nu \xi + 2\varepsilon, \tag{7.3.7}$$

in which the natural gauges are seen to be 1, ε^ν, and ε, in this order (since, again, $0 < \nu < 1$). Next, the outer solution is written in the intermediate variable by substituting $t = \varepsilon^\nu \xi$ into (7.3.3). This time the exponentials are not transcendentally small, but are expandable using the power series for the exponential function. The result is

$$y^o = A\left(1 - \varepsilon^\nu \xi + \cdots\right) + \varepsilon\left(-A\varepsilon^\nu \xi + B\right)\left(1 - \varepsilon^\nu \xi + \cdots\right),$$

where the dots denote terms of order $\varepsilon^{2\nu}$ and higher. If we now impose the additional requirement that

$$\tfrac{1}{2} < \nu < 1 \tag{7.3.8}$$

then the terms in $\varepsilon^{2\nu}$ are of higher order than the term in ε; then the terms represented by dots may be deleted, together with the terms of order $\varepsilon^{\nu+1}$ and $\varepsilon^{2\nu+1}$, giving

$$N_3 y^o = A - \varepsilon^\nu A\xi + \varepsilon B. \tag{7.3.9}$$

Now (7.3.7) and (7.3.9) are three-term expansions with respect to the same gauges and they become equal if

$$
\begin{aligned}
A &= 1, \\
B &= 2,
\end{aligned}
\tag{7.3.10}
$$

the same values found by the patching and Van Dyke methods. Finally the composite solution, found by adding (7.2.2) and (7.2.3) and subtracting (7.3.7), is

$$y^c = \left(e^{-t} - e^{-t/\varepsilon} - te^{-t/\varepsilon}\right) + \varepsilon\left(-te^{-t} + 2e^{-t} - 2e^{-t/\varepsilon}\right), \tag{7.3.11}$$

in agreement with (7.2.12). An examination of (7.3.7) and (7.3.9) shows why a three-term expansion is necessary in the intermediate variable although we are only matching two-term approximations: The natural gauges for the intermediate expansions contain an additional gauge inserted between the gauges 1 and ε used in the inner and outer solutions, and so three terms must be taken to give equal accuracy. Furthermore, the unknown B does not appear until the last of these three gauges. Once again (as in the Van Dyke method) A is overdetermined, but without inconsistency.

Now we consider the meaning of the matching condition $N_k y^i = N_k y^o$. Limit process expansions (in ε) of functions depending on ε and other variables are, like Taylor series, uniformly valid when the other variables are confined to compact subsets (provided that the limits taken in forming these expansions are uniformly valid on compact subsets, as is the case here and in most applications). Therefore given any compact interval $a \leq \xi \leq b$, the matching condition specifies that (if the constants are chosen correctly) y^i and y^o are asymptotically equivalent (up to their own order) on that interval. In terms of the original variable t, this means that y^i and y^o are nearly equal on the ε-dependent interval $a\varepsilon^\nu < t < b\varepsilon^\nu$ and differ there by a quantity which is of the same order as the expected error by which y^i and y^o should approximate the exact solution y. We say that the interval $a\varepsilon^\nu < t < b\varepsilon^\nu$ is a *formal overlap domain* for the formal solutions y^i and y^o, and we also say that y^i and y^o satisfy the *formal overlap condition* for scales ε^ν with $1/2 < \nu < 1$. (Often the interval $1/2 < \nu < 1$ is also called the *formal overlap domain*, since it specifies simultaneously all of the possible overlap domains in t: Just choose ν in this interval and choose any $0 < a < b$ and you have an overlap domain. These two usages will not cause confusion, since it is always clear whether one is referring to an interval of ν or an interval of t.)

The formal overlap condition defined in the last paragraph is a computable condition, not a hypothesis. That is, given y^i and y^o, it is generally a simple matter to check whether or not they overlap. The procedure for doing so is just to follow the IVMR; if the matching is possible (for ν in a suitable interval) then the overlap condition is satisfied. There is a stronger overlap condition, called *rigorous overlap*, that is not testable (without either knowing the exact solution or doing extensive error estimation for both y^i and y^o). Roughly speaking, rigorous overlap states that y^i is a valid approximation to the exact solution y on an inner domain that overlaps with an outer domain where y^o is valid. Thus, rigorous overlap makes reference to the exact solution, whereas formal overlap is merely a relationship between the two proposed approximate solutions. Formal overlap alone is not sufficient to justify the matching process, but rigorous overlap is.

In this book we will not work with the rigorous overlap condition. Partly, this is because it is not particularly useful in practice. The situation is this: One can compute the inner and outer solutions, attempt the matching (by the IVMR), and if this is successful one can form y^c. As a byproduct of this work, one knows that formal overlap holds. If rigorous overlap also holds, then the solution y^c is uniformly valid. But in order to know whether rigorous overlap holds, it is necessary to do error estimation for y^i and y^o using inner and outer domains which overlap. In the cases that we will encounter, this is more difficult than doing error estimation directly on y^c over the entire domain in which validity is expected. Thus the situation with the IVMR is not so different from that of the VDMR: Both offer a candidate for a composite solution y^c on the basis of conjectural hypotheses about the exact solution. The difference is that Van Dyke matching requires more hypotheses (in the form of the VDH) than does matching with an intermediate variable.

It is true that there exist certain theorems, called *extension theorems*, which are of some help in establishing rigorous overlap. These theorems are to the effect that if y^i and y^o are uniformly valid on inner and outer domains that do not overlap, then under certain circumstances they are actually valid on slightly larger inner and outer domains that do overlap. Although these theorems are important in theoretical discussions of the foundations of matching, they have found little application in concrete problem solving and will not be considered here. See Notes and References.

Exercises 7.3

1. Show that to leading order, the inner and outer solutions for the model problem overlap for all $0 < \nu < 1$ (and not only for $1/2 < \nu < 1$). Show that this fixes $A = 1$ but does not establish B.

2. Solve the problems in Exercise 7.2.7 using the IVMR.

3. Formulate and prove a theorem similar to Theorem 7.2.1 for the IVMR.

7.4. THE INITIAL CORRECTION LAYER METHOD

In this section, for the fourth (and last) time, a formal approximate solution will be found for the problem

$$\varepsilon \ddot{y} + b(t)\dot{y} + c(t)y = 0,$$
$$y(0) = \alpha, \tag{7.4.1}$$
$$\dot{y}(0) = \beta/\varepsilon + \gamma$$

for $t > 0$, under the assumption that $b(t) > 0$. The solution obtained will be the same as that found by matching (either by the VDMR or IVMR), but once again the heuristics will be different. For this specific problem (and for most ordinary differential equations problems with initial layers) this method is computationally the easiest of all of the methods in this chapter, and if we were only concerned with ordinary differential equations this method would have been presented at the beginning. However, this method requires in advance a clear idea of the structure of the solution. In complicated problems, especially in partial differential equations, it may not be clear in advance what scales should be used for the inner and outer variables, or what gauges should be used for the inner and outer expansions. In such a case, the matching methods are much more flexible than the correction layer method, because the inner and outer solutions may be calculated using undetermined scales and gauges, and the scales and gauges may then be fixed in the process of matching. (This will be illustrated in Section 8.2.) The correction layer method requires that all such decisions be made in advance, and if they are changed, the entire calculation must be repeated. Therefore the matching method has retained its popularity and is often used even in cases where the correction method is simpler. For these reasons, we have presented the matching methods first. However, the correction method will be used throughout the rest of this chapter.

The correction method will be developed "from scratch" without any mention of the matching method. However, for those who have been exposed to matching, it is helpful to motivate the correction method in the following way. The composite solution obtained by Van Dyke matching can be written in the following way:

$$y^c = y^i + y^o - y^{io}$$
$$= y^o(t, \varepsilon) + \left(y^i(\tau, \varepsilon) - y^{oi}(\tau, \varepsilon)\right)$$
$$= p(t, \varepsilon) + q(\tau, \varepsilon). \tag{7.4.2}$$

Here the "common part" $y^{io} = y^{oi}$ has been written in the latter form (so that it is a function of τ) and combined with the inner solution to form a function $q(\tau, \varepsilon)$ which (whenever Van Dyke matching is successful) will be significant only in the inner layer. (See the discussion following equation (7.2.12).) This fact can be expressed in the form

$$q(\infty, \varepsilon) := \lim_{\tau \to \infty} q(\tau, \varepsilon) = 0, \tag{7.4.3}$$

which we call the *correction layer condition* because it expresses that q is a "correction" to the outer solution $p = y^o$ that is negligible outside the thin initial layer. The aim of the correction method is to compute p and q directly, bypassing the matching step along with the calculation of y^{io} and y^{oi}. The correction layer condition (7.4.3) replaces the matching condition $y^{io} = y^{oi}$, to which it is equivalent. Each of the function p and q are to be found in the form of perturbation series in powers of ε.

The decomposition (7.4.2) suggests another comparison, one which to some extent bridges the gap between Parts II and III of this book. Namely, upon expanding (7.4.2) in the form

$$y \cong p(t, \varepsilon) + q(\tau, \varepsilon) \sim \{p_0(t) + q_0(\tau)\} + \varepsilon\{p_1(t) + q_1(\tau)\} + \cdots, \tag{7.4.4}$$

it would appear to be a special cases of a two-time-scale expansion

$$y \cong \tilde{y}(t, \tau, \varepsilon) \sim y_0(t, \tau) + \varepsilon y_1(t, \tau) + \cdots \tag{7.4.5}$$

of the type considered in Chapter 5. (As a warning to anyone who attempts detailed comparisons: Remember that in Chapter 5 $\tau = \varepsilon t$, whereas here $\tau = t/\varepsilon$.) In fact, this is a useful analogy to draw. Since each order in (7.4.4) is not an arbitrary function of t and τ, but is a sum of separate functions of each variable, the correction layer method is sometimes called the *additive two-scale method*. It is also possible to attack our problem (7.4.1) directly with a series of the form (7.4.5). One might claim that multiple scales is the most powerful and general asymptotic method, because it encompasses both oscillatory and transition layer phenomena and can be said to include the Lindstedt method, averaging, matching, and even regular perturbation theory (as an N-scale method with $N = 1$). On the other hand, the case can be made that no one has in fact developed a general theory of the multiple scale method, and it still seems necessary to treat oscillatory problems and transition layer problems separately. Furthermore the multiple scale method does not seem likely to duplicate the qualitative side of the averaging method (for instance, the material of Section 6.3) or the rigorous overlap theory (mentioned briefly in Section 7.3). If the multiple scale theory can accomplish these objectives and can present a unified theory covering oscillatory problems, transition layer problems, and problems exhibiting both phenomena at once and if the resulting method is flexible enough to handle partial differential equations when the correct gauges and scales are unknown in advance, then indeed the multiple scale method will supersede all the others. Until then, we still need them all.

Now it is time to develop the initial layer correction method for (7.4.1) from scratch, as promised. The first step is to write

$$y^c(t, \varepsilon) = p(t, \varepsilon) + q(\tau, \varepsilon), \tag{7.4.6}$$

with $\tau = t/\varepsilon$, and derive the differential equations and side conditions to be satisfied by p and q. The second step is to solve these equations by perturbation series of the form

$$\begin{aligned}
p(t, \varepsilon) &\sim p_0(t) + \varepsilon p_1(t) + \cdots, \\
q(\tau, \varepsilon) &\sim q_0(\tau) + \varepsilon q_1(\tau) + \cdots.
\end{aligned} \tag{7.4.7}$$

Sometimes these steps are performed as one step, that is, one might never mention p and q and instead work directly with the right hand side of (7.4.4). But while this may save some space, it entails a loss of conceptual clarity.

The differential equations and side conditions for p and q are derived from the following three requirements:

(i) The "outer" solution p should satisfy the differential equation in (7.4.1) but not the initial conditions.

(ii) The "composite" solution $p + q$ should satisfy the differential equation in (7.4.1) and also the initial conditions.

(iii) The "correction" q should satisfy the correction condition $q(\infty, \varepsilon) = 0$, that is, (7.4.3).

The condition expressed by (i) is simply

$$\varepsilon \ddot{p} + b(t)\dot{p} + c(t)p = 0. \tag{7.4.8a}$$

To derive the differential equation satisfied by q, condition (ii) says to substitute $y = p + q$ into (7.4.1). With $\dot{} = d/dt$ and $' = d/d\tau = \varepsilon d/dt$, this gives $\varepsilon(\ddot{p} + q''/\varepsilon^2) + b(\dot{p} + q'/\varepsilon) + c(p + q) = 0$. The terms in p vanish because of (7.4.8a), leaving

$$q'' + b(\varepsilon\tau)q' + \varepsilon c(\varepsilon\tau)q = 0. \tag{7.4.8b}$$

Condition (ii) also indicates that $y = p + q$ should be substituted into the initial conditions of (7.4.1), giving

$$p(0, \varepsilon) + q(0, \varepsilon) = \alpha, \tag{7.4.8c}$$

$$\dot{p}(0, \varepsilon) + \frac{1}{\varepsilon}q'(0, \varepsilon) = \frac{\beta}{\varepsilon} + \gamma. \tag{7.4.8d}$$

Finally, condition (iii) is that

$$\lim_{\tau \to \infty} q(\tau, \varepsilon) = 0. \tag{7.4.8e}$$

The next step is to insert the expansions (7.4.7) into (7.4.8) and equate powers of ε. Carrying the result to two terms, and taking proper care to expand the coefficients of (7.4.8b), the results are

$$b(t)\dot{p}_0 + c(t)p_0 = 0,$$
$$b(t)\dot{p}_1 + c(t)p_1 = -\ddot{p}_0, \tag{7.4.9a}$$

$$\vdots$$

$$q_0'' + b(0)q_0' = 0,$$
$$q_1'' + b(0)q_1' = -c(0)q_0 - \dot{b}(0)\tau q_0', \tag{7.4.9b}$$

$$\vdots$$

$$p_0(0) + q_0(0) = \alpha,$$
$$p_1(0) + q_1(0) = 0, \tag{7.4.9c}$$

$$\vdots$$

$$q_0'(0) = \beta,$$
$$\dot{p}_0(0) + q_1'(0) = \gamma, \tag{7.4.9d}$$

$$\vdots$$

$$q_0(\infty, \varepsilon) = 0,$$
$$q_1(\infty, \varepsilon) = 0, \tag{7.4.9e}$$

$$\vdots$$

Notice that the first equation in (7.4.9d) involves only one of the unknown functions, whereas the second and beyond each involve two unknown functions "at different levels." For instance, the third equation in this sequence will be $\dot{p}_1(0) + q_2'(0) = 0$. This can lead to some confusion about how these equations are to be used, and in particular, the order in which they are to be solved. The answer is that one should always *select the first k equations from each family*. To find a two-term solution ($k = 2$), all of the equations listed above in (7.4.9) will be used, and no others; this implies that $\dot{p}_1(0) + q_2'(0) = 0$ will not be used, since it refers to q_2 which is not among the functions being found. To be sure that the right number of equations is present, it is helpful to count constants of integration and side conditions. Equations (7.4.9a), being first order equations, contribute one unknown constant each, while (7.4.9b) contribute two, for a total of six constants (when $k = 2$). The side conditions (7.4.9c,d,e) provide exactly six equations to determine these constants. Another way to look at this is to take the first equation from each family; this forms a complete system with three constants and three side conditions. Once these are known, taking the second equation from each set gives another complete system. From this point of view, the second equation of (7.4.9d) can be written $q_1'(0) = \gamma - \dot{p}_0(0)$ and regarded as an initial condition for q_1 which is known after p_0 has been found.

Of course, the mere fact that the right number of side conditions are present does not guarantee that the system (7.4.9) is solvable, since this system does not present an initial value problem; the side conditions are a mixed set of initial and boundary values (with a boundary at infinity). But no existence theorem is necessary. The solution can be explicitly calculated, under the assumption that $b(t) > 0$ for $t > 0$. The crucial point here concerns (7.4.9b) and (7.4.9e). Consider the first equation of (7.4.9b). This is a second order homogeneous linear equation with constant coefficients having general solution $q_0 = c_1 + c_2 e^{-b(0)\tau}$. Since the exponential term decays when $b(0) > 0$, it is possible to satisfy (7.4.9e) by taking $c_1 = 0$, thus using one side condition and determining one constant. On the other hand if $b(0)$ were less than zero, it would not be possible to satisfy (7.4.9e). This pattern continues with the other equations in the family (7.4.9b): At the point when each equation needs to be solved, its right hand side is known, so it becomes an inhomogeneous second order linear equation with general solution equal to $c_1 + c_2 e^{-b(0)\tau}$ plus a particular solution. Since it can be seen recursively that the particular solutions decay at infinity, the correction layer condition (7.4.9e) can be met once again by taking $c_1 = 0$. Once it is known that $b(0)$ must be greater than zero, the necessity of $b(t) > 0$ follows from the need to divide by $b(t)$ in solving the rest of (7.4.9).

The complete solution of (7.4.9) to second order is left to the reader (Exercise 7.4.2). As stated before, the results coincide with those obtained by matching in (7.2.20) and Exercise 7.2.6. Error estimation for this solution will be carried out in the next section. One point should perhaps be mentioned: Although $y^c = p + q$ is required to satisfy the initial conditions exactly, the approximate solution obtained from the truncation of p and q does not satisfy the second initial condition exactly. This will become clear when we calculate the residuals in the next section. (The same is of course true for the matching method, but not for the patching method. This is in contrast to the initial value problems studied in Parts I and II, where the approximations always satisfied the initial conditions exactly.)

Exercises 7.4

1. Solve the problems in Exercise 7.2.7 by the correction method. Do not use (7.4.9), but begin with conditions (i), (ii), and (iii), and derive (and

solve) the specific equations of the form (7.4.9) for each problem. Compare your results with those previously obtained by matching.

2. Solve the system consisting of the first equation of each family in (7.4.9) and compare your result with (7.2.20). Then solve the system consisting of the second equation in each family and compare with your solution of Exercise 7.2.6.

3. Show that if $b(t) < 0$ for $t < 0$, the correction method can be used to obtain a formal approximate solution of (7.4.1) for $t < 0$. Then consider the problem given in Exercise 7.2.5. Solve this problem for $t < 0$ by the correction method (or by matching, or both) and compare the result with the exact solution. Observe that no problem can be solved for both $t > 0$ and $t < 0$, either by matching or correction.

7.5. ERROR ESTIMATION

Error estimation for the matching and correction methods is considerably more complicated than for the methods treated in previous chapters. There exist a considerable variety of results: uniform estimates on finite intervals, on infinite intervals, with a damping condition, with both "damping" and "restoring" conditions, and so forth. This section proves only the simplest of these results: an error estimate for the leading order approximation to a second order initial value problem on a finite interval, assuming a damping condition but no restoring condition. This much can be proved fairly easily by methods similar to those of Section 3.2: integral equations and Gronwall's inequality.

Of course, the arguments contain special features due to the presence of an initial layer. In particular, some of the steps on the way to the final result involve bounds which are a sum of two terms, one of order $\mathcal{O}(\varepsilon)$ everywhere, the other of order $\mathcal{O}(1)$ in the initial layer and exponentially small in the outer region. Nevertheless, the end result is that the error is uniformly $\mathcal{O}(\varepsilon)$ in $0 \leq t \leq T$. It is no more difficult (merely computationally tedious) to extend this result to higher order approximations. What does become more difficult is to extend the argument to the infinite interval $0 \leq t < \infty$, and this will not be attempted here. (See Notes and References.) To do so requires a "restoring" condition $c(t) > 0$ in addition to damping, and the argument uses "comparison equations" and not just Gronwall's inequality.

The problem to be considered is the same one treated by matching in Section 7.2 and by correction in Section 7.4, namely

$$\varepsilon \ddot{y} + b(t)\dot{y} + c(t)y = 0,$$
$$y(0) = \alpha, \qquad\qquad (7.5.1)$$
$$\dot{y}(0) = \beta/\varepsilon + \gamma.$$

The leading order solution by either method has been given in (7.2.20); see also Exercise 7.4.2. This solution is

$$y_0^c = -\frac{\beta}{b_0}e^{-b_0 t/\varepsilon} + \left(\alpha + \frac{\beta}{b_0}\right)\exp\left[-\int_0^t \frac{c(s)}{b(s)}ds\right]. \tag{7.5.2}$$

The result to be proved is:

Theorem 7.5.1. *If (7.5.1) satisfies the damping condition $b(t) > 0$ for $0 \leq t \leq T$, then the exact solution y and the approximate solution y^c given by (7.5.2) satisfy $y = y^c + \mathcal{O}(\varepsilon)$ uniformly for $0 \leq t \leq T$.*

(Remark: The condition $b(t) > 0$ for $0 \leq t \leq T$ implies the existence of a constant $k > 0$ such that $b(t) \geq 2k$, since the interval is compact. This will be used in the following proof. The reader may wonder whether strengthening $b(t) > 0$ to $b(t) \geq 2k > 0$ is sufficient to establish this theorem on $0 \leq t < \infty$, and the answer is that it is not. The restoring condition $c(t) > $ constant $ > 0$ also seems to be necessary.)

As usual, the first step in estimating the error of this approximate solution is to write the exact solution as

$$y = y^c + R, \tag{7.5.3}$$

where R is the error or remainder to be estimated. Since y satisfies (7.5.1) exactly, it follows immediately that

$$
\begin{aligned}
\varepsilon\ddot{R} + b(t)\dot{R} + c(t)R &= -p(t, \varepsilon), \\
R(0) &= \sigma_1(\varepsilon), \\
\dot{R}(0) &= \sigma_2(\varepsilon),
\end{aligned}
\tag{7.5.4}
$$

where

$$
\begin{aligned}
p(t, \varepsilon) &= \varepsilon\ddot{y}^c + b(t)\dot{y}^c + c(t)y^c, \\
\sigma_1(\varepsilon) &= \alpha - y^c(0, \varepsilon), \\
\sigma_2(\varepsilon) &= \beta/\varepsilon + \gamma - \dot{y}^c(0, \varepsilon)
\end{aligned}
\tag{7.5.5}
$$

are the residuals, that is, the amounts by which the approximate solution y^c fails to satisfy the differential equation and initial conditions (7.5.1). It is not difficult to compute these residuals exactly, by differentiating (7.5.2) twice (using the fundamental theorem of calculus) and putting the results into (7.5.5). The result will be given below. Once the residuals are known, equation (7.5.4) in principle determines R completely. Of course, this cannot be solved, since that would be equivalent to solving the original problem exactly. So (7.5.4) will be converted into an integral equation for R, and this will be used to obtain an upper bound for R.

It is at this point that one of the crucial differences arises between this section and Section 3.2. The equation for R in that section, namely (3.2.5), was of the form $a\ddot{R} + b\dot{R} + cR =$ a nonlinear function of R. In other words, the left-hand side was linear with constant coefficients, so that the associated homogeneous equation $a\ddot{y} + b\dot{y} + cy = 0$ was solvable. The two basic linearly independent solutions of that equation were used to construct the kernel function $K(t, \tau)$ by means of which the differential equation for R was converted to an integral equation. In the present situation, the left-hand side of (7.5.4) has nonconstant coefficients, and the associated homogeneous equation cannot be solved. The solution to this difficulty is astonishingly simple; it is just to move the last term of the left-hand side to the right:

$$\varepsilon\ddot{R} + b(t)\dot{R} = -c(t)R - \rho(t, \varepsilon). \tag{7.5.6}$$

Now the associated homogeneous equation $\varepsilon\ddot{y} + b(t)\dot{y} = 0$ is solvable, since it can be viewed as a first order linear equation for $\dot{y}$ which can be solved by separation of variables (or by an integrating factor). We will now solve this equation, construct a kernel $K(t, \tau)$, and use this kernel to convert (7.5.6), along with the initial conditions from (7.5.4), into an integral equation.

It is not hard to find (using either of the methods just mentioned, or checking the proposed solutions directly) that the solutions of $\varepsilon\ddot{y} + b(t)\dot{y} = 0$ satisfying $y_1(0) = 1$, $\dot{y}_1(0) = 0$, $y_2(0) = 0$, and $\dot{y}_2(0) = 1$ are

$$y_1(t) = 1 \quad \text{and} \quad y_2(t) = \int_0^t e^{-B(\sigma)/\varepsilon} d\sigma,$$

where

$$B(\sigma) := \int_0^\sigma b(s)ds.$$

From the formula $K(t, \tau) := (y_1(\tau)y_2(t) - y_1(t)y_2(\tau)) / a(\tau)W(\tau)$, given along with (C.26) in Appendix C, it follows that

$$K(t, \tau) = \frac{1}{\varepsilon} \int_\tau^t \exp\left[-\frac{1}{\varepsilon} \int_\tau^\sigma b(s)ds\right] d\sigma. \tag{7.5.7}$$

Observing that in fact $y_2(t) = \varepsilon K(t, 0)$, the equation (C.26) for the solution of the inhomogeneous linear equation $\varepsilon\ddot{y} + b(t)\dot{y} = f(t)$ with initial conditions $y(0) = \alpha$, $\dot{y}(0) = \beta$ is

$$y(t) = \alpha + \beta\varepsilon K(t, 0) + \int_0^t K(t, \tau)f(\tau)d\tau.$$

It then follows, as in the derivation of (C.30), that the solution of (7.5.4) satisfies the integral equation

$$R(t) = \sigma_1 + \sigma_2 \varepsilon K(t,0) - \int_0^t K(t,\tau)[c(\tau)R(\tau) + \rho(\tau)]d\tau. \qquad (7.5.8)$$

It is important to remember that in this equation almost everything depends upon ε, although this dependence has been "suppressed." That is, the arguments to follow will be carried out for fixed ε so that there is no need to list ε as an argument of every function. On the other hand the constants $c_1, \ldots, c_8$ which will be introduced below are independent of ε; this is necessary in order to obtain the final $\mathcal{O}(\varepsilon)$ error estimate. So it is helpful to look back over the last paragraph and notice just where ε appears. In particular, it is absent from the functions $b(t)$ and $c(t)$ and is present in $K(t,\tau)$ only through the factor $1/\varepsilon$ which occurs in two places: in front of the whole expression, and also in the exponent. An additional comment before beginning the detailed estimates: There is no need to prove the existence of a solution to the integral equation (7.5.8), as there was in Chapter 3. The reason is that (7.5.4) is a linear differential equation and therefore has a global solution (that is, one defined on the largest interval where all of the coefficients and the inhomogeneous term are defined).

As an aid to the reader, the following outline of the estimation process is given. First the residuals will be computed and estimated, with the result that

$$|\rho(t)| \le c_1 e^{-b_0 t/\varepsilon} + c_2 \varepsilon,$$
$$|\sigma_1| = 0, \qquad\qquad\qquad (7.5.9)$$
$$|\sigma_2| = c_3.$$

The first term in the estimate for ρ is significant only in the initial layer. Next the kernel will be estimated, with the result

$$K(t,\tau) \le c_4, \qquad (7.5.10)$$

in spite of the factor of the initial $1/\varepsilon$ occurring in K; the integrand decays exponentially as $\varepsilon \to 0$, overcoming the "blow-up" of the initial factor. Next all of these estimates will be substituted into (7.5.8) to obtain the integral inequality

$$|R(t)| \le \int_0^t c_5|R(\tau)|d\tau + c_6\varepsilon. \qquad (7.5.11)$$

Finally, a Gronwall argument will be applied to (7.5.11) giving

$$|R(t,\varepsilon)| \le c_8\varepsilon, \qquad (7.5.12)$$

the desired conclusion.

Equations (7.5.2) and (7.5.5) imply the following explicit expression for ρ, in which $b(t)$ and $c(t)$ are abbreviated to b and c:

$$\rho = \beta \frac{b - b_0}{\varepsilon} e^{-b_0 t/\varepsilon} - \frac{c\beta}{b_0} e^{-b_0 t/\varepsilon}$$

$$- \varepsilon \left(\alpha + \frac{\beta}{b_0} \right) \left(\frac{\dot{c}b - c\dot{b} - c^2}{b^2} \right) \exp \left[-\int_0^t \frac{c(s)}{b(s)} ds \right]. \tag{7.5.13}$$

The simplest term to estimate here is the second, containing a factor of ε; it is only necessary to check that the coefficient of ε is bounded on $0 \le t \le T$. Since $b(t) \ge 2k > 0$ the denominators pose no difficulty, and the coefficient of ε is a continuous function and therefore is bounded on the compact set. This yields the term $c_2 \varepsilon$ in (7.5.9). Obtaining the first term $c_1 e^{-kt/\varepsilon}$ in (7.5.9) is slightly more difficult. First, $b(t)$ has a Lipschitz constant L on $0 \le t \le T$, so $|b(t) - b_0| \le Lt$. Next,

$$\beta \frac{|b(t) - b_0|}{\varepsilon} e^{-b_0 t/\varepsilon} \le \beta \frac{Lt}{\varepsilon} e^{-2kt/\varepsilon}$$

$$= \beta \frac{Lt}{\varepsilon} e^{-kt/\varepsilon} \cdot e^{-kt/\varepsilon}$$

$$\le \beta \frac{L}{ke} \cdot e^{-kt/\varepsilon}.$$

The last inequality holds because the function $(Lt/\varepsilon)e^{-kt/\varepsilon}$ takes its maximum value (for $t > 0$) at $t = \varepsilon/k$. This estimates the first term in (7.5.13), and the second is clearly bounded by $(\max |c\beta|/b_0)e^{-kt/\varepsilon}$; these two terms together give the term $c_1 e^{-kt/\varepsilon}$ in (7.5.9). The rest of (7.5.9) is also obtained from (7.5.2) and (7.5.5). These imply that $\sigma_1 = 0$ and that

$$\sigma_2 = \gamma + \frac{(\alpha b_0 + \beta)c(0)}{b_0^2} =: c_3,$$

a constant independent of ε.

The next step is to estimate the kernel. Since $b(s) \ge 2k > 0$ it follows that $-b(s)/\varepsilon \le -2k/\varepsilon$. Integrating gives

$$-\frac{1}{\varepsilon} \int_\tau^\sigma b(s)ds \le -\frac{2k}{\varepsilon}(\sigma - \tau).$$

Using this in (7.5.7) implies

$$0 \le K(t, \tau) \le \frac{1}{\varepsilon} \int_\tau^t e^{-\frac{2k}{\varepsilon}(\sigma - \tau)} d\sigma$$

$$= \frac{1}{2k} \left[1 - e^{-\frac{2k}{\varepsilon}(t - \tau)} \right]$$

$$\le \frac{1}{2k} =: c_4.$$

This proves (7.5.10).

Substituting (7.5.9) and (7.5.10) into (7.5.8) gives

$$|R(t)| \leq c_3 c_4 \varepsilon + \int_0^t c_4 |c(\tau)| |R(\tau)| d\tau + \int_0^t c_4 \left(c_1 e^{-k\tau/\varepsilon} + c_2 \varepsilon \right) d\tau.$$

The final integral here can be evaluated and is easily seen to be less than $\varepsilon(c_1 c_4 / k + c_2 c_4 T)$ for $0 \leq t \leq T$. This term then combines with the first term $c_3 c_4 \varepsilon$ to form the term $c_6 \varepsilon$ in (7.5.11), while c_4 times the maximum value of $|c(\tau)|$ for $0 \leq \tau \leq T$ becomes c_5. This completes the proof of (7.5.11).

All that remains is a standard Gronwall argument. Let

$$S(t) := \int_0^t |R(\tau)| d\tau.$$

Then

$$\dot{S}(t) - c_5 S(t) \leq c_6 \varepsilon.$$

Multiplying by $e^{-c_5 t}$ and integrating, using $S(0) = 0$, gives

$$S(t) \leq \frac{c_6}{c_5} \varepsilon \left(e^{c_5 t} - 1 \right) \leq \frac{c_6}{c_5} \varepsilon \left(e^{c_5 T} - 1 \right) =: c_7 \varepsilon.$$

Finally

$$|R(t)| \leq c_5 S(t) + c_6 \varepsilon \leq (c_5 c_7 + c_6) \varepsilon = c_8 \varepsilon,$$

which proves (7.5.12) and completes the proof of Theorem 7.5.1.

In Chapter 3 (see the last paragraph of Section 3.3), it was found that if an approximate solution by regular perturbation theory of a second order differential equation satisfies an error estimate of a given order, then the first derivative of the approximate solution differs from the first derivative of the exact solution by an error of the same order. (In mechanical terms, the velocity is approximated just as well as the position.) This is not the case for singularly perturbed initial value problems with an initial impulse. Under the same hypotheses as Theorem 7.5.1, the following result is obtained if y^c is the k-term approximation (by the matching or correction method):

$$
\begin{aligned}
y &= y^c + \mathcal{O}(\varepsilon^k), \\
\dot{y} &= \dot{y}^c + \mathcal{O}(\varepsilon^{k-1}),
\end{aligned}
\tag{7.5.14}
$$

uniformly for $0 \leq t \leq T$. In particular, the derivative of the leading order solution (7.5.2) is essentially useless as an approximation to the velocity, since the error is not small with ε but is only $\mathcal{O}(1)$. This should not be surprising, since the initial velocity is $\beta/\varepsilon + \gamma$, and γ does not even appear in (7.5.2); that is, even initially there is an $\mathcal{O}(1)$ error in the velocity. Another way to say this is that because of the impulse β/ε, the sequence of gauges that must be used to expand the velocity is $1/\varepsilon, 1, \varepsilon, \varepsilon^2, \ldots$. Therefore a leading order approximation to the velocity will use the gauge $1/\varepsilon$, and the error should be of the order of the first omitted gauge 1. Looked at from this perspective, (7.5.14) appears quite natural.

The same result (7.5.14) can be proved uniformly for $0 \leq t < \infty$ under the additional hypothesis that $c(t)$ is positive and bounded away from zero. This "restoring condition" (so called because it implies that the term $c(t)y$ is a restoring force directed toward the origin) must hold in addition to the similar "damping condition" $b(t) \geq 2k > 0$. In proving such a result one does not "take the c term to the other side" as we did in going from (7.5.4) to (7.5.6). Instead a comparison technique is used, in which the solution of (7.5.4) is compared with the solution of a constant coefficient equation having a restoring force and damping less than that of the equation to be solved. In this way the arguments are related to the comparison methods used to prove the Sturm–Liouville theorems in ordinary differential equations.

7.6. TRIPLE DECKS

Both the matching and correction methods were successful for the simple initial value problem $\varepsilon \ddot{y} + b(t)\dot{y} + c(t)y = 0$, $y(0) = \alpha$, $\dot{y}(0) = \beta/\varepsilon + \gamma$ treated in Sections 7.1–7.5. In this section we discuss two problems of a more complicated type. These problems will be attacked formally, beginning with the ideas used so far in this chapter and adding others when failure is encountered. The matching method, rather than correction, will be used because of its flexibility: Encountering a "dead end" does not require starting over entirely, since it may only be necessary to change one part of the calculation (just the inner solution, for instance). The first problem is a third order equation with a computable exact solution and illustrates triple decks and matching of derivatives. The second problem is of second order, also involves triple decks, but turns out not to be solvable by matching (at least by the methods presented here; it is not clear if there is another way to solve it). The example is interesting both for calling attention to limitations of the method and because it is almost identical to a problem (presented in Example 8.1.1) that *can* be solved (at least formally) by matching.

Example 7.6.1

The initial value problem

$$\varepsilon^{3/2}\dddot{y} + (\varepsilon^{1/2} + \varepsilon + \varepsilon^{3/2})\ddot{y} + (1 + \varepsilon^{1/2} + \varepsilon)\dot{y} + y = 0,$$
$$y(0) = 3,$$
$$\dot{y}(0) = -1 - \frac{1}{\varepsilon^{1/2}} - \frac{1}{\varepsilon}, \qquad (7.6.1)$$
$$\ddot{y}(0) = 1 + \frac{1}{\varepsilon} + \frac{1}{\varepsilon^2}$$

is somewhat artificial, but is constructed to have the following simple exact solution:

$$y = e^{-t} + e^{-t/\sqrt{\varepsilon}} + e^{-t/\varepsilon}. \qquad (7.6.2)$$

The graph of this solution is shown in Fig. 7.6.1 and clearly illustrates the existence of what is called a "triple deck" solution structure: The solution decays very rapidly at first in an "inner deck" (as the third term in (7.6.2) loses significance), then more slowly in a "middle deck" (until the second term drops out), and finally more slowly still as it approaches the asymptote $y = 0$. The presence of three time scales (t, $t/\sqrt{\varepsilon}$, and t/ε) is clear analytically in (7.6.2). We will attempt to construct a leading order asymptotic approximation to the solution of (7.6.1), proceeding as if we did not know the exact solution (7.6.2) and did not anticipate the triple deck structure. The investigation will reach several dead ends before it achieves success (at a formal level). At the end, our final solution will be checked against the exact solution.

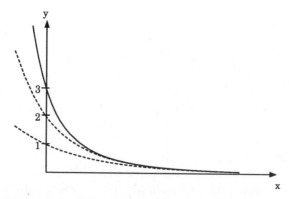

FIG. 7.6.1. A solution showing a triple-deck structure. The solid upper curve is $y = e^{-t} + e^{-t/\sqrt{\varepsilon}} + e^{-t/\varepsilon}$. The dotted curve below this omits the last term (the inner layer correction), and the one below that omits the last two terms (inner and intermediate layer corrections).

As a start, notice that when $\varepsilon = 0$ the differential equation (7.6.1) drops order. In earlier sections of this chapter, it was helpful to introduce an "inner variable" $\tau = t/\varepsilon$ in such cases, and we will naively do so now. (The reason why this is naive will appear before long.) Letting $'$ denote differentiation with respect to τ, it is found that (7.6.1) becomes

$$y''' + (1 + \varepsilon^{1/2} + \varepsilon)y'' + (\varepsilon^{1/2} + \varepsilon + \varepsilon^{3/2})y' + \varepsilon^{3/2}y = 0. \qquad (7.6.3)$$

Since this equation does not drop order, the strategy of introducing an inner variable appears successful. Therefore we will attempt to solve this "inner equation" (with initial conditions) and match the solution to the general solution of the "outer equation" (7.6.1) (without initial conditions). It will turn out that this attempt fails and the approach must be modified.

Before doing so, it will simplify the writing to set $\mu = \sqrt{\varepsilon}$. Then the original equations (7.6.1) become

$$\mu^3 \dddot{y} + (\mu + \mu^2 + \mu^3)\ddot{y} + (1 + \mu + \mu^2)\dot{y} + y = 0,$$
$$y(0) = 3,$$
$$\dot{y}(0) = -1 - \frac{1}{\mu} - \frac{1}{\mu^2}, \qquad (7.6.4)$$
$$\ddot{y}(0) = 1 + \frac{1}{\mu^2} + \frac{1}{\mu^4},$$

whereas the inner equations (7.6.3) and their initial conditions, with $\tau = t/\mu^2$, become

$$y''' + (1 + \mu + \mu^2)y'' + (\mu + \mu^2 + \mu^3)y' + \mu^3 y = 0,$$
$$y(0) = 3,$$
$$y'(0) = -1 - \mu - \mu^2, \qquad (7.6.5)$$
$$y''(0) = 1 + \mu^2 + \mu^4.$$

Notice that if we had started from the beginning with (7.6.4) instead of (7.6.1), the temptation would have been to use t and t/μ as the time scales instead of t and $t/\varepsilon = t/\mu^2$. This would also have been wrong (since all three variables, t, t/μ, and t/μ^2 are actually needed), but the nature of the failure is different. See Exercise 7.6.4.

The leading order outer solution will satisfy the reduced differential equation of (7.6.4), that is, $\dot{y} + y = 0$, and is not expected to satisfy the initial conditions. The outer solution, then, is

$$y^o = Ae^{-t}, \qquad (7.6.6)$$

where A is to be determined by matching with the inner solution. The leading order inner solution should satisfy the reduced problem of (7.6.5), including the initial condition. It is easy to calculate that

$$y^i = 2 + e^{-\tau}. \tag{7.6.7}$$

To match these, one would write y^i in the outer variable as $2 + e^{-t/\mu^2}$ and take the leading term, which gives $y^{io} = 2$ since the exponential is transcendentally small. Then one would write y^o in the inner variable as $Ae^{-\mu^2\tau}$ and take the leading order expansion. We have not specified the gauges for the expansion, but since only the leading order is needed the result is the same whether we use powers of μ (as seems natural here) or powers of $\mu^2 = \varepsilon$, which would seem natural if we were working from (7.6.1). In either case, the result is $y^{oi} = A$. The matching condition $y^{io} = y^{oi}$ then gives $A = 2$, and the matching appears to be successful. The composite solution $y^c = y^i + y^o - y^{io}$ is then

$$y^c = 2e^{-t} + e^{-\tau} = 2e^{-t} + e^{-t/\mu^2} \qquad \text{(incorrect)}.$$

Since the formal calculations went through without a hitch, how do we know that the result is incorrect? The exact solution (7.6.2) is $y = e^{-t} + e^{-t/\mu} + e^{-t/\mu^2}$. The difference between the exact solution and our proposed y^c is $y^c - y = e^{-t} - e^{-t/\mu}$. If y^c were a uniformly valid approximation on, say, $0 \le t \le 1$, then this difference would be $o(1)$, that is, it would be bounded (for all t in $0 \le t \le 1$ by a function $f(\mu)$ such that $f(\mu) \to 0$ as $\mu \to 0$. But at the point $t = \mu$ (which lies in $0 \le t \le 1$ for all small positive μ), $y^c - y = e^{-\mu} - e^{-1} \to 1 - e^{-1} \ne 0$. Therefore this y^c is not a uniformly valid approximation.

What went wrong? Why did matching give an incorrect result, without giving any warning? In Exercise 7.6.1, it will be shown that the first Van Dyke hypothesis fails for the exact solution; in fact, $I_1 O_1 y = 1$ and $O_1 I_1 y = 2$. Therefore Van Dyke matching, at least with the present choice of inner and outer variables, cannot be expected to work. Of course, this can only be known when the exact solution is known. Since in "real" perturbation problems the exact solution is not known, one is immediately prompted to ask: is there any computable *formal* test that would suggest there is a difficulty with our solution? And if the difficulty lies in an incorrect choice of variables, is there a way to find the correct variables? It turns out that both questions have the same answer. So we will now begin again, taking the problem as given in the form (7.6.4) but not assuming that τ is the correct inner variable. (The same methods would work if we began from (7.6.1).)

The student who has read Section 7.6.3 may wonder whether matching with the overlap condition gives any additional insights here. If y^o and y^i are given by (7.6.6) and (7.6.7), then it is easy to check that they overlap formally and that matching by the IVMR is possible and gives the same incorrect result as the VDMR. Since this result does not depend on the Van Dyke hypotheses, it cannot be completely explained by the failure of these hypotheses. In addition, the inner and outer solutions must fail the *rigorous overlap condition*, that is, the (μ-dependent) regions on which they are good approximations to the exact solution do not overlap. This indicates that there is (at least) a third region, between the inner and outer regions, on which a different approximation is necessary. This conclusion is not at all obvious from the Van Dyke heuristic and is an indication of the additional power of the overlap idea. Nonetheless we will quickly reach the same conclusion from the search for correct variables in the following paragraph.

The method to be used here may be called the *method of undetermined scales* and has much in common with the method of undetermined gauges in Section 1.5. It is important to be clear that there are both *scales* and *gauges* in problems of this type and that they are distinct: eh scales are the variables defining inner and outer (and intermediate) problems, the gauges are (in the present problem) the powers of μ used to expand the inner and outer solutions. In some problems it is necessary to take both the scales and the gauges as undetermined, but for this problem there is not much difficulty with the gauges (although we will shortly be forced to use some *negative* powers of μ as gauges!), so we will take only the scales as undetermined. Beginning with (7.6.4), then, we will substitute

$$\tau = t/\mu^\nu \tag{7.6.8}$$

and obtain (after using the chain rule)

$$\mu^{3-3\nu} y''' + (\mu^{1-2\nu} + \cdots) y'' + (\mu^{-\nu} + \cdots) y' + y = 0, \tag{7.6.9}$$

where the dotted terms are subdominant. Equating the powers of μ in pairs (including the μ^0 in the last term), it is found that the three choices $\nu = 0, 1, 2$ each cause a pair of terms in (7.6.9) to be dominant, while $\nu = 3/2$ causes a pair of terms to have the same order but not to be dominant. Therefore these define the *significant time scales*

$$t = t,$$
$$\sigma = t/\mu, \tag{7.6.10}$$
$$\tau = t/\mu^2.$$

It should be emphasized that we have not *proved* that these time scales will give a correct solution. But these are the time scales which focus attention on the interaction of two terms in the differential equation. Any other time scale will focus on only one term of the equation, giving a reduced solution that is almost trivial and does not exhibit the richness of structure needed to reflect a significant aspect of the behavior of the solutions. Since τ represents greatest "magnification" of the time variable at the origin, the behavior found on the τ scale should occur in the layer nearest the origin. The σ scale should reflect what is seen when the origin is examined with a magnifying glass of lower power that does not penetrate the innermost layer and make it visible; therefore it will describe the middle deck. And of course t itself is used away from the origin.

When using the three time scales given in (7.6.10), the inner and outer equations are the same as before, but there is now a middle equation using the σ variable. Let $y^* = dy/d\sigma$; then this middle equation is

$$\mu y^{***} + (1 + \mu + \mu^2)y^{**} + (1 + \mu + \mu^2)y^* + \mu y = 0. \qquad (7.6.11)$$

Since the middle layer is not in contact with the initial values, it will not be expected to satisfy any initial conditions. Rather, the arbitrary constants in the middle solution will be determined by matching with the inner solution and will in turn determine the constants in the outer solution by another matching. The matching is best done in the order "inner to middle" and then "middle to outer," since it is the inner solution that has no arbitrary constants and so provides the information needed to determine the constants in the other solutions. (In boundary value problems there can be information at both ends, and one must in effect do both matchings simultaneously. That is, each matching provides equations relating the constants, but these equations may be coupled in such a way that they cannot be solved until all the matchings have been done.) Since we are working only to leading order, the middle solution will satisfy the reduced equation of (7.6.11), namely, $y^{**} + y^* = 0$. The general solution of this equation is

$$y^m = B + Ce^{-\sigma}. \qquad (7.6.12)$$

Recall, from (7.6.7) and (7.6.6), that the inner and outer solutions are $y^i = 2 + e^{-\tau}$ and $y^o = Ae^{-t}$.

To match the inner and middle solutions, write the inner solution in the middle variable as $2 + e^{-\sigma/\mu}$ and take the one-term expansion; since the exponential is transcendentally small, this gives $y^{im} = 2$. Next write the middle solution in the inner variable as $B + Ce^{-\mu\tau}$ and expand to one term, giving $y^{mi} = B + C$. Putting $y^{im} = y^{mi}$ gives

$$B + C = 2. \qquad (7.6.13)$$

This does not completely determine B and C, so it is necessary to find another condition. This can be done by *matching the derivatives* of the inner and middle solutions. Choosing σ as the variable of differentiation (it is necessary to differentiate both solutions with respect to the same variable, either σ or τ), one has $y^i = 2 + e^{-\sigma/\mu}$ in the inner variable, so $dy^i/d\sigma = -(1/\mu)e^{-\sigma/\mu}$. This is exponentially small, so its leading order expansion is 0. On the other hand, $dy^m/d\sigma = -Ce^{-\sigma}$. Writing this in the inner variable and taking the leading order expansion gives $-C$. Matching these would give the result that $C = 0$, *but this is incorrect*. Once again, we have led you into a trap, a situation where matching leads, without warning, into an error. The solution constructed this way would not be an asymptotic approximation of the true solution. This time, the difficulty is not the existence of an additional layer, but is easier to remedy. It turns out that the correct result is obtained by calculating the inner solution to *two terms* instead of one. The two-term inner solution is easily found, from (7.6.5), to be

$$y^i = [2 + e^{-\tau}] + \mu[-\tau].\qquad(7.6.14)$$

Write this in the middle variable as $2 + e^{-\sigma/\mu} - \sigma$; then its derivative with respect to σ is found to be $-(1/\mu)e^{-\sigma/\mu} - 1$, with leading order expansion -1 instead of 0. In other words, *the one-term expansion of the derivative of y^i depends on the two-term expansion of y^i*. The reason for this is that the exact solution y satisfies the following rule of Van Dyke type:

$$I_1 \frac{d}{d\sigma} M_1 y = M_1 \frac{d}{d\sigma} I_2 y,\qquad(7.6.15)$$

where I_k and M_k are the inner and middle expansion operators (defined in an obvious way), but it does not satisfy the corresponding rule with I_2 replaced by I_1. See Exercise 7.6.3. We will not attempt to formulate Van Dyke rules for matching derivatives in any generality; our purpose is only to indicate the sort of difficulty that can arise. In the present problem, matching $-C$ with -1 gives the correct conclusion $C = 1$. (We will see that it is correct by comparing with the exact solution below.) Putting this together with (7.6.13), we have $B = C = 1$, and

$$y^m = 1 + e^{-\sigma}.\qquad(7.6.16)$$

The next step is to match the middle solution to the outer. This poses no difficulties. If we write y^m in the outer variable as $1 + e^{-t/\mu}$, the leading order expansion is 1; if we write the outer solution $y^o = Ae^{-t}$ in the middle variable as $Ae^{-\mu\sigma}$, the leading order is A; therefore $A = 1$ and

$$y^o = e^{-t}.\qquad(7.6.17)$$

Finally, the leading order inner, outer, and middle solutions (7.6.7), (7.6.16), and (7.6.17) can be combined into a composite solution according to the following formula, using the results $y^{im} = 2$ and $y^{mo} = 1$ which were obtained in the course of the matching:

$$\begin{aligned}y^c &= y^i + y^m + y^o - y^{im} - y^{mo}\\&= (2 + e^{-\tau}) + (1 + e^{-\sigma}) + e^{-t} - 2 - 1\\&= e^{-t/\mu^2} + e^{-t/\mu} + e^{-t}.\end{aligned}\qquad(7.6.18)$$

Comparison with (7.6.2) shows that the leading order composite solution for this problem is actually equal to the exact solution, so the error is zero. (This is entirely unexpected, since we have not even *used* the complete initial conditions given in (7.6.5) for the inner equations, but only their leading order. The explanation for it is that the exact solution to this problem is so simple: The parameter ε enters (7.6.2) only through the time scales and not through the presence of any higher order terms.) In a typical problem, of course, the error could only be expected to be $\mathcal{O}(\mu)$.

For ordinary differential equations, the best way to handle problems of the type considered in this example is probably to change the single higher order equation into a system of first order equations by the usual procedure of introducing variables $u_1 = y$, $u_2 = \dot{y}, \ldots$. In this way the derivatives become ordinary variables, and matching the derivatives is accomplished automatically by matching the variables. An introduction to such systems is given in Section 7.7, although we will not take the subject far enough to handle examples of the present type. For partial differential equations, it is not practicable to reduce everything to first order systems, and matching of derivatives is often necessary.

Example 7.6.2

Consider the problem

$$\varepsilon^3 \ddot{y} + t^3 \dot{y} + (t^3 - \varepsilon)y = 0,$$
$$y(0) = 0, \tag{7.6.19}$$
$$\dot{y}(0) = 1/\varepsilon.$$

Upon introducing $\tau = t/\varepsilon^\nu$, the significant scales are found to be $\nu = 0, 1/2, 1$ (with $1/3, 3/4, 3/5$ distinguished but not significant). Using the same notations as in the last example ($\mu = \sqrt{\varepsilon}$, scales given by (7.6.10), and $* = d/d\sigma$), the reduced equation in each scale (satisfied by the leading order approximation) is

$$y^{i''} - y^i = 0,$$
$$\sigma^3 y^{m*} - y^m = 0, \tag{7.6.20}$$
$$\dot{y}^o + y^o = 0.$$

This appears to pose a reasonable matching problem: First solve the inner equation, which is second order and will have a unique solution satisfying both initial conditions; then solve the middle equation and fix its single unknown constant by matching with y^i; then solve the outer equation and fix its constant by matching with y^m. Unfortunately, this does not work.

The solution for y^i, with initial conditions $y^i(0) = 0$, $y^{i'}(0) = 1$ obtained from (7.6.19), is

$$y^i = \sinh \tau = \tfrac{1}{2}(e^\tau - e^{-\tau}). \qquad (7.6.21)$$

The first step in matching would be to write this in the middle variable as $1/2(e^{\sigma/\mu} - e^{-\sigma/\mu})$ and expand to leading order (as a limit process expansion in powers of ε). But such an expansion is impossible, since one term of y^i is exponentially large. Therefore it is not possible to match. (This is clear without even computing y^m.)

It might seem that one could at least solve the problem if the initial conditions were changed so that the exponentially large term of (7.6.21) is eliminated. But in this case, an additional difficulty arises. Suppose the initial conditions in (7.6.19) are changed to

$$\begin{aligned} y(0) &= 1, \\ \dot{y}(0) &= -1/\varepsilon. \end{aligned} \qquad (7.6.22)$$

Then (7.6.21) is changed to

$$y^i = e^{-\tau}. \qquad (7.6.23)$$

When expressed in the middle variable, this is exponentially small (for fixed σ) and therefore $y^{im} = 0$. The general solution for y^m from (7.6.20) is

$$y^m = Ae^{-1/2\sigma^2}.$$

Written in the inner variable, $y^m = Ae^{-1/2\varepsilon\tau^2}$, which is also exponentially small, so that $y^{mi} = 0$. That is, the matching condition $y^{im} = y^{mi}$ holds for any A and therefore does not determine the constant A! It may be possible to find the "correct" value of A by imposing some requirement stronger than Van Dyke matching, but we will not pursue this question.

Exercises 7.6

1. Write the exact solution (7.3.2) of Example 7.6.1 in terms of $\mu = \sqrt{\varepsilon}$ and show that $I_1 O_1 y = 1$ and $O_1 I_1 y = 2$. (Here the inner variable is $\tau = t/\mu^2$ and the gauges for the inner and outer expansions are $1, \mu, \mu^2, \ldots$.)

2. Find the significant time scales for (7.6.1) in the same way that they were found in the text for (7.6.4).

3. Show that the exact solution (7.3.2) satisfies (7.6.15) and does not satisfy the same equation if I_2 is replaced by I_1 on the right hand side.

4. Attempt to solve (7.6.4) using only the time scales t and t/μ, and determine why this fails. (See the remark following (7.6.5).) Then try to

solve (7.6.4) using only the time scales t/μ and t/μ^2. Does this approach succeed? (Hint: What is the interval of uniform validity for the solution you get? A solution is not acceptable for this problem unless it is valid at least for a fixed finite interval $0 \leq t \leq T$.)

7.7. NONLINEAR INITIAL LAYER PROBLEMS AND SLOW CURVES

Up to this point, the problems discussed in this chapter have all been linear. We discussed an exactly solvable model problem $\varepsilon\ddot{y} + \dot{y} + y = 0$, and then generalized to $\varepsilon\ddot{y}+b(t)\dot{y}+c(t)y = 0$, which is not exactly solvable but can be solved approximately by the same methods developed for the model problem. This problem is linear but nonautonomous. In this section, the model problem will be generalized in a different way: we will study problems that are nonlinear but autonomous. Because they are autonomous, phase plane methods (which have not yet appeared in this chapter, although they were important in several previous chapters) can be used.

It is not much more difficult to consider problems that are both nonlinear and nonautonomous, but the geometry must be presented in three dimensions (as was also necessary in, for instance, Section 4.5). It is also possible to repeat the discussion in higher dimensions, by letting u and v in (7.7.2) become vectors. In the first case, the "slow curve" discussed below becomes a "slow surface," and in the second, it becomes a "slow manifold" of the appropriate dimension. But the ease of drawing pictures in two dimensions is a great advantage, and we will consider only the two-dimensional case. Sources for the more general cases are given in Notes and References.

When phase plane variables $u = y$, $v = \dot{y}$ are introduced into the equation for the model problem, the result is

$$\dot{u} = v,$$
$$\varepsilon\dot{v} = -u - v. \tag{7.7.1}$$

The natural generalization of this system to a nonlinear (by which we mean *not necessarily linear*) autonomous system is

$$\dot{u} = f(u, v),$$
$$\varepsilon\dot{v} = g(u, v). \tag{7.7.2}$$

This is the type of system to be studied in this section. In the spirit of phase plane work, we will often view (7.7.2) as defining a vector field

$$[\dot{u}, \dot{v}] = [f(u, v), \frac{1}{\varepsilon}g(u, v)] \tag{7.7.3}$$

in the phase plane; the solutions of (7.7.2) lie on orbits which are curves tangent to (7.7.3) at each point. (If you are reading this chapter without having studied previous phase plane work in this book, such as in Chapter 4, you may want to read Appendix D at this point to familiarize yourself with orbits and the notion of the flow of an autonomous system.) This point of view is most useful in studying the geometry of the entire family of solutions. For approximation purposes, it is necessary to consider a particular initial value problem for the system (7.7.2), and the one we will discuss is

$$u(0) = \alpha,$$
$$v(0) = \gamma. \tag{7.7.4}$$

In place of (7.7.4) it is also possible to study initial conditions of the form

$$u(0) = \alpha,$$
$$v(0) = \beta/\varepsilon + \gamma, \tag{7.7.5}$$

that is, problems with an initial impulse (as we did with the original model problem). But the method that must be used for (7.7.5) is less efficient in case $\beta = 0$ than the method we will present here for (7.7.4), so it is best to assume $\beta \neq 0$ in (7.7.5) and to treat (7.7.4) and (7.7.5) separately. For simplicity we will handle only (7.7.4). (See Notes and References.)

Our experience with the model problem in Section 7.1 suggests that there will be some difficulty in defining the problem to which (7.7.2) reduces when $\varepsilon = 0$. Setting $\varepsilon = 0$ in (7.7.2) gives

$$0 = g(u, v), \tag{7.7.6}$$

which is not a differential equation at all. The meaning of (7.7.6) will become clear as we proceed, but for the moment let us note that it defines a particular level curve of g in the phase plane which will be called the *slow curve*. It will be necessary to impose some hypotheses on the function g in order to guarantee that this is a smooth curve with nice properties. The simplest assumption that can be made here is the following hypothesis.

Hypothesis of Global Stability of the Slow Curve. *There exists a constant $k > 0$ such that $g_v(u, v) \leq -k < 0$ for all u and v.*

There are weaker (more local) versions of this hypothesis that are satisfactory but more difficult to state. The first consequence of this hypothesis is that along vertical lines (with $u = $ constant), g decreases monotonically as v increases, approaching infinity at the "bottom" of the phase plane and minus infinity at the top. Then for each u there is a single value of v for which $g = 0$, so that (7.7.6) is solvable for v to give a unique function

$$v = \phi(u) \tag{7.7.7}$$

defined for all u. (The implicit function theorem, applied at each point of the slow curve, implies that (7.7.7) is smooth everywhere.) Therefore the slow curve is a single curve crossing the plane from left to right without doubling back. Furthermore, g is negative above the slow curve and positive below. Now we can interpret (7.7.2) with $\varepsilon = 0$ as follows. The second equation, namely (7.7.6), can be written as (7.7.7) and substituted back into the first equation to obtain

$$\dot{u} = f(u, \phi(u)), \tag{7.7.8}$$

which is a differential equation for u as a function of t. After solving this for $u(t)$, the result may be substituted into (7.7.7) to obtain $v(t)$. This describes a situation in which the point (u, v) moves along the slow curve with time. From this point of view it does not seem as if (7.7.2) makes sense when $\varepsilon = 0$ if the initial point does not lie on the slow curve. This defect can be remedied (in a somewhat fanciful manner) by putting ε on the right-hand side of (7.7.2), as in (7.7.3), before setting it equal to zero. Since $g(u, v)$ is negative above the slow curve and positive below, letting $\varepsilon = 0$ (or better, letting $\varepsilon \to 0$) in (7.7.3) seems to give the result that

$$\dot{v} = \begin{cases} -\infty & \text{above the slow curve} \\ 0 & \text{on the slow curve} \\ +\infty & \text{below the slow curve.} \end{cases}$$

This suggests a way of interpreting the reduced problem when the initial conditions are not on the slow curve. Suppose that an initial point (α, γ) is given as in (7.7.4). If $\gamma \neq \phi(\alpha)$, that is, if the initial point is not on the slow curve, then the initial rate of change of v is infinite and is directed toward the slow curve. Thus the solution should make an instantaneous vertical jump from the initial point (α, γ) to the point $(\alpha, \phi(\alpha))$ on the slow curve directly above or below the initial point. From there, the motion can proceed along the slow curve according to (7.7.8) and (7.7.7). This *discontinuous solution* will be taken as the solution of the reduced problem (7.7.2) and (7.7.4) with $\varepsilon = 0$. Of course, this definition will have to be justified. It will be seen in the following discussion that the solution of (7.7.2) and (7.7.4) for small positive ε approaches this "discontinuous solution" as $\varepsilon \to 0$. Figure 7.7.1 shows an initial point, a slow curve, a discontinuous solution, and an actual solution for small positive ε. The actual solution rapidly approaches the slow curve and then moves slowly (by comparison) in a neighborhood of it. The first phase—the rapid transition—is the initial layer.

In order to show that this qualitative picture of the motion is correct, consider a compact region containing a portion of the slow curve, such as the box in Fig 7.7.2. Draw a small tube around the slow curve, and consider the vector field in the box outside of the tube. By continuity and compactness, there will be a constant $c > 0$ such that above the tube

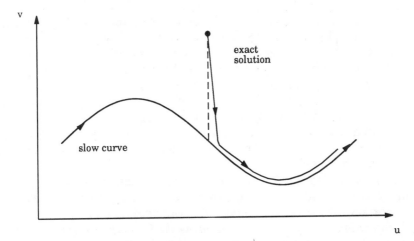

FIG. 7.7.1. Schematic illustration of the solution to a nonlinear autonomous initial layer problem in the phase plane. The solution drops rapidly from the initial point to the slow curve and then moves slowly along a path close to the slow curve. As $\varepsilon \to 0$ the initial motion approaches the vertical dashed line and the later motion follows the slow curve more closely.

$g(u, v) < -c$ while below the tube, $g(u, v) > c$. It follows from (7.7.3) that above the tube $\dot{v} < -c/\varepsilon$ and below the tube $\dot{v} > c/\varepsilon$; this assumes $\varepsilon > 0$. Thus for small positive ε, every solution will move rapidly towards the slow curve until it enters the tube. Now consider the vector field (7.7.3) on the boundary of the tube. The v component of this vector field is directed toward the slow curve and can be made as large as desired by decreasing ε. On the other hand the u component is bounded and does not change with ε. We will assume that the tube has been chosen so that its boundary curves have no vertical tangents; therefore the slope of these curves is bounded away from infinity. Therefore it is possible to choose ε_0 small enough that for all ε in $0 < \varepsilon < \varepsilon_0$ the vector field (7.7.3) points *into the tube* at

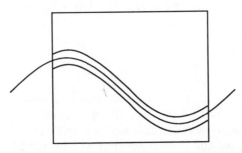

FIG. 7.7.2. Inside a box, but outside a tubular neighborhood of the slow curve, the speed of the fast vertical motion has a uniform lower bound.

each point of the boundary. This guarantees that once the solutions enter the tube, they do not leave (unless they leave the box itself through its right or left edges). Of course, the choice of ε_0 depends upon c, which in turn depends upon the choice of box and tube; in particular, narrowing the tube would tend to reduce c and therefore require a smaller ε_0. Such an argument could be carried out globally (that is, without the box) if the quantities involved in the argument have global bounds, but in general one cannot confine the solutions to an *arbitrary* tube as they move to infinity (to the right or left): It may be necessary to let the tube widen in order to achieve suitable bounds.

When we turn to the construction of approximate solutions for (7.7.2) in the next paragraph, one aspect of the previous qualitative study plays an important role: *In the initial layer the v-variable undergoes a substantial rapid transition, but the change in u during the initial phase of the motion is much smaller.* This feature will be build into the form of the approximate solution from the beginning. Specifically, the solution will be constructed on the assumption that v has a initial layer correction term with leading gauge $\mathcal{O}(1)$, but that the leading gauge for the correction layer for u is only $\mathcal{O}(\varepsilon)$. As in Chapter 4 and elsewhere in this book, an understanding of the qualitative features of the solutions is very helpful in setting up a suitable form of asymptotic solution.

It is here that a crucial difference occurs between initial conditions of the form (7.7.4) without an impulse and ones of the form (7.7.5) with an impulse. In the impulse case, the initial point is not fixed in the phase plane but moves off to infinity (in the v direction) as $\varepsilon \to 0$. This increases the distance that must be travelled to reach a neighborhood of the slow curve, giving an opportunity for u also to undergo a rapid change during the initial layer. The initial layer correction must now be constructed with a leading gauge of 1 for u and $1/\varepsilon$ for v. This in turn causes a rearrangement in the order in which the terms of the outer solution and correction must be computed. As mentioned above, we will not treat this case here.

It is time to begin the actual construction of the approximate solutions. Although it is possible to do this by matching, we will instead use the initial layer correction method. Let $\tau = t/\varepsilon$. The first step is to write the solution in the form

$$u(t, \varepsilon) = p(t, \varepsilon) + \varepsilon q(\tau, \varepsilon),$$
$$v(t, \varepsilon) = r(t, \varepsilon) + s(\tau, \varepsilon).$$

$$(7.7.9)$$

Here p and r are intended to form the "outer solution" and εq and s the "initial layer correction." (Actually, the quantities p, q, r, and s will be expanded only to a given order in ε, and it is these truncations that are the outer approximation and correction that will be used.) The inclusion of the factor ε with the q-term expresses the fact (discussed in the previous paragraph) that there is no $\mathcal{O}(1)$-term in the initial layer correction for

u. As in Section 7.4, three conditions are imposed on (7.7.9), which will suffice to fix the differential equations and side conditions satisfied by p, q, r, and s:

(i) The outer solution (p, r) should satisfy the differential equations (7.7.2) but not necessarily the initial conditions (7.7.4).

(ii) The composite solution (u, v) should satisfy both the differential equations (7.7.2) and the initial conditions (7.7.4).

(iii) The initial layer correction $(\varepsilon q, s)$ should satisfy the correction conditions, that is, should approach zero as τ approaches infinity.

The first condition simply says that

$$\dot{p} = f(p, r),$$
$$\varepsilon \dot{r} = g(p, r). \tag{7.7.10}$$

To derive the differential equations for q and s from condition (ii), let $' = d/d\tau$. Then

$$\dot{u} = \dot{p} + \varepsilon \cdot \frac{1}{\varepsilon} q' = f(p + \varepsilon q, r + s).$$

(Note the appearance and cancellation of the factor $1/\varepsilon$.) In view of (7.7.10), it follows that $q' = f(p + \varepsilon q, r + s) - f(p, r)$. Similarly,

$$\varepsilon \dot{v} = \varepsilon \dot{r} + \varepsilon \cdot \frac{1}{\varepsilon} s' = g(p + \varepsilon q, r + s).$$

Then (7.7.10) implies $s' = g(p + \varepsilon q, r + s) - g(p, r)$. These equations, together with the initial values implied by condition (ii) and (7.7.4), take the following form when written out in detail using $t = \varepsilon \tau$:

$$q'(\tau) = f(p(\varepsilon\tau) + \varepsilon q(\tau), r(\varepsilon\tau) + s(\tau)) - f(p(\varepsilon\tau), r(\varepsilon\tau)),$$
$$s'(\tau) = g(p(\varepsilon\tau) + \varepsilon q(\tau), r(\varepsilon\tau) + s(\tau)) - g(p(\varepsilon\tau), r(\varepsilon\tau)),$$
$$p(0, \varepsilon) + \varepsilon q(0, \varepsilon) = \alpha,$$
$$r(0, \varepsilon) + s(0, \varepsilon) = \gamma. \tag{7.7.11}$$

The dependence of each quantity on ε as a second argument has been suppressed in writing (7.7.11). It is interesting that both equations in (7.7.11) have exactly the same form, except that f in the first equation is replaced by g in the second, in spite of the fact that ε enters differently into the derivations of each. The third condition (iii) can be written as

$$\lim_{\tau \to \infty} q(\tau, \varepsilon) = 0,$$
$$\lim_{\tau \to \infty} s(\tau, \varepsilon) = 0. \tag{7.7.12}$$

Next we substitute the expansions

$$p(t, \varepsilon) = p_0(t) + \varepsilon p_1(t) + \cdots,$$
$$r(t, \varepsilon) = r_0(t) + \varepsilon r_1(t) + \cdots,$$
$$q(\tau, \varepsilon) = q_0(\tau) + \varepsilon q_1(\tau) + \cdots, \tag{7.7.13}$$
$$s(\tau, \varepsilon) = s_0(\tau) + \varepsilon s_1(\tau) + \cdots$$

into equations (7.7.10–(7.7.12) and derive the following sequence of equations, which are not to be taken entirely "at face value" for reasons that will be made clear soon. For the derivation of these equations see Exercise 7.7.1, and notice that in the equations for the correction terms one has $p_0(t) = p_0(\varepsilon\tau) = p(0) + \mathcal{O}(\varepsilon)$ and similarly for r; this explains the combination $r_0(0) + s_0(\tau)$ appearing in several of these equations. The functions $K_n(\tau)$ and $L_n(\tau)$ have expressions in terms of certain of the functions p_k, q_k, r_k and s_k; K_1 L_1 are computed explicitly in Exercise 7.7.1. These expressions become quite complicated beyond $n = 1$; the important thing about them is that at the point when these functions are needed in the recursive solution of the equations (which will be spelled out in detail below), the K_n and L_n are known functions of τ because the quantities on which they depend have already been computed.

$$\dot{p}_0 = f(p_0, r_0),$$
$$\dot{p}_1 = f_u(p_0, r_0)p_1 + f_v(p_0, r_0)r_1, \tag{7.7.14a}$$

$$\vdots$$

$$0 = g(p_0, r_0),$$
$$\dot{r}_0 = g_u(p_0, r_0)p_1 + g_v(p_0, r_0)r_1, \tag{7.7.14b}$$

$$\vdots$$

$$q'_0(\tau) = f(p_0(0), r_0(0) + s_0(\tau)) - f(p_0(0), r_0(0)),$$
$$q'_1(\tau) = f_v(p_0(0), r_0(0) + s_0(\tau)) s_1(\tau) + K_1(\tau), \tag{7.7.14c}$$

$$\vdots$$

$$s'_0(\tau) = g(p_0(0), r_0(0) + s_0(\tau)) - g(p_0(0), r_0(0)),$$
$$s'_1(\tau) = g_v(p_0(0), r_0(0) + s_0(\tau)) s_1(\tau) + L_1(\tau), \tag{7.7.14d}$$

$$\vdots$$

$$p_0(0) = \alpha,$$
$$p_1(0) + q_0(0) = 0,$$

$$\tag{7.7.14e}$$

$$\vdots$$

$$r_0(0) + s_0(0) = \gamma,$$
$$r_1(0) + s_1(0) = 0,$$

$$\tag{7.7.14f}$$

$$\vdots$$

$$q_0(\infty) = 0$$
$$q_1(\infty) = 0$$

$$\tag{7.7.14g}$$

$$\vdots$$

$$s_0(\infty) = 0$$
$$s_1(\infty) = 0$$

$$\tag{7.7.14h}$$

$$\vdots$$

When we said that these equations were not quite to be taken "at face value," we meant that the individual equations are not necessarily to be used in the way that would seem natural at first glance. For instance, consider the second equation of the set (7.7.14b). At first sight this would appear to be a differential equation for r_0. But this does not make sense, because the right hand side depends upon r_1, which will not be known at the time that r_0 is sought.

It turns out that this equation *should be regarded as an equation for* r_1, and therefore is not a differential equation at all, but merely an "algebraic" equation. (This is not meant in the technical sense of algebraic function theory.) This, in turn, raises an additional perplexity: Since all the equations of the family (7.7.14b) are "algebraic" in this sense, there seem to be too many side conditions. That is, the equations of the four families (e) through (h) give side conditions for the four families (a) through (d), but since one of these families is algebraic, it needs no side conditions. The resolution of this difficulty is that one family of side conditions, namely (7.7.14h), is not to be used.

Instead, this family of equations will be proved to hold as an automatic consequence of the "hypothesis of the stability of the slow curve." Because of these difficulties, it is necessary to rearrange the equations in (7.7.14) and spell out precisely the order in which they are to be solved.

The first step is to determine the leading order outer solution. For this purpose, the first equations of families (a), (b), and (e) are used, namely

$$\dot{p}_0 = f(p_0, r_0),$$
$$p_0(0) = \alpha, \tag{7.7.15}$$
$$0 = g(p_0, r_0).$$

The last of these can be solved for r_0 in the form

$$r_0 = \phi(p_0) \tag{7.7.16}$$

in view of (7.7.6) and (7.7.7). Next (7.7.16) is substituted into the first equation of (7.7.15) to obtain

$$\dot{p}_0 = f(p_0, \phi(p_0)),$$
$$p_0(0) = \alpha, \tag{7.7.17}$$

an initial value problem that completely determines $p_0(t)$. Substituting this back into (7.7.16) gives $r_0(t) = \phi(p_0(t))$, completing the first order outer solution. Notice that (7.7.17) is the same as (7.7.8), describing the motion of the reduced system on the slow curve.

Next the leading order correction is determined in two stages. From the last paragraph, p_0 and r_0 are known and $g(p_0, r_0) = 0$. Therefore the first equations of the (d) and (f) families can be written as

$$s_0' = g(p_0(0), r_0(0) + s_0(\tau)),$$
$$s_0(0) = \gamma - r_0(0), \tag{7.7.18}$$

an initial value problem that completely determines $s_0(\tau)$. Notice that this problem is nonlinear, but since the right-hand side is known it can be solved by integrating both sides with respect to τ. Since $s_0(\tau)$ is completely determined, with no remaining arbitrary constants, it is not possible to "use" the first equation of the (h) family, that is, $s_0(\infty) = 0$, but instead it must be proved to hold automatically as a consequence of the stability of the slow curve. To do so, let $G(\xi) := g(p_0(0), r_0(0) + \xi)$. Since $G(0) = 0$, the mean value theorem implies $G(\xi) = G'(\eta)\xi$ for some η (depending on ξ) in the interval $0 < \eta < \xi$. But from the hypothesis of global stability of the slow curve, $G' = g_v < -k$, so $G(\xi) < -k\xi$ for all ξ. Applying this to $\xi = s_0(\tau)$, (7.7.18) implies $s_0' < -ks_0$, so that

$$s_0(\tau) \le s_0(0)e^{-k\tau}. \tag{7.7.19}$$

Thus not only is $s_0(\infty) = 0$, but the decay toward zero as $\tau \to \infty$ is exponential.

Once s_0 is known, the first equations of the (c) and (g) families

$$q_0' = f(p_0(0), r_0(0) + s_0(\tau)) - f(p_0(0), r_0(0)),$$
$$q_0(\infty) = 0$$
(7.7.20)

become solvable. The differential equation here is solvable by integration, with an undetermined additive constant. In order to use the condition at infinity to determine this constant, it is necessary to know that the solutions of the differential equation have a limit at infinity. But in view of (7.7.19), $s_0(\tau)$ remains in a compact set, so that the right hand side of (7.7.20) can be bounded using a Lipschitz constant; that is, $|q_0'(\tau)| \le L|s_0(\tau)| \le Ls_0(0)e^{-k\tau}$. It follows from the comparison test for the convergence of improper integrals that

$$\int_0^\infty q'(\sigma)d\sigma$$

exists. If we write the general solution of (7.7.20) as

$$q_0(\tau) = \text{const} + \int_0^\tau q_0'(\sigma)d\sigma,$$

it is now clear that $\lim_{\tau \to \infty} q_0(\tau)$ exists, and the constant can be chosen to make it equal zero. This completes the determination of the leading order correction.

The pattern established for the leading order calculations now repeats itself for each successive order. That is, the next order of the outer solution is found from the next equation in the (a), (b), and (e) families; then (d) and (f) are used to find the next order in s, after which (h) is proved (not used); and finally, (c) and (g) are used to find the next order in q. The only difference from the leading order occurs in handling the (b) family. All of the equations in the (b) family are interpreted as "algebraic" equations, not as differential equations. But while the first equation in the (b) family is nonlinear, and requires a hypothesis (the existence of the function ϕ) for its solution, the remaining equations are linear and can be solved easily. Thus, the second equation in the (b) family is to be used in the form

$$r_1 = \frac{\dot{r}_0 - g_u(p_0, r_0)p_1}{g_v(p_0, r_0)}.$$
(7.7.21)

The denominator is not zero, because of the stability hypothesis for the slow curve. The remaining details are left to the reader.

Exercises 7.7

1. Give a complete discussion of the model problem in the form (7.7.1) by the methods of this section. Draw the slow curve, calculate the solutions

(to two terms) for the initial value problem $u(0) = \alpha$, $v(0) = \beta$, sketch these solutions on your graph, and compare your solution for u with the two-term solution of $\varepsilon\ddot{y} + \dot{y} + y = 0$, $y(0) = \alpha$, $\dot{y}(0) = \beta$ found by the methods of Section 7.2 or 7.4. (Remember that these initial conditions are not the same as those used for the model problem in earlier sections, since the present section did not deal with impulses, so you will have to solve the problem from scratch.) Also solve (7.7.1) exactly by matrix methods and compare the exact solution with your approximation.

2. Express the singularly perturbed Duffing equation $\varepsilon\ddot{y} + \dot{y} + y + y^3 = 0$ in the form (7.7.2). Locate the slow curve and any rest points, check the stability hypothesis for the slow curve, and describe the motion in the plane for arbitrary initial conditions. Carry out the solution to leading order, noticing that (7.7.18) is in fact linear for this problem. Repeat the problem for $\varepsilon\ddot{y} + \dot{y} + y - y^3$.

7.8* RELAXATION OSCILLATIONS AND CANARDS

The general Van der Pol equation (without forcing) may be written

$$\ddot{y} + \lambda(y^2 - y)\dot{y} + y = 0. \tag{7.8.1}$$

In Part II, this equation was studied by several methods under the assumption that λ was a small parameter, using the notation $\lambda = \varepsilon$. In particular, it was shown in Section 4.4 that for sufficiently small λ there is a stable periodic orbit (a limit cycle in the phase plane) which is nearly sinusoidal and approximately computable by the Lindstedt method. In fact, equation (7.8.1) continues to have a stable limit cycle for all positive values of λ, and all solutions except the rest point are attracted to this limit cycle; we will not prove this here, since the proof depends on ideas from differential equations that are not of direct relevance to this book. Our purpose here is to discuss the limit cycle of (7.8.1) in the case of *large* λ, the extreme opposite case from that treated in Part II. The oscillations in this case are not at all sinusoidal, and instead are of a type called *relaxation oscillations*. This type of oscillation is characterized by a jerky motion in which a "slow phase" alternates with a "fast phase." A simple example of a relaxation oscillation (not governed by the Van der Pol equation) is a mass, attached by a spring to the wall, resting on a slowly moving conveyor belt. As the belt moves the mass slowly away from the wall, the tension in the spring increases until it is sufficient to overcome the friction between the mass and the belt, at which point the mass suddenly slides back toward the wall, then begins again to move slowly away. The name "relaxation" apparently comes from the phase in which the spring tension suddenly "relaxes" after its slow buildup.

We will not give a complete treatment of the relaxation oscillations of the Van der Pol equation, since that becomes quite technical and lengthy, but we will indicate the main features and how they relate to the ideas of this chapter. One of the central questions is the shape of the limit cycle. From Section 4.4 we know that the limit cycle approaches a circle as $\lambda \to 0$. As λ increases from zero, the limit cycle becomes more and more distorted, and as $\lambda \to \infty$ it approaches a limiting shape which (in a different set of coordinates) is easy to describe. It is possible to find this limiting shape in a simple intuitive manner; nevertheless, the issues involved are somewhat more subtle than they appear. In fact, there are equations very similar to the Van der Pol equation having solutions (called *canards*) that behave in a manner contrary to one's intuitive expectations. This discovery is relatively new and is not yet mentioned in most textbooks. Our discussion of canards will not be complete, but will serve as an introduction to the literature. (See Notes and References.)

It is interesting that canards were first discovered by a technique called "nonstandard analysis" that is very different from any method studied in this book; there now exists a treatment using "standard analysis." Nonstandard analysis is a mathematical theory that adds to the real number system certain infinitely small and infinitely large numbers to obtain what is sometimes called the "hyperreal" number system. Extending the real number system this way is fraught with dangers, as anyone knows who has studied the early history of calculus where the idea of infinitely small numbers was sometimes used in an imprecise way. But by the use of extremely sophisticated mathematical logic (model theory and ultrafilters) it has become possible to define and work with infinitesimals rigorously. Although there are a few results (such as the existence of canards) that were first discovered by nonstandard analysis, it seems unlikely that the method will ever be simplified enough to become popular.

The Van der Pol equation (7.8.1), with large λ, is equivalent to the following singularly perturbed system of equations which has the form (7.7.2), with small ε:

$$\dot{u} = -v - 1,$$
$$\varepsilon \dot{v} = u - \frac{1}{3}v^3 - v^2. \tag{7.8.2}$$

Although the dot is used to denote the time derivative in both (7.8.1) and (7.8.2), the time variables involved are not the same. In order to derive (7.8.2), write (7.8.1) using s as the time variable:

$$\frac{d^2y}{ds^2} + \lambda(y^2 - 1)\frac{dy}{ds} + y = 0.$$

Then introduce the following new variables:

$$u := \frac{1}{\lambda} \cdot \frac{dy}{ds} + \frac{1}{3}y^3 - y + \frac{2}{3},$$
$$v := y - 1,$$
$$t = s/\lambda, \tag{7.8.3}$$
$$\varepsilon = 1/\lambda^2.$$

The result (using $' = d/dt$) is (7.8.2). Here u and v define a "phase plane" somewhat different from the one used in Chapter 4, but which is particularly suited for the present problem. Observe that ε is small when λ is large, and that $\varepsilon = 0$ does not actually correspond to any instance of (7.8.1) but in some sense represents the limit of (7.8.1) as $\lambda \to \infty$. We will make no further use of (7.8.1) in this section but will instead study (7.8.2) and the following slight generalization:

$$\dot{u} = -v - \alpha,$$
$$\varepsilon \dot{v} = u - \frac{1}{3}v^3 - v^2. \tag{7.8.4}$$

Notice that (7.8.2) has a rest point at $(-2/3, -1)$, whereas the rest point of (7.8.4) is at $(\alpha^3/3 + \alpha^2, -\alpha)$.

Following the ideas of Section 7.7, the first step in studying (7.8.2) is to locate the slow curve $u - \frac{1}{3}v^3 - v^2 = 0$, which is drawn in Fig. 7.8.1. The motion on the slow curve (for the limiting system $\varepsilon = 0$) is governed by $\dot{u} = -v - 1$, and therefore the motion is to the left above the rest point $(-2/3, -1)$ and to the right below it; at the fold points there is a paradoxical situation in which it appears that the solution is expected to keep moving to the right or left and also to stay on the slow curve, which is impossible. However, the case $\varepsilon = 0$ does not correspond to any actual physical situation (since λ cannot actually equal infinity). When ε is very small (but not zero), the solution is not required to follow the slow curve exactly. So it is reasonable to expect that solutions following near the slow curve will in fact move away from the slow curve near the fold points and fall under the influence of the fast vector field. This intuition will now be developed more carefully.

It is clear that the slow curve fails to satisfy the stability hypothesis of Section 7.7 because of the vertical tangents and the "fold." If the plane is cut along the slow curve, it falls into two pieces, one of which is (generally speaking) "above" the slow curve and the other "below" it; for small ε the vector field "below" the slow curve is directed (basically) upward, and the vector field "above" it is directed downward, with the result that the folded-over portion of the slow curve is unstable. Since we will be especially interested in the motion near the vertical tangents (which we will call the

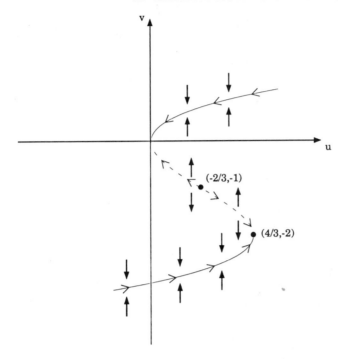

FIG. 7.8.1. The slow curve and vertical field for the Van der Pol equation with large parameter.

fold points), we have placed one of these points at the origin in Fig. 7.8.1 by including the constant terms $+2/3$ and -1 in (7.8.3). Now imagine a solution starting at the point A in Fig. 7.8.2. The reasoning of Section 7.7 (see Figure 7.7.2) will apply in a box containing a portion of the upper branch of the slow curve and excluding the fold point at C. Therefore the solution should rapidly drop to a point B near the slow curve and then follow (a neighborhood of) the slow curve to (a neighborhood of) the fold point C at the origin. (In the language of Section 7.7, the "discontinuous solution," which is the limit as $\varepsilon \to 0$ of the actual solutions, will instantaneously jump to the slow curve at a point B directly under A and travel exactly along the slow curve to the fold point; the actual solution will lie close to this discontinuous solution and may either lie slightly above the slow curve, as shown, or slightly below it.) As the solution approaches C, the reasoning in Section 7.7 loses its validity. But it is natural to expect that the solution will "fall off" the slow curve at C and move under the influence of the fast vertical field to a point D almost directly under C on the bottom branch of the slow curve and then proceed to points E, F, and so forth. (Notice that at point F we have shown the solution crossing the slow curve. The slow curve is not an orbit of the exact system, except in the limiting case $\varepsilon = 0$, and there is no reason solutions for small ε cannot cross it. As suggested above, the solution beginning at A could have

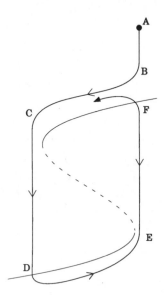

FIG. 7.8.2. Schematic illustration of the solutions of the Van der Pol equation with large parameter.

crossed the slow curve at B and then followed along beneath it. Of course, the solution cannot cross itself or any other solution.)

The solution starting at an arbitrary point A will not, of course, be the limit cycle, but it should approach the limit cycle as $t \to \infty$. It is clear that if the arguments of the last paragraph are correct, the solution in Fig. 7.8.2 will essentially retrace itself endlessly from F to C, D, E, and back to F, so this must be the approximate shape of the limit cycle (for a fixed small ε). The reasoning should be more accurate the smaller ε is. Therefore as $\varepsilon \to 0$, the segments CD and EF will become more and more nearly vertical, and "in the limit," one expects the limit cycle to lie close to the curve shown in Fig. 7.8.3, which has been called "the limit of the limit cycles." This curve is a "discontinuous solution" in which the vertical jumps are instantaneous.

The weak point in the above analysis is clearly the behavior in a neighborhood of the fold point. To examine this neighborhood more closely, it is reasonable to "enlarge" the neighborhood by introducing stretched coordinates. To do so we will stretch the v-coordinate by an undetermined factor $1/\varepsilon^\mu$ and stretch the horizontal distance to the slow curve by a different undetermined factor $1/\varepsilon^\nu$. That is, new coordinates ξ and η will be defined by

$$\varepsilon^\nu \xi = u - \left(\tfrac{1}{3}v^3 + v^2 \right),$$

$$\varepsilon^\mu \eta = v,$$

$$(7.8.5)$$

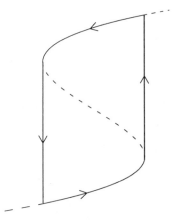

FIG. 7.8.3 The limit of the limit cycles for the Van der Pol equation.

leading (after some calculation) to the equations

$$\dot{\xi} = -\varepsilon^{-\nu} - \varepsilon^{\mu-1}2\xi\eta + \cdots,$$
$$\dot{\eta} = \varepsilon^{\nu-\mu-1}\xi, \tag{7.8.6}$$

where the dots denote terms that are subdominant for any positive μ and ν. The unknown scales are now fixed by requiring $-\nu = \mu - 1 = \nu - \mu - 1$ so as to make all terms in (7.8.6) equal in dominance and therefore obtain the "richest" possible approximate system. Then (7.8.6) becomes

$$\varepsilon^{2/3}\dot{\xi} = -1 - 2\xi\eta + \cdots,$$
$$\varepsilon^{2/3}\dot{\eta} = \xi, \tag{7.8.7}$$

which upon suitably scaling time (we leave this to you) and passing to the reduced problem (by dropping the dotted terms) becomes

$$\xi' = -1 - 2\xi\eta,$$
$$\eta' = \xi. \tag{7.8.8}$$

This is equivalent to $\eta'' + 2\eta\eta' + 1 = 0$, which is solvable in terms of Bessel functions or Airy functions. From here the remainder of the problem becomes very difficult. Suffice it to say that the solution of (7.8.8) confirms the conclusion that near the fold points, the solution moves away from the slow curve and into the fast vertical field. It is possible to obtain an approximation to the limit cycle for small ε (large λ) by matching the three types of solutions: fast solutions for the jumps (these are "inner" solutions in the sense of Section 7.7), slow ("outer") solutions along the slow curve, and solutions of (7.8.8) near the fold points. This is quite difficult because the transition layers do not occur near an initial point, but occur at points in time which themselves depend on the solution; furthermore it is necessary to use logarithmic gauges; and to make it still worse, Van Dyke matching fails and it is necessary to use matching by intermediate variables or overlap domains.

Now let us consider what happens in the modified Van der Pol equation (7.8.4). This equation has the same slow curve as before, and (roughly) the same fast vertical field, so it is described by Fig. 7.8.1 with the one difference being the position of the rest point on the slow curve. As long as the rest point is on the unstable branch of the slow curve, the discussion will proceed as before and the "limit of the limit cycles" will be as in Fig. 7.8.3. This will be the case no matter how close the rest point is to a fold point, as long as it is strictly on the unstable branch. On the other hand, when the rest point is on a stable branch there is no limit cycle at all. (Exercise 7.8.1.) This leads to the question: How does the limit cycle cease to exist as the rest point crosses the fold point?

The first thing to notice is that we have not paid sufficient attention to questions of uniformity in the last paragraph. We said that for any position of the rest point on the unstable branch, the limit cycle approaches Fig. 7.8.3 for small ε; but we said nothing about how small ε must be. The fact is that the closer the rest point is to the fold point, the smaller ε must be in order to obtain a picture close to Fig. 7.8.3. Therefore if ε is held at a fixed small value and α is moved so that the rest point crosses the fold point, there will be a neighborhood of the fold point in which the picture can look quite different from Fig. 7.8.3, and it is in this region that one must look in order to see how the limit cycle disappears.

The difficulty that presents itself is this: In order to use perturbation methods, one must let $\varepsilon \to 0$; but in order to see the desired behavior, we cannot let $\varepsilon \to 0$ with α constant. The solution is a rescaling of the parameter. (Compare Section 1.7, and the discussion leading to equation (4.5.11).) That is, it is necessary to consider perturbation families in which α depends upon ε in such a way that $\alpha(0) = 0$, that is, the rest point approaches the fold point as $\varepsilon \to 0$.

In order to carry out such a program in detail, it is necessary to select the right scaling, that is, to choose $\alpha(\varepsilon)$ correctly to exhibit the behavior being sought. This will not be done here, since we will confine ourselves to looking qualitatively at the reduced system that results from (7.8.4) by taking both ε and α equal to zero; this is independent of the scaling. In this case Fig. 7.8.1 is replaced by Fig. 7.8.4. Notice that in this figure the motion on the slow curve is towards the fold/rest point from above, and away from it below. What is important here is to think about the vector field just below the unstable branch of the slow curve near the fold/rest point. For the actual (impossible) reduced system, the vector field is discontinuous, and off the slow curve the vectors are vertical and infinite. For actual systems close to this one the vector field must be continuous, so the vectors near the slow curve must point more or less along the slow curve to the right.

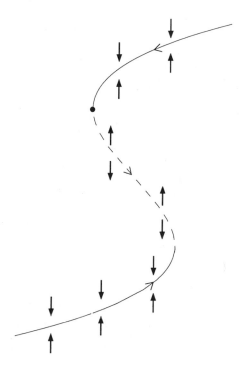

FIG. 7.8.4 The slow manifold and vertical field for a problem with canards.

On the other hand the slow curve is unstable, and so the vectors must point at least a little bit away from the slow curve (downwards). This downward component must become rapidly greater as one moves away from the slow curve.

Now consider a solution (of an actual system close to Fig. 7.8.4) that closely follows the stable branch of the slow curve down to (a neighborhood of) the fold point (which will not exactly coincide with the rest point, but almost). Instead of "falling off" the slow curve, a new possibility arises: The solution may appear to follow the unstable branch of the slow curve for a time. Because of the continuity of the vector field, there must exist solutions very close to the unstable branch, which diverge gradually from the slow curve and then, once they are far enough from the slow curve, begin to move rapidly downward. Several such solutions are shown in Fig. 7.8.5; one among these will be the limit cycle. A limit cycle of this type, which follows a portion of the unstable curve, is known as a *canard*. It is all a matter of the subtle interplay of time scales, the central question being how rapidly the solutions diverge from the unstable curve in comparison to how rapidly they travel along it.

As we have hinted, by taking suitable functions $\alpha(\varepsilon)$ and then letting $\varepsilon \to 0$ it is possible to find "discontinuous solutions" that represent the "limit of the limit cycles" when these limit cycles are of the type shown in Fig. 7.8.5. There are two possibilities, indicated in Fig. 7.8.6. Now it

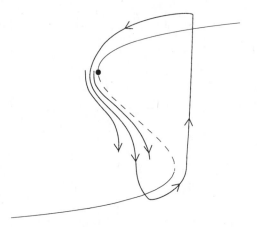

FIG. 7.8.5. A canard limit cycle and some nearby solutions. Notice that if these solutions are continued they approach the limit cycle, which is still stable.

is possible to answer the question of how the limit cycle disappears as the rest point crosses the fold point while ε is held at a small constant value. Namely, when the rest point is far from the fold, the limit cycle will be close to that shown in Fig. 7.8.3. As the rest point approaches the fold point, the vertical drop will begin to be delayed, giving a limit cycle like that of Fig. 7.8.6a but with the drop-off point at first very close to the fold. Then as the rest point is moved closer to the fold,

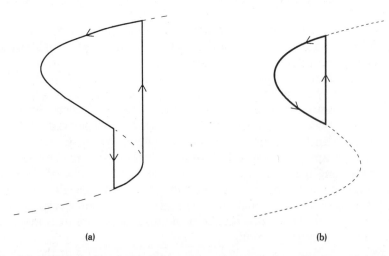

(a)   (b)

FIG. 7.8.6. The limit of the limit cycles for canard oscillations. (a) and (b) show two cases.

the delay will become longer and the drop-off point will approach the other (right-hand) fold point, after which the limit cycle will take the form of Fig. 7.8.6b. Now the "drop-up" point will move back to the left, and the limit cycle will shrink to a point (in fact, to the rest point) and disappear. This is in fact a reverse Hopf bifurcation (Section 4.4); the rest point changes from instability to stability (as seen in Exercise 7.8.1). All of this will take place while the rest point is in an extremely small neighborhood of the fold point; as a result, it is quite difficult actually to find (for instance, on a computer) solutions having the shape shown in Fig. 7.8.6.

Exercises 7.8

1. Show by a qualitative discussion (similar to the discussion of Fig. 7.8.1) that when the rest point is above the fold point, all solutions of (7.8.4) tend to the rest point.

7.9. NOTES AND REFERENCES

The term "patching" was apparently first used for a method of switching from an inner to an outer solution at a point determined by experimental data. What I have called the patching method is my own creation, as a simplification (for pedagogical purposes) of matching; it is most closely related to matching by an intermediate variable, as explained in Section 7.3. I claim no originality here, since anyone with the same goal could obtain the same result (and very likely has).

The Van Dyke method is presented with many computational examples and exercises in the books of Nayfeh. The book by Van Dyke containing the original formulation is focused exclusively on fluid flow problems, treated briefly in Section 8.2 below; the reference is given in the Notes and References for Chapter 8. The Fraenkel theorem, asserting that the first Van Dyke hypothesis often holds, as well as examples for which it does not hold, appear in

L. E. Fraenkel, On the method of matched asymptotic expansions, Part I: A matching principle, *Proc. Camb. Phil. Soc.* **65** (1969), 209–231.

The most thorough treatment of the method of matching by intermediate variables and overlap domains is probably

Wiktor Eckhaus, *Asymptotic Analysis of Singular Perturbations*, North-Holland, Amsterdam, 1979.

An elementary discussion of the extension theorem sometimes used for proving rigorous overlap is given in

R. Spigler, Double limits and matched asymptotic expansions, *American Mathematical Monthly* **91** (October, 1984), 501–505.

The correction method for both linear and nonlinear problems, together with error estimates (going far beyond the material presented here), can be found in

Donald R. Smith, *Singular-perturbation Theory*, Cambridge University Press, Cambridge, 1985.

A discussion of matching for the limit cycle of Van der Pol's equation for large λ is given in

H. Bavinck and J. Grasman, The method of matched asymptotic expansions for the periodic solution of the Van der Pol equation, *Int. J. Non-Linear Mechanics* **9** (1974), 421–434.

A treatment of canards by standard analysis, with references to the nonstandard treatment, is given in

Wiktor Eckhaus, Relaxation oscillations including a standard chase on French ducks, in *Lecture Notes in Mathematics* **985**, Springer-Verlag, New York, 1983.

CHAPTER 8

BOUNDARY LAYERS

8.1. SECOND ORDER LINEAR BOUNDARY VALUE PROBLEMS

This chapter is concerned with *boundary layer problems*, that is, boundary value problems exhibiting one or more transition layers near the boundary. The present section treats problems of the form

$$
\varepsilon \frac{d^2 y}{dx^2} + b(x)\frac{dy}{dx} + c(x)y = 0,
$$
$$
y(0) = \alpha, \tag{8.1.1}
$$
$$
y(1) = \beta,
$$

with $b(x) \neq 0$ for $0 \leq x \leq 1$. This has the same differential equation as the primary problem discussed in the last chapter (see (7.2.13) or (7.4.1)). As in Chapter 2, the independent variable is denoted by x rather than t, because boundary value problems most often (although not always) concern functions of a spatial rather than temporal variable. Since the techniques are similar to those developed at great length in Chapter 7, this problem can be handled much more rapidly. Therefore the Van Dyke method, the correction method, and a triple-deck example will all be presented within a single section. Section 8.2 will introduce the original example from which all boundary layer study began, the partial differential equations for fluid flow past a flat plate.

There are two fundamental differences between the treatment of (8.1.1) and the analogous initial value problem. The first of these concerns the location of the boundary layer. The initial value problem could be solved (by an asymptotic approximation) only if $b(t) > 0$, and in this case the transition layer always occurred close to the initial conditions. But (8.1.1) can be solved as long as $b(x) \neq 0$. If $b(x) > 0$ the boundary layer occurs at the left end of the interval, $x = 0$; if $b(x) < 0$ it occurs at the right end, $x = 1$. (There is, of course, a situation parallel to this in Chapter 7, namely that if $b(t) < 0$ then a "terminal-condition problem" on an interval such as $T \leq x \leq 0$ can be solved, with the transition layer at the right hand end of the interval; see Exercise 7.4.3.) Cases in which $b(x)$ changes sign inside the interval, or vanishes everywhere, can exhibit considerably more complex behavior, such as boundary layers at both ends or transition layers at internal points of the interval.

The second difference between boundary and initial value problems for singularly perturbed equations is the location of the undetermined constants that are to be fixed by matching. Consider the case $b > 0$, so that the boundary layer occurs at the left hand end. Then the inner solution is valid near the left hand end, and the outer solution is valid elsewhere, including at the right hand end. As in Chapter 7, each term of the outer solution satisfies a first order equation, while each term of the inner solution satisfies a second order equation. With initial value problems, there are two conditions imposed at the left end, which fully determines the inner solution, while the outer solution has an undetermined constant at each order. But with boundary value problems, one condition is imposed at each end. The condition at the right end is sufficient to fully determine the outer solution, since it is found from first order equations, but the inner solution, which would require two conditions for its full determination, satisfies only one, leaving one constant (at each order) in the inner solution to be determined by matching. (If the correction method is used in place of the matching method, there is a similar change in the way the outer solution and correction are characterized.)

These remarks, together with Chapter 7, are probably enough to enable the reader to develop everything in this section without reading any further, except for some remarks about the existence of the exact solution. Existence of solutions is always a question when boundary value problems are concerned. Since we have not dealt with boundary value problems since Chapter 2, it might be worthwhile to review the remarks about existence made there and in particular in Section 2.1, beginning with equation (2.1.8). At that point a distinction was made between oscillatory and nonoscillatory boundary value problems, and it was claimed that a unique solution always exists for nonoscillatory problems. The definition of "nonoscillatory" given there was formulated only for constant coefficient problems: The equation $ay'' + by' + cy = 0$ is nonoscillatory if and only if $b^2 - 4ac > 0$. It happens that a similar condition is valid for problems with nonconstant coefficients: The problem $ay'' + b(x)y' + c(x)y = 0$ (with constant a) is nonoscillatory if $b(x)^2 - 4ac(x) + 2ab'(x) > K$ for some $K > 0$, for all x in the domain of interest. Since we are concerned with the compact interval $0 \leq x \leq 1$, it is enough to have $b(x)^2 - 4ac(x) + 2ab'(x) > 0$;

the existence of K follows by compactness. Now for (8.1.1), $a = \varepsilon$, and the nonoscillatory condition is $b^2 + \varepsilon(2b' - 4c) > 0$ for $0 \le x \le 1$. This is automatic for sufficiently small ε. Since asymptotic solutions are only valid for small ε, (8.1.1) may be regarded for present purposes as nonoscillatory. In fact, as this brief discussion suggests, there always exists a unique solution to the boundary value problem (8.1.1) for ε in some interval $0 < \varepsilon < \varepsilon_0$. No proof of this fact will be attempted here.

Unfortunately, the error estimation for the problems of this section is more difficult than that of Section 7.5 and will be omitted here. (A reference is given in Section 8.3.) Briefly, the reason for the difficulty is as follows. After computing the approximate solution (8.1.15) below, it is possible to find an integral equation for the remainder, which is similar to (7.5.8), the primary difference being that the integral is from zero to one in view of (C.37). This means that Gronwall's technique cannot be used to estimate the remainder. The algebraic technique used in Section 3.1 does not work either, since the terms do not satisfy strong enough estimates. One must use techniques suitable for Fredholm (instead of Volterra) integral equations; in this way one can simultaneously prove existence of R (which implies existence of the exact solution y) and an estimate for $|R|$. The result is that y^c has an error that is of the order of the first omitted term, uniformly for $0 \le x \le 1$. As "evidence," one may consider an exactly solvable problem such as the "model problem" of Chapter 7 (with the initial values replaced by boundary values). See Exercise 8.1.4.

In order to find an approximate solution of (8.1.1) by the method of matching (by the Van Dyke matching rules), we will first assume

$$b(x) > 0. \tag{8.1.2}$$

We have claimed, without proof, that in this case the boundary layer occurs at the left end of the interval. Our actual course of action is to simply assume that this is the location of the layer and then observe that (8.1.2) makes the matching possible under this assumption, whereas if $b(x) < 0$ the matching is not possible. Even this, of course, does not prove that the exact solution has a boundary layer at the left end, but that follows once the approximate solution is shown to be asymptotically valid (by direct error estimation, which, as mentioned above, is omitted).

Assuming, then, that the boundary layer occurs near the left end, we introduce a stretched or inner variable near the left end by setting

$$\xi := x/\varepsilon. \tag{8.1.3a}$$

(If the boundary value problem were formulated on an interval $A \le x \le B$, this would have the form

$$\xi := \frac{x - A}{\varepsilon} \tag{8.1.3b}$$

instead.) The inner and outer variables are ξ and x in this chapter, τ and t in the last chapter. In the last chapter, $\dot{} = d/dt$ and $' = d/d\tau$, that is, dot is differentiation with respect to the outer variable and prime is differentiation with respect to the inner variable. In order to facilitate comparison between this chapter and the last, the same convention will be adopted here, that is,

$$\dot{} = d/dx,$$
$$' = d/d\xi. \tag{8.1.4}$$

(Notice that this conflicts with the convention in Chapter 2 that prime denotes d/dx.) In this notation, the *outer equations* consist of (8.8.1) together with the right hand boundary condition,

$$\varepsilon \ddot{y} + b(x)\dot{y} + c(x)y = 0,$$
$$y(1) = \beta, \tag{8.1.5}$$

while the inner equations are found from (8.1.1) and (8.1.3a), keeping the left hand boundary condition:

$$y'' + b(\varepsilon\xi)y' + \varepsilon c(\varepsilon\xi)y = 0,$$
$$y(0) = \alpha. \tag{8.1.6}$$

(Notice that even if the original problem is posed on $A \le x \le B$ with $y(A) = \alpha$, the inner solution will satisfy $y(0) = \alpha$ (and not $y(A) = \alpha$) in view of (8.1.3b).)

The perturbation series solutions of (8.1.5) and (8.1.6), taken to a specified number of terms, are denoted $y^o = y_0^o + \varepsilon y_1^o + \cdots$ and $y^i = y_0^i + \varepsilon y_1^i + \cdots$, respectively. For the present we will determine only the leading order approximation, so we take $y^o = y_0^o$ and $y^i = y_0^i$. These will satisfy (8.1.5) and (8.1.6) with $\varepsilon = 0$, that is,

$$b(x)\dot{y}^o + c(x)y^o = 0,$$
$$y^o(1) = \beta \tag{8.1.7}$$

and

$$y^{i\prime\prime} + b_0 y^{i\prime} = 0,$$
$$y^i(0) = \alpha, \tag{8.1.8}$$

where

$$b_0 = b(0). \tag{8.1.9}$$

The solutions of these equations are

$$y^o = \beta \exp \int_x^1 \frac{c(u)}{b(u)} du \tag{8.1.10}$$

and

$$y^i = Ae^{-b_0\xi} + (\alpha - A), \tag{8.1.11}$$

where A is a constant to be determined by matching. Notice that in agreement with our preliminary discussion, the outer solution is fully determined whereas it is the inner solution that contains an undetermined constant. In order to match the inner and outer solutions, express y^o from (8.1.10) in the inner variable as

$$y^o = \beta \exp \int_{\varepsilon\xi}^1 \frac{c(u)}{b(u)} du$$

and expand to leading order in ε to find

$$y^{oi} = \beta \exp \int_0^1 \frac{c(u)}{b(u)} du. \tag{8.1.12}$$

Similarly, write y^i from (8.1.11) in the outer variable as

$$y^i = Ae^{-b_0 x/\varepsilon} + (\alpha - A)$$

and expand to leading order as

$$y^{io} = \alpha - A, \tag{8.1.13}$$

using the fact that $b_0 > 0$ (from (8.1.2) and (8.1.9)) which implies that the exponential is transcendentally small. Equating (8.1.12) and (8.1.13) identifies the constant A as

$$A = \alpha - \beta \exp \int_0^1 \frac{c(u)}{b(u)} du. \tag{8.1.14}$$

Finally,

$$y^c = y^i + y^o - y^{io} = \left(\alpha - \beta \exp \int_0^1 \frac{c(u)}{b(u)} du\right) e^{-b_0 x/\varepsilon} + \beta \exp \int_x^1 \frac{c(u)}{b(u)} du. \tag{8.1.15}$$

Notice that the composite solution does not satisfy the boundary conditions exactly, but satisfies

$$y^c(0) = \alpha,$$
$$y^c(1) = \beta + \text{transcendentally small.} \tag{8.1.16}$$

If $b_0 < 0$ then y^i has no limit process expansion, and the matching fails. Intuitively, boundary layer behavior is characterized by exponential decay of the inner solution toward the outer, and (8.1.11) decays only if $b_0 > 0$. Under the opposite hypothesis, (8.1.11) grows exponentially, but an inner solution constructed "backwards" from the right hand end of the interval will decay. In fact (see Exercise 8.1.1) it can be shown that an inner variable

$$\xi := \frac{1-x}{\varepsilon} \tag{8.1.17}$$

allows an approximate solution to be constructed in this case.

The same composite solution (8.1.15) can be found by the correction method. To do so, write $y(x) = p(x) + q(\xi)$, where p is the outer solution and q the boundary layer correction (so that $q = y^i - y^{io}$). Then $\dot{y} = \dot{p} + q'/\varepsilon$ and $\ddot{y} = \ddot{p} + q''/\varepsilon^2$. Require that p satisfies the differential equation in (8.1.1) and the right hand boundary condition; that $p + q$ satisfies the differential equation and the left hand boundary condition; and that $q \to 0$ as $\xi \to \infty$. These conditions imply

$$\varepsilon \ddot{p} + b(x)\dot{p} + c(x)p = 0,$$
$$p(1) = \beta \tag{8.1.18}$$

and

$$q'' + b(\varepsilon\xi)q' + \varepsilon c(\varepsilon\xi)q = 0,$$
$$p(0) + q(0) = \alpha \tag{8.1.19}$$

and

$$q(\infty) = 0. \tag{8.1.20}$$

Now (8.1.18) is the same as (8.1.5), and (8.1.19) is the same as (8.1.6) except for the initial condition, so their solutions proceed very similarly. One sets $p \cong p_0 + \varepsilon p_1 + \cdots$ to a fixed number of terms, and similarly for q; for a leading order solution, it is only necessary to set $\varepsilon = 0$ in both (8.1.18) and (8.1.19). Solve (8.1.18) first, since it is a self-contained system; then find $p(0)$ to determine the initial condition for (8.1.19). The solution of (8.1.19) will contain an unknown constant that can be evaluated from (8.1.20), provided that $\beta_0 > 0$. See Exercise 8.1.2.

The solutions to (8.1.1) have only been carried to the first order, because it is complicated to handle the general case to higher orders. It is much simpler to treat individual problems when proceeding to, say, two terms. See Exercise 8.1.3.

We will conclude this section with a triple-deck example with an unusual feature. This example is to be compared with Example 7.6.2. See Exercise 8.1.5 for a problem related to Example 7.6.1.

Example 8.1.1

The problem in question is

$$\varepsilon^3 \ddot{y} + x^3 \dot{y} + (x^3 - \varepsilon)y = 0,$$
$$y(0) = \alpha, \qquad (8.1.21)$$
$$y(1) = \beta,$$

having the same differential equation as Example 7.6.2 but with boundary conditions in place of initial conditions. The preliminary work to determine the scales and the differential equations in each layer are the same here as in Example 7.6.2. For this reason we will continue to use t as the independent variable here (instead of the x which is usually our signal indicating "boundary value problem"). Therefore the scales are t, $\sigma = t/\mu$, and $\tau = t/\mu^2$, where $\mu = \sqrt{\varepsilon}$, and the leading order inner, middle, and outer equations are

$$y^{i''} - y^i = 0,$$
$$\sigma^3 y^{m*} - y^m = 0, \qquad (8.1.22)$$
$$\dot{y}^o + y^o = 0,$$

as in (7.6.20). The solution procedure will be to work from both ends. The outer solution will have a single arbitrary constant, which will be fixed by the right hand boundary condition. The middle solution will likewise have a single constant, fixed by matching with y^o. The inner solution has two constants, one to be fixed by the left hand boundary condition, and the other by matching with y^m. It is here that the unique feature of this example occurs. It was seen in Example 7.6.2 that there are difficulties in matching y^i with y^m which cannot be resolved in the case of an initial value problem. Therefore it is interesting to see how these difficulties resolve themselves for the boundary value problem.

The general solutions of the equations in (8.1.22) are

$$y^i = Ae^\tau + Be^{-\tau},$$
$$y^m = Ce^{-1/2\sigma^2}, \qquad (8.1.23)$$
$$y^o = De^{-t}.$$

The right hand boundary condition $y^o(1) = \beta$ determines $D = \beta e$. Matching this to y^m goes as follows. Write y^o in the middle variable as $De^{-\mu\sigma}$ and expand to leading order as $y^{om} = D$. Write y^m in the outer variable as $Ce^{-\mu^2/2t^2}$ and expand to leading order as $y^{mo} = C$. The matching condition $y^{om} = y^{mo}$ then implies $C = D = \beta e$. The left hand boundary condition applied to y^i gives $A + B = \alpha$, requiring one more condition to fix A and B (or, equivalently, leaving one unknown quantity). At this point we must face the question of matching y^i to y^m. It has been seen in Example 7.6.2 that y^{im} does not even exist unless $A = 0$ (so that y^i does not contain an exponentially growing term). Therefore we must take $A = 0$ in order for matching to be even a possibility. But if we do this, there is no constant remaining to be fixed by matching! How can the matching condition $y^{im} = y^{mi}$ be achieved when there is no adjustable constant available? Miraculously, $y^{im} = 0$ as long as $A = 0$ regardless of the value of B, and $y^{mi} = 0$ regardless of the value of C, so the matching takes place automatically. The very fact which made matching useless in Example 7.6.2 for the case of the initial values (7.6.22) makes matching succeed (at least formally) in the present example. The final conclusion, using $y^c = y^i + y^m + y^o - y^{im} - y^{mo}$, is that a leading order formal approximation to the solution of (8.1.21) is

$$y^c = \beta e^{1-t} + \beta e^{1-\frac{t}{2t^2}} + \alpha e^{-t/\varepsilon} - \beta e. \qquad (8.1.24)$$

It seems that an error analysis for this solution has not been carried out.

Exercises 8.1

1. Use (8.1.17) to obtain a leading order approximation for (8.1.1) when $b_0 < 0$ by Van Dyke matching. Warning: The variable ξ runs "backwards," that is, $\xi = 0$ at the right end and is positive for $x < 1$. How are the calculations affected if you choose $\xi = (x - 1)/\varepsilon$, an equally valid choice?

2. Complete the details of the correction method, in particular the derivation of (8.1.19) from the requirement imposed on $p + q$, and the calculation that $p + q$ (to leading order) agrees with (8.1.15). Also use the correction method to handle the case $b_0 < 0$ and check that your results agree with the previous exercise.

3. Find two-term approximations to the problem (8.1.1) in each of the following special cases: $\varepsilon \ddot{y} + (1 + x)\dot{y} = 0$, $\varepsilon \ddot{y} + \dot{y} - xy = 0$, $\varepsilon \ddot{y} + (2 - x)\dot{y} + xy = 0$, and $\varepsilon \ddot{y} - (1 + x)\dot{y} = 0$. (The first three of these have appeared in Exercise 7.2.1 as initial value problems. Of course, here you are to use the boundary values given in (8.1.1).)

4. Find a two-term approximation to the "model problem" of Chapter 7, that is, $\varepsilon \ddot{y} + \dot{y} + y = 0$, with the boundary values of (8.1.1). Also find the exact solution. Compare these, following the pattern of Exercise 7.2.4.

5. Create and solve (formally, to leading order) a third order triple-deck boundary value problem using the differential equation of Example 7.6.1. Hint: How many boundary conditions should be imposed at each end?

8.2* PARTIAL DIFFERENTIAL EQUATIONS: FLOW PAST A FLAT PLATE

The problem to be addressed in this section was first treated in 1905, and yet remains a subject of current research. (See Notes and References.) The results obtained here concern only the leading order approximation and coincide with the 1905 results which were obtained by ad hoc physical reasoning. These results are obtained here by a combination of the methods of undetermined gauges and Van Dyke matching. It does not seem to be possible to construct higher order approximations in the same way, without introducing additional features such as triple decks.

Mathematically, the problem is easy to state: to find a function $\psi(x,y)$ defined in the upper half-plane $y \geq 0$ satisfying

$$\varepsilon(\psi_{xxxx} + 2\psi_{xxyy} + \psi_{yyyy}) - \psi_y(\psi_{xxx} + \psi_{xyy}) + \psi_x(\psi_{xxy} + \psi_{yyy}) = 0,$$

$$(8.2.1a)$$

$$\psi(x,0) = 0 \qquad \text{for} \quad -\infty < x < \infty, \qquad (8.2.1b)$$

$$\psi_y(x,0) = 0 \qquad \text{for} \quad 0 \leq x \leq 1, \qquad (8.2.1c)$$

$$\psi(x,y) \to y \qquad \text{as} \quad x^2 + y^2 \to \infty. \qquad (8.2.1d)$$

The physical meaning of the equations (8.2.1) will now be described, without attempting to derive these equations from the physical laws governing fluid flow. (It is not necessary to understand the physical meaning of (8.2.1) in order to follow the solution presented below.)

Imagine a fluid flowing steadily in three-dimensional space with each molecule having velocity vector $\hat{\mathbf{i}}$, that is, unit velocity in the positive x direction. What we have just described is called the *velocity field* of the steady flow. Now suppose that an infinitely thin flat plate is introduced into the flow, occupying the region $0 \leq x \leq 1$, $y = 0$, $-\infty < z < \infty$. Our aim is to find the velocity field of the resulting steady flow. Since the plate is infinitely long in the z direction, nothing in the resulting flow pattern will depend upon z, and we may ignore the z variable, imagining the flow to take place in the (x,y) plane with the plate occupying the line segment $0 \leq x \leq 1$ on the x-axis. Furthermore, since the domain of the flow is symmetrical around the x-axis we will look for a symmetrical solution, and therefore we may confine our attention to the upper half-plane. (Asymmetrical solutions are possible; the symmetry of the domain does not preclude these, but implies that for every asymmetrical solution there will be another which is its reflection in the x-axis. So the assumption that the solution is symmetrical is a hypothesis which, it is hoped, helps to single out a unique solution.) The assumption that the flow is "steady" means that the velocity field does not depend on time. In other words, there is a fixed velocity vector attached at every point of the upper half plane, and each molecule of fluid that passes through such a point has that velocity at the moment it occupies that point.

Now it has been found convenient to describe velocity fields in the plane by their *stream functions* $\psi(x, y)$. Such a description is possible whenever the fluid in question is incompressible. For an incompressible fluid, the velocity vector $u\hat{i} + v\hat{j}$ satisfies $u_x + v_y = 0$ (that is, the divergence vanishes); it follows that there is a function ψ such that $u = \psi_y$ and $v = -\psi_x$. The single function ψ then replaces u and v as the unknown function of x and y. This is the function which satisfies (8.2.1). Notice that if $\psi = y$ then $u = 1$ and $v = 0$; that is, $\psi = y$ corresponds to the flow which we assumed (at the beginning of the previous paragraph) to take place in the absence of the flat plate. Equation (8.2.1d) then states that far away from the plate, the flow approaches this laminar condition. This corresponds to the observation that in a river, water can appear to flow in a steady parallel pattern until it reaches the neighborhood of a rock, remain nearly parallel at a distance from the rock, and return to a parallel pattern after passing the rock; it is only in some neighborhood of the rock that the flow is disrupted. An infinitely thin flat plate is of course a gross oversimplification of a rock, but the mathematical methods devised for the flat plate can also be applied to other geometries.

Next it is necessary to say something about the parameter ε in (8.2.1a). It is a nondimensional parameter proportional to the "viscosity" of the water; physicists are accustomed to using $R := 1/\varepsilon$, called the *Reynolds number*. When $\varepsilon = 0$ (or $R = \infty$) the fluid is called "inviscid" (pronounced "in-viss'-id" even though the "c" in "viscous" is hard). The concept of viscosity is somewhat mysterious, but it has two important effects. The first is the difference between water and the proverbial molasses in January: The latter is more viscous. But the second effect of viscosity is the crucial one in the present discussion and appears to be exactly the same for water and molasses: If a fluid has any viscosity at all, it sticks to things, whereas if it does not, it flows past them as though they were not there. So, for $\varepsilon > 0$, the x component of velocity must be zero along the flat plate, to which the fluid sticks. Therefore (considering the meaning of the stream function) condition (8.2.1c) expresses the effect of the flat plate when $\varepsilon > 0$. If $\varepsilon = 0$, on the other hand, condition (8.2.1c) should be dropped. In this case the remaining equations are satisfied by $\psi = y$; that is, the flow when $\varepsilon = 0$ is the same as the flow which we have postulated when the plate is not present.

Nothing further will be said here about (8.2.1a), which is a consequence of the Navier–Stokes equations for fluid flow. Condition (8.2.1b) is a consequence of the assumption of symmetry about the x-axis. Now we will return to the question of solving (8.2.1).

The first remark to make is that, essentially, nothing whatever is known about the rigorous existence or uniqueness of an exact solution to (8.2.1). Some are inclined to assume the solution must exist on physical grounds, while others point out that (8.2.1) might not be an adequate model of the physical situation and therefore its mathematical solvability is a separate question from the existence of a physical flow past an obstructing object. In any case, our explorations here cannot be more than formal, because the framework for an existence proof, not to mention an error analysis, does not exist.

Heuristically, therefore, we begin with the guess that an approximation can be created by matching two solutions: an outer solution valid away from the x-axis (and therefore away from the plate) and an inner solution valid near the x-axis. The experience of Chapters 7 and 8 suggests that the outer solution should satisfy the reduced equation of (8.2.1a), that is, the equation with $\varepsilon = 0$.

Since this reduces the order of the equation, it should also reduce the number of boundary conditions that may be imposed, and the discussion about inviscid flow suggests that it is condition (8.2.1c) that should be dropped. Under these circumstances we have observed that

$$\psi^o(x,y) = y \qquad (8.2.2)$$

is a solution, and without further ado we accept this as the leading order outer solution. As a slightly more sophisticated approach (comparable with our approach to the inner problem below) one might assume an outer solution in the form

$$\psi^o = \delta_0(\varepsilon)\psi_0(x,y) + \cdots, \qquad (8.2.3)$$

substitute this into (8.2.1), and argue that $\delta_0 = 1$ and $\psi_0 = y$. See Exercise 8.2.1.

To find an inner solution, we expect to have to dilate or stretch the y variable near the x-axis, that is, around $y = 0$. To this end we introduce

$$\eta := y/\mu(\varepsilon), \qquad (8.2.4)$$

where μ is an undetermined scale function. Now in Section 8.1, it was enough to assume a solution in the form of an asymptotic power series such as $\psi^i(x,y,\varepsilon) = \psi_0(x,\eta) + \varepsilon\psi_1(x,\eta) + \cdots$. For the present problem this does not work, and it is necessary to seek the solution in the form

$$\psi^i = \Delta_0(\varepsilon)\Psi_0(x,\eta) + \cdots. \qquad (8.2.5)$$

That is, the matching method does not work unless it is combined with the method of undetermined gauges. (It would suffice, in this problem but not all problems, to take the gauges in the form ε^ν for fractional ν.)

Upon differentiating (8.2.5) using (8.2.4) and substituting into (8.2.1a), one obtains

$$
\Delta_0 \left\{ \varepsilon \Psi_{0xxxx} + \frac{2\varepsilon}{\mu^2} \Psi_{0xx\eta\eta} + \frac{\varepsilon}{\mu^4} \Psi_{0\eta\eta\eta\eta} \right\}
$$
$$
+ \Delta_0^2 \left\{ -\frac{1}{\mu} \Psi_{0\eta} \Psi_{0xxx} - \frac{1}{\mu^3} \Psi_{0\eta} \Psi_{0x\eta\eta} + \frac{1}{\mu} \Psi_{0x} \Psi_{0xx\eta} + \frac{1}{\mu^3} \Psi_{0x} \Psi_{0\eta\eta\eta} \right\}.
$$

$$(8.2.6)$$

In the first set of braces, the term in ε/μ^4 dominates the others, while in the second set of braces, the two terms in $1/\mu^3$ dominate. Therefore the dominant part of (8.2.6) is (after simplifying)

$$
\frac{\varepsilon}{\mu} \Psi_{0\eta\eta\eta\eta} + \Delta_0 \left[\Psi_{0x} \Psi_{0\eta\eta\eta} - \Psi_{0\eta} \Psi_{0x\eta\eta} \right] = 0. \tag{8.2.7}
$$

The relative significance of the two gauges appearing in (8.2.7) is not yet determined. At this point, on the principle that the leading order equation should be kept as "rich" as possible so that the solution contains the maximum possible information (recall the discussion following equation (1.7.16)), we make these two gauges equal so that all of the terms in (8.2.7) are retained. That is, we set

$$
\frac{\varepsilon}{\mu} = \Delta_0 \tag{8.2.8}
$$

and conclude that

$$
\Psi_{0\eta\eta\eta\eta} + \Psi_{0x} \Psi_{0\eta\eta\eta} - \Psi_{0\eta} \Psi_{0x\eta\eta} = 0. \tag{8.2.9}
$$

Equation (8.2.9) is the leading order inner equation. (It would have been more in keeping with previous sections to have derived a "full" inner equation from (8.2.1) and (8.2.4), and then to substitute (8.2.5) into this equation to get (8.2.9), but we have done all of this in one step.)

In previous sections, the inner equation was solved prior to matching. In the present case, equation (8.2.9) is not actually solvable without some computer assistance, as will be discussed below. But it is possible to carry out the matching without knowing the full solution; the result of such a matching process is to complete the determination of the unknown gauge and scale function Δ_0 and μ and to derive a side condition on the solution of (8.2.9). But this cannot be done without some assumption about the behavior of Ψ_0 at infinity. In order to see what is involved, we first express the outer solution $\psi^o = y$ in the inner variable as $\mu\eta$; this is already its own leading order expansion, so $\psi^{oi} = \mu\eta = y$. Next the unknown inner solution $\psi^i = \Delta_0 \Psi_0(x, \eta)$ is written in the outer variable as $\Delta_0 \Psi_0(x, y/\mu)$. This is to be expanded to leading order in ε (recalling that Δ_0 and μ are unknown positive monotone functions of ε that approach zero as $\varepsilon \to 0$) to obtain ψ^{io}, which is then to be equated to $\psi^{oi} = y$. The difficulty concerns how to expand $\Psi_0(x, y/\mu)$. Clearly $y/\mu \to \infty$ as $\varepsilon \to 0$ for fixed x and y. That is, we are essentially concerned with approximating $\Psi_0(x, \eta)$ for large η.

In this book we have usually been concerned with approximations using a small parameter, but approximations for a large value of a variable are also important. Consider for a moment a function $y = f(x)$ in elementary calculus. If this function has a horizontal asymptote $y = y_0$ as $x \to \infty$ (that is, if the limit $y_0 = f(\infty) = \lim_{x\to\infty} f(x)$ exists), then $f(x) \cong y_0$ is the leading order approximation to $f(x)$ for large x. But if the function has a slant asymptote, then $f(\infty)$ does not exist but $m = f'(\infty) = \lim_{x\to\infty} f'(x)$ does exist (and is not zero). In this case the leading order approximation is $f(x) \cong mx$ for large x; this is the leading term in an expansion of the form $mx + b_0 + b_1/x + \cdots$ in powers of $1/x$, with $mx + b_0$ being the equation of the slant asymptote. Often such problems can be reduced to small-parameter problems by setting $x = 1/t$ and examining $t^n f(1/t)$ for various integers n. If for some n this function has a limit process expansion for small t as an asymptotic power series $a_0 + a_1 t + a_2 t^2 + \cdots$, then $f(x) \sim a_0 x^n + a_1 x^{n-1} + \cdots$ for large x; the case of a horizontal asymptote corresponds to $n = 0$ and the slant asymptote to $n = 1$.

It turns out (see Exercise 8.2.2) that the correct assumption about the behavior of $\Psi_0(x, \eta)$ as $\eta \to \infty$ is that $\Psi_{0\eta}(x, \infty)$ exists and $\Psi_0(x, \eta) \cong \eta \Psi_{0\eta}(x, \infty)$ to leading order; that is, for each x the function Ψ_0 has a slant asymptote as $\eta \to \infty$. Under this hypothesis the leading order expansion of $\Psi_0(x, y/\mu)$ is $(y/\mu)\Psi_{0\eta}(x, \infty)$, so the leading order expansion of $\Delta_0 \Psi_0(x, y/\mu)$ is

$$\psi^{io} = \frac{\Delta_0}{\mu} \Psi_{0\eta}(x, \infty) \cdot y. \tag{8.2.10}$$

Now the matching condition $\psi^{io} = \psi^{oi}$ gives

$$\frac{\Delta_0}{\mu} \Psi_{0\eta}(x, \infty) \cdot y = y, \tag{8.2.11}$$

from which we have

$$\frac{\Delta_0}{\mu} = 1 \tag{8.2.12}$$

and

$$\Psi_{0\eta}(x, \infty) = 1. \tag{8.2.13}$$

Finally, from (8.2.8) and (8.2.12) we conclude

$$\Delta_0(\varepsilon) = \mu(\varepsilon) = \sqrt{\varepsilon}, \tag{8.2.14}$$

which fixes both the scaling factor for the inner variable η and the leading gauge for the inner solution. Equation (8.2.13) becomes a side condition to be imposed on the solution of (8.2.9).

It remains only to find the solution of (8.2.9) satisfying $\Psi_0(x,0) = 0$ for $-\infty < x < \infty$, $\Psi_{0\eta}(x,0) = 0$ for $0 \le x \le 1$, and (8.2.13); the first two of these are derived from (8.2.1b,c). Once Ψ_0 is found, the solution is completed by

$$\psi^c = \psi^o + \psi^i - \psi^{oi} = y + \sqrt{\varepsilon}\Psi_0(x, y/\sqrt{\varepsilon}) - y$$
$$= \sqrt{\varepsilon}\Psi_0(x, y/\sqrt{\varepsilon}). \qquad (8.2.15)$$

Thus it happens that for this problem, to leading order, the composite solution coincides with the inner solution.

> It might appear from this that the outer solution was unnecessary in this problem, but this is not the case; in fact the inner solution could only be found by matching with the outer solution. It might also seem that there is no boundary layer, since the inner solution is valid everywhere. This also is not correct. The fact that there is a boundary layer is clear because the outer solution does not have the required behavior near the plate. The true explanation of why $y^c = y^i$ in this problem is simply that the inner solution accidentally happens to have the correct behavior (to leading order) away from the plate; that is, the inner solution already exhibits the boundary layer effect by itself. This does not happen at higher orders, so it is truly "accidental" and does not reflect a deeper property of the solution.

We will leave the problem at this point, except for a few further remarks. In attempting to solve (8.2.9) to find the function Ψ_0, one first notices that it is equivalent to

$$\frac{\partial}{\partial \eta}\left[\Psi_{0\eta\eta\eta} + \Psi_{0x}\Psi_{0\eta\eta} - \Psi_{0\eta}\Psi_{0x\eta}\right] = 0, \qquad (8.2.16)$$

which asserts that the quantity in brackets is a function of x alone. Next, this quantity is "evaluated" at $\eta = \infty$ using (8.2.13) and assuming (heuristically) that all higher derivatives of $\Psi_{0\eta}$ vanish at (x, ∞). (This assumption is based on interpreting (8.2.13) as $\Psi_{0\eta}(x, \eta) \cong 1$ for large η and differentiating the approximation. Mathematically this amounts to assuming that differentiation commutes with taking the limit as $\eta \to \infty$, which is an assumption about the smoothness of Ψ_0 at infinity.) One finds that the bracket vanishes at $\eta = \infty$, and hence everywhere; that is,

$$\Psi_{0\eta\eta\eta} + \Psi_{0x}\Psi_{0\eta\eta} - \Psi_{0\eta}\Psi_{0x\eta} = 0, \qquad (8.2.17)$$

a partial differential equation of one less order. It is possible, by some further analysis, to express the solution of (8.2.17) in the form

$$\Psi_0(x, \eta) = \sqrt{2x}\,f(\eta/\sqrt{2x}), \qquad (8.2.18)$$

where f is the solution of the following ordinary differential equation and side conditions:

$$f''' + ff'' = 0,$$
$$f(0) = 0,$$
$$f'(0) = 0,$$
$$f'(\infty) = 1,$$

(8.2.19)

but this problem is not solvable analytically and must be handled by numerical methods. It may appear that little has been gained by using asymptotic methods if a numerical solution is needed at the end; we might as well have solved the original problem (8.2.1) numerically. But this feeling is not correct. In fact, it is much easier to solve (8.2.19) numerically than (8.2.1). Furthermore, the leading order solution is given completely by (8.2.15) together with (8.2.18), as soon as the function f is known. This is quite similar to the fact that $y'' + k^2 y = 0$ is solved as $y = A \cos(kt + \phi)$ once the function cos is known. Now cos is known from its power series, but most importantly, we have a good feel for it from its familiar graph. A computer can easily make a graph of f, and then the meaning of (8.2.18) becomes clear. One advantage of (8.2.18) is that it shows how ε, x, and η enter into the solution; these things would not be clear at all from a numerical solution of (8.2.1).

Exercises 8.2

1. Substitute (8.2.3) into (8.2.1) and obtain an equation for ψ_0. Show that (8.2.1d) implies $\delta_0 = 1$, and that $\psi_0 = y$ is a solution.

2. Show that if one were to assume the existence of $\Psi_0(x, \infty)$ (that is, a horizontal rather than a slant asymptote) then matching would lead to $\psi^{io} = \Delta_0 \Psi_0(x, \infty)$, $\mu = \Delta_0 = \sqrt{\varepsilon}$, and $\Psi_0(x, \infty) = y$. The latter is impossible, since the left hand side does not depend on y.

8.3. NOTES AND REFERENCES

The computational aspect of Section 8.1 is presented, using the matching method with worked examples, in Nayfeh's *Introduction to Perturbation Techniques*, Chapter 12. Example 8.1.1 is taken from this book. The theoretical aspects are covered quite fully in Chapter 8 of Smith's *Singular Perturbation Theory*. Smith's treatment uses the correction method and gives complete proofs of the existence of the exact solution and of the error estimates. In addition to the correction method outlined here, Smith gives a version of the correction method using a different stretched variable ξ that is not a linear function of x. This method has some advantages, and is related to the WKB method it presented in Chapter 9 below. Other references for Section 8.1 are

P. A. Lagerstrom, *Matched Asymptotic Expansions*, Springer-Verlag, New York, 1988

and

Robert E. O'Malley, Jr., *Introduction to Singular Perturbations*, Academic Press, New York, 1974.

The standard reference for the material of Section 8.2 is Chapter 7 of

Milton Van Dyke, *Perturbation Methods in Fluid Mechanics*, annotated edition, The Parabolic Press, Stanford, Calif., 1975

or the original version without the annotations, published in 1964 by Academic Press. (The annotations are quite helpful and bring the book somewhat more up to date.) Unfortunately this book is quite difficult to read without a good grounding in fluid mechanics. For surveys of more recent results including the triple-deck solution for the wake, see

F. T. Smith, On the high Reynolds number theory of laminar flows, *IMA Journal of Applied Mathematics* **28** (1982), 207–281

or

Keith Stewartson, D'Alembert's Paradox, *SIAM Review* **23** (1981), 308–343.

CHAPTER 9

METHODS OF THE WKB TYPE

9.1. INTRODUCTION: THREE LINEAR PROBLEMS

In the class of second order linear differential equations, those with constant coefficients are exactly solvable and therefore do not require perturbation theory. There are two ways to get unsolvable second order problems: introduce nonlinearities, or allow the coefficients to become nonconstant. For the most part, this book has dealt with nonlinear perturbations of linear equations (or systems), although linear problems with nonconstant coefficients have also appeared. For instance, Mathieu's equation in Example 6.1.1 was linear, and many of the problems in Chapters 7 and 8 have been linear. But until this chapter, the methods presented have been useful for both linear and nonlinear problems. In this chapter, we address ourselves to a group of methods (loosely known as WKB methods) that are applicable only to linear problems. The term WKB (sometimes seen as WKBJ) stands for Wentzel–Kramers–Brillouin–(Jeffreys), names which are associated with the method particularly in regard to its applications in quantum mechanics. Within mathematics, these techniques are often referred to under the name *Liouville–Green method* and (for turning point cases) *Langer transformation.*

Our primary focus will be the following three equations:

$$\varepsilon^2 y'' + a(x)y = 0, \tag{9.1.1}$$

$$\ddot{y} + a(\varepsilon t)y = 0, \tag{9.1.2}$$

and

$$y'' + (\lambda^2 p(x) + q(x))y = 0. \tag{9.1.3}$$

433

In these equations ε is a small parameter as usual, and λ is a large parameter; that is, in considering (9.1.3) we will be concerned with approximate solutions which improve in accuracy as $\lambda \to \infty$. The functions a, p, and q will be assumed to be smooth and defined on the entire real line, although many of the results will also be valid for functions defined on an open interval. In this introduction we will discuss the background of each of these equations and the relationships between them. In the remainder of the chapter we will solve them asymptotically, first formally and then rigorously, and answer the most important questions that one wishes to know about each equation from the point of view of its applications.

From one point of view, the material in this chapter is also applicable to the equation

$$\varepsilon u'' + b(x, \varepsilon)u' + c(x, \varepsilon)u = 0.$$

Namely, the transformation $u = P(x)y$, where $P(x) := e^{-B(x)/2\varepsilon}$ with $B' = b$, carries this equation into (9.1.1) with

$$a(x, \varepsilon) = -\frac{b^2(x)}{4} - \varepsilon \left(\frac{b'(x)}{2} - c(x) \right).$$

This change of variables is called the *Sturm transformation*. A special case of it has already been given in Exercise 2.1.1. The point to be made here is that *if one is only interested* in finding two linearly independent solutions of $\varepsilon u'' + bu' + cu = 0$, then the Sturm transformation is useful: It carries this equation into one of the form (9.1.1), and then the solutions obtained in this chapter can be transformed back into the original variable u. But the Sturm transformation drastically changes the nature of any initial or boundary values applied to this equation; in particular, when boundary values are transformed from u to y, they may approach infinity as $\varepsilon \to 0$. Therefore the Sturm transformation is almost entirely useless for asymptotic approximation of solutions to boundary values problems for small ε. That is why it was never used in Chapters 7 and 8.

The first thing to notice about the three equations above is that no initial or boundary values have been given. This is closely related to the fact that the methods presented here are valid only for linear equations. Instead of posing an initial or boundary value problem at the beginning, we will instead look for (approximations to) two linearly independent solutions of the equations, without regard to any side conditions which they may satisfy. Since the general solution of a second order linear equation is a linear combination of two linearly independent solutions, specific boundary or initial value problems can be addressed afterward by attempting to choose the constants appropriately.

In looking at equation (9.1.1), it is not important that the small parameter is squared; this is merely a matter of convenience to avoid a lot of square roots later. What is important is that when $\varepsilon = 0$ the differential equation "drops order" from 2 to 0; that is, it ceases to be a differential equation at all! Chapters 7 and 8 concerned equations that dropped from second to first order in the reduced problem, but at least one was still dealing with a differential equation. Here, when $\varepsilon = 0$ we are left merely with the algebraic equation $a(x)y(x) = 0$. Ignoring for a moment that $y(x)$ should be continuous, this algebraic equation implies that $y(x) = 0$ wherever $a(x) \neq 0$, but at any point where $a(x) = 0$, $y(x)$ is arbitrary. It does in fact turn out that even for $\varepsilon \neq 0$, solutions of (9.1.1) exhibit different behavior at points where $a(x) = 0$. Therefore the study of (9.1.1) is broken down into several subcases. The simplest cases are

$$\varepsilon^2 y'' - k^2(x)y = 0 \tag{9.1.4}$$

and

$$\varepsilon^2 y'' + k^2(x)y = 0, \tag{9.1.5}$$

where $k(x) > 0$. These are called the *nonoscillatory* and *oscillatory* cases, respectively. Next in simplicity come problems of the form (9.1.1) where $a(x)$ has one or more simple zeroes (and therefore changes sign at these zeroes). These zeroes are called *simple turning points*. We will not consider any other cases.

Equation (9.1.1) becomes equation (9.1.2) upon substituting $x = \varepsilon t$, which is a reasonable change of independent variable as long as $\varepsilon \neq 0$. That is, (9.1.1) and (9.1.2) differ only by a rescaling of the independent variable. (It happens that the primary physical applications leading to the form (9.1.1) involve a space variable whereas those leading to (9.1.2) involve time; that is why we have used x in (9.1.1) and t in (9.1.2), rather than (say) x and ξ, or t and τ, which are familiar from previous chapters as pairs of rescaled variables.) As the reader should expect by now, such a rescaling totally changes the reduced problem, which in the case of (9.1.2) becomes

$$\ddot{y} + a(0)y = 0. \tag{9.1.6}$$

It is a little misleading to write the reduced equation this way, since to see $a(0)$ suggests that there is something special about time zero; what actually happens in (9.1.2) is that as $\varepsilon \to 0$ the coefficient $a(\varepsilon t)$ varies more and more slowly in time, and for $\varepsilon = 0$ it becomes constant, so that $a(0)$ is actually the value of the coefficient for all time. When written in the form (9.1.2), then, the reduced problem does not drop order, and it is this reduced problem (rather than $a(x)y = 0$) which suggests the right way to go about constructing formal solutions for both (9.1.1) and (9.1.2). For instance, consider the case (9.1.4). Under the transformation $x = \varepsilon t$ this becomes $\ddot{y} - k^2(\varepsilon t)y = 0$ with reduced problem $\ddot{y} - k^2(0)y = 0$. This reduced problem has the two linearly independent solutions $y = e^{\pm k(0)t}$. In the original independent variable, this reduced solution takes the form $y = e^{\pm k(0)x/\varepsilon}$. This last expression may not seem very reasonable, since it contains one part (namely $k(0)$) which comes from $\varepsilon = 0$ and another part (x/ε) which requires $\varepsilon \neq 0$. But this is only heuristic, and it does suggest looking for solutions of (9.1.4) in the form

$$y(x, \varepsilon) = e^{u(x,\varepsilon)/\varepsilon} = \exp\left\{\frac{u_0(x)}{\varepsilon} + u_1(x) + \varepsilon u_2(x) + \cdots\right\}. \qquad (9.1.7)$$

This procedure will prove successful (at least formally) in Section 9.2.

Equation (9.1.3) is also essentially equivalent to (9.1.1). Namely, if we set $\varepsilon := 1/\lambda$, then ε is small when λ is large and (9.1.3) becomes $\varepsilon^2 y'' + (p(x) + \varepsilon^2 q(x))y = 0$; this is only a slight generalization of (9.1.1), which amounts to replacing $a(x)$ by $a(x, \varepsilon)$. Since the three problems are basically the same, why do we give nearly equal weight to all of them in this chapter? Because the three problems arise in different contexts which lead to different questions about the solutions. Although the same approximate solutions are applicable in all three problems, they call forth different interpretations. Equation (9.1.1) arises in quantum mechanics as the Schrödinger equation in one dimension; it is the solutions themselves which are of interest, since they explain such things as the "tunnelling effect." Equation (9.1.2) arises as the equation of a linearized pendulum of slowly varying length, and one is not interested so much in the solution as in the so-called "adiabatic invariance of the action," which is a prototype for many other adiabatic invariants in physics. Finally, equation (9.1.3) arises (together with boundary conditions) as a Sturm–Liouville eigenvalue problem. It is known that such a problem has an infinite sequence of eigenvalues λ_n approaching infinity, and the interest here is to use approximate solutions of (9.1.3) for large λ to obtain asymptotic approximations to the eigenvalues λ_n for large n. We will turn now to the physical description of these various problems.

Example 9.1.1. Elementary Quantum Mechanics

Quantum mechanics is the physical theory developed in the twentieth century by Plank, Heisenberg, Bohr, Schrödinger, and others to explain the behavior of matter on the atomic and molecular level; it is sometimes called wave mechanics and is associated with such themes as "wave-particle duality," "the uncertainty principle," and "complementarity." The following brief description of the quantum mechanics of a one-dimensional system can be omitted, without loss of mathematical continuity but probably with some loss in motivation. Consider a system consisting of a particle of mass m moving on a line (the x-axis) in a force field with potential energy $V(x)$. The classical (that is, non-quantum-mechanical) description of this motion is given by the nonlinear differential equation

$$m\frac{d^2 x}{dt^2} + F(x) = 0,$$

where $F = V'$. Equations of this form were discussed in Section 4.3, and it is helpful to recall Figure 4.3.1, which shows a typical "potential well" which produces a regime of periodic solutions. (The coordinate labeled u in Fig. 4.3.1 is now our x.) For each value of the total energy E between the maximum and minimum heights of the well, there exists a periodic orbit which repeatedly changes direction at two "turning points" determined by the intersection of the line $V = E$ with the graph of the function $V(x)$.

The description of such a system in quantum mechanics is rather different. First of all, the particle does not have a definite position at any time. Instead, the most that one can say about the position at time t is to give a function $\psi(x, t)$ called the *wave function*, which has the following interpretation: If you look within the interval $a \leq x \leq b$, the probability of finding the particle there at time t is

$$\int_a^b |\psi(x, t)|^2 dx.$$

In classical mechanics the state of the system at time t is given by the position and velocity (or momentum). In quantum mechanics the wave function completely specifies the state: It not only tells the most that can be told about the position (as just explained), but it also tells all that can be told about the momentum, although we will not describe how that information can be extracted from ψ. Among all possible states of the system, there are some for which the concept of energy is well-defined, in the sense that a measurement of energy will give a unique value, and others for which the energy has only a probability distribution (which, as for position and momentum, can be derived from ψ). We will concern ourselves only with states having a definite energy, since these are the simplest and also they are the ones which correspond to the classical solutions drawn in Fig. 4.3.1 (as much as quantum-mechanical solutions can be said to correspond to classical solutions at all). For solutions having a definite energy E, the wave function is given by

$$\psi(x, t) = y(x)e^{-2\pi i E t/h}$$

where h is a specific very small number called *Plank's constant* and where $y(x)$ satisfies

$$\frac{h^2}{8\pi m}y'' + (E - V(x))y = 0, \tag{9.1.8}$$

which is called the *time-independent Schrödinger equation*. Now (9.1.8) has the form of (9.1.1), since the coefficient of y'' is small due to the smallness of Plank's constant. (We will skip the process of nondimensionalizing to save time.) Comparing (9.1.8) and (9.1.1) we see that $a(x) = E - V(x)$, so that the "turning points" of the classical motions in Fig. 4.3.1 (where $E = V$) are just the "turning points" of (9.1.1) which we have defined as isolated zeroes of $a(x)$. This explains the origin of the expression "turning point." It turns out that the solutions of (9.1.8) corresponding to values of E between the top and bottom of a potential well are not exactly equal to zero outside the interval between the turning points. In view of our discussion of the meaning of ψ, it follows that there is a nonzero probability of finding the particle outside of the region (the periodic regime) to which the classical solutions are confined. This is the famous "tunnelling effect" in quantum mechanics: There is a nonzero probability that a dollar bill in my pocket will suddenly be found in yours, even if you are a totally honest human being. (Of course such things are reasonably likely only on an atomic scale, which may explain why electrons don't carry money.) As a final remark, we observe that a phenomenon which is classically described by a *nonlinear* differential equation is treated in quantum mechanics by a *linear* differential equation. Of course this does not mean that the quantum mechanical treatment is easier. In particular, we have discussed only the solutions having a definite energy. In order to consider all solutions, (9.1.8) must be replaced by a partial differential equation, which, however, is still linear.

This discussion of elementary quantum mechanics has omitted one important feature: In order for the wave function to represent a probability, it is necessary that the integral of $|\psi(x, t)|^2$ over the x-axis be equal to 1, which implies that for states with definite energy, the integral of $|y(x)|^2$ must also equal 1. What is most important here is that this improper integral converge (that is, take a finite value), since in this case it is always possible to normalize the integral to 1 by including a constant factor in y (which is permissible since (9.1.8) is linear). This convergence condition is equivalent to requiring that $y(x)$ decay sufficiently rapidly as $x \to \pm\infty$, which may be thought of as a pair of boundary conditions at $\pm\infty$. It turns out that solutions of (9.1.8) can only satisfy these conditions for certain discrete values of E. In other words, these boundary conditions make (9.1.8) into an eigenvalue problem, and E must be one of a discrete set of *energy eigenvalues*. This does not invalidate any of our previous remarks; the principal problem (from our standpoint) is still to solve (9.1.8) for arbitrary E, since the eigenvalues can only be determined (approximately) once the solutions are known (approximately). For any E, (9.1.8) is an equation of the form (9.1.1). We do not wish to stress the eigenvalue aspect of this problem, in order to avoid confusing it with problem (9.1.3), where eigenvalues (of a different boundary value problem) will play a major role. In (9.1.8), as in (9.1.1), the coefficient of the second derivative is the perturbation parameter and E is the eigenvalue; these are two distinct quantities. In (9.1.3), the eigenvalue λ^2 *is at the same time the (large) perturbation parameter.*

Example 9.1.2. Adiabatic Invariance of Action

Consider a pendulum of mass m hanging from the ceiling by a string (not a rod) which passes through a small hole in a horizontal plate which can be raised and lowered so as to change the effective length of the pendulum string. Suppose now that the height of the plate is slowly varied. Since the equation of motion of a pendulum of (fixed) length l is $\ddot{y} + (g/l)\sin y = 0$, where y is the angle from the vertical, and since moving the plate does no work on the system, the equation of motion is $\ddot{y} + (g/l(\varepsilon t))\sin y = 0$, where ε is a parameter introduced to express the slowness of the plate's motion. This equation is not exactly (9.1.5) since it is nonlinear, but the approximation for small oscillations (setting $\sin y = y$) is of the form (9.1.5). A similar example which is "exactly" linear could be obtained in the form of an RC circuit (an electric circuit containing a battery, a resistor, and a capacitor) having a variable capacitor subjected to a slow variation. A similar mechanical example does not seem possible, since once cannot "tune" the Hooke's law constant of a linear spring. (Of course no physical system is exactly linear, but at least in the electrical example the neglected nonlinearities are not as obvious, and probably not as large (over a reasonable range of y), as in the pendulum example.)

From a physical point of view, great attention is paid to conserved quantities, since they often represent fundamental aspects of the underlying reality. A time-dependent system such as (9.1.5) does not conserve energy, so it is natural to look for other quantities that might remain constant. (We have remarked that moving the plate in the pendulum problem does no work on the system, since any force exerted by the plate against the string is only a constraint force and does not act through any distance. Therefore one might wonder how moving the plate can change the energy of the system. The answer is that the position of the plate affects the relationship between the angle y and the height of the bob, and therefore affects how the potential energy is related to y.) It turns out that there is no *exactly* conserved quantity associated with (9.1.5), but there is a very interesting *approximately* conserved quantity (or *adiabatic invariant*) called the "action." Action is a concept which arises first in the Hamilton–Jacobi theory (in advanced classical mechanics), and it is rather difficult to give an intuitive meaning to this notion. In the case of (9.1.5), however, the action $J(t)$ associated with a given solution $y(t)$ at time t can be defined as follows: Consider the orbit in the $(y, \dot{y})$ phase plane which *would be obtained* if the value of k were frozen at its value at time t and if $(y(t), \dot{y}(t))$ were allowed to evolve beginning from its actual value at time t (as an initial value). The (fictitious) orbit obtained in this way is an ellipse, and the area inside this ellipse is the action. It is not difficult to show that this definition is equivalent to π times the quantity

$$J(t) := \frac{1}{k}(\dot{y}^2 + k^2 y^2). \tag{9.1.9}$$

In order to avoid factors of π we will adopt (9.1.9) as our definition of the action in the sequel.

There are two senses in which this action is an "adiabatic invariant."

The first and simplest is that it varies slowly on a time scale of $1/\varepsilon$, that is, given $T > 0$ there exist constants c and ε such that

$$|J(t) - J(0)| \leq c\varepsilon \qquad \text{for } 0 \leq t \leq T/\varepsilon \text{ and } 0 \leq \varepsilon \leq \varepsilon_0. \qquad (9.1.10)$$

The WKB method is not well suited to proving (9.1.10); the proof uses Hamilton–Jacobi theory together with the method of averaging and will not be given in this book, although a sketch of the proof is given in the Notes and References for those familiar with Hamiltonian mechanics.

The second sense in which $J(t)$ is an adiabatic invariant is somewhat more remarkable and is provable by WKB methods. Suppose that the function $k(x)$ is independent of x except for x between zero and one; that is, $k(x)$ is equal to one constant for $x < 0$ and a possibly different constant for $x > T$. Then $k(\varepsilon t)$ is constant for $t < 0$, varies slowly for $0 \leq t \leq T/\varepsilon$, and then remains constant for the rest of time. According to (9.1.10), the action never varies by more than $c\varepsilon$ during this entire time. The more remarkable fact is that if the action is compared at the beginning and at the end of the period of variation, the difference is much smaller even than $c\varepsilon$; in fact it is exponentially small with ε. In other words, the pendulum seems in some way to "remember" what the amount of its action had been prior to the slow movement of the plate and restores its action to this value extremely closely when this movement comes to an end. The variations in action which occur during the movement are themselves small (of order ε), but may be much larger than the "final difference," that is, the difference between the initial and final values of the action. In this book we will not prove the exponential smallness of the final difference, but only that it is smaller than any power of ε: that is, there exists a constant c such that

$$|J(T/\varepsilon) - J(0)| \leq c\varepsilon^2 \qquad (9.1.11)$$

under the hypothesis that $k(\varepsilon t)$ varies only between times 0 and T/ε.

Example 9.1.3. Large Eigenvalues of Sturm–Liouville Problems

A *Sturm–Liouville problem* is a boundary value problem of the form

$$
\begin{aligned}
y'' + (\lambda^2 p(x) + q(x))y &= 0, \\
y(0) &= 0, \\
y(1) &= 0.
\end{aligned}
\qquad (9.1.12)
$$

More general types of boundary conditions are also allowed, but we will consider only this type. Here $p(x)$ is assumed to be positive and smooth (usually two continuous derivatives are enough), and λ is a parameter which is allowed to take any real value. Boundary value problems of the Sturm–Liouville type arise from separation of variables in linear partial differential equations such as the wave equation or heat equation with (possibly) nonconstant coefficients. Any value of λ for which (9.1.12) has a nontrivial solution is called an *eigenvalue*. The principal result in Sturm–Liouville theory is the existence of an infinite sequence of eigenvalues λ_n, $n = 1,\ldots,\infty$, with $\lambda_1 < \lambda_2 < \lambda_3 < \cdots$ and $\lambda_n \to \infty$ as $n \to \infty$. Since the sequence of eigenvalues approaches infinity, it follows that the large eigenvalues (and the corresponding solutions for y, called eigenfunctions) can be computed approximately from an approximate solution for (9.1.3) for large λ. In particular, we will see from these calculations that the eigenvalues approach a uniform spacing as $n \to \infty$.

9.2. THE FORMAL WKB EXPANSION WITHOUT TURNING POINTS

In this section we will construct formal approximate solutions to the equations

$$\varepsilon^2 y'' - k^2(x)y = 0 \tag{9.2.1}$$

and

$$\varepsilon^2 y'' + k^2(x)y = 0 \tag{9.2.2}$$

where $k(x) > 0$, and use these solutions to discuss (formally) the questions raised in Examples 9.1.2 and 9.1.3 in the last section. Example 9.1.1, the tunnelling effect in quantum mechanics, cannot be handled until we have discussed turning points, but since turning points are handled by combining solutions of (9.2.1) and (9.2.2), the work done here is also essential to that problem.

It has already been suggested that the way to solve (9.2.1) is to use (9.1.7). It is possible to substitute (9.1.7) directly into (9.2.1) (see Exercise 9.2.1), but we will obtain the same results by a (very slightly) different method. First, observe that if k in (9.2.1) is independent of x, then the problem is exactly solvable, and two linearly independent solutions are given by $y = e^{kx/\varepsilon}$ and $y = e^{-kx/\varepsilon}$. This suggests that the structure of the solution in the general case will be exponential and that the exponent should contain ε in the denominator. (This is reasonable since the manner in which (9.2.1) depends on ε is not affected by whether or not k is constant.) Therefore we will introduce a new dependent variable u in (9.2.1) by the equation

$$y = e^{u/\varepsilon}. \tag{9.2.3}$$

(Recall that a change of variable needs no justification; if a heuristic insight suggests a change of variables, it is generally worthwhile to try the change and see if it puts the problem in a more tractable form.) The result of applying (9.2.3) to (9.2.1) is (see Exercise 9.2.2)

$$\varepsilon u'' + (u')^2 - k^2(x) = 0. \tag{9.2.4}$$

This equation somewhat resembles those studied in Chapters 7 and 8, but it is nonlinear (because the term in u' is squared) and the last term does not have a factor of u. Since the equation does not explicitly involve u (except in the form of u' and u''), it can be reduced to a first order equation by setting

$$v := u' \qquad (9.2.5)$$

to obtain

$$\varepsilon v' + v^2 - k^2(x) = 0. \qquad (9.2.6)$$

This is a type of differential equation known as a *Riccati equation*. The presence of an ε multiplying the highest derivative suggests that an ordinary asymptotic expansion for the solution will not be uniformly valid, and appears to call for inner and outer solutions; on the other hand, no initial or boundary conditions are specified, and so an "inner" solution would seem to have no meaning. We will therefore find two linearly independent straightforward perturbation expansions (or "outer" solutions) for (9.2.6) and leave the question of uniformity for the next section (the "rigorous" section, as opposed to the present formal analysis). Substituting

$$v \sim v_0 + \varepsilon v_1 + \varepsilon^2 v_2 + \cdots \qquad (9.2.7)$$

into (9.2.6) and equating corresponding powers of ε yields the following sequence of equations:

$$
\begin{aligned}
v_0^2 &= k^2(x), \\
2v_0 v_1 &= -v_0', \\
2v_0 v_2 &= -v_1' - v_1^2, \\
&\vdots
\end{aligned}
\qquad (9.2.8)
$$

The first equation in (9.2.8) has exactly two solutions, namely

$$v_0(x) = \pm k(x). \qquad (9.2.9)$$

The remaining equations have the form $2v_0 v_n =$ terms involving $v_0, v_1, \ldots, v_{n-1}$. Each of these equations is solvable for v_n merely by dividing by $2v_0$, once the previous v_i are known. Therefore the equations in (9.2.8) are not differential equations at all (even though they involve derivatives of the v_i), but merely a sequence of algebraic equations. The two solutions for v_0 give rise to two sequences $v_0, v_1, \ldots$ which then give two solutions (9.2.7) for (9.2.6). These are easily computed to be

$$v^{(1)}(x) = k(x) - \varepsilon \frac{k'(x)}{2k(x)} + \varepsilon^2 \left\{ \frac{2k''(x)k(x) - 3(k'(x))^2}{8k^3(x)} \right\} + \mathcal{O}_F(\varepsilon^3)$$

$$(9.2.10a)$$

and

$$v^{(2)}(x) = -k(x) - \varepsilon \frac{k'(x)}{2k(x)} - \varepsilon^2 \left\{ \frac{2k''(x)k(x) - 3(k'(x))^2}{8k^3(x)} \right\} + \mathcal{O}_F(\varepsilon^3).$$
$$(9.2.10b)$$

To the indicated order, these solutions differ only in the signs of the first and third terms. Since (9.2.6) is nonlinear, these two solutions cannot be combined with arbitrary constants to give a general solution of that equation; but on the other hand we do not require a general solution of (9.2.6). Each of the solutions in (9.2.10) can be integrated once (see (9.2.5)) to obtain a solution for u, which in turn can be substituted into (9.2.3) to give a solution for y. Since the equation for y is linear, these two solutions can then be combined with arbitrary constants to give a general solution of (9.2.1).

In writing these solutions out, we will at first drop the terms of order ε^2 from (9.2.10), changing the error term to $\mathcal{O}_F(\varepsilon^2)$. The solutions obtained in this way are the ones most frequently seen under the name "WKB approximation." Therefore we begin with

$$v(x) \cong \pm k(x) - \varepsilon \frac{k'(x)}{2k(x)}$$

and integrate to obtain

$$u(x) \cong \int \pm k(x)dx - \varepsilon \int \frac{k'(x)}{2k(x)}dx$$
$$= \int \pm k(x)dx - \frac{\varepsilon}{2} \ln k(x). \qquad (9.2.11)$$

Using (9.2.3) and basic identities for the exponential and logarithm functions, (9.2.11) yields

$$y \cong \frac{1}{\sqrt{k(x)}} \exp \frac{1}{\varepsilon} \int \pm k(x)dx. \qquad (9.2.12)$$

Equation (9.2.12) is what is usually called the WKB approximation for the nonoscillatory problem (9.2.1).

It is quite remarkable that the second term in (9.2.11) interacts with the exponential function in just such a way as to produce the simple factor of $1/\sqrt{k}$ in (9.2.12). It is not possible to integrate the third term of (9.2.10) in closed form in terms of k, so we will not make any explicit use of the higher order WKB approximations. Of course the integrals can be evaluated numerically whenever this is useful, and it will be seen that the mere existence of the higher order approximations is useful for certain theoretical purposes.

Since this section is concerned only with formal calculations, nothing can be said at this point about rigorous error estimates. But there is something to say about the *formal structure* to be expected in an error estimate for the WKB approximation. The WKB approximation arises in the form $y \cong \exp(u_0/\varepsilon + u_1)$, which is a truncation of $y = \exp(u_0/\varepsilon + u_1 + \mathcal{O}_F(\varepsilon)) = \exp(u_0/\varepsilon + u_1) \cdot \exp \mathcal{O}_F(\varepsilon)$. Since $\exp \mathcal{O}_F(\varepsilon) = 1 + \mathcal{O}_F(\varepsilon)$, we have the following formal error estimate for the WKB approximation $\tilde{y}$:

$$y(x, \varepsilon) = \tilde{y}(x, \varepsilon) \cdot (1 + \mathcal{O}_F(\varepsilon)). \qquad (9.2.13)$$

Instead of an *error term* of specified order (close to zero), the error estimate appears (formally) to come out as an *error factor* close to one. In other words, what is being estimated here is not the *absolute error* but the *relative error*: If (9.2.13) is multiplied out, one has $y = \tilde{y} + \tilde{y} \cdot \mathcal{O}_F(\varepsilon) = \tilde{y} + R$, which reveals the absolute error R as a certain fraction $\mathcal{O}_F(\varepsilon)$ of the approximate solution. A little thought shows that this is the only useful form that an error estimate could take for this problem. Both the actual solution y and the WKB approximation $\tilde{y}$ are either exponentially small as $\varepsilon \to 0$ (when (9.2.12) has the negative sign) or exponentially large (for the positive sign). In the former case, the solution itself is smaller than any possible additive error term using polynomial gauges $\mathcal{O}_F(\varepsilon^n)$, so the presence of such an error would make the approximation useless. In the latter case, the solution is so large for small ε that it is unreasonable to expect the absolute error to be of polynomial order. In view of this formal observation, when we come to rigorous error estimation in Section 9.3, we will naturally look for an error factor rather than an error term. (On the other hand for the oscillatory problem we will find an absolute error estimate.)

To obtain the WKB approximation for the oscillatory problem (9.2.2), we will use a slightly different approach which leads to a linear equation rather than a Riccati equation. The difference is not crucial; both (9.2.1) and (9.2.2) can be solved by either method. (See Exercise 9.2.3.) Once again, the motivation comes from considering the solvable case of (9.2.2) which arises when k is constant. In this case (9.2.2) has the following two linearly independent complex solutions: $y = e^{\pm ikx/\varepsilon}$. This suggests looking for complex solutions of the general case in the form $y = e^{iu/\varepsilon}$ where $u(x, \varepsilon) \sim u_0(x) + \varepsilon u_1(x) + \cdots$. But instead of following this suggestion exactly, we first observe the identity

$$e^{i(u_0/\varepsilon + u_1 + \varepsilon u_2 + \cdots)} = e^{iu_0/\varepsilon} \cdot e^{i(u_1 + \varepsilon u_2 + \cdots)}. \qquad (9.2.14)$$

Now the second factor on the right hand side of (9.2.14) appears to be a regular function at $\varepsilon = 0$, that is, it does not appear to "blow up" there (as the first factor clearly does). Of course the series $u_1 + \cdots$ in the exponent of the second factor may not converge, but at the present stage our work is formal and this will not bother us; we will happily write this factor as a function $A(x, \varepsilon)$ and proceed to look for a solution of (9.2.2) in the form

$$y = e^{iu_0(x)/\varepsilon} A(x, \varepsilon), \qquad (9.2.15)$$

where $u_0(x)$ and $A(x,\varepsilon)$ are functions to be determined. Substituting (9.2.15) into (9.2.2) leads to

$$\left\{k^2 - (u_0')^2\right\} A + \varepsilon i u_0'' A + 2\varepsilon i u_0' A' + \varepsilon^2 A'' = 0. \qquad (9.2.16)$$

Setting $\varepsilon = 0$ (and assuming that $A(x,0) \neq 0$) implies that $u_0' = \pm k(x)$. Since we are ultimately interested in two linearly independent *real* solutions of (9.2.2), and since these can be found as the real and imaginary parts of a single complex solution, we will take only the plus sign, so that

$$u_0(x) = \int k(x)dx. \qquad (9.2.17)$$

Now the first term of (9.2.16) drops out, and the rest becomes

$$\varepsilon A'' + 2ikA' + ik'A = 0, \qquad (9.2.18)$$

which is the linear equation that must be solved in place of a nonlinear Riccati equation. Once again, this is a singular-looking problem without initial or boundary conditions, for which we will find only an "outer" solution in the form

$$A(x,\varepsilon) \sim A_0(x) + \varepsilon A_1(x) + \cdots . \qquad (9.2.19)$$

The coefficients are easily found to satisfy

$$2ikA_0' + ik'A_0 = 0,$$
$$2ikA_1' + ik'A_1 = -A_0'', \qquad (9.2.20)$$
$$\vdots$$

The equation for A_0 is readily solvable to give (as one particular solution) $A_0 = 1/\sqrt{k(x)}$. The equations for A_1 and higher order coefficients are inhomogeneous first order linear equations with nonconstant coefficients which are solvable by quadrature after multiplying by an integrating factor; that is, the solutions are expressible as integrals, with integrands which involve only k and its derivatives. However, the integrals do not seem to be computable in closed form. But the higher order approximations do exist, and their mere existence (together with the error estimates given in the next section) are crucial to proving the higher order adiabatic invariance of action (in the sense of (9.1.11)), as will be seen below. Confining ourselves, then, to the first term in (9.2.19), and using (9.2.15) and (9.2.17), we obtain the WKB approximation for the oscillatory case in complex form,

$$y(x,\varepsilon) \cong \frac{1}{\sqrt{k(x)}} \exp\frac{i}{\varepsilon} \int k(x)dx, \qquad (9.2.21)$$

which should be compared with (9.2.12). Taking real and imaginary parts gives two linearly independent real approximate solutions in the form

$$y^{(1)} \cong \frac{1}{\sqrt{k(x)}} \cos \frac{1}{\varepsilon} \int k(x) dx \qquad (9.2.22a)$$

and

$$y^{(2)} \cong \frac{1}{\sqrt{k(x)}} \sin \frac{1}{\varepsilon} \int k(x) dx. \qquad (9.2.22b)$$

If the right hand side of (9.2.21) is denoted $\tilde{y}$, then just as in the nonoscillatory case one has the formal result (9.2.13) which suggests that the error estimate should be relative (once the "formal" big-oh symbol is replaced by a "rigorous" big-oh in the next section). But in taking the real and imaginary parts of this expression there arise "cross terms": If the $\mathcal{O}_F(\varepsilon)$ term is written $\mu + i\nu$, and if the right hand sides of (9.2.22) are denoted $\tilde{y}^{(1)}$ and $\tilde{y}^{(2)}$, then for instance $y^{(1)} = \tilde{y}^{(1)}(1 + \mu) - \tilde{y}^{(2)}\nu$, which does not give a relative error estimate for $y^{(1)}$. However, it is clear that both $\tilde{y}^{(1)}$ and $\tilde{y}^{(2)}$ are bounded, so an absolute estimate (still formal for now) does follow:

$$y^{(j)} = \tilde{y}^{(j)} + \mathcal{O}_F(\varepsilon). \qquad (9.2.23)$$

It will be seen below that this, together with a similar estimate for the derivatives, is just what is needed for the adiabatic invariance question.

Example 9.2.1. Adiabatic Invariance of Action

Recall from Example 9.1.2 that for the equation $\ddot{y} + k^2(\varepsilon t)y = 0$, the action is defined by $J = \dot{y}^2/k + ky^2$. We will show that the approximations developed above, taken to any order of accuracy, predict that if $k(x)$ varies only during the interval $0 \le x \le 1$ then the action after the change in k returns exactly to its value prior to the change in k. Since this is based on an approximation, the actual value of J does not return to its exact prior value, but since the approximation can be taken to any order (and since error estimations done in the next section confirm that the result is valid to any order) it follows that the difference between the values of J before and after the change in k is "zero to any order," that is, of smaller order than any power of ε.

The first step is to introduce slow time $x = \varepsilon t$ with $' = d/dx$ so that the equation of motion becomes $\varepsilon^2 y'' + k^2(x)y = 0$ and the action becomes

$$J = \varepsilon^2(y')^2/k + ky^2. \qquad (9.2.24)$$

Next, an approximation for y must be substituted into (9.2.24). For a first approximation, a linear combination of (9.2.22a) and (9.2.22b) could be used for this purpose; it is not sufficient to use each of these solutions separately, since we wish to prove the adiabatic invariance of J for all solutions, not merely for the two given in (9.2.22), and J is not a linear function of y. But it is simpler to go back to the form (9.2.15), taking u_0 to be given by (9.2.17) and A to be a truncation (at any order n) of the series (9.2.19), with coefficients satisfying the first n equations of (9.2.20). It turns out that there is no need to have an explicit expression for A. Instead we will use directly the fact that A (even when truncated) satisfies (9.2.18). However, we will put these equations in real form by setting $A = \alpha - i\beta$ and taking y to be the real part of (9.2.15). Then

$$y \cong \alpha(x, \varepsilon) \cos \frac{u_0}{\varepsilon} + \beta(x, \varepsilon) \sin \frac{u_0}{\varepsilon}, \qquad (9.2.25)$$

and (9.2.18) becomes

$$\varepsilon \alpha'' - 2k\beta' - k'\beta = 0,$$
$$\varepsilon \beta'' + 2k\alpha' + k'\alpha = 0. \qquad (9.2.26)$$

It follows from (9.2.26) that α and β are constant outside $0 \le x \le 1$, where k is constant.

With these preliminaries, (9.2.25) can be differentiated and substituted into (9.2.24) to obtain a rather complicated expression for J that is valid for all x. We will not need this expression, but will use instead the following much simpler expression that results from differentiating (9.2.25) *as though $\alpha' = 0$ and $\beta' = 0$* and substituting the result into (9.2.24):

$$K(x, \varepsilon) = k(\alpha^2 + \beta^2). \qquad (9.2.27)$$

Of course $K = J$ outside of $0 \le x \le 1$, where α' and β' are actually zero, but within that interval they are not equal. Nevertheless, for any $a < 0$ and $b > 1$ we have, by the fundamental theorem of calculus, suppressing the dependence on ε,

$$J(b) - J(a) = K(b) - K(a) = \int_a^b K'(x)dx. \qquad (9.2.28)$$

That is, the change in the action can be computed just as well by following the simpler function K across the interval, as by following the actual action J. This is the crucial observation. It only remains to differentiate (9.2.27) and use (9.2.26) to obtain

$$\begin{aligned} K'(x) &= (k'\alpha^2 + 2k\alpha\alpha') + (k'\beta^2 + 2k\beta\beta') \\ &= \alpha(k'\alpha + 2k\alpha') + \beta(k'\beta + 2k\beta') \\ &= \alpha(-\varepsilon\beta'') + \beta(\varepsilon\alpha'') \\ &= \varepsilon(\alpha'\beta - \alpha\beta')'. \end{aligned} \qquad (9.2.29)$$

Putting (9.2.29) into (9.2.28) gives

$$J(b) - J(a) = \varepsilon \left[(\alpha'\beta - \alpha\beta')|_{x=b} - (\alpha'\beta - \alpha\beta')|_{x=a} \right] = 0. \qquad (9.2.30)$$

The last equality in (9.2.30) is because α' and β' are zero at a and b since these lie outside $0 \le x \le 1$. Now (9.2.30) says that the action associated with our approximate solution returns exactly to its previous value after the slow variation in k is over. Since this result holds for an approximation of any order, it follows that the action associated with the actual solution returns to its original value with a discrepancy that is $\mathcal{O}(n)$ for all n. As remarked in the last section, it can be shown that the discrepancy is actually exponentially small.

Of course, to make the above argument rigorous requires error estimates, not only for the approximation of y but also for y' (since y' enters into J). That is, if y denotes an actual solution and $\tilde{y}$ denotes one of the approximations (obtained by truncating A at some particular n) one must have estimates for the errors of $y \cong \tilde{y}$ and $y' \cong \tilde{y}'$, and the latter will not follow from the former (since inequalities cannot be differentiated). Once such estimates are available, the error of $J \cong \tilde{J}$ can be estimated, where $\tilde{J}$ is the action of the approximation. In the next section it will be shown that for the basic WKB approximation (9.2.22) (and therefore for linear combinations of these) one has

$$
\begin{aligned}
y &\cong \tilde{y} + \mathcal{O}(\varepsilon), \\
y' &\cong \tilde{y}' + \mathcal{O}(1)
\end{aligned}
\tag{9.2.31}
$$

uniformly for $0 \le x \le 1$ (that is, in the present problem, for t in expanding intervals). Now (9.2.24) and (9.2.31) imply that $J = \tilde{J} + \mathcal{O}(\varepsilon)$, and since we have just proved that the change in $\tilde{J}$ across the interval is zero, it follows that the change in J is $\mathcal{O}(\varepsilon)$. In the same way, estimates for the higher order approximations imply that the change in J is $\mathcal{O}(\varepsilon^n)$ for any n.

Example 9.2.2. Large Eigenvalues of Sturm–Liouville Problems

The eigenvalue problem

$$
\begin{aligned}
y'' &+ (\lambda^2 p(x) + q(x))y = 0, \\
y(0) &= 0, \\
y(1) &= 0.
\end{aligned}
\tag{9.2.32}
$$

becomes

$$
\begin{aligned}
\varepsilon^2 y'' &+ (k^2(x) + \varepsilon^2 q(x))y = 0, \\
y(0) &= 0, \\
y(1) &= 0
\end{aligned}
\tag{9.2.33}
$$

under the substitutions $\varepsilon = 1/\lambda$ and $p(x) = k^2(x)$; recall that in Example 9.1.3 it was assumed that $p > 0$. Now when (9.2.15) is substituted into (9.2.33) the result is the same as (9.2.16) with an additional term of $\varepsilon^2 q A$;

this adds $\varepsilon q A$ to (9.2.18), which does not affect the differential equation for A_0, the first equation of (9.2.20). Therefore the approximations (9.2.22) are not changed. The first boundary condition $y(0) = 0$ picks out (9.2.22b) and its multiples, after which the second boundary condition implies

$$\sin \frac{1}{\varepsilon} \int_0^1 k(x)dx \cong 0$$

or

$$\frac{1}{\varepsilon} \int_0^1 k(x)dx = n\pi.$$

Since $\lambda = 1/\varepsilon$, this means that the nth eigenvalue is given by

$$\lambda_n \cong n\pi/K \tag{9.2.34}$$

where

$$K = \int_0^1 k(x)dx.$$

Since this holds for small ε, which means large λ, it follows that (9.2.34) should be increasingly accurate for larger n, that is, the eigenvalues will approach equal spacing as $n \to \infty$.

Exercises 9.2

1. Obtain (9.2.12), the WKB approximation to the solution of (9.2.1), by directly substituting (9.1.7) into (9.2.1) and obtaining equations for u_0 and u_1.

2. Derive (9.2.4) by substituting (9.2.3) into (9.2.1).

3. Derive the WKB approximation (9.2.12) once again by the procedure used in the text for the oscillatory case. Begin with the analog of (9.2.14).

9.3. RIGOROUS THEORY WITHOUT TURNING POINTS

In the previous section, the WKB approximation was derived formally for the second order differential equations $\varepsilon^2 y'' \pm k^2(x)y = 0$. There are four cases altogether: First the $\pm$ sign in the differential equation must be specified (thereby selecting the nonoscillatory or the oscillatory case), and then the $\pm$ sign in the WKB approximation itself must be specified (or, in the case of the real form of the solution in the oscillatory case, the sine or cosine solution must be chosen). In the present section, error estimates will be derived for each of these cases. In the oscillatory case, there is no significant difference between the plus and minus (or sine and cosine) subcases, and both can be handled by the same proof. In the nonoscillatory case, the plus subcase is exponentially large and the minus subcase is exponentially small (for fixed x), leading to certain differences in the proofs. Therefore three distinct arguments are needed in total, although they all follow the same basic pattern.

The most direct way to attack the problem of error estimation is simply to write

$$y = \tilde{y}z, \tag{9.3.1}$$

where $\tilde{y}$ is one of the WKB approximations and z is an error factor (taking the hint from (9.2.13) that an error factor rather than an error term is correct for this problem). This can be substituted into the differential equation for y, and a new second order differential equation satisfied by z can be derived. Finally, once can prove that $z = 1 + \mathcal{O}(\varepsilon)$ uniformly for compact subsets of the x-axis. More precisely, what one wishes to prove is that *there exists* a solution of the differential equation for z which satisfies $z = 1 + \mathcal{O}(\varepsilon)$; then (9.3.1) can be taken as the definition of a particular solution for y which is well approximated by $\tilde{y}$. (After all, one does not expect $\tilde{y}$ to approximate *all* solutions for y.) These are, in fact, the fundamental steps that will be carried out in this section. However, the entire process is made much simpler by an initial transformation of both the dependent and independent variables (y and x) in the problem, and therefore this change of variables (called the *Liouville–Green transformation*) will be introduced before getting down to the business of error estimation. The Liouville–Green transformation also leads to a very direct formal derivation of the WKB approximation itself, and it could have been used for this purpose in the last section. The only drawback to this is that the Liouville–Green transformation is not obvious, and we prefer to motivate it from the WKB approximation (which arises fairly naturally, as we have seen) rather than the other way around. When we come to turning points, the other direction is more natural: Having seen what the Liouville–Green transformation can do in the absence of turning points, one is motivated to look for a similar transformation that can handle a simple turning point. The result is the Langer transformation (in Section 9.4).

Consider the nonoscillatory case

$$\varepsilon^2 y'' - k^2(x)y = 0 \tag{9.3.2}$$

and its WKB approximation

$$\tilde{y}(x,\varepsilon) = \frac{1}{\sqrt{k(x)}} \exp \pm \frac{1}{\varepsilon} \int k(x)dx. \tag{9.3.3}$$

Now (9.3.3) can be written as

$$\tilde{y} = \frac{1}{\sqrt{k(x)}}\tilde{w}, \tag{9.3.4a}$$

where

$$\tilde{w} = e^{\pm s/\varepsilon} \tag{9.3.4b}$$

with

$$s = \int_a^x k(\xi)d\xi. \tag{9.3.4c}$$

The lower limit a of the last integral is arbitrary, but some choice must be made for definiteness; the error estimates to be developed will be valid on an arbitrary compact interval $a \leq x \leq b$, and it is convenient to use the left endpoint of this interval as the lower limit in (9.3.4c). Now (9.3.4b) is a pair of linearly independent solutions of the differential equation

$$\varepsilon^2 \frac{d^2 \tilde{w}}{ds^2} - \tilde{w} = 0. \tag{9.3.5}$$

In other words, (9.3.4a) and (9.3.4c) define a change of independent and dependent variables $(x, \tilde{y}) \rightarrow (s, \tilde{w})$ which transforms the WKB approximation $\tilde{y}$ into (9.3.4b), which satisfies (9.3.5). Since $\tilde{y}$ is intended as an approximation to y, the same change of variables applied to y in the form $(x, y) \rightarrow (s, w)$ ought to produce a function w which satisfies an equation close to (9.3.5), and therefore (perhaps) simpler (in some respects) than (9.3.2).

With this as motivation (and recalling once again that any coordinate transformation is justified by its results) we introduce new variables s and w defined by

$$w = \sqrt{k(x)} \cdot y, \tag{9.3.6a}$$

$$s = \int_a^x k(\xi)d\xi \tag{9.3.6b}$$

into (9.3.2). After a good deal of grubby calculation (outlined in Exercise 9.3.1), the following equation is obtained:

$$\varepsilon^2 \frac{d^2 w}{ds^2} - w = \varepsilon^2 \delta(s)w \tag{9.3.7}$$

where

$$\delta(s) := -\left\{ \frac{3(k')^2}{4k^4} - \frac{k''}{2k^3} \right\}. \tag{9.3.8}$$

The definition of (9.3.8) is awkward in that the right hand side is a function of x rather than s; the meaning is that x is to be regarded as a function of s by inverting (9.3.6b). In other words each occurrence of k in (9.3.7) is to be understood as $k(x(s))$, and similarly for k' and k''. The expression for (9.3.8) is quite impossible to work with, but it is not necessary to do so. The argument to follow will only use (9.3.7), together with the fact that $\delta(s)$ is continuous and hence bounded on compact subsets. This fact follows from (9.3.8), as long as k has two continuous derivatives.

The coordinate transformation leading to (9.3.7) was motivated by the WKB approximation. If this transformation had occurred to us in some other way, we could now use (9.3.7) to suggest the WKB approximation. For suppose that the right hand side of (9.3.7) has only a small influence on the solutions. Then each solution w should be close to a solution $\tilde{w}$ of (9.3.5). Two solutions of (9.3.5) are given by (9.3.4b), and these transform into the WKB approximation when expressed in the variables y and x.

It remains now to prove that in fact the right hand side of (9.3.7) does have only a small influence on the solution. This is not entirely obvious even intuitively, since one might ask why it is permissible to ignore one term of order ε^2 (the right hand side) while retaining another (the first term of the left hand side). Therefore it is necessary to argue carefully, using the principles of error estimation introduced in Chapter 3. By this method we will obtain an error estimate for $\tilde{w}$, which will immediately imply an estimate for $\tilde{y}$, that is, for the WKB approximation. Of the three cases described at the beginning of this section, the following lemma will be used for the exponentially large solution of the nonoscillatory equation. (Notice the minus signs in (9.3.9) and (9.3.10).)

Lemma 9.3.1. *Let $\delta(s)$ be continuous. Then there exists a smooth family $w(s, \varepsilon)$ of solutions of the differential equation*

$$\varepsilon^2 \frac{d^2 w}{d s^2} - w = \varepsilon^2 \delta(s) w \tag{9.3.9}$$

which satisfies

$$w(s, \varepsilon) = e^{s/\varepsilon} \left(1 + \mathcal{O}(\varepsilon)\right) \tag{9.3.10}$$

uniformly on compact subsets of the s-axis.

Proof. Let $Q(s, \varepsilon)$ denote the term that is to be estimated as $\mathcal{O}(\varepsilon)$ in (9.3.10); that is, introduce

$$w = e^{s/\varepsilon}(1 + Q) \tag{9.3.11}$$

as an s-dependent change of the dependent variable (from w to Q) in (9.3.9). It is now a routine calculation (Exercise 9.3.2) to obtain the following differential equation for Q:

$$\varepsilon \frac{d^2 Q}{d s^2} + 2 \frac{dQ}{ds} = \varepsilon \delta(s)(1 + Q). \tag{9.3.12}$$

The differential equation $\varepsilon d^2 Q/ds^2 + 2dQ/ds = 0$ has the linearly in-dependent solutions $Q_1 = 1$ and $Q_2 = e^{-2s/\varepsilon}$, with Wronskian $W = (-2/\varepsilon)e^{-2s/\varepsilon}$. If $f(s)$ is any continuous function, it follows from (C.24) that the general solution of the inhomogeneous linear equation

$$\varepsilon \frac{d^2 Q}{ds^2} + 2\frac{dQ}{ds} = f(s) \tag{9.3.13}$$

is given by

$$Q(s) = c_1 + c_2 e^{2s/\varepsilon} + \frac{1}{2}\int_0^s \{1 - e^{2(\sigma-s)/\varepsilon}\}f(\sigma)d\sigma. \tag{9.3.14}$$

As usual, the *solution* (9.3.14) of the inhomogeneous equation (9.3.13) becomes an *integral equation* for the solution of the perturbed equation (9.3.12). (See Appendix C.) But there is a slight variation in the argu-ment because we have not fixed c_1 and c_2 to satisfy any specific initial or boundary conditions, so we will work through the reasoning again. Let $Q(s)$ be any solution of (9.3.12). Then $Q(s)$ also satisfies the dif-ferential equation obtained from (9.3.12) by substituting $Q = Q(s)$ into the right hand side (but leaving Q unknown on the right hand side). The equation obtained in this way has the form (9.3.13) for fixed ε. It follows that *for some choice of* c_1 *and* c_2, Q satisfies (9.3.14), with, of course, $f(\sigma)$ replaced by $\varepsilon\delta(\sigma)(1 + Q(\sigma))$. Notice that c_1 and c_2 are no longer arbitrary in this equation: One has to first choose a solution Q, then the function f appearing in (9.3.14) becomes defined, and finally c_1 and c_2 must be chosen so that (9.3.14) holds. If ε is regarded as a variable rather than as a fixed quantity, so that $Q(s, \varepsilon)$ is a family of solutions of (9.3.12), then this argument must be carried out separately for each value of ε, with the result that c_1 and c_2 become functions of ε. But in spite of this complication in the derivation of the integral equation, it remains true (as can be checked directly) that any solution of the integral equation for arbitrary c_1 and c_2 satisfies the differential equation. Our purpose is to find a solution of the differential equation that is $\mathcal{O}(\varepsilon)$, which clearly requires taking $c_1 = c_2 = 0$. Therefore the desired solution (if it exists at all) will be found by solving the integral equation

$$Q(s) = \frac{\varepsilon}{2}\int_a^s \{1 - e^{2(\sigma-s)/\varepsilon}\}\delta(\sigma)(1 + Q(\sigma))d\sigma, \tag{9.3.15}$$

suppressing the dependence of Q on ε.

As usual, we will omit the proof (by iteration) that the solution of (9.3.15) exists and proceed to the proof that it is in fact $\mathcal{O}(\varepsilon)$ uniformly on any compact interval, as hoped. Choose the interval $a \leq x \leq b$ (the left endpoint of which already appears in (9.3.15)) and let $\| \cdot \|$ denote the maximum of the absolute value of a function over $a \le x \le b$. Then

$$
\begin{aligned}
|Q(s)| &\le \frac{\varepsilon}{2} \int_a^s \|\{1 - e^{2(\sigma - s)/\varepsilon}\}\delta(\sigma)\|(1 + \|Q(\sigma)\|)d\sigma \\
&\le \frac{\varepsilon}{2} \int_a^b \|\delta\|(1 + \|Q\|)d\sigma \\
&\le \frac{\varepsilon}{2}\|\delta\|(1 + \|Q\|)(b - a) \\
&\le c_1\varepsilon\|Q\| + c_2\varepsilon.
\end{aligned}
\tag{9.3.16}
$$

(For the second inequality, notice that the kernel of the integral equation, that is, the quantity in braces, is bounded by 1. This will become a critical issue in the other cases treated below.) Maximizing over s in (9.3.16) gives $\|Q\| \le c_1\varepsilon\|Q\| + c_2\varepsilon$, so that

$$
\|Q\| \le \frac{c_2\varepsilon}{1 - c_1\varepsilon} = \mathcal{O}(\varepsilon). \quad \blacksquare
\tag{9.3.17}
$$

The next case to be considered is the exponentially small solution of the nonoscillatory problem. For this, one seeks a solution of (9.3.9) which satisfies

$$
w = e^{-s/\varepsilon}(1 + \mathcal{O}(\varepsilon))
\tag{9.3.18}
$$

in place of (9.3.10). To this end, introduce Q by $w = e^{-s/\varepsilon}(1+Q)$ (in place of (9.3.11)) and deduce the following equation for Q:

$$
\varepsilon\frac{d^2Q}{ds^2} - 2\frac{dQ}{ds} = \varepsilon\delta(s)(1 + Q).
\tag{9.3.19}
$$

Now the presence of the minus sign in the second term, as opposed to the plus sign in (9.3.13), appears to give a good deal of difficulty, since upon deriving the modified version of the integral equation (9.3.15), the kernel (the expression in braces) is exponentially large instead of being bounded. But in fact the sign of the second term of (9.3.19) can be changed by reversing the direction of the s axis, that is, by introducing a new independent variable $t = -s$. Alternatively, one can work out a form of the integral equation in which the upper limit of the integration is fixed and the lower limit is variable; in this form, the quantity in braces is bounded. The details are left to the reader in Exercise 9.3.3.

The last case to be considered is the oscillatory case. Most of the details are left to the reader in Exercise 9.3.4. One first checks that the Liouville–Green transformation carries

$$
\varepsilon^2 y'' + k^2(x)y = 0
\tag{9.3.20}
$$

into

$$\varepsilon^2 \frac{d^2 w}{ds^2} + w = \varepsilon^2 \delta(s) w \tag{9.3.21}$$

for a suitable function $\delta(s)$. Then one looks for a complex solution having the form

$$w = e^{is/\varepsilon}(1 + Q), \tag{9.3.22}$$

where $Q = \mathcal{O}(\varepsilon)$. The differential equation for Q is

$$\varepsilon \frac{d^2 Q}{ds^2} + 2i \frac{dQ}{ds} = \varepsilon \delta(s)(1 + Q). \tag{9.3.23}$$

The kernel of the integral equation for Q is found from the solutions of the homogeneous equation obtained by deleting the right hand side of (9.3.23); these solutions are 1 and $e^{-2is/\varepsilon}$. It follows easily that the kernel is bounded, and the rest of the argument goes through as before. Therefore $w = e^{is/\varepsilon}(1 + \mathcal{O}(\varepsilon))$. This transfers to the WKB approximation (using the boundedness of $k(x)$) to give $y = \tilde{y}(1 + \mathcal{O}(\varepsilon))$, which upon taking real parts becomes the absolute error estimate

$$y^{(j)} = \tilde{y}^{(j)} + \mathcal{O}(\varepsilon) \tag{9.3.24}$$

as discussed in connection with (9.2.23).

For the adiabatic invariance of the action in Example 9.2.1, the two estimates (9.2.30) were required. The first of these follows from (9.3.24) by taking arbitrary linear combinations of the two solutions. The second estimate of (9.2.30) is $y' \cong \tilde{y}' + \mathcal{O}(1)$ uniformly on compact subsets. This is shown in the following way. First, the exact solution for $y^{(1)}$ may be written

$$y^{(1)}(x, \varepsilon) = k(x)^{-1/2} \operatorname{Re} e^{is/\varepsilon}(1 + Q). \tag{9.3.25}$$

If Q is omitted, what remains is the WKB approximation. Now differentiate (9.3.25), remembering that s is a function of x and that the derivative may be taken inside the "real part" sign. The resulting terms may be grouped into three parts. The terms without Q or Q' are the derivative of the WKB approximation. The terms with a factor of Q can be estimated from the previous discussion, since $Q = \mathcal{O}(\varepsilon)$; most of these terms are $\mathcal{O}(\varepsilon)$, but there is one term that is $\mathcal{O}(1)$ because Q is multiplied by $1/\varepsilon$. Finally there are the terms that contain Q'. By differentiating the integral equation for $Q(s)$, it is easy to see that $Q' = \mathcal{O}(1)$; see Exercise 9.3.5.

Exercises 9.3

1. Derive (9.3.7) and (9.3.8) by carrying out the following steps. Step 1: Write (9.3.6a) in the form $y = k^{-1/2}w$ and differentiate twice with

respect to x. Step 2: Use (9.3.6b) to derive the operator equation $d/ds = k^{-1}d/dx$. Then apply this operator twice to w to obtain

$$\frac{d^2w}{ds^2} = \frac{1}{k^2}\left[\frac{d^w}{dx^2} - \frac{k'}{k}\frac{dw}{dx}\right].$$

Step 3: Substitute the result of step 2 into the result of step 1 to obtain a formula for d^2y/dx^2 in terms of d^2w/ds^2 and w; your formula should not involve dw/ds. Step 4: Substitute $y = k^{-1/2}w$ and the result of step 3 into $\varepsilon^2 d^2y/dx^2 + k^2y = 0$. After dividing by $k^{3/2}$ you should have (9.3.7), with δ appearing in the form of (9.3.8).

2. Derive equation (9.3.12).

3. Check that the integral equation for (9.3.19) does not lead to an estimate for Q because of the unbounded kernel. Remedy this by one of the methods mentioned in the text and complete the proof that the relative error for the exponentially small solution in the nonoscillatory case is $\mathcal{O}(\varepsilon)$.

4. Carry out all of the details leading from (9.3.20) to (9.3.24).

5. Show that in the oscillatory case $Q' = \mathcal{O}(1)$. Hint: Since the integral equation for Q (which you found in the previous exercise) has the form

$$Q(s) = \varepsilon \int_a^s K(s,\sigma)(1 + Q(\sigma))d\sigma$$

for a certain kernel function K, its derivative will be

$$Q'(s) = \varepsilon K(s,s)(1 + Q(s)) + \varepsilon \int_a^s K_s(s,\sigma)(1 + Q(\sigma))d\sigma.$$

Find K_s and notice that an ε cancels.

9.4. INTRODUCTION TO TURNING POINTS

In the previous sections of this chapter we have studied the equation $\varepsilon^2 y'' + a(x)y = 0$ in the cases $a(x) = k^2(x) > 0$ and $a(x) = -k^2(x) < 0$. In this section, $a(x)$ will be allowed to have simple zeroes, called *simple turning points*. At first we will assume that there is only one such zero, that it is located at the origin, and that $a'(0) < 0$, so that $a(x)$ is negative for positive x and positive for negative x. Such an equation may be written in the form

$$\varepsilon^2 y'' - xh^2(x)y = 0, \tag{9.4.1}$$

where $h(x) \neq 0$ (and we will assume $h(x) > 0$, that is, h denotes the positive square root of the quantity $-a(x)/x$).

In contrast to all other topics treated in this book, it is not possible to solve (9.4.1), even asymptotically, in terms of the elementary functions of calculus. It is necessary to have at hand the solutions of a special differential equation, the *Airy equation*

$$\ddot{w} - tw = 0. \tag{9.4.2}$$

This equation does not contain a small parameter, but it does contain the factor $-t$ multiplying w, so that the equation is oscillatory for negative t and nonoscillatory for positive t; therefore it serves as a "model" for the phenomenon of transition (at the origin) from one type to the other, which is the characteristic feature of (9.4.1), and its solutions are able to capture the behavior necessary to express solutions of (9.4.1). Of course, (9.4.2) is a second order linear equation and has two linearly independent solutions. They are denoted

$$w = \text{Ai}\ (t),$$

(because Ai stands for Airy), and

$$w = \text{Bi}\ (t)$$

(because—what else is there to call it?). Of course these are not just any two solutions of (9.4.2); they are specific solutions that have been thoroughly studied, and their properties are discussed in books on special functions. (They can also be written in terms of Bessel functions of order 1/3.)

For our purposes we will not need to know much about these functions, only that they have the following leading order asymptotic expansions for large positive and negative values of t:

$$\begin{aligned}
&\text{Ai}\ (t) \sim \frac{1}{2\sqrt{\pi}} t^{-1/4} \exp\left(-\frac{2}{3} t^{3/2}\right) && \text{as}\quad t \to \infty, \\
&\text{Ai}\ (t) \sim \frac{1}{\sqrt{\pi}} (-t)^{-1/4} \sin\left(\frac{2}{3}(-t)^{3/2} + \frac{\pi}{4}\right) && \text{as}\quad t \to -\infty, \\
&\text{Bi}\ (t) \sim \frac{1}{\pi} t^{-1/4} \exp\left(\frac{2}{3} t^{3/2}\right) && \text{as}\quad t \to \infty, \\
&\text{Bi}\ (t) \sim \frac{1}{\pi} (-t)^{-1/4} \cos\left(\frac{2}{3}(-t)^{3/2} + \frac{\pi}{4}\right) && \text{as}\quad t \to -\infty.
\end{aligned} \tag{9.4.3}$$

These are not asymptotic expansions in a small parameter, but in the large "parameter" t; that is to say, the difference between the right and left hand side of each equation in (9.4.3), divided by $t^{-1/4}$, approaches zero as t approaches either plus or minus infinity, according to the case. (In other words, as for asymptotic expansions in general, "the error is little-oh of the last term," and here only one term is given.)

In a manner similar to the Liouville–Green transformation of the last section, it is possible to transform the equation (9.4.1) into a form close to Airy's equation. Namely, introduce a new independent variable s and a new dependent variable w by the following equations which define the *Langer transformation*:

$$s(x) = \left[\frac{3}{2}\int_0^x \sqrt{|\xi|}h(\xi)d\xi\right]^{2/3},$$

$$w = \left(\frac{s(x)}{xh^2(x)}\right)^{1/4}.$$

(9.4.4)

In the latter equation, the function in parentheses is understood to be defined at $x = 0$ by its limit, which must be shown to exist. (Exercise 9.4.1.) Then (9.4.1) is transformed into

$$\varepsilon^2\frac{d^2w}{ds^2} - sw = \varepsilon^2\delta(s)$$

(9.4.5)

for a suitable function $\delta(s)$. The solutions of (9.4.5) can be approximated by the solutions of

$$\varepsilon^2\frac{d^2w}{ds^2} - sw = 0,$$

(9.4.6)

just as the solutions of (9.3.7) are approximated by those of (9.3.5). Now the further transformation $s = \varepsilon^{2/3}t$ carries (9.4.6) into (9.4.2). Therefore the solutions of (9.4.6) are $w = \text{Ai}\,(\varepsilon^{-2/3}s)$ and $w = \text{Bi}\,(\varepsilon^{-2/3}s)$; these are approximate solutions of (9.4.5); and feeding this back into the Langer transformation we arrive at the following approximate general solution of (9.4.1):

$$y = \left(\frac{xh^2(x)}{s(x)}\right)^{1/4}\left(c_1\,\text{Ai}\,(\varepsilon^{-2/3}s(x)) + c_2\,\text{Bi}\,(\varepsilon^{-2/3}s(x))\right),$$

(9.4.7)

where $s(x)$ is given by (9.4.4).

The representation (9.4.7) of the solution in terms of Airy functions may not seem to be very useful if the Airy functions are not familiar. Actually, it is not necessary to use this representation except in a neighborhood of the origin. In fact, as $\varepsilon \to 0$ for fixed x, the arguments of the Airy functions in (9.4.7) become large, so that the Airy functions may be replaced by their approximations as given by (9.4.3). This must be done separately for positive x and negative x, leading to an expression in exponentials on the right half-line and in sines and cosines on the left. (See Exercise 9.4.2.) These expressions are uniformly valid in compact sets excluding the origin and in fact are valid outside of suitable shrinking neighborhoods of the origin as well.

If one is only concerned about solutions away from the origin on the positive or negative side, there is no need to be concerned about the turning point at all. In fact the methods of Section 9.2 are valid on any interval in which $a(x)$ is strictly positive or strictly negative, so there exist exponential WKB solutions for (9.4.1) on the positive axis and sinusoidal ones on the negative, which are valid on compact sets excluding the origin. These solutions agree with those obtained from the large-argument approximations of the Airy functions. (See Exercise 9.4.3). Frequently it is the case that one is interested in solutions on the whole real line, but an inaccuracy in a small neighborhood of the origin is not particularly important. In such a case it would seem that the WKB solutions would be adequate, but they have a severe drawback: It is not clear how a particular linear combination of the sine and cosine solutions on the left is to be associated with a particular linear combination of the exponential solutions on the right. This is called the *connection problem*, and it is solved in a simple way by the use of Airy functions. Namely, for a given choice of c_1 and c_2 in (9.4.7), the use of (9.4.3) gives the correct choice of WKB solutions on either side of the turning point.

These remarks make it possible to handle problems with several turning points. Suppose for instance that $a(x)$ has simple zeroes at $x = 0$ and $x = 1$, and is oscillatory between them, nonoscillatory outside. There exist easy WKB approximations valid in compact subsets of the three intervals into which the real line is divided by the turning points. The problem is to connect them.

The turning point at zero is "reversed" from the one in (9.4.1), but it is easy to modify the Langer transformation to treat this case (or else introduce $\xi = -x$). The only serious change from the case of (9.4.1) is that the Langer transformation is not defined everywhere, because the quantity (s/a) is undefined at the second turning point $x = 1$. Similarly, the turning point at $x = 1$ can be shifted to the origin by introducing $\xi = x - 1$, but the resulting Langer transformation is undefined at the first turning point. But this is not a serious drawback, because the Langer transformation for each turning point need only be used in a neighborhood of that turning point, in order to establish the connection between particular WKB expansions on either side.

In particular, in quantum mechanics one is concerned with problems of the type discussed in the last paragraph, in which the desired solution approaches zero as $x \to \infty$ and also as $x \to \infty$. Such solutions can be found in the following way. First find the general oscillatory WKB solution in the interval between the turning points; this will depend on two arbitrary constants. Then use the Airy approximations near each turning point to find the exponential WKB solutions on the outside of each turning point, expressed in terms of the same two constants. Finally, the requirement that the solution decay in both directions fixes the constants. See Exercise 9.4.4.

Exercises 9.4

1. Let the derivative of $h^2(x)$ at the origin be denoted by C. Show that the function $(xh^2(x)/s(x))^{1/4}$, which appears in the Langer transformation,

has the limiting value $C^{1/6}$ as $x \to 0$. Hint: Expand $h^2(x) = C + \cdots$ and substitute this into the definition of $s(x)$.

2. Substitute the asymptotic expressions for the Airy functions (9.4.3) into (9.4.7), treating $x > 0$ and $x < 0$ separately.

3. Solve (9.4.1) separately for $x > 0$ and $x < 0$ by the WKB approximations of Section 9.2, and show that the results agree with the previous exercise except that there is no obvious way to relate the constants for the cases $x > 0$ with those for $x < 0$.

4. Find the WKB approximation in each region for the solution of $\varepsilon^2 y'' + x(1 - x)y = 0$ which decays to zero as $x \to \pm\infty$.

9.5. NOTES AND REFERENCES

The WKB approximation and turning points are studied formally in Chapter 14 of Nayfeh's *Introduction* and in Chapter 7 of his *Perturbation Methods*. A somewhat more careful, but still introductory, presentation is given in Chapter 2 of

William D. Lakin and David A. Sanchez, *Topics in Ordinary Differential Equations*, Dover, New York, 1970.

A thorough and up-to-date treatment (requiring extensive knowledge of differential equations) may be found in

W. Wasow, *Linear Turning Point Theory*, Springer-Verlag, New York, 1985.

The argument given in Example 9.1.2 I learned from Jürgen Moser in some lectures that he gave at the Courant Institute of Mathematical Sciences in the late 1960s; this is the result that the action of a linearized pendulum is an adiabatic invariant, in the sense that *after* a slow variation of its length the action returns to its original value with an error that is $\mathcal{O}(\varepsilon^n)$ for all n.

The proof that the action of a linearized pendulum of variable length is an adiabatic invariant, in the sense that its variation is of order ε *during* a t-interval of length $1/\varepsilon$ (or τ-interval of length $\mathcal{O}(1)$), is not done by the WKB method and so was not included in this chapter. One might think that this might be proved by showing that J', that is, $dJ/d\tau$, is of order ε. This would certainly imply the result, but it is not true. To see this it is only necessary to differentiate $J = \varepsilon^2(y')^2/k + ky^2$ and use $\varepsilon^2 y'' + ky = 0$; the result is $J' = (2/k - 1)y^2 k' - \varepsilon^2(y'/k)^2 k'$. The first term here is of order 1, not ε, but is rapidly oscillating because of the y^2 term; the WKB approximation shows that y is approximately sinusoidal with very large frequency. Now the change in J over a finite τ-interval is the integral of J', and if this is small it will not be because J' is small but because it is oscillating. In fact the easiest way to carry out the proof is to use a different set of ideas altogether. It is possible to express the equations of motion of the linearized pendulum in the form

$$\dot{J} = \varepsilon H_\theta(J, \tau, \theta, \varepsilon),$$

$$\dot{\tau} = \varepsilon,$$

$$\dot{\theta} = 1 - \varepsilon H_J(J, \tau, \theta, \varepsilon),$$

where H is a so-called Hamiltonian function and J (which equals the action) and θ are called action/angle variables. This is accomplished by the "canonical transformations" of Hamilton–Jacobi theory. Now these equations are in angular standard form for the method of averaging, with θ as a fast angle. (See Section 6.6.) Since the equation for $\dot{J}$ involves the derivative of H with respect to an angular variable, its average will be zero (since H can be expanded in a Fourier series and differentiating will eliminate the constant term). Therefore the averaged equation reads $\dot{J} = 0$, so that the approximation to J obtained from this equation will be constant. But this approximation is valid with error $\mathcal{O}(\varepsilon)$ for intervals of length $\mathcal{O}(1/\varepsilon)$. The adiabatic invariance of J (in the sense being discussed here) follows immediately.

APPENDIX A

TAYLOR'S THEOREM

Functions of a real variable are often classified according to the number of continuous derivatives which they possess. The most commonly encountered functions, such as $\sin x$ are *infinitely differentiable*, that is, all of the derivatives $f', f'', \ldots, f^{(n)}, \ldots$ exist (and it follows that they are all continuous). Other functions such as $|x|$ are not differentiable even once. (Of course $|x|$ is differentiable everywhere except at the origin. But there exist continuous functions that are not differentiable anywhere.) The function

$$f(x) = \begin{cases} 0 & \text{for } x \leq 0 \\ x^2 & \text{for } x > 0 \end{cases}$$

has one continuous derivative

$$f'(x) = \begin{cases} 0 & \text{for } x \leq 0 \\ 2x & \text{for } x > 0, \end{cases}$$

but the second derivative does not exist (as a function on the whole real line) because it is not defined at the origin; loosely speaking, one says the second derivative is discontinuous (more accurately, it is continuous wherever it exists but cannot be extended continuously to zero). Most functions that appear in this book can be differentiated any number of times. However, functions with discontinuities in the derivatives do arise in applied mathematics (for instance, in the theory of shock waves), and it is therefore sometimes useful to know exactly how many continuous derivatives are required for a particular mathematical result to be valid. In the early chapters of this book, some care is taken to "count derivatives" in the statement of theorems in order to make the reader aware of the occasional need to do this. In later chapters we tend merely to assume that the functions that appear are "smooth" (or "sufficiently smooth"), meaning that by this time readers should be able to look over the proof and count for themselves how many derivatives are required. A useful fact from advanced calculus is that if the nth derivative of a function exists then all lower order derivatives, and the function itself, are continuous.

Let $f(\varepsilon)$ be an infinitely differentiable real-valued function of a real variable ε. For each integer $k \geq 0$ the kth *Taylor polynomial* of f is defined as

$$p_k(\varepsilon) := a_0 + a_1\varepsilon + \cdots + a_k\varepsilon^k, \tag{A.1}$$

where

$$a_n := f^{(n)}(0)/n!. \tag{A.2}$$

If f is only r times differentiable, the Taylor polynomials can only be defined for $0 \leq k \leq r$. The *remainder* of the kth Taylor polynomial is defined by

$$R_k(\varepsilon) := f(\varepsilon) - p_k(\varepsilon). \tag{A.3}$$

Taylor's theorem states that the remainder is small for small ε, that is, that $p_k(\varepsilon)$ is a good approximation to $f(\varepsilon)$ when ε is sufficiently small. There is no reason to expect the approximation to be good for large ε, since the only data from the function f which is used in forming p_k is data from near $\varepsilon = 0$ (namely, the derivatives of f at zero, which by definition are limits constructed using values of f near zero).

One version of Taylor's theorem is as follows:

Theorem A.1. *Let $f(\varepsilon)$ be defined for $|\varepsilon| < \varepsilon_0$. If f has r continuous derivatives, then for each $k \leq r - 1$ the remainder R_k satisfies*

$$R_k(\varepsilon) = \int_0^\varepsilon f^{(k+1)}(\eta)\frac{(\varepsilon - \eta)^k}{k!}\, d\eta, \tag{A.4}$$

and for each ε_1 with $0 < \varepsilon_1 < \varepsilon_0$,

$$|R_k(\varepsilon)| \leq \frac{M_k(\varepsilon_1)}{(k+1)!}|\varepsilon|^{k+1} \quad \text{for} \quad |\varepsilon| \leq \varepsilon_1, \tag{A.5}$$

where

$$M_k(\varepsilon_1) := \max\{|f^{(k+1)}(\varepsilon)| : |\varepsilon| \leq \varepsilon_1\}. \tag{A.6}$$

Proof. Use the fundamental theorem of calculus to write $f(\varepsilon) = f(0) + \int_0^\varepsilon f'(\eta)d\eta$ and then integrate by parts repeatedly to obtain

$$f(\varepsilon) = f(0) + f'(\eta)(\eta - \varepsilon)\big|_{\eta=0}^\varepsilon - \int_0^\varepsilon f''(\eta)(\eta - \varepsilon)d\eta$$

$$= f(0) + f'(0)\varepsilon + \int_0^\varepsilon f''(\eta)(\varepsilon - \eta)d\eta$$

$$= f(0) + f'(0)\varepsilon - f''(\eta)\frac{(\varepsilon - \eta)^2}{2}\bigg|_{\eta=0}^\varepsilon + \int_0^\varepsilon f'''(\eta)\frac{(\varepsilon - \eta)^2}{2}d\eta$$

$$= f(0) + f'(0)\varepsilon + \frac{f''(0)}{2}\varepsilon^2 + \int_0^\varepsilon f'''(\eta)\frac{(\varepsilon - \eta)^2}{2}d\eta.$$

(In the first integration by parts, take $u = f'(\eta)$, $v = \eta - \varepsilon$; in the second, $u = f''(\eta)$, $v = (\varepsilon - \eta)^2/2$; and so on.) Continuing in this way

$$f(\varepsilon) = p_k(\varepsilon) + \int_0^\varepsilon f^{(k+1)}(\eta)\frac{(\varepsilon - \eta)^k}{k!}d\eta.$$

This gives (A.4). Given ε_1 as in the theorem, the maximum value $M = M_k(\varepsilon_1)$ of $|f^{(k+1)}(\eta)|$ for $|\eta| \le \varepsilon_1$ exists because $f^{(k+1)}$ is assumed to be continuous. (A continuous function on a closed bounded interval has a maximum.) Then for $|\varepsilon| < \varepsilon_1$,

$$|f(\varepsilon) - p_k(\varepsilon)| = \left| \int_0^\varepsilon f^{(k+1)}(\eta)\frac{(\varepsilon - \eta)^k}{k!}d\eta \right|$$

$$\le \int_0^\varepsilon \left| f^{(k+1)}(\eta)\frac{(\varepsilon - \eta)^k}{k!} \right| d\eta$$

$$\le M \int_0^\varepsilon \frac{|\varepsilon - \eta|^k}{k!}d\eta = \frac{M\varepsilon^{k+1}}{(k+1)!}.$$

This is (A.5). ∎

For fixed k and ε_1, the right hand side of (A.5) always decreases when ε is decreased. It decreases at a faster rate if k is larger, since higher powers of ε decrease faster as $\varepsilon \to 0$. (Recall that $|\varepsilon| > |\varepsilon|^2 > |\varepsilon|^3 > \cdots$ for $|\varepsilon| < 1$.) However, this gain from increasing k (taking more terms in the Taylor polynomial) may be offset by an increase in M_k. For instance, a certain function $f(\varepsilon)$ may satisfy $|f(\varepsilon) - p_1(\varepsilon)| \le 5\varepsilon^2$ and $|f(\varepsilon) - p_2(\varepsilon)| \le 100\varepsilon^3$. For $\varepsilon = 0.1$ the error bound for $p_1(\varepsilon)$ is 0.05, while the error bound for $p_2(\varepsilon)$ is 0.1; thus p_1 is better than p_2 (unless by accident the actual error for p_2 is much less than its known upper bound). But for $\varepsilon = 0.01$ the error bound for p_1 is 0.0005, and for p_2, it is 0.0001, so p_2 is better. This is because ε^3 decreases more rapidly than ε^2 as $\varepsilon \to 0$, and this eventually overcomes the advantage of the smaller constant 5 (as opposed to 100).

The error estimate (A.5) is most often used in the following form: Given ε_1 with $0 < \varepsilon_1 < \varepsilon_0$, there exists a constant $c > 0$ such that

$$|R_k(\varepsilon)| \leq c|\varepsilon|^{k+1} \qquad \text{for} \quad |\varepsilon| \leq \varepsilon_1. \tag{A.7}$$

The mere *existence* of a constant c for which (A.7) holds is a useful piece of information, since it "justifies" the use of the Taylor approximation "for sufficiently small ε." When c is not specified, (A.7) is often written as

$$R_k(\varepsilon) = \mathcal{O}(\varepsilon^{k+1}). \tag{A.8}$$

This notation, called the big-oh symbol, is explained at greater length in Section 1.8 and is called an *asymptotic error estimate*.

Of course in (A.7) $c = M_k(\varepsilon_1)/(k+1)!$. It follows from (A.6) that c is likely to increase if ε_1 is increased, and may decrease if ε_1 is decreased. For this reason it may seem advantageous to take ε_1 as small as possible for each ε: Namely, if $\varepsilon_1 = |\varepsilon|$, (A.5) becomes

$$|R_k(\varepsilon)| \leq \frac{M_k(|\varepsilon|)}{(k+1)!} |\varepsilon|^{k+1}.$$

This is in fact the sharpest form of (A.5) for any given ε, but is not very useful since M_k is difficult to compute. When it is computable at all, it is best to do so only once for some suitable ε_1 and then to use the simple form (A.7).

In calculus, Taylor polynomials are usually introduced in connection with the infinite Taylor series $\sum_{n=0}^{\infty} a_n \varepsilon^n$. Each Taylor polynomial is a partial sum of this series. The infinite series is said to *converge* to a function $g(\varepsilon)$ if $\lim_{k \to \infty} p_k(\varepsilon)$ exists and equals $g(\varepsilon)$ for each ε in an interval around zero. In this case one writes $\sum_{n=0}^{\infty} a_n \varepsilon^n = g(\varepsilon)$; if the series does not converge, the expression $\sum_{n=0}^{\infty} a_n \varepsilon^n$ does not have a value, although it is still used to denote the infinite series itself. When the infinite series converges, the sum $g(\varepsilon)$ is not always the same as the function $f(\varepsilon)$ from which the series was constructed (that is, the function appearing in (A.2)); in other words, the Taylor series of a function may converge to a different function, not the one intended. If the infinite Taylor series of a function $f(\varepsilon)$ not only converges (for ε in some interval about zero), but converges to the "correct" function f, then f is called an *analytic function* or more specifically a *real-analytic function*, (since "analytic" by itself usually means "complex-analytic"). If the Taylor series converges at all, one has

$$f(\varepsilon) = p_k(\varepsilon) + R_k(\varepsilon),$$

$$g(\varepsilon) = p_k(\varepsilon) + \sum_{n=k+1}^{\infty} a_n \varepsilon^n,$$

and if f is not analytic, that is, if $g \neq f$, then the remainder $R_k(\varepsilon)$ is not equal to the "tail" of the infinite series (consisting of the terms after p_k). If the series is divergent, neither g nor the "tail" have a value at all, although $R_k(\varepsilon)$ is always well defined. To find the condition under which f is analytic, write $g = \lim_{k \to \infty} p_k = f - \lim_{k \to \infty} R_k$; it follows that $g = f$ if and only if $R_k \to 0$ as $k \to \infty$. For example, if $f(\varepsilon) = \sin \varepsilon$ it is easy to see that $M_k(\varepsilon_1) \leq 1$ regardless of ε_1, and then $|R_k(\varepsilon)| \leq |\varepsilon|^{k+1}/(k+1)!$ Since this approaches zero as $k \to \infty$, $\sin \varepsilon$ equals the sum of its infinite Taylor series and so is analytic.

This review of the infinite Taylor series has been given in order to contrast the use of Taylor's theorem in calculus with its use in perturbation theory. In this book we are not concerned with the convergence of the infinite Taylor series. Instead what matters is the error estimate (A.7) for a fixed partial sum of the series, usually with a low value of k such as 1 or 2. This error estimate holds regardless of the convergence or divergence of the series as $k \to \infty$. Some of the series constructed in this book converge, and many do not, but we will not even pause to consider the issue.

Several extensions of Taylor's theorem will be needed. If $\mathbf{f}(\varepsilon)$ is vector-valued, that is, $\mathbf{f}(\varepsilon) = (f_1(\varepsilon), f_2(\varepsilon), \ldots, f_N(\varepsilon))$, the coefficients $\mathbf{a}_n$ in (A.2) are vectors, and so $\mathbf{p}_k$ is also vector-valued. The proof of Theorem A.1 goes through exactly as stated, the integration in (A.4) being componentwise; in (A.5) and (A.6) the absolute value of $R_k(\varepsilon)$ and $f^{(k+1)}(\varepsilon)$ should be replaced by the Euclidean norms $\|\mathbf{R}_k(\varepsilon)\|$ and $\|\mathbf{f}^{(k+1)}(\varepsilon)\|$, where $\|\mathbf{y}\| := \sqrt{y_1^2 + y_2^2 + \cdots + y_N^2}$. (It is equally permissible to use another vector norm such as $\max\{|y_1|, \ldots, |y_N|\}$ or $|y_1| + |y_2| + \cdots + |y_N|$. For further information about vector norms, see Appendix F.)

Next, let $\mathbf{f}(\mathbf{x}, \varepsilon)$ depend continuously on an additional vector variable $\mathbf{x} = (x_1, \ldots, x_M)$ of any dimension M; ε still denotes a scalar. The Taylor polynomials of $\mathbf{f}$ with respect to ε are defined by

$$\mathbf{p}_k(\mathbf{x}, \varepsilon) := \sum_{n=0}^{k} \mathbf{a}_n(x)\varepsilon^k, \qquad (A.9)$$

where

$$\mathbf{a}_n(\mathbf{x}) := \frac{1}{n!} \frac{\partial^n}{\partial \varepsilon^n} \mathbf{f}(\mathbf{x}, \varepsilon) \Big|_{\varepsilon=0}. \qquad (A.10)$$

For each fixed **x**, (A.9) and (A.10) are the same as (A.1) and (A.2). The remainder is defined by

$$\mathbf{R}_k(\mathbf{x}, \varepsilon) = \mathbf{f}(\mathbf{x}, \varepsilon) - \mathbf{p}_k(\mathbf{x}, \varepsilon).$$

A *compact set* is a closed and bounded set such as a finite closed interval or a solid sphere or box including its boundary. (The actual definition is more technical, but this is equivalent in Euclidean spaces according to the Heine–Borel theorem.) By a theorem of Weierstrass, every continuous real-valued function on a compact set has a maximum; we have already used this for closed bounded intervals in the proof of Theorem A.1.

Theorem A.2. Let $\mathbf{f}(\mathbf{x}, \varepsilon)$ be defined for $|\varepsilon| < \varepsilon_0$ and have r continuous partial derivatives with respect to ε, each of which depends continuously on **x**. Then for each compact set K in the **x**-space $(K \subset \mathbf{R}^M)$ and for each ε_1 with $0 < \varepsilon_1 < \varepsilon_0$ there exists a constant $c > 0$ such that

$$\|\mathbf{R}_k(\mathbf{x}, \varepsilon)\| \le c|\varepsilon|^{k+1} \quad \text{for} \quad \mathbf{x} \text{ in } K \quad \text{and} \quad |\varepsilon| \le \varepsilon_1.$$

Proof. Theorem A.1 applies for each fixed x, so that

$$\mathbf{R}_k(\mathbf{x}, \varepsilon) = \int_0^\varepsilon \left\{ \frac{\partial^{k+1}}{\partial \varepsilon^{k+1}} \mathbf{f}(\mathbf{x}, \varepsilon) \Big|_{\varepsilon=\eta} \right\} \frac{(\varepsilon - \eta)^k}{k!} d\eta.$$

By hypothesis, the expression in braces is a continuous function of **x** and ε, and therefore its norm has a maximum value M for x in K and $|\varepsilon| \le \varepsilon_1$. It follows that

$$\|\mathbf{R}_k(\mathbf{x}, \varepsilon)\| \le M \int_0^\varepsilon \frac{|\varepsilon - \eta|^k}{k!} d\eta = \frac{M|\varepsilon|^{k+1}}{(k+1)!}. \quad \blacksquare$$

In the terminology of Section 1.8 this result can be stated as

$$\mathbf{R}_k(\mathbf{x}, \varepsilon) = \mathcal{O}(\varepsilon^{k+1}) \quad \text{uniformly for} \quad \mathbf{x} \text{ in } K \quad \text{and} \quad |\varepsilon| \le \varepsilon_1. \quad (A.11)$$

A final extension of Taylor's theorem is possible to cover the case in which ε becomes a vector. As we will not use this, we will state only the second order Taylor polynomial of $f(\varepsilon_1, \varepsilon_2)$ as an example:

$$f(\varepsilon_1, \varepsilon_2) \cong f(0, 0) + \{f_{\varepsilon_1}(0, 0)\varepsilon_1 + f_{\varepsilon_2}(0, 0)\varepsilon_2\}$$

$$+ \frac{1}{2!}\{f_{\varepsilon_1\varepsilon_1}(0, 0)\varepsilon_1^2 + 2f_{\varepsilon_1\varepsilon_2}(0, 0)\varepsilon_1\varepsilon_2 + f_{\varepsilon_2\varepsilon_2}(0, 0)\varepsilon_2^2\}. \quad (A.12)$$

The next term is 1/3! times a collection of cubic terms with coefficients made from third derivatives of f. Equation (A.12) is intended to be a good approximation to $f(\varepsilon_1, \varepsilon_2)$ when both ε_1 and ε_2 are close to zero.

APPENDIX B

THE IMPLICIT FUNCTION THEOREM

The equation $x^2 + y^2 - 1 = 0$ describing the unit circle can be solved for y to yield the two functions $y = +\sqrt{1 - x^2}$ (the upper half of the circle) and $y = -\sqrt{1 - x^2}$ (the lower half). These two functions are said to be *implicit* in the equation $x^2 + y^2 - 1 = 0$. (The original Latin meaning of "implicit" is "folded in," and the dictionary definition is "suggested or understood but not plainly expressed." When the equation is solved for y the two concealed functions are rendered *explicit*.) If (x_0, y_0) is any point on the circle other than $(1,0)$ or $(-1,0)$, there is an arc of the circle which contains (x_0, y_0) and which belongs entirely to the graph of one of the two explicit functions. The "bad" points $(1,0)$ and $(-1,0)$ are points where the circle "doubles back" and has a vertical tangent. It is not possible in practice to solve every "implicit function" $\varphi(x, y) = 0$ to obtain (one or more) explicit functions $y = f(x)$, because the necessary functions $f(x)$ may not be expressible using the well-known elementary functions of calculus. But it is possible to prove that even if it cannot be written down, a function $y = f(x)$ does exist in a neighborhood of any point on the graph of $\varphi(x, y) = 0$ not having a vertical tangent, provided that φ is smooth. The condition of "no vertical tangent" is called *nondegeneracy*, and analytically takes the form $\varphi_y \neq 0$. A precise statement follows.

Theorem B.1. *Let $\varphi(x, y)$ be a real-valued function having continuous partial derivatives. Suppose $\varphi(x_0, y_0) = 0$ and $\varphi_y(x_0, y_0) \neq 0$, that is, the tangent to the curve $\varphi(x, y) = 0$ at (x_0, y_0) is not vertical. Then in a sufficiently small open interval around x_0, there is a unique function $y = f(x)$ such that $f(x_0) = y_0$ and $\varphi(x, f(x)) = 0$; that is, the graph of f is an arc of the curve $\varphi = 0$ containing (x_0, y_0). In addition, f is continuously* *differentiable (and hence of course also continuous), and its derivative is given by*

$$f'(x) = -\frac{\varphi_x(x, f(x))}{\varphi_y(x, f(x))}.$$

If φ has continuous partial derivatives (including mixed partials) of all orders $\leq r$, where $r \geq 1$, then f has r continuous (ordinary) derivatives.

A proof of the implicit function theorem can be found in most advanced calculus or real analysis texts. Since any proof is rather long and technical and since there are several proofs that illuminate the theorem in different ways, we have chosen to sketch several proofs at the end of this appendix instead of giving the complete details of any one proof. The discussion given here should therefore complement that given in a good analysis text.

A useful way to think of the theorem is to write down the equation of the tangent line to the curve $\varphi = 0$ at (x_0, y_0). This equation is

$$\varphi_x(x_0, y_0)(x - x_0) + \varphi_y(x_0, y_0)(y - y_0) = 0. \tag{B.1}$$

The tangent line is a linear approximation to the curve near (x_0, y_0). Equation (B.1) can be solved for y if and only if $\varphi_y(x_0, y_0) \neq 0$; thus the implicit function theorem says that the nonlinear equation $\varphi(x, y) = 0$ can be solved for y near (x_0, y_0), provided that the linear equation (B.1) can be solved for y. Since the left hand side of (B.1) can be regarded as the total differential

$$d\varphi(x_0, y_0) := \varphi_x(x_0, y_0)dx + \varphi_y(x_0, y_0)dy \tag{B.2}$$

with $dx = x - x_0$, $dy = y - y_0$, another way to say this is that $\varphi(x, y) = 0$ can be solved for y near (x_0, y_0) provided $d\varphi(x_0, y_0) = 0$ can be solved for dy. This formulation is useful for remembering the implicit function theorem when other variables are used. For instance, in Chapter 1 the problem arises of solving $\varphi(x, \varepsilon) = 0$ for $x = f(\varepsilon)$ given a starting solution x_0 such that $\varphi(x_0, 0) = 0$. This will be possible provided $d\varphi = \varphi_x dx + \varphi_\varepsilon d\varepsilon = 0$ can be solved for dx, when the coefficients φ_x and φ_ε are evaluated at $(x_0, 0)$. So the condition is $\varphi_x(x_0, 0) \neq 0$. This is Theorem 1.4.1. The differential notation is also useful for remembering the formula for $f'(x)$: From $\varphi = 0$ one has $d\varphi = 0$, that is, $\varphi_x dx + \varphi_y dy = 0$; therefore $dy/dx = -\varphi_x/\varphi_y$. This is, of course, just the procedure called "implicit differentiation" in elementary calculus.

Now suppose that x and y are allowed to become vector variables $\mathbf{x} = (x^1, x^2, \ldots, x^M)$ and $\mathbf{y} = (y^1, y^2, \ldots, y^N)$. Notice that the dimensions M and N are not necessarily equal and that we have written the indices as superscripts rather than subscripts; this allows subscripts to be used to designate particular fixed values of $\mathbf{x}$ or $\mathbf{y}$, so that (for instance)

$\mathbf{x}_0$ denotes the vector $(x_0^1, \ldots, x_0^M)$. (In this book, components of a vector will usually be indicated by subscripts, but whenever it is convenient the present notation will be used.) It would seem reasonable to consider solving $\varphi(\mathbf{x}, \mathbf{y}) = 0$ for $\mathbf{y} = \mathbf{f}(\mathbf{x})$, provided that φ is a vector-valued function with N components. In this case $\varphi(\mathbf{x}, \mathbf{y}) = 0$ stands for the following system of equations when written out in full:

$$
\begin{aligned}
\varphi^1(x^1, \ldots, x^M, y^1, \ldots, y^N) &= 0, \\
\varphi^2(x^1, \ldots, x^M, y^1, \ldots, y^N) &= 0, \\
&\vdots \\
\varphi^N(x^1, \ldots, x^M, y^1, \ldots, y^N) &= 0.
\end{aligned}
\tag{B.3}
$$

Since there are exactly N equations and N "unknowns" (that is, the components of $\mathbf{y}$), it is plausible that (B.3) can be solved for $\mathbf{y}$, at least under certain conditions. Our experience with Theorem B.1 suggests that the solution can only be done "locally," that is, given one point satisfying (B.3) it should be possible to find a function $\mathbf{y} = \mathbf{f}(\mathbf{x})$ that solves (B.3) in a neighborhood of that point. Theorem B.1 also suggests that some nondegeneracy condition involving the derivatives of φ should be imposed in order to guarantee the existence of a solution. No doubt any reader of this book has one other piece of relevant experience: the solution of systems of *linear* algebraic equations. A system of N linear equations in N unknowns can be solved (uniquely), provided that a certain determinant is not zero. In the present case, it turns out that the nondegeneracy condition is expressed as the nonvanishing of a determinant made from derivatives of components of φ.

In stating the following theorem, certain common notations are used which we now review. To say $f : A \to B$ means that f is a function (or mapping or transformation) which to each element of the set A associates an element of B. The symbol $\mathbf{R}^N$ denotes the set of N-tuples of real numbers, that is, N-dimensional space. The Cartesian product $A \times B$ of two sets A and B is the set of ordered pairs (a, b) of elements a of A and b of B. In particular, $\mathbf{R}^M \times \mathbf{R}^N$ is more or less the same thing as $\mathbf{R}^{(M+N)}$, since $(\mathbf{x}, \mathbf{y})$ with $\mathbf{x}$ in $\mathbf{R}^M$ and $\mathbf{y}$ in $\mathbf{R}^N$ is essentially just an $M + N$-tuple of real numbers; but the notation $\mathbf{R}^M \times \mathbf{R}^N$ suggests that the first M numbers play a different role than the last N and are denoted by a different letter.

Theorem B.2. *Suppose* $\varphi : \mathbf{R}^M \times \mathbf{R}^N \to \mathbf{R}^N$ *has continuous first partial derivatives. Suppose that* $\mathbf{x}_0$ *and* $\mathbf{y}_0$ *satisfy* $\varphi(\mathbf{x}_0, \mathbf{y}_0) = 0$ *and that the following Jacobian determinant of* φ *at* $(\mathbf{x}_0, \mathbf{y}_0)$ *is nonzero:*

$$
J\varphi(\mathbf{x}_0, \mathbf{y}_0) := \frac{\partial(\varphi^1, \ldots, \varphi^N)}{\partial(y^1, \ldots, y^N)} \bigg|_{(\mathbf{x}_0, \mathbf{y}_0)} \neq 0.
\tag{B.4a}
$$

Then for a sufficiently small neighborhood U of $\mathbf{x}_0$ in $\mathbf{R}^M$ there exists a unique continuous function $\mathbf{f} : U \rightarrow \mathbf{R}^N$ satisfying $\mathbf{f}(\mathbf{x}_0) = \mathbf{y}_0$ and $\boldsymbol{\varphi}(\mathbf{x}, \mathbf{f}(\mathbf{x})) = \mathbf{0}$ for $\mathbf{x}$ in U. Furthermore $\mathbf{f}$ has continuous first partial derivatives, and its matrix of partial derivatives $\mathbf{f}'(\mathbf{x}) := [\partial f^i / \partial x^j]$ is given by

$$\mathbf{f}'(\mathbf{x}) = -\boldsymbol{\varphi}_\mathbf{y}(\mathbf{x}, \mathbf{f}(\mathbf{x}))^{-1} \, \boldsymbol{\varphi}_\mathbf{x}(\mathbf{x}, \mathbf{f}(\mathbf{x})), \tag{B.4b}$$

where $\boldsymbol{\varphi}_\mathbf{y}$ and $\boldsymbol{\varphi}_\mathbf{x}$ are the matrices of partial derivatives $[\partial \varphi^i / \partial y^k]$ and $[\partial \varphi^k / \partial x^l]$. If $\boldsymbol{\varphi}$ has continuous partial derivatives of all orders $\leq r$ then $\mathbf{f}$ does also.

In the equation for $\mathbf{f}'(\mathbf{x})$ it should be observed that $\boldsymbol{\varphi}_\mathbf{y}$ is an $N \times N$ matrix that is invertible for $\mathbf{x}$ near $\mathbf{x}_0$ because of (B.4), and $\boldsymbol{\varphi}_\mathbf{x}$ is $N \times M$; therefore the matrix product is $N \times M$, which is the correct size for $\mathbf{f}'$. It is also possible to write this rule as a formula for specific partial derivatives which are components of $\mathbf{f}'$. This formula may be written as follows, although the notation requires some comment (given immediately below):

$$\frac{\partial y^i}{\partial x^j} = -\frac{\partial(\varphi^1, \ldots, \varphi^i, \ldots, \varphi^N)}{\partial(y^1, \ldots, x^j, \ldots, y^N)} \bigg/ \frac{\partial(\varphi^1, \ldots, \varphi^i, \ldots, \varphi^N)}{\partial(y^1, \ldots, y^i, \ldots, y^N)}. \tag{B.4c}$$

In this formula, i and j are understood as any specific integers in the range from 1 to N. Then the Jacobian determinants on the right hand side are identical except for the ith column. The denominator is a standard Jacobian, while in the numerator, the ith column, instead of having derivatives with respect to y^i, has derivatives with respect to x^j. In other words, the expression $(y^1, \ldots, x^j, \ldots, y^N)$ is short for $(y^1, \ldots, y^{i-1}, x^j, y^{i+1} \ldots, y^N)$, indicating that x^j replaces y^i in the ith position. It is not hard to see that (B.4c) follows from (B.4b) and the formula for the inverse of a matrix.

The intuition behind Theorem B.2 is that the system (B.3) can be solved for $y^1, \ldots, y^N$ near $(\mathbf{x}_0, \mathbf{y}_0)$, provided that the linearized system

$$\frac{\partial \varphi^1}{\partial y^1} dy^1 + \cdots + \frac{\partial \varphi^1}{\partial y^N} dy^N = -\frac{\partial \varphi^1}{\partial x^1} dx^1 - \cdots - \frac{\partial \varphi^1}{\partial x^M}, dx^M,$$

$$\vdots \quad + \quad + \quad = \quad \vdots \tag{B.5}$$

$$\frac{\partial \varphi^N}{\partial y^1} dy^1 + \cdots + \frac{\partial \varphi^N}{\partial y^N} dy^N = -\frac{\partial \varphi^N}{\partial x^1} dx^1 - \cdots - \frac{\partial \varphi^N}{\partial x^M}, dx^M$$

can be solved for $dy^1, \ldots, dy^N$ when the coefficients are evaluated at $(\mathbf{x}_0, \mathbf{y}_0)$. This is possible when the coefficient matrix is nonsingular, that is, if its determinant is nonzero, which is condition (B.4a).

Instead of proving the implicit function theorem, we will briefly sketch the main ideas of several proofs of Theorem B.1. Each of these proofs requires a good deal of technical detail to complete, and yet the main ideas are simple and informative.

The first proof is geometrical. Assume $\varphi(x_0, y_0) = 0$ and $\varphi_y(x_0, y_0) > 0$. By continuity, $\varphi_y(x, y) > 0$ for (x, y) near (x_0, y_0). That is, $\varphi(x, y)$ increases as y increases along vertical lines near (x_0, y_0). In particular, there is a vertical line segment PQ through (x_0, y_0) on which φ is negative at P (below (x_0, y_0)) and positive at Q. (See Fig. B.1.) By continuity, φ is negative near P and positive near Q. It follows that there is a box around (x_0, y_0) with the following properties: φ is negative at the bottom, positive at the top, and increasing along vertical lines in the box. Therefore there is a unique point on each vertical line where $\varphi = 0$. These points make up the graph of f. From here the proof becomes technical, because it is necessary to prove that f is continuous and continuously differentiable.

The second proof uses the method of iteration, or successive approximation. The equation $\varphi(x, y) = 0$ is equivalent, since $\varphi_y(x_0, y_0) \neq 0$, to the equation

$$y = y_0 - \frac{\varphi(x, y) - \varphi_y(x_0, y_0)(y - y_0)}{\varphi_y(x_0, y_0)}. \tag{B.6}$$

As a first approximation to $y = f(x)$ for any x near x_0, take $y = y_0$. Suppose that an nth approximation y_n has been found; define the next approximation by

$$y_{n+1} := y_0 - \frac{\varphi(x, y_n) - \varphi_y(x_0, y_0)(y_n - y_0)}{\varphi_y(x_0, y_0)}. \tag{B.7}$$

It is possible to show that the sequence y_n defined in this way converges for x sufficiently near x_0; it is then easy to see by letting $n \to \infty$ in (B.7) that the limit $y := \lim_{n \to \infty} y_n$ satisfies (B.6) and hence $\varphi(x, y) = 0$. This defines $f(x)$, and again it can be shown to be smooth.

The third proof uses differential equations. Assuming for a moment that there is a function $y = f(x)$ such that $\varphi(x, f(x)) = 0$ and $f(x_0) = y_0$, it follows by differentiating $\varphi(x, y) = 0$ with respect to x that

$$y' = -\frac{\varphi_x(x, y)}{\varphi_y(x, y)}. \tag{B.8}$$

Since $\varphi_y(x_0, y_0) \neq 0$ it follows by continuity that $\varphi_y(x, y) \neq 0$ for (x, y) near (x_0, y_0). Thus (B.8) can be thought of as a differential equation for $y = f(x)$ defined in a neighborhood of (x_0, y_0), and the solution with initial condition $y = y_0$ when $x = x_0$ exists for x near x_0 and is the desired function $f(x)$. This proof depends upon the existence theorem for differential equations, which in turn is usually proved by an iteration argument.

The last proof uses power series and requires the additional assumption that $\varphi(x, y)$ has a convergent double power series for (x, y) near (x_0, y_0):

$$\varphi(x, y) = \sum_{m=0}^{\infty} \sum_{n=0}^{\infty} a_{mn}(x - x_0)^m (y - y_0)^n. \tag{B.9}$$

This is a series of the same type as (A.12). In this case a solution of $\varphi(x, y) = 0$ may be sought in the form

$$y - y_0 = \sum_{k=0}^{\infty} b_k (x - x_0)^k. \tag{B.10}$$

Substituting (B.10) into (B.9), the coefficients b_k can be determined recursively and (B.10) can be shown to converge, defining $y = f(x)$ for x near x_0. The calculations are essentially the same as those carried out in Section 1.2, except that the focus is on convergence of infinite series rather than approximation by partial sums. The first three proofs are better than this one, because for our purposes the convergence of (B.9) is too restrictive an assumption.

We conclude this section with the statement of a modified implicit function theorem needed for the first time in Section 1.7. Since this does not appear in most advanced calculus books, a rather technical proof is included for the interested reader.

Theorem B.3. *Let $\varphi(x, \tau, \varepsilon)$ be a real-valued function having continuous partial derivatives of all orders $\leq r$, with $r \geq 1$. Suppose for each $\tau \in [\tau_1, \tau_2]$ there is an $x_0 = h(\tau)$ such that $\varphi(h(\tau), \tau, 0) = 0$ and $\varphi_x(h(\tau), \tau, 0) \neq 0$: That is, the hypotheses of the implicit function theorem are satisfied for each τ. Suppose also that $h(\tau)$ is continuous. Then there exists a unique continuous function $x = f(\tau, \varepsilon)$ defined for $\tau \in [\tau_1, \tau_2]$ and for ε in an interval $|\varepsilon| < \varepsilon_0$, satisfying $f(\tau, 0) = h(\tau)$ and $\varphi(f(\tau, \varepsilon), \tau, \varepsilon) = 0$. Furthermore f has continuous partial derivatives of all order $\leq r$, and f has a Taylor expansion*

$$f(\tau, \varepsilon) = x_0(\tau) + x_1(\tau)\varepsilon + \cdots + x_k(\tau)\varepsilon^k + R_{k+1}(\tau, \varepsilon)$$

with $x_0(\tau) = h(\tau)$, in which the remainder satisfies the following error bound: Given ε_1 with $0 < \varepsilon_1 < \varepsilon_0$, there exists a constant c independent of τ and ε such that $|R_{k+1}(\tau, \varepsilon)| \leq c\varepsilon^{r+1}$ for all $\tau \in \tau_1, \tau_2$ and $|\varepsilon| \leq \varepsilon_1$.

Proof. Once the existence and smoothness of the function $f(\tau, \varepsilon)$ is established, the Taylor expansion and uniform error estimate follow from Theorem A.2. The difficult part is to establish the existence of f. By a standard version of the implicit function theorem, if $\varphi(x_0, \tau_0, 0) = 0$ and $\varphi_x(x_0, \tau_0, 0) \neq 0$ then there is a function $x = f(\tau, \varepsilon)$ defined on a disk about $(\tau_0, 0)$ whose graph coincides with the graph of $\varphi = 0$ in a ball about $(x_0, \tau_0, 0)$. Applying this theorem with $\tau_0 = \sigma$, $x_0 = h(\sigma)$ for each $\sigma \in [\tau_1, \tau_2]$, we obtain functions $x = f_\sigma(\tau, \varepsilon)$, each defined in a disk about $(\sigma, 0)$ and each satisfying $f_\sigma(\tau, 0) = h(\tau)$ for all τ for which the left hand side is defined. Since the set K of $(\sigma, 0)$ for $\sigma \in [\tau_1, \tau_2]$ is compact, there are finitely many points $\sigma_1, \ldots, \sigma_k$ such that the domains of $f_i = f_{\sigma_i}$, $i = 1, \ldots, k$, cover K. We wish to amalgamate these functions into a single function by defining $f(\tau, \varepsilon) = f_i(\tau, \varepsilon)$ whenever (τ, ε) belongs to the domain of f_i.

The difficulty is that f_i and f_j need not agree on the full intersection of their domains (although they do agree wherever they intersect on K). We claim that there is a neighborhood of K on which no conflicts arise. Consider first any pair of functions f_i and f_j. Either there is a distance $d_{ij} > 0$ such that f_i and f_j never conflict within distance d_{ij} of K, or there is a point $\sigma \in K$ having arbitrarily near points at which $f_i \neq f_j$. The latter possibility violates the fact that in a neighborhood of $(h(\sigma), \sigma, 0)$ the graph of $\varphi = 0$ coincides with that of f_σ. The finite set of d_{ij}, $(i, j = 1, \ldots, k)$ has a positive minimum, so there is a neighborhood of K in which no conflicts arise. On this neighborhood, $f(\tau, \varepsilon)$ can be defined and meets the requirements of the theorem. ∎

APPENDIX C

SECOND ORDER DIFFERENTIAL EQUATIONS

The homogeneous linear second order ordinary differential equation with constant coefficients,

$$ay'' + by' + cy = 0 \quad (a \neq 0), \tag{C.1}$$

is solved by first solving the quadratic equation $a\lambda^2 + b\lambda + c = 0$. If the roots λ_1 and λ_2 are real and distinct the general solution is

$$y = c_1 e^{\lambda_1 x} + c_2 e^{\lambda_2 x}. \tag{C.2}$$

If $\lambda_1 = \lambda_2 = \lambda$ then

$$y = (c_1 + c_2 x) e^{\lambda x}. \tag{C.3}$$

If $\lambda_1 = \mu + i\nu$ and $\lambda_2 = \mu - i\nu$ with μ and ν real then

$$y = e^{\mu x} (c_1 \cos \nu x + c_2 \sin \nu x). \tag{C.4}$$

If $c/a > 0$ the most convenient form of (C.1) is

$$\ddot{y} + 2h\dot{y} + k^2 y = 0. \tag{C.5}$$

We have used the symbol $\dot{} = d/dt$ rather than $' = d/dx$ since this form usually arises in problems where the independent variable is time, such as the oscillations of a mass suspended by a spring. For (C.5), $\lambda = -h \pm \sqrt{h^2 - k^2}$. The following four cases are particularly important; the letters

477

A and δ denote arbitrary constants playing the same role as c_1 and c_2 in alternate forms of the solutions.

$h = 0$ *(Undamped):*

$$y = c_1 \cos kt + c_2 \sin kt = A \cos(kt + \delta). \tag{C.6}$$

$0 < h < k$ *(Undercritically Damped):*

$$y = e^{-ht}(c_1 \cos \nu t + c_2 \sin \nu t) = Ae^{-ht} \cos(\nu t + \delta), \qquad \nu = \sqrt{k^2 - h^2}. \tag{C.7}$$

$h = k$ *(Critically Damped):*

$$y = (c_1 + c_2 t)e^{-ht}. \tag{C.8}$$

$h > k$ *(Overcritically Damped):*

$$y = c_1 e^{\lambda_1 t} + c_2 e^{\lambda_2 t}, \qquad \lambda_{1,2} = -h \pm \sqrt{h^2 - k^2}. \tag{C.9}$$

A linear differential equation is called *inhomogeneous* or *forced* if its right hand side is a nonzero function of t or x. The fundamental fact about such equations is that their general solution is equal to the general solution of the associated homogeneous equation plus any particular solution of the inhomogeneous equation. Inhomogeneous second order linear differential equations with constant coefficients can be solved by the method of *undetermined coefficients* if the inhomogeneous part is a simple combination of polynomials, sine, cosine, and exponential functions; otherwise, by the method of *variation of parameters* or *variation of constants*. Variation of parameters will be reviewed later in this appendix. The method of undetermined coefficients involves making an educated guess of the form taken by a solution, substituting the guess into the equation, and choosing the coefficients so that it works. The "guessing" can be done either by "intuition" or according to precise rules which will not be reviewed here. The most important cases for this book arise in Part II and concern the forced oscillator

$$\ddot{y} + 2h\dot{y} + k^2 y = F \cos \omega t. \tag{C.10}$$

The following general solutions may be checked by substituting into (C.10):

$h = 0$, $\omega \neq k$ *(Undamped and Nonresonant):*

$$y = A \cos(kt + \delta) + \frac{F}{k^2 - \omega^2} \cos \omega t. \tag{C.11}$$

$h = 0$, $\omega = k$ *(Undamped and Resonant):*

$$y = A \cos(kt + \delta) + \frac{F}{2k} t \sin kt. \tag{C.12}$$

$0 < h < k$ *(Undercritically Damped):*

$$y = Ae^{-ht} \cos(\nu t + \delta) + B \cos(\omega t + \psi), \tag{C.13}$$

with

$$\nu = \sqrt{k^2 - h^2},$$

$$B^2 = \frac{F^2}{(k^2 - \omega^2)^2 + 4h^2\omega^2},$$

$$\tan \psi = \frac{-2h\omega}{k^2 - \omega^2}, \qquad -\frac{\pi}{2} \le \psi \le \frac{\pi}{2}.$$

(C.12) shows that in the undamped case, resonance (forcing frequency ω equal to free frequency k) produces unbounded solutions due to the $t \sin kt$ term. In the damped case (C.13), solutions remain bounded even if $\omega = k$. In (C.13) the constants A and B do not appear on an equal footing; A is arbitrary whereas B is determined by the forcing. The same remark applies to δ and ψ. If $\omega = k$ the expression for $\tan \psi$ is infinite, and one should take $\psi = \pi/2$ and $B = -F/2h\omega$ or else $\psi = -\pi/2$ and $B = F/2h\omega$; the two are equivalent. If $\omega \ne k$ the sign of B should be taken equal to the sign of $F/(k^2 - \omega^2)$.

Each solution of these second order equations contains two arbitrary constants (c_1 and c_2 or A and δ) which can be determined if two *initial conditions* are given in the form $y(t_0) = \alpha$, $\dot{y}(t_0) = \beta$. Notice that *three* real numbers (t_0, α, and β) are required to fix the two constants. Frequently it is assumed that the initial time t_0 equals zero, so that only two numbers are required. A fundamental existence theorem guarantees that for every choice of t_0, α, and β there is a unique solution (and hence a unique choice of c_1 and c_2). Since the equations so far considered are linear, another theorem guarantees that the solutions exist for all time. These facts can be seen, of course, from the explicit solutions, but the theorems in question apply more generally to linear equations with nonconstant coefficients such as

$$a(t)\ddot{y} + b(t)\dot{y} + c(t)y = f(t), \tag{C.14}$$

provided $a(t) \ne 0$ and a, b, c, and f have continuous first derivatives for all t. In this case the unique solution $y = \varphi(t, t_0, \alpha, \beta)$ satisfying $y = \alpha$, $\dot{y} = \beta$ at $t = t_0$ has continuous first partial derivatives with respect to t, t_0, α, and β, and exists for all t. If a, b, c, and f have continuous derivatives of all orders $\le r$, then φ has continuous partial derivatives of all order $\le r$.

Another way to restrict the choice of arbitrary constants is by imposing *boundary conditions* in place of initial conditions. There are no general theorems guaranteeing the existence or uniqueness of solutions to boundary value problems. For instance if the general solution is (C.2) then boundary conditions $y(0) = \alpha$, $y(1) = \gamma$ determine c_1 and c_2 uniquely. But if the general solution is (C.4) with $\mu = 0$, then the boundary conditions $y(0) = 0$ and $y(1) = 0$ determine the unique solution $c_1 = c_2 = 0$ unless $\nu = n\pi$ for some integer n, in which case $c_1 = 0$ and c_2 is arbitrary; that is, solutions exist but are not unique. In the same case $\nu = n\pi$ the boundary value problem $y(0) = 0$, $y(1) = \gamma \neq 0$ has no solution.

A second order equation can be written as a *system of two first order equations* (confusingly, this is also called a *second order system*) in several ways. The simplest is to set $u_1 = y$, $u_2 = \dot{y}$. For the case of (C.10) this leads to

$$\dot{u}_1 = u_2,$$
$$\dot{u}_2 = -k^2 u_1 - 2hu_2 + F \cos \omega t. \tag{C.15}$$

A more symmetrical formulation is $u_1 = y$, $u_2 = \dot{y}/k$, giving

$$\dot{u}_1 = k u_2,$$
$$\dot{u}_2 = -ku_1 - 2hu_2 + \frac{F}{k} \cos \omega t. \tag{C.16}$$

Equations (C.15) and (C.16) are most useful in the unforced case $F = 0$, because then the solutions lie on curves (called *orbits*) in the (u_1, u_2) plane (called the *phase plane*) that fill the plane without intersecting. A graph showing a number of these orbits is called a *phase portrait*. (See Appendix D for more information about orbits of autonomous systems.) If $F = 0$ and $h = 0$, system (C.16) takes the following form in polar coordinates ($u_1 = r \cos \theta$, $u_1 = r \sin \theta$):

$$\dot{r} = 0,$$
$$\dot{\theta} = -k. \tag{C.17}$$

The procedure for obtaining (C.17) from (C.16) is to substitute $\dot{u}_1 = \dot{r} \cos \theta - r\dot{\theta} \sin \theta$, $\dot{u}_2 = \dot{r} \sin \theta + r\dot{\theta} \cos \theta$ into (C.16) and solve for $\dot{r}$ and $\dot{\theta}$. The solutions of (C.17) are circles $r = A$, $\theta = -kt - \delta$ for arbitrary A and δ; since $\cos \theta = \cos(-\theta)$, this implies $u_1 = A \cos(kt + \delta)$ as we already knew (equation (C.6)). The phase portrait of the system is simply the collection of circles centered at the origin. If the unsymmetrical form (C.15) were used, the polar coordinate version would be more complicated, because the solutions would lie on ellipses rather than circles. These equations (C.15)–(C.17) will be used frequently as the starting point for various perturbations. In the case $F = 0$, $h > 0$ corresponding to (C.7), the phase portrait consists of spirals tending toward the origin.

The concept of *variation of parameters* or *variation of constants* first appears as a general method of solution for second order equations of the form

$$a(t)\ddot{y} + b(t)\dot{y} + c(t)y = f(t), \qquad (C.18)$$

but its significance extends far beyond this. Two approaches to this method will be outlined here. The presentation in an elementary differential equations text usually runs something like this. The associated homogeneous problem $a\ddot{y} + b\dot{y} + cy = 0$ has basic solutions y_1 and y_2, general solution $c_1y_1 + c_2y_2$. Look for a solution of the inhomogeneous problem in the form

$$\begin{aligned} y &= v_1y_1 + v_2y_2, \\ \dot{y} &= v_1\dot{y}_1 + v_2\dot{y}_2, \end{aligned} \qquad (C.19)$$

with v_1 and v_2 being unknown functions of t. Since the derivative of the first equation in (C.19) is actually $\dot{y} = \dot{v}_1y_1 + v_1\dot{y}_1 + \dot{v}_2y_2 + v_2\dot{y}_2$, the second equation implies that

$$\dot{v}_1y_1 + \dot{v}_2y_2 = 0. \qquad (C.20)$$

Substituting (C.19) into (C.18) leads to

$$\dot{v}_1\dot{y}_2 + \dot{v}_2\dot{y}_2 = f(t)/a(t). \qquad (C.21)$$

Then (C.20) and (C.21) imply

$$\dot{v}_1 = -y_2f/aW, \qquad \dot{v}_2 = y_1f/aW, \qquad (C.22)$$

where

$$W(t) = \begin{vmatrix} y_1(t) & y_2(t) \\ \dot{y}_1(t) & \dot{y}_2(t) \end{vmatrix} \qquad (C.23)$$

is called the *Wronskian*. Equations (C.22) can be integrated and substituted into (C.19) to obtain the solution. The result can be written

$$y(t) = c_1y_1(t) + c_2y_2(t) + \int_0^t \frac{y_1(\tau)y_2(t) - y_1(t)y_2(\tau)}{a(\tau)W(\tau)} f(\tau)d\tau, \qquad (C.24)$$

where c_1 and c_2 are constants of integration. If y_1 and y_2 are chosen so that

$$\begin{aligned} y_1(0) &= 1 & y_2(0) &= 0, \\ \dot{y}_1(0) &= 0 & \dot{y}_2(0) &= 1, \end{aligned} \qquad (C.25)$$

then the constants c_1, c_2 equal the initial conditions $y(0) = \alpha$, $\dot{y}(0) = \beta$. In this case (C.24) can be written

$$y(t) = \alpha y_1(t) + \beta y_2(t) + \int_0^t K(t, \tau)f(\tau)d\tau, \qquad (C.26)$$

where $K(t, \tau) := (y_1(\tau)y_2(t) - y_1(t)y_2(\tau))/a(\tau)W(\tau)$. The last term in (C.26) is called the *Volterra integral transform* of f with kernel K.

The preceding derivation may seem somewhat mysterious; in particular, why is it legitimate to impose condition (C.20) or equivalently, the second equation of (C.19)? The usual answer is that it requires two conditions to determine the two unknown functions v_1 and v_2; one condition is that $y = v_1y_1 + v_2y_2$ solve the differential equation (C.18) and the second may be imposed arbitrarily. Then condition (C.20) is chosen in order to simplify $\dot{y}$, so that $\ddot{y}$ will contain four terms rather than eight. But there is another point of view which makes the derivation much more transparent. First write (C.18) as a system of two equations, setting $u_1 = y$, $u_2 = \dot{y}$:

$$\dot{u}_1 = u_2,$$
$$\dot{u}_2 = -\frac{c(t)}{a(t)}u_1 - \frac{b(t)}{a(t)}u_2 + \frac{f(t)}{a(t)}. \qquad (C.27)$$

Next introduce new variables v_1, v_2 in place of u_1, u_2 by the equations

$$u_1 = v_1y_1 + v_2y_2,$$
$$u_2 = v_1\dot{y}_1 + v_2\dot{y}_2, \qquad (C.28)$$

where y_1 and y_2 are as before. (Thus $y_1, y_2, \dot{y}_1$, and $\dot{y}_2$ are known functions of t, whereas u_1, u_2, v_1, and v_2 are simply variables.) A short calculation shows that v_1 and v_2 satisfy (C.22). From this viewpoint (C.28), which is the same as (C.19), is just a change of variables, and there is no need to explain why the second equation of (C.28) is permissible.

Part of the importance of variation of parameters lies in its use for changing differential equations into integral equations. This usage occurs in this book in Chapter 3. Consider the equation

$$a(t)\ddot{y} + b(t)\dot{y} + c(t)y = g(y, \dot{y}, t), \qquad (C.29)$$

which is nonlinear (in general) because g need not be a linear function of y and $\dot{y}$. This resembles (C.18), except that the right hand side is not a known function of t since it involves the unknown function y and its derivative. But there is a clever way to exploit (C.26) even though it does not apply directly. Namely, suppose $y(t)$ is a solution of (C.29) and define a function f by $f(t) := g(y(t), \dot{y}(t), t)$. Then it follows from (C.29) that $y(t)$ is also a solution of (C.18) with this particular forcing function $f(t)$.

Therefore $y(t)$ also satisfies (C.26), and therefore, using the definition of f, we have

$$y(t) = \alpha y_1(t) + \beta y_2(t) + \int_0^t K(t, \tau) g(y(\tau), \dot{y}(\tau), \tau) d\tau. \qquad \text{(C.30)}$$

It is assumed in this derivation that y_1 and y_2 are solutions of the homogeneous problem satisfying (C.25)). (C.30) cannot be regarded as a *formula* for the solution $y(t)$, as can (C.26), since this unknown function also occurs on the right hand side as an argument of g. Instead it is an *integral equation* for y, that is, an equation in which the unknown function appears both inside and outside of an integral. An integral equation sometimes has advantages over a differential equation. As a simple example of how an integral equation can be used, if the function $|g|$ is bounded by a constant c, an upper bound for $|y(t)|$ can be derived from (C.30); if the initial conditions are $y(0) = \dot{y}(0) = 0$, we have $|y(t)| \leq c \int_0^t K(t, \tau) d\tau$. Such techniques are important in error estimation for perturbation methods, and play a crucial role beginning with Chapter 3.

There exist formulas similar to (C.26) and (C.30) for boundary value problems instead of initial value problems. Suppose that the equation

$$a(x)y'' + b(x)y' + c(x)y = 0, \qquad 0 \leq x \leq 1 \qquad \text{(C.31)}$$

has two solutions $y_1(x)$ and $y_2(x)$ satisfying

$$\begin{aligned} y_1(0) &= 1 & y_2(0) &= 0, \\ y_1(1) &= 0 & y_2(1) &= 1 \end{aligned} \qquad \text{(C.32)}$$

in place of (C.25). (The existence of these solutions is not automatic, as it is for (C.25), because (C.32) gives boundary conditions rather than initial conditions.) These solutions y_1 and y_2 can be used in (C.24) to write the general solution of

$$a(x)y'' + b(x)y' + c(x)y = f(x) \qquad \text{(C.33)}$$

in the form

$$y(x) = c_1 y_1(x) + c_2 y_2(x) + \int_0^x \frac{y_1(\xi)y_2(x) - y_1(x)y_2(\xi)}{a(\xi)W(\xi)} f(\xi) \, d\xi, \qquad \text{(C.34)}$$

where $W = y_1 y_2' - y_1' y_2$. To look for a solution of (C.33) with boundary conditions

$$y(0) = \alpha, \qquad y(1) = \gamma, \qquad \text{(C.35)}$$

substitute $x = 0$ and $x = 1$ into (C.34) and solve for c_1 and c_2 using (C.32). The result is $c_1 = \alpha$, $c_2 = \gamma + \int_0^1 (y_2(\tau)/a(\tau)W(\tau))f(\tau)d\tau$. Putting these back into (C.34) gives

$$y(0) = \alpha y_1(x) + \gamma y_2(x) + \int_0^x \frac{y_1(x)y_2(\xi)}{a(\xi)W(\xi)} f(\xi)\, d\xi + \int_x^1 \frac{y_1(\xi)y_2(x)}{a(\xi)W(\xi)} f(\xi)\, d\xi,$$
$$\text{(C.36)}$$

which can be written as

$$y(x) = \alpha y_1(x) + \gamma y_2(x) + \int_0^1 G(x,\xi)f(\xi)d\xi, \qquad \text{(C.37)}$$

where

$$G(x,\xi) = \begin{cases} y_1(x)y_2(\xi)/a(\xi)W(\xi) & \text{for } 0 \le \xi \le x \\ y_1(\xi)y_2(x)/a(\xi)W(\xi) & \text{for } x \le \xi \le 1. \end{cases} \qquad \text{(C.38)}$$

Equation (C.37) is the analog of (C.26) for boundary value problems, and $G(x,\xi)$ is called Green's function. Notice that these calculations in fact prove the existence and uniqueness of the solution to (C.33) with boundary conditions (C.35), under the condition that (C.31) has solutions satisfying (C.32).

The analog of (C.30) for boundary value problems follows immediately. Namely, any solution of the nonlinear boundary value problem

$$a(x)y'' + b(x)y' + c(x)y = g(x, y, y'),$$
$$y(0) = \alpha, \qquad \text{(C.39)}$$
$$y(1) = \gamma$$

must also satisfy the integral equation

$$y(x) = \alpha y_1(x) + \gamma y_2(x) + \int_0^1 G(x,\xi)g(\xi, y(\xi), y'(\xi))d\xi, \qquad \text{(C.40)}$$

provided of course that (C.31) has solutions satisfying (C.32). Again, the integral equation (C.40) does not provide an immediate solution, but is often a useful alternative formulation of the problem.

The last term of (C.37) is called a *Fredholm integral transform* of f, and (C.40) is called a *Fredholm integral equation*. The name "Fredholm" indicates that the integrals are over a fixed interval (in the present case [0,1]), as opposed to the *Volterra transform* in (C.26) and the *Volterra integral equation* (C.30), where the integration is over [0, t].

APPENDIX D

SYSTEMS OF DIFFERENTIAL EQUATIONS

The introduction of second order systems in Appendix C brings us to the subject of *Nth order systems* of differential equations, meaning systems of N first order equations, which can be written in vector form as

$$\dot{\mathbf{u}} = \mathbf{f}(\mathbf{u}, t), \tag{D.1}$$

where $\mathbf{u} = (u_1, \ldots, u_N)$. The fundamental facts about the existence of solutions to such systems are as follows. If $\mathbf{f}$ has continuous first partial derivatives with respect to $\mathbf{u}$ (that is, with respect to the components u_i of $\mathbf{u}$) and t, then there is a unique solution

$$\mathbf{u} = \boldsymbol{\varphi}(t, t_0, \boldsymbol{\alpha}), \tag{D.2}$$

defined for t in some open interval containing t_0, which satisfies $\mathbf{u} = \boldsymbol{\alpha}$ when $t = t_0$, or $\boldsymbol{\varphi}(t_0, t_0, \boldsymbol{\alpha}) = \boldsymbol{\alpha}$. This solution has continuous first partial derivatives with respect to t, t_0, and $\boldsymbol{\alpha}$, and if $\mathbf{f}$ has continuous partial derivatives of all orders $\leq r$, so does $\boldsymbol{\varphi}$. If the interval of t on which the solution is defined is not infinite, then $\|\boldsymbol{\varphi}(t, t_0, \boldsymbol{\alpha})\| \to \infty$ as t approaches any finite endpoint of the interval. That is, solutions can only cease to exist by becoming infinite. (This is assuming $\mathbf{f}$ is defined for all $\mathbf{u}$ and t. If not, solutions can also cease to exist by running into a point where $\mathbf{f}$ is not defined.)

One refinement of the existence theorem for differential equations is very helpful in perturbation problems.

Theorem D.1. *Let* $\mathbf{f}(\mathbf{u}, t, \varepsilon)$ *be defined for all* $(\mathbf{u}, t, \varepsilon)$ *and have continuous first partial derivatives, and suppose that for* $\varepsilon = 0$ *the solution of the initial value problem*

$$\dot{\mathbf{u}} = \mathbf{f}(\mathbf{u}, t, \varepsilon),$$
$$\mathbf{u}(0) = \alpha \tag{D.3}$$

exists on the interval $|t| \leq T$. *Then there exists* $\varepsilon_0 > 0$ *such that for* $|\varepsilon| \leq \varepsilon_0$, *the solution of (D.3) also exists for* $|t| \leq T$.

The idea of the proof is that for ε sufficiently small the solution is close to the solution when $\varepsilon = 0$ and hence cannot become unbounded; since it does not become unbounded, it cannot cease to exist.

A linear homogeneous system of differential equations is a system of the form

$$\dot{\mathbf{u}} = A(t)\mathbf{u}, \tag{D.4}$$

where $A(t)$ is a continuously differentiable $N \times N$ matrix-valued function of t. (In all equations using matrices, it is understood that vectors such as $\mathbf{u}$ are expressed as column vectors.) Solutions of such a system cannot become unbounded and hence must exist for all time. If $A(t) = A$ is a constant matrix, the system

$$\dot{\mathbf{u}} = A\mathbf{u} \tag{D.5}$$

can be solved explicitly. The simplest case is when A is diagonalizable. If $C^{-1}AC = D$ is a diagonal matrix with diagonal entries (eigenvalues) equal to $\lambda_1, \ldots, \lambda_N$, and if a new vector variable $\mathbf{v}$ is introduced by setting $\mathbf{v} := C^{-1}\mathbf{u}$, then $\dot{\mathbf{v}} = D\mathbf{v}$, that is, $\dot{v}_i = \lambda_i v_i$. The system of N equations decouples into N first order scalar equations. It follows that $v_i(t) = e^{\lambda_i t}v_i(0)$, or $\mathbf{v}(t) = e^{Dt}\mathbf{v}(0)$, where e^{Dt} is the diagonal matrix with diagonal entries $e^{\lambda_i t}$. The solution of (D.5) is then

$$\mathbf{u}(t) = Ce^{Dt}C^{-1}\mathbf{u}(0). \tag{D.6}$$

Complex eigenvalues can occur in these calculations and are handled by the formula $e^{\alpha + i\beta} = e^{\alpha}(\cos \beta + i \sin \beta)$. In this case C will be complex but $Ce^{Dt}C^{-1}$ will be real, and (D.6) will give real solutions for real vectors $u(0)$. When A is not diagonalizable, D is replaced by the Jordan canonical form of A, and (D.6) continues to hold. In this case the general definition of the exponential of a matrix must be used, according to which $e^B = I + B + B^2/2! + \cdots$, that is, the usual power series for the exponential except that the leading 1 is replaced by the identity matrix I. The components of $u(t)$ then turn out to involve linear combinations of $t^k e^{\lambda_i t}$ for various powers k and eigenvalues λ_i.

All linear homogeneous systems (D.4) have the "trivial solution" $\mathbf{u}(t) \equiv \mathbf{0}$, and it is often important to decide whether this solution is "stable." In the case (D.5) with constant A, there is a simple criterion depending only on the eigenvalues λ_i. If all eigenvalues have negative real part, then every solution decays toward the trivial solution as $t \to \infty$, and the trivial solution is called *asymptotically stable*. This should be thought of as damping. If even one eigenvalue has positive real part, some solutions move away from the trivial solution at an exponential rate, and it is unstable. The case in which some eigenvalues lie on the imaginary axis but none lie in the right half-plane is more subtle, and is not very important for this book.

In this case there are solutions which oscillate around the origin without damping out, and the trivial solution is at best *neutrally stable*. But even this neutral stability need not hold. If there is a repeated pure imaginary eigenvalue occurring in a Jordan block, then a term $te^{\lambda t}$ with λ imaginary will occur in the solution. This term becomes unbounded as $t \to \infty$ and causes the origin to be unstable. This should be thought of as a resonance phenomenon: Pure imaginary eigenvalues are frequencies, and two equal frequencies in the absence of damping can cause a resonant growth of amplitude if they interact with each other in certain ways indicated by the presence of a Jordan block. If A is diagonalizable, or more generally if there are no pure imaginary eigenvalues in Jordan blocks, then neutral stability will hold, and all solutions will either decay or oscillate. Such neutral stability is very sensitive to any small change in the differential equation such as adding a perturbation term or a nonlinear term, whereas both asymptotic stability and instability are resistant to such changes. This resistance is what makes those cases much more significant in this book, as will be seen below.

The "solution function" φ of (D.1), defined by (D.2), satisfies a fundamental identity, which can be derived as follows. The solution $\varphi(t, t_0, \alpha)$ is the unique solution taking the value α at time t_0. Let $\beta = \varphi(t_1, t_0, \alpha)$. Then $\varphi(t, t_0, \alpha)$ is also the unique solution taking the value β at time t_1. In symbols, $\varphi(t, t_0, \alpha) = \varphi(t, t_1, \beta)$, or, substituting the definition of β,

$$\varphi(t, t_0, \alpha) = \varphi\left(t, t_1, \varphi(t_1, t_0, \alpha)\right). \tag{D.7}$$

If all solutions of (D.1) exist for all time, then (D.7) holds for all t, t_0, and t_1. Otherwise it holds whenever both sides are well defined. In words, (D.7) says that if you begin a solution at α at time t_0 and follow that solution until time t, it will reach the same point as if you follow the solution until time t_1, "restart" it at that time (by solving a new initial value problem), and follow the new solution until time t.

Many of the systems of the form (D.1) which occur in this book, especially in Part II, are either independent of t or periodic in t. Each of these cases has its own special results and concepts: *orbits* and *flows* for the first case, *period maps* for the second. Each case also has its own simplified form of the fundamental identity (D.7). We will now turn to these topics.

The first special case is when f in (D.1) is independent of time, that is,

$$\dot{\mathbf{u}} = \mathbf{f}(\mathbf{u}). \tag{D.8}$$

This is called the *autonomous* case. In this case the function $\mathbf{f}(\mathbf{u})$ may be regarded as defining a vector field in N-dimensional space, and the solutions of (D.8) are just the curves $\mathbf{u}(t)$ which have $\mathbf{f}(\mathbf{u}(t))$ as their velocity vector at each point; in particular, a solution passing through $\mathbf{u}$ is tangent to $\mathbf{f}(\mathbf{u})$. Since the vector field $\mathbf{f}(\mathbf{u})$ is independent of time, the solution passing through α at time t_0 will be exactly the same as the solution passing through α at time 0, except that the solution will be delayed by t_0; that is,

$$\boldsymbol{\varphi}(t, t_0, \alpha) = \boldsymbol{\varphi}(t - t_0, 0, \alpha). \tag{D.9}$$

An analytical proof of (D.9) consists of checking that both sides satisfy (D.8) (the reader should be sure to check that this is *not* in general true if the equation is *not* autonomous) and that both sides equal α when $t = t_0$; therefore, by the uniqueness of solutions with a given initial condition, they are equal for all t in their domain. Because of (D.9) all expressions involving φ can be replaced by expressions in which the second argument of φ is zero. Upon defining a shorter form $\boldsymbol{\varphi}(t, \alpha) := \boldsymbol{\varphi}(t, 0, \alpha)$ for convenience and writing t_2 for t, (D.7) takes the form $\boldsymbol{\varphi}(t_2 - t_0, \alpha) = \boldsymbol{\varphi}(t_2 - t_1, \boldsymbol{\varphi}(t_1 - t_0, \alpha))$. Putting $t_2 - t_1 = t$, $t_1 - t_0 = s$ gives

$$\boldsymbol{\varphi}(t + s, \alpha) = \boldsymbol{\varphi}(t, \boldsymbol{\varphi}(s, \alpha)), \tag{D.10}$$

the so-called flow property of autonomous systems. The term "flow" is meant to suggest a fluid, whose molecules occupy various positions α at time zero and move to positions $\boldsymbol{\varphi}(s, \alpha)$ at time s. But because of (D.9), it is possible to think of $\boldsymbol{\varphi}(s, \alpha)$ as the result of starting in a state α at any time (not necessarily time zero) and "flowing forward" for a time interval of length s. Then (D.10) says that if the resulting state $\boldsymbol{\varphi}(s, \alpha)$ is continued for an additional time t, the result is the same as if the state α has been advanced through time $t + s$. (In the language of fluid mechanics, this situation is called a *steady flow*, but in differential equations the word "steady" is omitted.) The set of points occupied by $\boldsymbol{\varphi}(t, \alpha)$ for fixed α and for all t is called the *orbit* of α. (In fluid mechanics the same thing is called a *stream line*.) All *solutions* $\boldsymbol{\varphi}(t - t_0, \alpha)$ passing though α at different times t_0 travel along the same *orbit*, so each *orbit* carries infinitely many *solutions*. Every point belongs to a unique orbit; orbits cannot cross. It is important to be clear about the distinction between orbits and solutions. As an example, in the case of (C.16) with $h = 0$ and $F = 0$ (which is autonomous since $F = 0$), or its equivalent polar form (C.17), the *orbits* are the circles $r = A$, while specifying a *solution* requires a choice of δ in $\theta = -kt - \delta$.

An autonomous system can have orbits of an important special type called *rest points* or *equilibrium points*. (Sometimes these are called *singular points*, but since "singular" has so many other meanings in mathematics this terminology will not be used.) A rest point is an orbit consisting of only one point, which does not move with time. Clearly, a point α is a rest point if and only if $\mathbf{f}(\alpha) = \mathbf{0}$, and in this case $\varphi(t, \alpha) = \alpha$ for all t. A rest point is *asymptotically stable* if all solutions beginning near the rest point remain near it for all future time and approach it as $t \to \infty$. (Technical point: "Remaining near it" and "approaching it" are separate issues. Nearby solutions could move far away, then return and approach the rest point. This would not be asymptotic stability.) The behavior of solutions near the rest point can often be determined from the linearization of the system at the rest point. To find the linearization, make the change of variables $\mathbf{v} := \mathbf{u} - \alpha$, so that in the new variables the rest point is at the origin; then expand the new equation in the form $\dot{\mathbf{v}} = \mathbf{f}(\alpha + \mathbf{v}) = A\mathbf{v} + \cdots$, where A is the matrix of partial derivatives of $\mathbf{f}$ evaluated at a rest point α and the dots denote terms of quadratic and higher orders in the components of $\mathbf{v}$; then the *linearization* of (D.8) at the rest point is the linear system $\dot{\mathbf{v}} = A\mathbf{v}$, which is of the form (D.5). Since the omitted terms are small compared to the linear terms in a neighborhood of the origin, the behavior of (D.8) near α can be shown to be the same as the behavior of the linear system (D.5) near its rest point at $\mathbf{0}$, provided that (D.5) is either asymptotically stable or is unstable. (It has already been pointed out that neutrally stable systems are sensitive to small influences, so nothing can be said about the rest point of the nonlinear system when its linearization is neutrally stable.) Therefore, if the eigenvalues of A lie in the left half-plane, then the rest point α of (D.8) is asymptotically stable; if A has an eigenvalue in the right half-plane, α is unstable.

The second special case of (D.1) of particular importance is the case in which $\mathbf{f}(\mathbf{u}, t)$ is periodic in t with some period $T > 0$, in other words

$$\mathbf{f}(\mathbf{u}, t + T) = \mathbf{f}(\mathbf{u}, t) \qquad (D.11)$$

for all t. Whenever $\mathbf{f}$ in (D.1) depends on t, solutions starting at the same point α at different times will not in general follow the same path; that is, solutions do not lie on orbits. It is still possible to think of $\mathbf{f}$ as a vector field, but it is now a *changing* vector field, and each solution passing through a point has a velocity vector equal to what $\mathbf{f}$ happens to be at that time. If we think of $\mathbf{f}(\mathbf{u}, t)$ as specifying the "environment" experienced by a solution as it evolves in time, then solutions starting at the same point at different times experience different environments. However, when $\mathbf{f}$ is periodic with period T, the solution starting at any point (α, nT) (n an integer) will experience the same environment as the solution beginning there at time 0, and will therefore follow the same path, with a constant time delay of nT. Thus

$$\varphi(t, nT, \alpha) = \varphi(t - nT, 0, \alpha). \qquad (D.12)$$

This equation replaces (D.9), which does not hold in this setting except for $t_0 = nT$.

The rigorous proof of (D.12) is to observe that both sides are solutions of (D.1) when (D.11) holds, and that both sides equal α when $t = nT$. For the former, begin with

$$\frac{d}{dt}\boldsymbol{\varphi}(t, 0, \boldsymbol{\alpha}) = \mathbf{f}\big(\boldsymbol{\varphi}(t, 0, \boldsymbol{\alpha}), t\big).$$

Therefore

$$\frac{d}{dt}\boldsymbol{\varphi}(t - nT, 0, \boldsymbol{\alpha}) = \mathbf{f}\big(\boldsymbol{\varphi}(t - nT, 0, \boldsymbol{\alpha}), t - nT\big)$$
$$= \mathbf{f}\big(\boldsymbol{\varphi}(t - nT, 0, \boldsymbol{\alpha}), t\big),$$

which is the assertion that $\boldsymbol{\varphi}(t - nT, 0, \boldsymbol{\alpha})$ is a solution of (D.1).

From (D.12) one proves, in exactly the manner in which (D.10) is derived from (D.9), that

$$\boldsymbol{\varphi}(nT + mT, 0, \boldsymbol{\alpha}) = \boldsymbol{\varphi}\big(nT, 0, \boldsymbol{\varphi}(mT, 0, \boldsymbol{\alpha})\big). \qquad \text{(D.13)}$$

In particular at the *stroboscopic times* $0, T, 2T, 3T, \ldots$ the solution occupies the following sequence of positions:

$$\boldsymbol{\alpha},$$
$$\boldsymbol{\varphi}(T, 0, \boldsymbol{\alpha}),$$
$$\boldsymbol{\varphi}(2T, 0, \boldsymbol{\alpha}) = \boldsymbol{\varphi}\big(T, 0, \boldsymbol{\varphi}(T, 0, \sigma))\big), \qquad \text{(D.14)}$$
$$\boldsymbol{\varphi}(3T, 0, \boldsymbol{\alpha}) = \boldsymbol{\varphi}\big(T, 0, \boldsymbol{\varphi}(2T, 0, \boldsymbol{\alpha})\big),$$
$$\text{etc.}$$

The term "stroboscopic time" refers to the fact that if the solution, considered as a point moving through N-dimensional space, were illuminated by a stroboscopic light at time $0, T, 2T, \ldots$, the sequence of points (D.14) would be observed. This sequence of points can be obtained as the iterates of a function $\boldsymbol{\Phi}$ mapping points to points, called the *period map*. Namely, if

$$\boldsymbol{\Phi}(\boldsymbol{\alpha}) := \boldsymbol{\varphi}(T, 0, \boldsymbol{\alpha}), \qquad \text{(D.15)}$$

then the sequence (D.14) is simply

$$\boldsymbol{\alpha}, \boldsymbol{\Phi}(\boldsymbol{\alpha}), \boldsymbol{\Phi}^2(\boldsymbol{\alpha}), \boldsymbol{\Phi}^3(\boldsymbol{\alpha}), \ldots . \qquad \text{(D.16)}$$

Here the powers do not denote products but rather repetitions of the operation $\boldsymbol{\Phi}$, that is, $\boldsymbol{\Phi}^2(\boldsymbol{\alpha}) = \boldsymbol{\Phi}\big(\boldsymbol{\Phi}(\boldsymbol{\alpha})\big) = \boldsymbol{\varphi}\big(T, 0, \boldsymbol{\varphi}(T, 0, \boldsymbol{\alpha})\big) = \boldsymbol{\varphi}(2T, 0, \boldsymbol{\alpha})$, and similarly for higher powers. The values of solutions at points between times nT are not obtainable from the function $\boldsymbol{\Phi}$ but only via the complete solution $\boldsymbol{\varphi}$.

APPENDIX E

FOURIER SERIES

The Fourier series of a periodic function $f(t)$ with period $2a$ is defined to be

$$f(t) = \frac{1}{2}a_0 + \sum_{n=1}^{\infty} a_n \cos \frac{n\pi t}{a} + b_n \sin \frac{n\pi t}{a}, \qquad (E.1)$$

with

$$a_n = \frac{1}{a} \int_{-a}^{a} f(t) \cos \frac{n\pi t}{a} \, dt \qquad n = 0, 1, 2, \ldots$$

$$b_n = \frac{1}{a} \int_{-a}^{a} f(t) \sin \frac{n\pi t}{a} \, dt \qquad n = 1, 2, \ldots. \qquad (E.2)$$

If f and f' exist and are continuous for all t, the Fourier series converges uniformly to $f(t)$ for all t. Since in this book most functions can be assumed to be this smooth, there is no need for any of the more sophisticated convergence theories for Fourier series (convergence to the midpoint of a jump discontinuity; Gibbs' phenomenon; Caesaro means; L_1 and L_2 convergence).

The integrals in (E.2) can be changed to integrals from 0 to $2a$. If f is even, $f(-t) = -f(t)$, the equations can be replaced by

$$a_n = \frac{2}{a} \int_{0}^{a} f(t) \cos \frac{n\pi t}{a} \, dt$$

$$b_n = 0; \qquad (E.3)$$

if f is odd, $f(-t) = -f(t)$, they become

$$a_n = 0$$

$$b_n = \frac{2}{a} \int_0^a f(t) \sin \frac{n\pi t}{a} \, dt. \tag{E.4}$$

These are called half-range formulas (although "half-domain" would be a more appropriate term). If the function $f(t)$ is defined only for $0 \le t \le a$, either (E.3) or (E.4) can be used, resulting in a *Fourier cosine series* and a *Fourier sine series* for $f(t)$. If the original f (on $0 \le t \le a$) is continuously differentiable, the cosine series converges to f on $0 \le t \le a$ and the sine series on $0 < t < a$; the sine series converges to zero at $t = 0$ and $t = 1$, and hence converges to $f(t)$ there, provided $f(0) = f(a) = 0$. One way to think of these series is to first restrict $f(t)$ to $0 < t < a$, then extend it to an even or odd periodic function of period $2a$. The cosine and sine series of f are the Fourier series of these extended functions. (The reason for restricting to $0 < t < a$ is to avoid conflicts in the odd extension which arise unless $f(0) = f(a) = 0$. The odd periodic extension should be defined to be zero at all points na, for integers n, and will be discontinuous unless $f(0) = f(a) = 0$.)

Certain trigonometric functions have Fourier series which can be calculated without integration by use of identities. For instance the half-angle formula $\sin^2 x = \frac{1}{2} - \frac{1}{2} \cos 2x$ is actually the Fourier series of $\sin^2 x$. Note that $\sin^2 x$ is even and so only cosine terms appear. In some problems (an example occurs in Chapter 2) one wants to represent $\sin^2 x$ on the half-range $0 \le x \le \pi$ by a sine series; the result is an infinite series that cannot be found by trig identities. Table E.1 is a list of important Fourier series obtainable by identities. The simplest way to obtain these is to use $\cos x = (e^{ix} + e^{-ix})/2$ and $\sin x = (e^{ix} - e^{-ix})/2i$. Another useful set of identities is

$$\cos nx \cos mx = \frac{1}{2} \cos(n + m)x + \frac{1}{2} \cos(n - m)x$$

$$\cos nx \sin mx = \frac{1}{2} \sin(n + m)x - \frac{1}{2} \sin(n - m)x \tag{E.5}$$

$$\sin nx \sin mx = -\frac{1}{2} \cos(n + m)x + \frac{1}{2} \cos(n - m)x$$

For many purposes the complex form of the Fourier series (E.1) is more convenient. When the period is 2π the complex form is

$$f(t) = \sum_{n=-\infty}^{+\infty} c_n e^{int}, \tag{E.6}$$

with

$$c_n = \frac{1}{2\pi} \int_0^{2\pi} f(t) e^{-int} \, dt. \tag{E.7}$$

TABLE E.1. Fourier Series of $\sin^n x \cos^m x$

$$n+m=2 \begin{cases} \sin^2 x & = \tfrac{1}{2} - \tfrac{1}{2}\cos 2x \\ \sin x \cos x & = \tfrac{1}{2}\sin 2x \\ \cos^2 x & = \tfrac{1}{2} + \tfrac{1}{2}\cos 2x \end{cases}$$

$$n+m=3 \begin{cases} \sin^3 x & = \tfrac{3}{4}\sin x - \tfrac{1}{4}\sin 3x \\ \sin^2 x \cos x & = \tfrac{1}{4}\cos x - \tfrac{1}{4}\cos 3x \\ \sin x \cos^2 x & = \tfrac{1}{4}\sin x + \tfrac{1}{4}\sin 3x \\ \cos^3 x & = \tfrac{3}{4}\cos x + \tfrac{1}{4}\cos 3x \end{cases}$$

$$n+m=4 \begin{cases} \sin^4 x & = \tfrac{3}{8} - \tfrac{1}{2}\cos 2x + \tfrac{1}{8}\cos 4x \\ \sin^3 x \cos x & = \tfrac{1}{4}\sin 2x - \tfrac{1}{8}\sin 4x \\ \sin^2 x \cos^2 x & = \tfrac{1}{8} \qquad\quad - \tfrac{1}{8}\cos 4x \\ \sin x \cos^3 x & = \tfrac{1}{4}\sin 2x + \tfrac{1}{8}\sin 4x \\ \cos^4 x & = \tfrac{3}{8} + \tfrac{1}{2}\cos 2x + \tfrac{1}{8}\cos 4x \end{cases}$$

$$n+m=5 \begin{cases} \sin^5 x & = \tfrac{5}{8}\sin x - \tfrac{5}{16}\sin 3x + \tfrac{1}{16}\sin 5x \\ \sin^4 x \cos x & = \tfrac{1}{8}\cos x - \tfrac{3}{16}\cos 3x + \tfrac{1}{16}\cos 5x \\ \sin^3 x \cos^2 x & = \tfrac{1}{8}\sin x + \tfrac{1}{16}\sin 3x - \tfrac{1}{16}\sin 5x \\ \sin^2 x \cos^3 x & = -\tfrac{1}{8}\cos x - \tfrac{1}{16}\cos 3x - \tfrac{1}{16}\cos 5x \\ \sin x \cos^4 x & = \tfrac{1}{8}\sin x + \tfrac{3}{16}\sin 3x + \tfrac{1}{16}\sin 5x \\ \cos^5 x & = \tfrac{5}{8}\cos x + \tfrac{5}{16}\cos 3x + \tfrac{1}{16}\cos 5x \end{cases}$$

In order for f to be real, the complex coefficients c_n must satisfy $c_{-n} = \bar{c}_n$ for each n; the bar denotes complex conjugate. In particular, c_0 must be real. The constant term in either the real or the complex forms (c_0 or $1/2a_0$) is the average of f over one period.

If $f(\theta_1,\ldots,\theta_M)$ is a real-valued function of M angles which is 2π-periodic in each angle when the others are fixed, then the *multiple Fourier series* of f is

$$f(\theta_1,\ldots,\theta_M) = \sum_{\nu_1=-\infty}^{+\infty} \cdots \sum_{\nu_M=-\infty}^{+\infty} c_{\nu_1\ldots\nu_M} e^{i(\nu_1\theta_1 + \cdots + \nu_M\theta_M)}, \tag{E.8}$$

with

$$c_{\nu_1 \ldots \nu_M} = \frac{1}{(2\pi)^M} \int_0^{2\pi} \cdots \int_0^{2\pi} f(\theta_1, \ldots, \theta_M) e^{-i(\nu_1 \theta_1 + \cdots + \nu_M \theta_M)} \, d\theta_1 \ldots d\theta_M.$$

(E.9)

Abbreviated notations for multiple Fourier series are developed in Section 6.6. The smoothness conditions for uniform convergence of (E.8) are more stringent than for single Fourier series. The easiest theorem is that if f has $M+2$ continuous derivatives then the coefficients decay sufficiently rapidly to give uniform convergence.

If $f(t)$ or $f(\theta_1, \ldots, \theta_M)$ are vector-valued functions which are periodic in their arguments, everything said in this Appendix E holds exactly as stated, except that the coefficients of the various Fourier series are vectors (and become boldface).

APPENDIX F

LIPSCHITZ CONSTANTS AND VECTOR NORMS

The aim of this appendix is to prove the following result of advanced calculus:

Theorem F.1. *Let* **f** *be a continuously differentiable mapping from a compact convex set* $D \subset \mathbf{R}^N$ *to* $\mathbf{R}^M$. *Then there exists a constant* L *such that*

$$\|\mathbf{f}(\mathbf{u}) - \mathbf{f}(\mathbf{v})\| \le L\|\mathbf{u} - \mathbf{v}\| \tag{F.1}$$

for all **u** *and* **v** *in* D.

In the applications, M usually equals 1 or N; that is, **f** is usually either a scalar-valued function (in which case it is denoted by a lightface f) or else a mapping from $\mathbf{R}^N$ to itself. The constant L is called a *Lipschitz constant* for **f** on D.

The vector norm $\|\mathbf{z}\|$ appearing in (F.1) is the standard or Euclidean norm defined by

$$\|\mathbf{z}\| := \sqrt{z_1^2 + \cdots + z_N^2}; \tag{F.2}$$

of course, on the left hand side of (F.1) there are M terms inside the square root instead of N, and if $M = 1$ the norm reduces to the absolute value. In the course of the argument it will be useful to have two additional vector norms, defined by

$$\|\mathbf{z}\|_{\max} := \max\{|z_1|, \ldots, |z_N|\} \tag{F.3}$$

and

$$\|\mathbf{z}\|_{\text{sum}} := |z_1| + \cdots + |z_N|. \tag{F.4}$$

Any of these three is bounded both above and below by a constant times any of the others; the constants depend only on the dimension. We will need the following easily checked instances:

$$\|\mathbf{z}\|_{\text{sum}} \leq N\|\mathbf{z}\| \qquad \text{for} \quad \mathbf{z} \in \mathbf{R}^N \tag{F.5}$$

and

$$\|\mathbf{w}\| \leq \sqrt{M}\|\mathbf{w}\|_{\text{max}} \qquad \text{for} \quad \mathbf{w} \in \mathbf{R}^M. \tag{F.6}$$

Proof of Theorem F.1. A *convex set* is one which contains the straight line segment joining any two of its points; the interior of an ellipse or rectangle in the plane is convex, while an annulus or a "heart-shaped" region is not. The line segment joining $\mathbf{u}$ to $\mathbf{v}$ can be parameterized as $t\mathbf{u} + (1 - t)\mathbf{v}$ with $0 \leq t \leq 1$. Define a vector-valued function of the real variable t by

$$\mathbf{g}(t) := \mathbf{f}(t\mathbf{u} + (1 - t)\mathbf{v}). \tag{F.7}$$

This is well-defined because D is convex. The mean value theorem does not apply directly to vector functions, but can be applied to each component of $\mathbf{g}$; this shows that for each $i = 1, \ldots, n$ there exists a t_i between 0 and 1 such that

$$f_i(\mathbf{u}) - f_i(\mathbf{v}) = g_i(1) - g_i(0) = g_i'(t_i). \tag{F.8}$$

Differentiating the ith component of (F.7) by the chain rule gives

$$g_i'(t) = \sum_{j=1}^{n} f_{ij}(t\mathbf{u} + (1 - t)\mathbf{v})(u_j - v_j), \tag{F.9}$$

where f_{ij} is the partial derivative of f_i with respect to its jth argument. Let

$$B := \max|f_{ij}(\mathbf{x})|, \tag{F.10}$$

where the maximum is taken over all i and j and over all $\mathbf{x}$ in D. (The maximum over $\mathbf{x}$ exists because D is compact.) It follows from (F.8), (F.9), and (F.10) that

$$|f_i(\mathbf{u}) - f_i(\mathbf{v})| \leq B \sum_{j=1}^{n} |u_j - v_j| = B\|\mathbf{u} - \mathbf{v}\|_{\text{sum}}$$

for each i; taking the maximum over i,

$$\|\mathbf{f}(\mathbf{u}) - \mathbf{f}(\mathbf{v})\|_{\text{max}} \leq B\|\mathbf{u} - \mathbf{v}\|_{\text{sum}}. \tag{F.11}$$

Finally, (F.1) follows from (F.5), (F.6), and (F.11), with $L = BN\sqrt{M}$. ∎

APPENDIX G

LOGICAL QUANTIFIERS AND UNIFORMITY

The term *quantifier* is used in mathematical logic to refer to the expressions "for every ... " and "there exists ... such that," as well as various equivalent expressions such as "given any ... " and "one can find ... for which." These expressions are the fundamental building blocks of almost any rigorous mathematical statement. For instance, the definition of continuity begins "for every $\varepsilon > 0$ there exists a $\delta > 0$ such that ... ." The precise meaning of a sentence depends very critically on the order in which the quantifiers occur. The meaning of continuity would be entirely changed if the definition were written "there exists a $\delta > 0$ such that for every ε ... ." Therefore it is absolutely essential for the reader of this or any other mathematics book to understand thoroughly the meaning of quantifiers and of the order in which they appear. In this book, it is necessary to understand quantifiers in order to appreciate the meaning of uniformity. The entire subject of uniformity of approximations hinges on the order in which two quantifiers appear; this will be explained in detail below. Although every mathematics student should be grounded in logic somewhere in the undergraduate curriculum, this task is often left to a course in real analysis or topology and even then, the student is frequently left to pick up the material "by osmosis" without any formal exposition. The purpose of this appendix is to fill in this background for students who may not have seen it or who may not have realized its importance until encountering its practical significance in the concept of uniformity.

Consider the graph of $y^2 = x$. From this graph, it is easy to see that

$$\text{for every } x \geq 0 \text{ there exists a } y \text{ such that } y^2 = x. \tag{G.1a}$$

In symbolic logic, this statement would be written

$$(\forall x \geq 0)(\exists y) \quad y^2 = x. \qquad \text{(G.1b)}$$

The symbol $\forall$ is read "for every," and $\exists$ means "there exists ... such that." English is a flexible language, and it is possible to say the same thing in a variety of ways. Symbolic logic is very rigid and allows only the version (G.1b). Among the possible English expressions, (G.1a) should be considered the most important, because the quantifiers are clearly arranged in order at the beginning of the sentence exactly as in the symbolic form (G.1b). Another way to say the same thing in English words is

Given any $x \geq 0$ it is possible to choose a y for which $y^2 = x$.   (G.1c)

Still another way, somewhat less formal, is

Every nonnegative x is equal to y^2 for some y.   (G.1d)

This last form illustrates the fact that the word "some" (or "for some") is often equivalent to "there exists." It is important for the student to practice putting such informal mathematical statements into the "standard form" illustrated in (G.1a) in order to achieve absolute clarity. At the same time, anyone who intends to do mathematical writing should become fluent in the alternate expressions in order to avoid a cramped style.

The singular expressions "there exists a" in (G.1a) or "it is possible to chose a" in (G.1c) does not imply that there exists *only* one value of y. In fact, for every $x > 0$ there are two choices for such a y, namely $+\sqrt{x}$ and $-\sqrt{x}$. The expression "there exists a ... " should be understood as "there exists *at least* one ... ." On the other hand, if y is required to be positive, it is the case that

for every $x \geq 0$ there exists a unique $y \geq 0$ such that $y^2 = x$.   (G.2a)

The word "unique" indicates that there is exactly one (one and only one) y. The symbolic notation for (G.2a) is

$$(\forall x \geq 0)(\exists! y \geq 0) \quad y^2 = x. \qquad \text{(G.2b)}$$

In any version of (G.1), the fact that "for every" occurs before "there exists" is absolutely crucial to the meaning. To emphasize this, we will repeat (G.1a) with parentheses around the two quantifiers:

(For every $x \geq 0$) (there exists a y such that) $y^2 = x$.   (G.1e)

If these expressions in parentheses were reversed, we would have

(There exists a y such that) (for every $x \geq 0$) $y^2 = x$.   (FALSE)

$$(G.3)$$

This statement would say that the same value of y works for every value of x in the formula $y^2 = x$, in other words, that the graph of $y^2 = x$ is a horizontal line, which is of course untrue.

Consider now the formal definition of a uniform big-oh symbol stated in Definition 1.8.1: To say that $f(x, \varepsilon) = \mathcal{O}(\delta(\varepsilon))$ uniformly for x in $a \leq x \leq b$ means that

there exist constants $c > 0$ and $\varepsilon_1 > 0$ such that $|f(x, \varepsilon)| \leq c\delta(\varepsilon)$

for every x in $a \leq x \leq b$ and for every ε in $0 < \varepsilon \leq \varepsilon_1$.   (G.4a)

In this definition one concession has been made to fluent English: The last two quantifiers have been moved to the end of the sentence. In more precise logical order, (G.4a) should read

there exist constants $c > 0$ and $\varepsilon_1 > 0$ such that for every x in $a \leq x \leq b$

and for every ε in $0 < \varepsilon \leq \varepsilon_1$ one has $|f(x, \varepsilon)| \leq c\delta(\varepsilon)$.   (G.4b)

In symbolic logic, using symbols from set theory as well as logical symbols, this would be written as

$$(\exists c > 0)(\exists \varepsilon_1 > 0)(\forall x \in [a, b])(\forall \varepsilon \in (0, \varepsilon_1]) \quad |f(x, \varepsilon)| \leq c\delta(\varepsilon).$$

$$(G.4c)$$

In any form, the important point is that the existence of c and ε_1 *precedes* the "for all x and ε" (suitably restricted to intervals). In particular, the values of c and ε_1 do not depend upon x. This fact (which is the essence of uniformity) would be lost if the quantifiers were written in the wrong order.

Now suppose that $f(x, \varepsilon)$ is defined for all real x, and that it is uniformly $\mathcal{O}(\delta(\varepsilon))$ on every compact subset of the real line but (perhaps) not uniformly $\mathcal{O}(\delta(\varepsilon))$ on the entire real line. This is a very common situation, and it is important to understand its meaning precisely. We begin with the statement in the commonly used short form

$f(x, \varepsilon) = \mathcal{O}(\delta(\varepsilon))$ uniformly for x in compact subsets of the real line.

$$(G.5a)$$

Here the phrase "in compact subsets" is a short way of saying "in any (that is, every) compact subset." Therefore we have

$f(x, \varepsilon) = \mathcal{O}(\delta(\varepsilon))$ uniformly for x in any compact subset of the real line.

(G.5b)

To bring this into standard logical form, the quantifier should be made more explicit and brought to the beginning of the sentence:

For every compact subset K of the real line, $f(x, \varepsilon) = \mathcal{O}(\delta(\varepsilon))$

uniformly for $x \in K$. (G.5c)

(It is permissible here to say "for every closed bounded interval $[a, b]$" in place of "for every compact subset K of the real line," since these are the compact subsets of most importance and since, in fact, one statement implies the other by some elementary topology.) Next we can substitute into (G.5c) the definition of uniformity in the manner of (G.4), to obtain

For every K there exist $c > 0$ and $\varepsilon_1 > 0$ such that for all $x \in K$

and for all $\varepsilon \in (0, \varepsilon_1]$, $|f(x, \varepsilon)| \leq c\delta(\varepsilon)$. (G.5d)

Notice that (G.5d) has been shortened by assuming, without an explicit statement, that K denotes a compact subset of the real line. By analyzing the order of quantifiers in (G.5d), one can determine the real meaning of uniformity on compact subsets: Since "for every K" comes before "there exist c and ε_1," it is clear that these two constants depend upon the choice of K. Since "for all x and ε" comes after "there exist c and ε_1," these constants do not depend upon x and ε. The fact that c and ε_1 depend upon K is related to the fact that f need not be uniformly $\mathcal{O}(\delta)$ on the entire real line: If K is allowed to expand (say by taking $K = [-n, n]$ for successively larger integers n), the value of c may increase and that of ε_1 may decrease (these are the "bad" directions of change for these two constants), so that there may be no values of c and ε_1 (finite and positive) that work on the entire real line.

Finally, it is important to know how to write down the negation of a statement having many quantifiers. For instance, when attempting to "prove or disprove" a conjecture, one must know the precise meaning of the conjecture, and also of its denial, in order to know what one must look for in the form of a counterexample. The fundamental idea is illustrated by the following examples: To deny that "*every* odd number is prime" is to affirm that "*there exists* an odd number (for instance, 9) that is *not* prime." To deny that "*there exists* a negative number that is a square (that is, the square of a real number)" is to claim that "*every* negative number is a *non-square*." These examples show that the negation of a statement that begins with "for every" is a statement that begins with "there exists," and vice versa; also, the rest of the statement after the quantifier must be negated. In symbolic logic, the symbol for "not" or negation is either $\neg$ or $\sim$. The rules for negation then take the form

$$\neg(\forall \dots) = (\exists \dots)\neg,$$
$$\neg(\exists \dots) = (\forall \dots)\neg. \tag{G.6}$$

Warning: English is frustratingly ambiguous in regard to the word order for negated quantifiers. If someone says "every snake is not a rattlesnake," she probably means "not every snake is a rattlesnake" (which is true), rather than "every snake is a non-rattlesnake" (which is false). To avoid this ambiguity, we had to write "every negative number is a non-square" in the preceding discussion, rather than the more natural "every negative number is not a square," which is prone to be misunderstood.

As an illustration of the negation rules given in (G.6), let us "compute" the negation of "$f(x, \varepsilon) = \mathcal{O}(\delta(\varepsilon))$ uniformly on compact subsets." First we will express (G.5d) in symbolic notation. For simplicity we will suppress (that is, not mention) all of the restrictions within each quantifier (such as $c > 0$ and $\varepsilon \in (0, \varepsilon_1]$). Then (G.5d) becomes

$$(\forall K)(\exists c)(\exists \varepsilon_1)(\forall x)(\forall \varepsilon) \quad |f| \leq c\delta. \tag{G.5e}$$

To negate this, we place the symbol $\neg$ in front and then carry it across each quantifier in turn. First, $\neg(\forall K)$ becomes $(\exists K)\neg$. Next, $\neg(\exists c)$ becomes $(\forall c)\neg$. Finally, when $\neg$ crosses the last quantifier, we encounter $\neg |f| \leq c\delta$; the negation of this inequality is simply $|f| > c\delta$. The final result is

$$(\exists K)(\forall c)(\forall \varepsilon_1)(\exists x)(\exists \varepsilon) \quad |f| > c\delta. \tag{G.7a}$$

Written out in full, restoring all of the suppressed assumptions, this says

There is a compact subset K of the real line such that given any c and ε_1, one can find an $x \in K$ and an $\varepsilon \in (0, \varepsilon_1)$ for which $|f(x, \varepsilon)| > c\delta(\varepsilon)$.

$$\tag{G.7b}$$

Thus, to prove that $f(x, \varepsilon)$ is *not* uniformly $\mathcal{O}(\delta)$ on compact subsets, it suffices to find a *single* compact subset on which it is not uniformly $\mathcal{O}(\delta)$. This in turn means that for *every* c and ε_1, the inequality $|f| < c\delta$ is violated somewhere.

SYMBOL INDEX

INDEX

When a topic has several page numbers, a number in boldface indicates a definition. Page numbers may indicate the **first page** or the **high point** of a discussion; further information on a topic may be found on nearby pages. **Phrases** are usually only indexed once; "pitchfork bifurcation" is under "bifurcation'" whereas "expansion operator" is under "expansion."